Polymer Science and Technology

Polymer Science and Technology

Editor: Bethany Ferguson

MURPHY & MOORE
www.murphy-moorepublishing.com

Published by Murphy & Moore Publishing
1 Rockefeller Plaza,
New York City, NY 10020, USA
www.murphy-moorepublishing.com

Polymer Science and Technology
Edited by Bethany Ferguson

International Standard Book Number: 978-1-63987-441-5 (Hardback)

Cataloging-in-Publication Data

Polymer science and technology / edited by Bethany Ferguson.
p. cm.
Includes bibliographical references and index.
ISBN 978-1-63987-441-5
1. Polymers. 2. Polymerization. 3. Polymer engineering. I. Ferguson, Bethany.
TA455.P58 P65 2022
620.192--dc23

Contents

Preface

This book has been an outcome of determined endeavour from a group of educationists in the field. The primary objective was to involve a broad spectrum of professionals from diverse cultural background involved in the field for developing new researches. The book not only targets students but also scholars pursuing higher research for further enhancement of the theoretical and practical applications of the subject.

A polymer consists of macromolecules which are made up of many repeated subunits. Polymer science is a multi-disciplinary science that involves characterization, designing and fabrication of various polymers. Plastics and elastomers are the primary focus of this discipline. Macromolecular chemistry, polymer physics and polymer characterization are the three primary sub-disciplines of this field. The chemical synthesis and chemical properties of polymers are studied within macromolecular chemistry. Polymer physics focuses on the engineering applications and bulk properties of polymer materials. Polymer characterization delves into the morphology and chemical structures of polymers. This book is a compilation of chapters that discuss the most vital concepts and emerging trends in the field of polymer science and technology. It brings forth some of the most innovative concepts and elucidates the unexplored aspects of this field. This book includes contributions of experts and scientists which will provide innovative insights into this field.

It was an honour to edit such a profound book and also a challenging task to compile and examine all the relevant data for accuracy and originality. I wish to acknowledge the efforts of the contributors for submitting such brilliant and diverse chapters in the field and for endlessly working for the completion of the book. Last, but not the least; I thank my family for being a constant source of support in all my research endeavours.

Editor

1

Investigation on Dynamic Propagation Characteristics of In-Plane Cracks in PVB Laminated Glass Plates

Xiaoqing Xu, Bohan Liu, and Yibing Li

State Key Laboratory of Automotive Safety & Energy, Department of Automotive Engineering, Tsinghua University, Beijing 100084, China

Correspondence should be addressed to Yibing Li; liyb@mail.tsinghua.edu.cn

Academic Editor: Majid R. Ayatollahi

Polyvinyl butyral (PVB) laminated glass has been widely used as an important component of mechanical and construction materials. Cracks on PVB laminated glass are rich in impact information, which contribute to its impact resistance design. In this paper, a three-dimensional (3D) numerical simulation model describing PVB laminated glass under impact loading is firstly established and validated qualitatively and quantitatively compared with the corresponding experimental results recorded by the high-speed photography system. In the meantime, the extended finite element method (XFEM) is introduced to analyze the crack propagation mechanism of laminated glass based on dynamic stress intensity factors (DSIFs) and propagations of stress waves. Parametric studies are then carried out to investigate the influence of five critical parameters, that is, plate dimension, crack length, impact energy, glass properties, and PVB properties, on crack propagation characteristics of laminated glass. Results show that the interaction between crack tip and stress waves as well as the propagations of stress waves corresponds to the fluctuations of DSIFs at crack tip. Both the structure and material variables are proven to play a very important role in glass cracking DSIFs and thus govern the crack propagation behavior. Results may provide fundamental explanation to the basic crack propagation mechanism on radial cracks in PVB laminated glass under impact loading conditions, thus to instruct its impact design improvement.

1. Introduction

Due to its excellent transparency, formability, and impact resistance, polyvinyl butyral (PVB) laminated glass becomes great potential transparent protective material and has been widely used in both automotive and construction industry. The impact of foreign objects on laminated glass can generate microcracks, which will gradually extend and eventually lead to the failure of laminated glass. Regardless of whether fracture failure occurs, the fragility of glass layer and the good adhesion of PVB interlayer ensure the integrity of sandwiched structure and make the structure have certain energy absorption characteristics and toughness, which effectively improve the security of laminated structure. In order to utilize laminated glass as transparent protective material more effectively, its cracking propagation behavior under impact conditions needs to be comprehensively studied. However, dynamic failure characteristics of laminated glass are very complex; therefore studies on its dynamic cracking propagation characteristics become increasingly necessary.

In recent years, investigations on mechanical properties of PVB laminated glass have been thoroughly conducted, ranging from quasi-static mechanical properties [1–4] to dynamic impact response [4–10]. In general, the impact test in accordance with its impact velocity can be divided into three categories, that is, low-speed drop-weight test [11–13], high-speed impact test with air gun [3, 14, 15], and explosion shock wave impact test [6, 9, 10, 16], where low-speed drop-weight test is one of the most widely used testing methods to evaluate the material critical failure conditions and its fracture mechanism. Lee et al. [17] and Park and Chen [14], respectively, studied the static and dynamic crack propagation behavior in layered glass cross adhesive interlayers. Failure wave was introduced to account for their observations. Zhang et al. [18] and Hu et al. [19] experimentally and numerically studied failure conditions and crack morphologies in brittle plate with a thin polymer backing under various impact velocities and concluded the determining factors to predict its dynamic fracture patterns, that is, elastic properties, fracture energy, and accurate stress wave tracing model. Meanwhile, Vandenberghe et al. [20] investigated crack patterns in brittle plate upon controlled transverse impact using a global scaling law to predict the number of radial cracks. Also, the coexistence of radial and circumferential cracks was observed. As for a further step, Xu et al. [21] and Chen et al. [22] conducted out-of-plane dynamic loading test on PVB laminated glass based on drop-weight combined with high-speed photography system. Velocity and acceleration time history for both radial and circular cracks were calculated to understand the crack initiation and propagation mechanism. It is found that radial cracks appear earlier than circular ones with higher propagation speed. Furthermore, Chen et al. [23, 24] investigated in-plane crack propagation behavior in PVB laminated glass plate via captured in-plane and interplane cracking process and found that the supported glass layer would always initiate before the loaded layer with the same final radial crack morphologies. The quantity effect of radial cracks on its propagation behavior was also thoroughly investigated.

Numerical simulation attracts more and more attentions to investigate the fracture mechanism of PVB laminated glass for its simplicity and efficiency [5, 8, 25–32]. Bennison et al. [33] studied the stress distribution and fracture sequence in PVB laminated glass upon biaxial flexure loading using finite element method (FEM) and found the relationship between crack initiation and maximum stress location. Meanwhile, Xu and Li [34] numerically investigated the crack evolutions on PVB laminated windshield plate subjected to pedestrian head. It was found that the plastic deformation, radial cracks, and circular cracks were mainly caused by shear stress, compressive stress, and tensile stress, respectively. Furthermore, a series of newly developed numerical simulation methods are increasingly applied to researches on fracture characteristics of PVB laminated glass for their unique advantages in cracking simulation. Xu and Zang [35], Zang et al. [36], and Gao and Zang [37] investigated the impact fracture mechanism of PVB laminated glass on the basis of discrete element method (DEM). Chen et al. [38] numerically investigated the cracking behavior of PVB laminated glass beam under low-velocity impact in the framework of cohesive zone modeling with the glass-ply cracking modeled by extrinsic cohesive model. The propagations of stress waves during cracking process were also primarily analyzed. Based on extended finite element method (XFEM), Xu et al. [39] and Xu et al. [40] established PVB laminated glass plate model subjected to impact loading validated by both radial crack velocity and crack morphology and deeply investigated its radial and circular crack propagation mechanism.

However, as a key parameter to indicate the cracking driving force and interpret the crack propagation behavior in cracking mechanism analysis, dynamic stress intensity factor (DSIF) cannot be easily obtained via experimental methods. The propagations of stress waves during cracking can also have significant effect on cracking response of PVB laminated glass under impact loading. Besides, seldom numerical studies are available in exploring the cracking behavior of PVB laminated glass during out-plane impact loading from the perspective of DSIFs and stress waves. It is therefore necessary to investigate the dynamic crack propagation behavior of

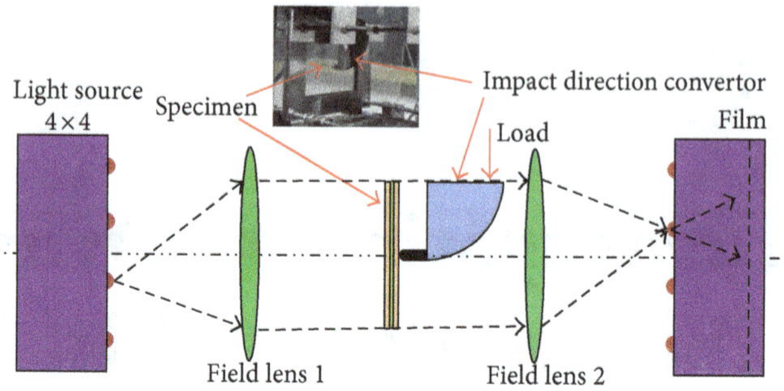

FIGURE 1: The schematic diagram of experimental system combing drop-weight test device with high-speed photography system.

PVB laminated glass on the basis of DSIFs and stress waves. Hence, the numerical model of PVB laminated glass plate with a layered structure is established in this study. The numerical model in simulations with propagation is then validated via the comparison between simulation and experimental results of radial crack propagation process, which are recorded by the high-speed photography system. Further, the time history of DSIFs is calculated in simulations with fixed crack, and the interaction between stress waves and crack tip is also introduced to interpret the relationship between crack propagation behavior and its driving force. Finally, parametric studies on five critical parameters, that is, plate dimension, crack length, impact energy, glass properties, and PVB properties, are conducted based on XFEM, which reveals the laws of crack propagation behavior on PVB laminated glass plates.

2. Experimental and Numerical Details

2.1. Experiment Setup. The drop-weight device combined with high-speed photography experimental system [21–23] is sketched in Figure 1. The 150 mm × 200 mm rectangular specimen of laminated glass is used in the experiment, which consists of a 0.76 mm PVB interlayer sandwiched by two 2 mm soda-lime glass sheets and is fabricated with the same processing technology of industrial automotive windshield. The drop-weight device is used to apply vertical load via free fall with different height. Meanwhile, the impact direction convertor with hemispherical end transfers the vertical load to horizontal load at the specimen center, that is, a load direction perpendicular to specimen plane. The high-speed photography system is adopted to record the real time crack propagation process on glass layer of specimen. The radial crack morphology is recorded on the film and crack propagation speed is calculated [22].

2.2. Computational Model. In this study, the three-dimensional (3D) finite element (FE) model consisting of an impactor, a prefabricated crack, and a PVB laminated glass plate is shown in Figure 2. One-quarter of the laminated plate and impactor are modeled to take advantage of the structure symmetry. The cylinder impactor has a mass of

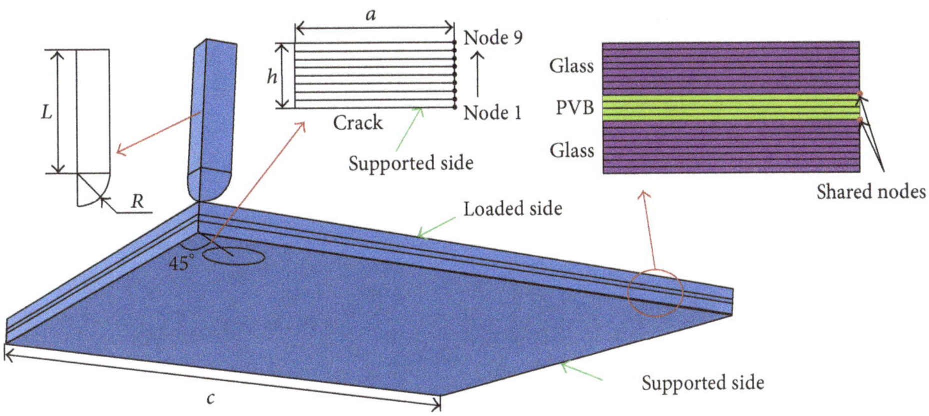

FIGURE 2: The overview of PVB laminated glass computational model subjected to impact loading.

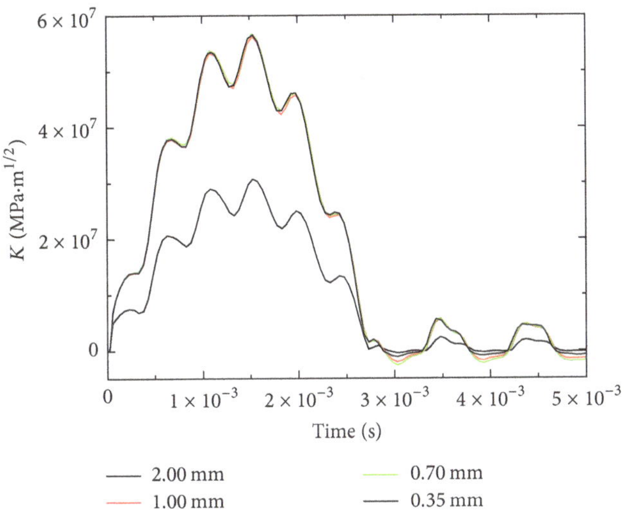

FIGURE 3: The mode I DSIFs of different mesh sizes at node 1 with fixed crack length of 0.02 m and impact speed of 5 m/s.

2 kg via equivalent density with the same mass of experimental impactor. The 2 mm thick crack with the length of a (shown in Figure 2) is placed at the center of laminated plate in the glass layer of supported side and the angle between crack and structure symmetry line is 45 degrees. In order to eliminate the influence of structural asymmetry, square laminated plate model with dimensions of 200 mm × 200 mm on plane direction is introduced to study its crack propagation characteristics. The laminated plate model is meshed with 2 mm × 2 mm (in plane direction) 3D eight-node reduced integration solid (C3D8R) elements. Since stress concentration exists and the stress gradients are very large at crack tip, the crack tip area needs to be meshed with sufficient fine elements to achieve stress and strain accurately. For this purposes, four mesh sizes at crack tip area in plane direction, that is, 0.35 mm × 0.35 mm, 0.7 mm × 0.7 mm, 1 mm × 1 mm, and 2 mm × 2 mm, are applied to achieve a stable convergence of the computational model. The results are shown in Figure 3, where the mode I DSIFs of node 1 (node 1 to node 9 at crack tip are defined in Figure 2) with fixed crack length $a = 0.02$ m under the impact speed of 5 m/s are illustrated. As can be seen, all simulations show the same oscillation frequencies and the mesh refinement has only effect on oscillation amplitudes. Therefore, the configuration of 0.35 mm × 0.35 mm can be used for further simulations sufficiently. In longitudinal direction, the laminated structure of specimen is modeled with 20 element layers, wherein the outside 8 element layers represent glass layers and the inner 4 element layers stand for the PVB interlayer (shown in Figure 2). Therefore, each layer of the laminated structure has more than three integration points to guarantee the accuracy of calculation and analysis in longitudinal direction.

Clamped boundary conditions of the laminated plate model and frictionless surface-to-surface contact between impactor and specimen are adopted to simulate the real experimental conditions. As typical viscoelastic material, PVB interlayer is simulated with a linear elastic model with a shear modulus $G = G_0$ and bulk modulus K, combined with a viscoelastic model. The elastic modulus E_{PVB} and Poisson's ratio ν are described as [42]

$$E_{PVB} = \frac{9KG_0}{3K + G_0}, \quad \nu = \frac{3K - 2G_0}{6K + 2G_0}, \tag{1}$$

where K and G_0 are bulk modulus and short time shear modulus of PVB, respectively. The generalized Maxwell model using Prony series with one term is introduced to simulate the viscoelasticity of PVB [42]:

$$G(t) = G_\infty + (G_0 - G_\infty) e^{-\beta t}, \tag{2}$$

where $G(t)$ is the relaxation modulus which is the function of time t, G_∞ is long time shear modulus, and β is decay factor.

As typical brittle material, glass layers are defined by linear elastic model. Reference parameters of glass layer, PVB interlayer, and impactor are listed in Table 1.

The computational model with the same dimensions of experimental system is set up using a cylinder impactor and a laminated plate model [39] to verify the effectiveness of the

TABLE 1: Basic parameters for numerical model.

	Properties	Reference parameters
Glass layer [40, 41]	Density	2500 kg/m^3
	Young's modulus	70 GPa
	Poisson's ratio	0.22
	Maximum allowable principal stress	60 MPa
	Energy release rate: G_{IC}, G_{IIC}, G_{IIIC}	$G_{IC} = 10\,J/m^2$, $G_{IIC} = G_{IIIC} = 50\,J/m^2$
	Thickness t_{glass}	2 mm
PVB interlayer [41, 42]	Density	1100 kg/m^3
	Shear modulus: G_0, G_∞	$G_0 = 0.33$ GPa, $G_\infty = 0.69 \times 10^{-3}$ GPa
	Decay factor β	12.6 s^{-1}
	Bulk modulus K	20 GPa
	Thickness t_{PVB}	0.76 mm
Impactor	Mass	2 kg
	Young's modulus	207 GPa
	Poisson's ratio	0.22
	Length L	20 mm
	Radius R	5 mm

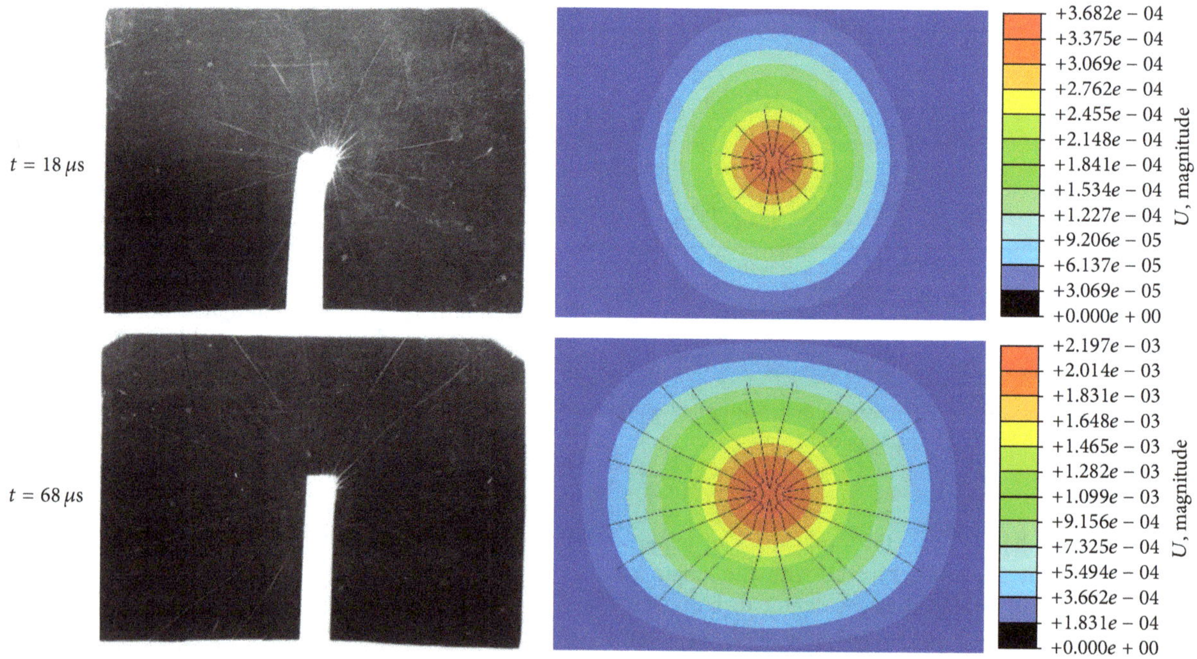

(a) Experiment results (b) Simulation results (unit: m)

FIGURE 4: Qualitative verification: comparison of radial crack morphologies between simulation results and experimental results.

abovementioned model. The experimental system model is then verified in simulations with propagation via qualitative method (radial crack morphology) and quantitative method (crack propagation velocity) shown in Figures 4 and 5, respectively. As can be seen from Figure 4, a total of 22 radial cracks exist in experimental images and most of them have propagated to the border of the specimen shown in Figure 4(a). In comparison, 20 radial cracks in all have propagated close to the border in simulation images shown in Figure 4(b). Thus, both the crack propagation process and the morphology of cracks in simulation agree well with the experimental results. Figure 5 shows the comparison of crack velocity time history curves between experiment and simulation results. As can be observed, the cracks propagation time history in simulation is recorded from the increment of the original crack, for initial flaws are preembedded in the model. Another difference between the experimental and simulation results is that the gradient of the descending stage of the velocity curves in simulation is smaller than that in experiment. The responsible reason should be the rubber pad used between the mental clamp and glass sheet to avoid possible damage during screwing in experiment. The rebound velocity in simulation is higher in simulation where no cushion pad is provided. So the peak value as well as the time history of the simulation results is in good agreement with the experimental results. The computational results are in good agreement with the corresponding experimental results from both qualitative perspective and quantitative perspective. Therefore, a reliable FE model with fairly high quality is established and can be adopted for further studies on crack behavior of PVB laminated glass.

2.3. Fundamentals of XFEM. The extended finite element method XFEM, which is a numerical technique based on generalized finite element method and the partition of unity method, extends the piecewise polynomial function space of conventional finite element methods with extra enrichment functions. XFEM does not need to be updated to track the crack path and can obtain accurate solution with coarse mesh [43, 44].

The displacement interpolation of XFEM is described as below [43]:

$$\mathbf{u}^h(\mathbf{x}) = \sum_{I\in N_\Gamma} N_I(\mathbf{x})\left[\mathbf{u}_I + H(\mathbf{x})\,\mathbf{a}_I + \sum_{\alpha=1}^{4}\sum_{J\in N_\Lambda} F_\alpha(\mathbf{x})\,\mathbf{b}_J^\alpha\right], \tag{3}$$

where N_Γ is the number of nodes in elements cut by crack; N_Λ is the number of nodes in elements with crack tip; $N_I(\mathbf{x})$ are the conventional shape functions; $\mathbf{u}_I$ are the nodal DOFs for conventional shape functions; $\mathbf{a}_I$ and $\mathbf{b}_J^\alpha$ are the nodal DOFs associated with the Heaviside function $H(\mathbf{x})$, with the value above the crack and −1 below the crack. The crack tip enrichment functions $F_\alpha(\mathbf{x})$ are expressed as [43]

$$[F_\alpha(\mathbf{x}), \alpha = 1 \sim 4] = \left[\sqrt{r}\sin\frac{\theta}{2}, \sqrt{r}\cos\frac{\theta}{2}, \sqrt{r}\sin\theta \cdot \sin\frac{\theta}{2}, \sqrt{r}\sin\theta\cos\frac{\theta}{2}\right], \tag{4}$$

where $\mathbf{x}$ is an integration point and (r, θ) denote the coordinate values from a polar coordinate system located at the crack tip.

Details of XFEM can be found in [43–45].

2.4. Calculation of Dynamic Stress Intensity Factors. In linear fracture mechanics, for example, soda-lime glass, the dynamic stress intensity factors ($K_\mathrm{I}^\mathrm{Dyn}, K_\mathrm{II}^\mathrm{Dyn}, K_\mathrm{III}^\mathrm{Dyn}$) are widely used to characterize the stress and displacement fields near the crack tip, where $K_\mathrm{I}^\mathrm{Dyn}$, $K_\mathrm{II}^\mathrm{Dyn}$, and $K_\mathrm{III}^\mathrm{Dyn}$ are mode I, II, and III DSIFs. DSIFs relevant to energy release rate (J-integral) can be described as

$$J = \frac{1}{8\pi}\overline{K}^T \cdot \overline{B}^{-1} \cdot \overline{K}, \tag{5}$$

where $\overline{K} = [K_\mathrm{I}^\mathrm{Dyn}, K_\mathrm{II}^\mathrm{Dyn}, K_\mathrm{III}^\mathrm{Dyn}]^T$ is SIF vector and $\overline{B}$ is the prelogarithmic energy factor matrix, which is diagonal for homogeneous and isotropic materials. Therefore the equation can be simplified to

$$J = \frac{1}{E^*}\left[\left(K_\mathrm{I}^\mathrm{Dyn}\right)^2 + \left(K_\mathrm{II}^\mathrm{Dyn}\right)^2\right] + \frac{1}{2G}\left(K_\mathrm{III}^\mathrm{Dyn}\right)^2, \tag{6}$$

where $E^* = E$ for plane stress and $E^* = E/(1-\nu^2)$ for plane strain.

The interaction integral method [46] is introduced to calculate DSIFs for cracks under mixed-mode loading, instead of according to a known J-integral. The J-integral can be described as

$$J = \frac{1}{8\pi}\left[K_\mathrm{I}^\mathrm{Dyn}B_{11}^{-1}K_\mathrm{I}^\mathrm{Dyn} + 2K_\mathrm{I}^\mathrm{Dyn}B_{12}^{-1}K_\mathrm{II}^\mathrm{Dyn} + 2K_\mathrm{I}^\mathrm{Dyn}B_{13}^{-1}K_\mathrm{III}^\mathrm{Dyn} + \left(\text{terms not involving } K_\mathrm{I}^\mathrm{Dyn}\right)\right]. \tag{7}$$

With regard to an auxiliary pure mode I crack tip field, the DSIF $k_\mathrm{I}^\mathrm{Dyn}$ can be defined as

$$J_\mathrm{aux}^\mathrm{I} = \frac{1}{8\pi}k_\mathrm{I}^\mathrm{Dyn}B_{11}^{-1}k_\mathrm{I}^\mathrm{Dyn}. \tag{8}$$

Then overlay the auxiliary field on the actual field yields

$$J_\mathrm{tot}^\mathrm{I} = \frac{1}{8\pi}\left[\left(K_\mathrm{I}^\mathrm{Dyn} + k_\mathrm{I}^\mathrm{Dyn}\right)B_{11}^{-1}\left(K_\mathrm{I}^\mathrm{Dyn} + k_\mathrm{I}^\mathrm{Dyn}\right) + 2\left(K_\mathrm{I}^\mathrm{Dyn} + k_\mathrm{I}^\mathrm{Dyn}\right)B_{12}^{-1}K_\mathrm{II}^\mathrm{Dyn} + 2\left(K_\mathrm{I}^\mathrm{Dyn} + k_\mathrm{I}^\mathrm{Dyn}\right)B_{13}^{-1}K_\mathrm{III}^\mathrm{Dyn} + \left(\text{terms not involving } K_\mathrm{I}^\mathrm{Dyn} \text{ or } k_\mathrm{I}^\mathrm{Dyn}\right)\right]. \tag{9}$$

Since

$$\left(\text{terms not involving } K_\mathrm{I}^\mathrm{Dyn}\right)\Big|_J = \left(\text{terms not involving } K_\mathrm{I}^\mathrm{Dyn} \text{ or } k_\mathrm{I}^\mathrm{Dyn}\right)\Big|_{J_\mathrm{tot}^\mathrm{I}} \tag{10}$$

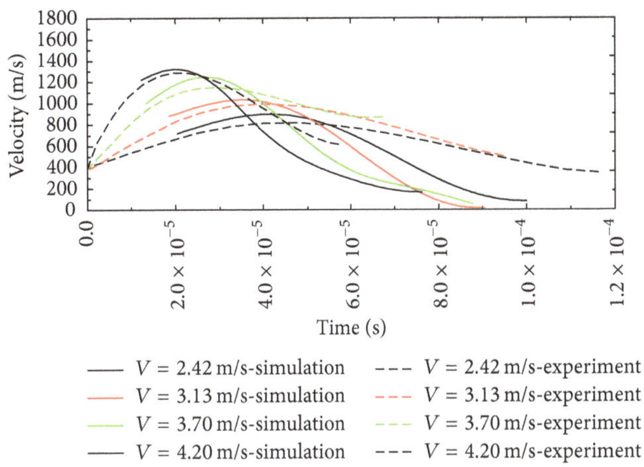

FIGURE 5: Quantitative verification: comparison of crack propagation velocities between simulation results and experimental results.

the interaction integral for mode I field can be expressed as

$$J_{\text{int}}^{\text{I}} = J_{\text{tot}}^{\text{I}} - J - J_{\text{aux}}^{\text{I}} = \frac{k_{\text{I}}^{\text{Dyn}}}{4\pi}\left(B_{11}^{-1}K_{\text{I}}^{\text{Dyn}} + B_{12}^{-1}K_{\text{II}}^{\text{Dyn}} + B_{13}^{-1}K_{\text{III}}^{\text{Dyn}}\right). \tag{11}$$

For mode II and mode III, similar results can be obtained:

$$\begin{aligned} J_{\text{int}}^{\text{II}} &= \frac{k_{\text{II}}^{\text{Dyn}}}{4\pi}\left(B_{21}^{-1}K_{\text{I}}^{\text{Dyn}} + B_{22}^{-1}K_{\text{II}}^{\text{Dyn}} + B_{23}^{-1}K_{\text{III}}^{\text{Dyn}}\right) \\ J_{\text{int}}^{\text{III}} &= \frac{k_{\text{III}}^{\text{Dyn}}}{4\pi}\left(B_{31}^{-1}K_{\text{I}}^{\text{Dyn}} + B_{32}^{-1}K_{\text{II}}^{\text{Dyn}} + B_{33}^{-1}K_{\text{III}}^{\text{Dyn}}\right). \end{aligned} \tag{12}$$

2.5. Crack Propagation Criteria. In static case, the open mode crack propagation criteria are generally expressed as $K_{\text{I}} \geq K_{\text{IC}}$, where K_{I} and K_{IC} (for glass: $K_{\text{IC}} \approx 0.7\,\text{MP}\sqrt{\text{m}}$) are static stress intensity factor and static fracture toughness, respectively. Similarly, the dynamic open mode crack propagation criteria are defined as $K_{\text{I}}^{\text{Dyn}}(t) > K_{\text{IC}}^{\text{Dyn}}$ [47], where $K_{\text{I}}^{\text{Dyn}}(t)$ related to time t is DSIF and $K_{\text{IC}}^{\text{Dyn}}$ associated with strain rate is dynamic fracture toughness. $K_{\text{I}}^{\text{Dyn}}(t)$ is determined by experiments, while $K_{\text{IC}}^{\text{Dyn}}$ is determined through dynamic structural analysis. Since the cracks are generated in linear elastic glass layer, the influence of loading rate on $K_{\text{IC}}^{\text{Dyn}}$ can be neglected within the scope of our discussion impact velocity [48, 49]. Therefore, the rate sensitivity of $K_{\text{IC}}^{\text{Dyn}}$ for glass layer can be ignored in the analysis of laminated glass fracture mechanism, that is, $K_{\text{IC}}^{\text{Dyn}} \approx K_{\text{IC}} \approx 0.7\,\text{MP}\sqrt{\text{m}}$.

2.6. Cracking Propagation Direction. The maximum tangential stress criterion is adopted to compute the cracking propagation direction, where $K_{\text{III}}^{\text{Dyn}}$ is not taken into consideration in these criteria. In the case of $\partial\sigma_{\theta\theta}/\partial\theta = 0$ (or $\tau_{r\theta} = 0$, where σ and τ are normal stress and tangential stress, resp., and r and θ are polar coordinates at the crack tip in a plane perpendicular to the crack plane), the crack propagation angle relative to the crack plane can be obtained [46]:

$$\widehat{\theta} = \cos^{-1}\left(\frac{3\left(K_{\text{II}}^{\text{Dyn}}\right)^2 + \sqrt{\left(K_{\text{I}}^{\text{Dyn}}\right)^4 + 8\left(K_{\text{I}}^{\text{Dyn}}\right)^2\left(K_{\text{II}}^{\text{Dyn}}\right)^2}}{\left(K_{\text{I}}^{\text{Dyn}}\right)^2 + 9\left(K_{\text{II}}^{\text{Dyn}}\right)^2}\right), \tag{13}$$

where $\widehat{\theta} = 0$ indicates the crack propagation in a straight direction. $\widehat{\theta} > 0$ corresponds to $K_{\text{II}}^{\text{Dyn}} < 0$, while $\widehat{\theta} < 0$ stands for $K_{\text{II}}^{\text{Dyn}} > 0$.

2.7. Simulations with Fixed Crack. According to our previous experimental studies [23], there are three types of cracks in PVB laminated plate under impact loading, that is, radial cracks in supported side, radial cracks in loaded side, and circumferential cracks in supported side, where supported side and loaded side are defined in Figure 2. Besides, radial cracks appear before circumferential cracks, while radial cracks in supported side are generated earlier than radial cracks in loaded side and radial cracks in both sides overlap completely. Therefore, the study on radial cracking propagation behavior in supported side is fairly important for investigating the fracture properties of PVB laminated glass.

Then, a series of numerical simulations on dynamic response of laminated glass with a fixed crack are conducted under different impact conditions using commercial finite element software ABAQUS. In our simulations, the crack is specified to be stationary and the material in crack tip should be in pristine condition to predict the crack propagation. Five structure and material variables, that is, plate dimension, crack length, impact energy, glass properties, and PVB properties, are introduced to conduct parametric studies in Section 4, and the detailed variable settings in parametric study are listed in Table 2 with other parameters kept at their reference values shown in Table 1.

3. Analysis of the Time History of DSIFs

In the condition of dynamic fracture, DSIFs are functions of time and the in-depth analysis of the time history of DSIFs helps to fully understand the crack propagation characteristics. Figure 6(a) shows the velocity and acceleration time history of impactor with the initial velocity of 5 m/s, and the corresponding mode I, II, and III DSIFs of node 1 at crack tip are shown in Figure 6(b). It can be seen that both the acceleration curve and DSIF curves exhibit regular fluctuations in the same law before 3×10^{-3} s, while the acceleration curve almost has no fluctuation and DSIF curves still have slight regular fluctuations after 3×10^{-3} s. Since the impactor has been away from the plate after 3×10^{-3} s, the existence of stress waves leads to the fluctuations of DSIF curves in the entire time course and the crest appears when crack tip interacts with stress waves, indicating that the interaction between crack tip and stress waves accounts for the DSIFs' changes. Therefore, the time of interaction between the crack tip and stress waves corresponds to the fluctuation crests of DSIFs.

Since $|K_{\text{I}}^{\text{Dyn}}|$ is much larger than $|K_{\text{II}}^{\text{Dyn}}|$ and $|K_{\text{III}}^{\text{Dyn}}|$ (Figure 6), the crack tip field is mixed-mode crack tip filed which mode I is in dominance [21, 22], that is, the mixed-mode DSIF $K^{\text{Dyn}} \approx K_{\text{I}}^{\text{Dyn}}$, and mode I DSIF can be introduced to represent mixed-mode DSIF. Figure 7 shows the mode I DSIF time history of node 1 to node 8 at crack tip under the impact velocity of 5 m/s (Figure 7(b) shows the data before 5×10^{-5} s). As can be seen, mode I DSIFs of node 1 to node 8 at crack tip begin to rise rapidly after a time lag which is defined as lag time, then fluctuate in the same regularity under the impact stress waves, and finally fluctuate slightly with the crest decreasing under the residual stress waves, so the entire process can be divided into three stages: lag stage, interaction stage, and residual stage. The propagation process of stress waves is introduced to explain this phenomenon (shown in Figure 8): the impact generates the compressive stress waves, which propagate from the loaded side to the supported side in the thickness direction.

TABLE 2: The detailed variable settings in parametric numerical study.

Numerical variables		Reference parameters
Plate dimension (m × m)		0.05 × 0.05, 0.1 × 0.1, 0.2 × 0.2
Crack length (m)		0.01, 0.02, 0.04
Impact energy	Impact velocity (m/s)	2, 5, 8, 10, 15, 20
	Impact mass (kg)	0.32, 2, 8
Glass properties	Young's modulus (GPa)	7, 70, 700
	Density (kg/m^3)	250, 2500, 25000
PVB properties	Young's modulus	$0.1E_{\text{PVB}}$, E_{PVB}, $10E_{\text{PVB}}$
	Density (kg/m^3)	110, 1100, 11000
	Shear modulus ratio	2.1×10^{-4}, 2.1×10^{-3}, 2.1×10^{-2}
	Delay factor (s^{-1})	0.126, 12.6, 1260

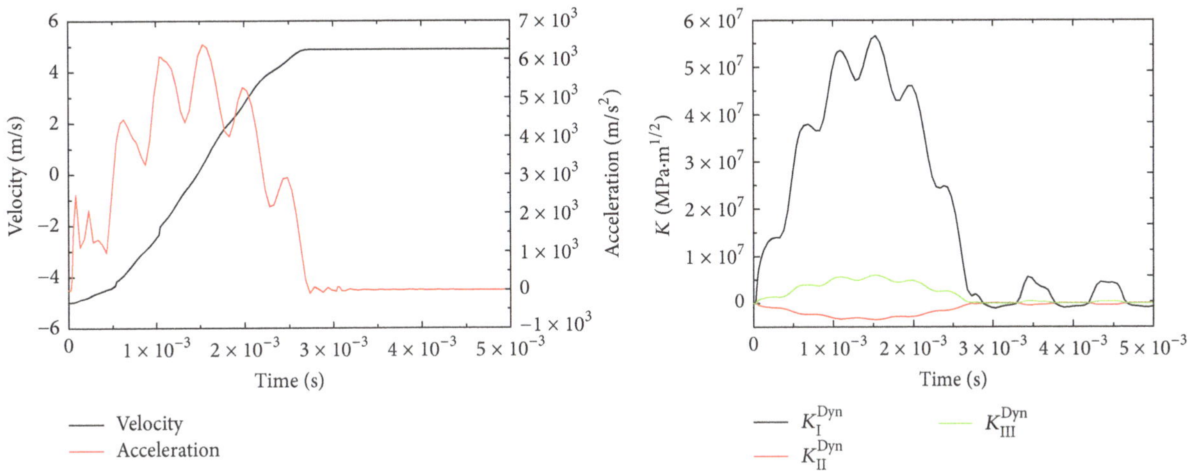

(a) Velocity and acceleration time history of impactor

(b) Mode I, II, and III DSIFs of node 1 at crack tip

FIGURE 6: Simulation results with the impact velocity of 5 m/s and crack length of 0.02 m.

Then the compressive stress waves reflect from the supported side (free surface) into tension stress waves, which consist of shear wave, longitudinal wave, and Rayleigh wave. The tension stress waves further propagate to the boundaries and reflect to the crack tip (shear wave and longitudinal wave through glass layer, while Rayleigh wave through free surface of supported side), which results in the crests of DSIFs. Besides, mode I DSIFs of node 1 to node 8 decrease one by one for the weakening of reflected tension stress waves during propagation, while there is almost no difference in peak times of these nodes for the extremely high propagation velocity of stress waves. Therefore, the round-trip propagation of tension stress waves leads to the fluctuations of DSIFs.

The velocities of involved stress waves are expressed as follows [50]:

$$c_l = \sqrt{\frac{E(1-\nu)}{(1-2\nu)(1+\nu)\rho}}$$
$$c_s = \sqrt{\frac{E}{2(1+\nu)\rho}} \tag{14}$$
$$c_R \approx 0.495 c_l,$$

where c_l, c_s, and c_R stand for the velocities of longitudinal wave, shear wave, and Rayleigh wave, respectively, besides $c_l = \max\{c_l, c_s, c_R\}$; E, ν, and ρ represent Young's modulus, Poisson's ratio, and density, respectively. The layer interfaces can cause large amounts of stress wave reflection and further dramatically delay the stress wave propagation. Similar delays have also been discovered in other experimental researches [14]. Therefore, the lag time of DSIF, which is the shortest propagation time of compressive stress waves from loaded side to supported side, can be expressed as

$$\begin{aligned} t_l &\approx 2\cdot t_{\text{Glass}} + t_{\text{PVB}} + 2\cdot t_{\text{Interface}} \\ &= 2\cdot \left.\frac{d}{c_l}\right|_{\text{Glass}} + \left.\frac{d}{c_l}\right|_{\text{PVB}} + 2\cdot t_{\text{Interface}}, \end{aligned} \tag{15}$$

where t_{Glass}, t_{PVB}, and $t_{\text{Interface}}$ represent propagation time of longitudinal wave through glass layers, PVB interlayer, and layer interfaces, respectively. Analogously, the interval of propagation time between node 1 and node 8 should be

$$t_{\text{Interval}} \approx \left.\frac{d}{c_l}\right|_{\text{Glass}} \approx 3.54\times10^{-7}\,\text{s}. \tag{16}$$

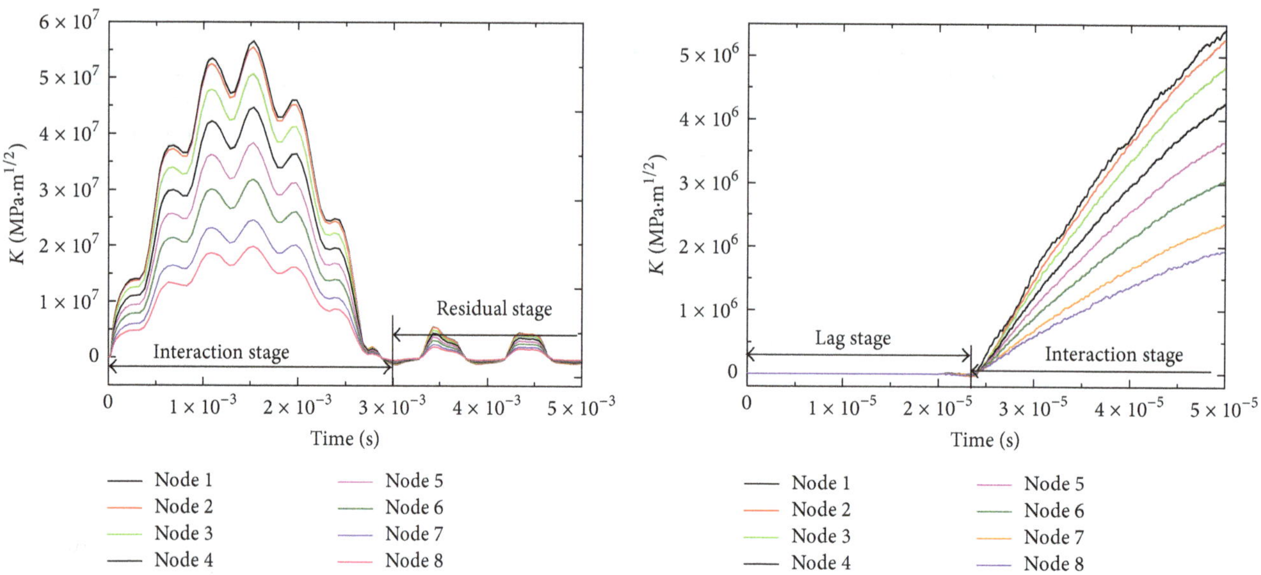

(a) Mode I DSIF curves of entire time course

(b) Mode I DSIF curves before 5×10^{-5} s

FIGURE 7: The mode I DSIF time history of node 1 to node 8 at crack tip with the impact velocity of 5 m/s and crack length of 0.02 m.

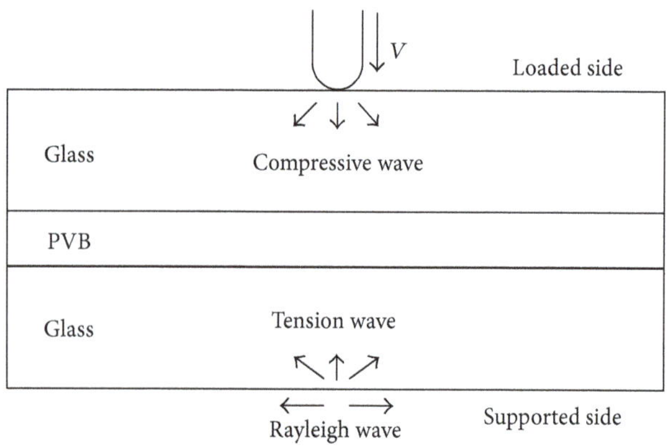

FIGURE 8: The propagation process of stress waves.

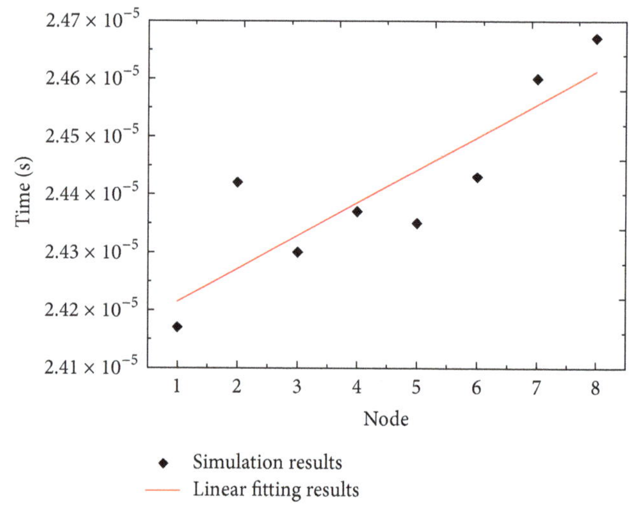

FIGURE 9: The simulated crack propagation times of node 1 to node 8 and the corresponding linear fitting curve.

Furthermore, the time when DSIF increases to dynamic fracture toughness with the stress waves accumulation, that is, $K^{\mathrm{Dyn}} \approx K_{\mathrm{IC}}^{\mathrm{Dyn}} \approx K_{\mathrm{IC}} \approx 0.7\,\mathrm{MP}\sqrt{\mathrm{m}}$, is considered as the simulated propagation time. The simulated crack propagation times of node 1 to node 8 and the corresponding linear fitting results are illustrated in Figure 9; then the simulated propagation time interval between node 1 and node 8 can be calculated as 4.55×10^{-7} s which is in good agreement with theoretical calculation.

According to the stress wave propagation theory, crack propagation firstly occurs at node 1, so parametric studies on crack propagation characteristics focus on mode I DSIF at node 1 in Section 4.

4. Discussions

4.1. Effects of Plate Dimension. Three different plate dimensions (0.05 m × 0.05 m, 0.1 m × 0.1 m, and 0.2 m × 0.2 m) in plane direction are considered to study the dimension effect on cracking propagation characteristics of glass layer. Figure 10 gives the effect of plate dimension on mode I DSIF of node 1. As can be seen, the interaction stage is getting longer and the peak DSIF would be smaller with the increase of plate size under the same impact energy due to the decrease of structure stiffness. The time that the stress wave propagation needed from the crack tip to the plate borders and return to the crack tip becomes longer due to the increase of plate edge length; therefore the time interval of DSIF peaks, that is, fluctuation cycle, increases with the increase of plate size which is fully consistent with the simulation results as shown in Figure 10.

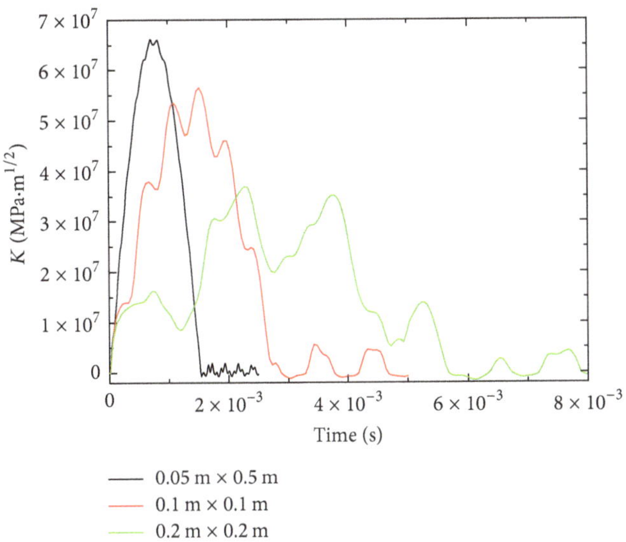

FIGURE 10: Mode I DSIF curves at node 1 of different plate dimensions with the impact velocity of 5 m/s and crack length of 0.02 m.

4.2. Effects of Crack Length. Figure 11 shows the mode I DSIF curves at node 1 with different fixed crack lengths (0.01 m, 0.02 m, and 0.04 m) under the impact velocity of 5 m/s. As can be seen, with the increase of fixed crack length, the lag time increases proportionally (Figure 11(b)) and the peak times are postponed with fluctuation frequency (reciprocal of fluctuation cycle) changing slightly (Figure 11(a)). Since the pre-crack length increases, the stress wave propagation path from impact point to crack tip increases, which leads to the increase of lag time, and the propagation path from crack tip to proximal border decreases, while it increases to distal border, which results in the changes of fluctuation frequency. In addition, the DSIF decreases with the increase of pre-crack length, due to the increase of propagation path from impact point to crack tip which leads to the increased stress wave weakening.

4.3. Effects of Impact Energy. Two aspects, that is, impact velocity and impact mass, are introduced to investigate the influence of impact energy on cracking propagation characteristics of glass layer. The mode I DSIFs of node 1 with different impact velocities and impact masses are given in Figures 12(a) and 12(b), respectively. As shown in Figure 12, both the increase of impact velocity and impact mass lead to the increase of peak mode I DSIF at all stages and almost have no effect on fluctuation frequency of DSIF curve. Besides, the interaction time between impactor and plate increases with the increase of impact mass but is not affected by impact velocity. Since the acceleration of impactor is mainly affected by impact velocity, the rebound time of impactor increases with the impact mass and almost has no change with impact speed. In summary, the increase of impact energy results in the increase of stress wave accumulation, that is, peak DSIF, while it has little influence on stress wave propagation velocity, that is, fluctuation frequency.

4.4. Effects of Glass Properties. In static case, stress intensity factors have nothing to do with material's Young's modulus and density. However, under impact loading conditions, the stress wave accumulation at crack tip leads to the crack propagation; thus stress wave propagation speed plays a critical role in crack propagation characteristics. As can be seen from (14), longitudinal wave, shear wave, and Rayleigh wave velocities are all functions of Young's modulus and density. Besides, stress wave velocities are proportional to Young's modulus and inversely proportional to density.

Figures 13(a) and 13(b) show the effects of glass Young's modulus E_{Glass} and density ρ_{Glass} on mode I DSIF at node 1. As can be observed, larger Young's modulus corresponds to the shorter interaction stage and larger peak DSIF owing to the increase of structure stiffness, while density changes have little effect on the length of interaction stage and peak DSIF with structure stiffness unchanged. Besides, the lag time increases and the peak times are postponed, for the decrease of Young's modulus or the increase of density leads to the decrease of stress wave velocities. Furthermore, the stress wave propagation time becomes longer with the decrease of Young's modulus or the increase of density; therefore the fluctuation frequency increases with the increase of Young's modulus or the decrease of density, which agrees well with the simulation results shown in Figure 13.

4.5. Effects of PVB Properties. As longitudinal boundaries of cracks in glass layers, PVB interlayer has an important influence on DSIFs of glass layer. Figures 14(a) and 14(b) show the effects of PVB Young's modulus E_{PVB} and density ρ_{PVB} on mode I DSIF curves at node 1. As can be seen, the interaction stage and peak DSIF decrease with the increase of PVB Young's modulus due to the increase of structure stiffness, while density changes have little effect on the length of interaction stage and peak DSIF with structure stiffness unchanged. Moreover, the fluctuation frequency increases with the increase PVB Young's modulus or the decrease of PVB density. According to (14), the stress wave propagation velocities in glass layer are far higher than that in PVB interlayer; thus the existence of PVB interlayer has an inhibiting effect on the stress wave propagation in glass layer, which leads to the increase of stress wave propagation time in glass layer.

The viscosity η in the one-term generalized Maxwell model can be expressed as

$$\eta = (G_0 - G_\infty)\tau = \frac{(G_0 - G_\infty)}{\beta}. \quad (17)$$

As seen, shear modulus ratio G_∞/G_0 and delay factor β are two main viscosity factors, which are inversely proportional to viscosity. Figures 14(c) and 14(d) give the effects of shear modulus ratio and delay factor on mode I DSIF curves at node 1. One may draw conclusion that the interaction stage, peak DSIF, and fluctuation cycle decrease with the increase of PVB viscosity. Besides, shear modulus ratio has a greater influence than delay factor. The relaxation occurs in PVB interlayer during the impact, which leads to the decrease of PVB modulus. PVB modulus decreases faster with the increase

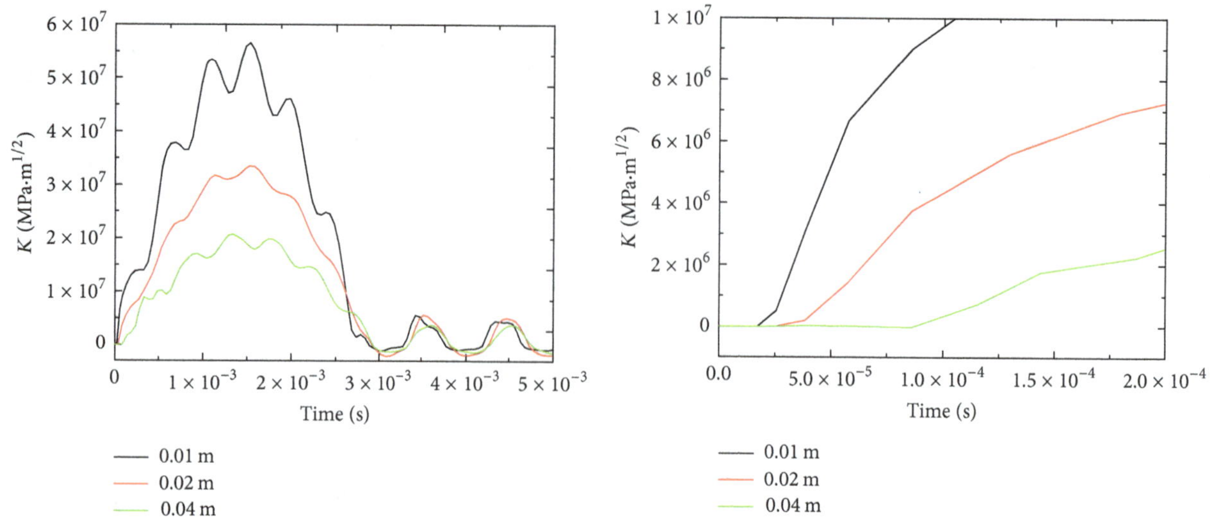

(a) Mode I DSIF curves of entire time course

(b) Mode I DSIF curves before 2×10^{-4} s

FIGURE 11: Mode I DSIF curves at node 1 with different fixed crack lengths under the impact velocity of 5 m/s.

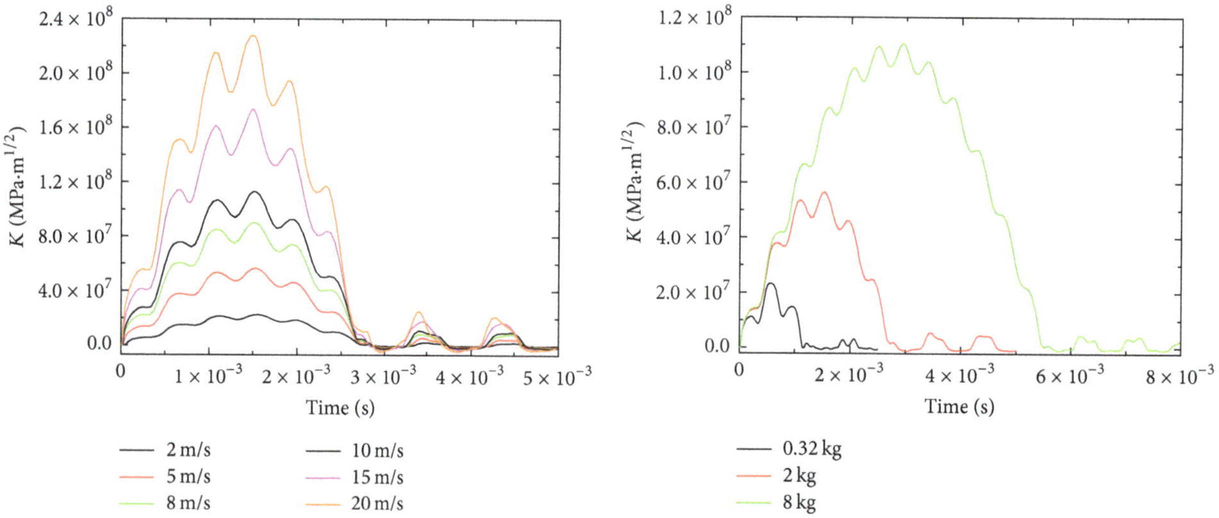

(a) Impact velocities

(b) Impact masses

FIGURE 12: Mode I DSIF curves at node 1 with crack length of 0.02 m under different impact energy.

of viscosity, which explains the viscosity effect. However, the stress wave propagation time is far less than the PVB relaxation time and only a small amount of relaxation occurs during the impact. Therefore, the influence of delay factor on DSIF is less obvious than shear modulus ratio.

5. Concluding Remarks

In this paper, the crack propagation characteristics of PVB laminated glass under low-speed impact are investigated in the framework of XFEM. A PVB laminated glass 3D model is proposed, which has been qualitatively and quantitatively validated via comparing with the experimental results captured by high-speed photography system. The verified numerical model with fixed crack is then used to systematically investigate the crack propagation behavior. The time history of brittle DSIFs as well as the propagations of stress waves is derived to interpret the crack propagation mechanism.

Further, five critical structure and material variables, that is, plate dimension, crack length, impact energy, glass properties, and PVB properties, are introduced to conduct parametric studies. It is found that the interaction between the crack tip and stress waves and the round-trip propagations of tension stress waves lead to the fluctuations of DSIFs, which corresponds to the influence of plate dimension and crack length. The impact energy and glass properties result in the changes of DSIF peak values and fluctuation frequencies, respectively. As longitudinal boundaries of cracks in glass

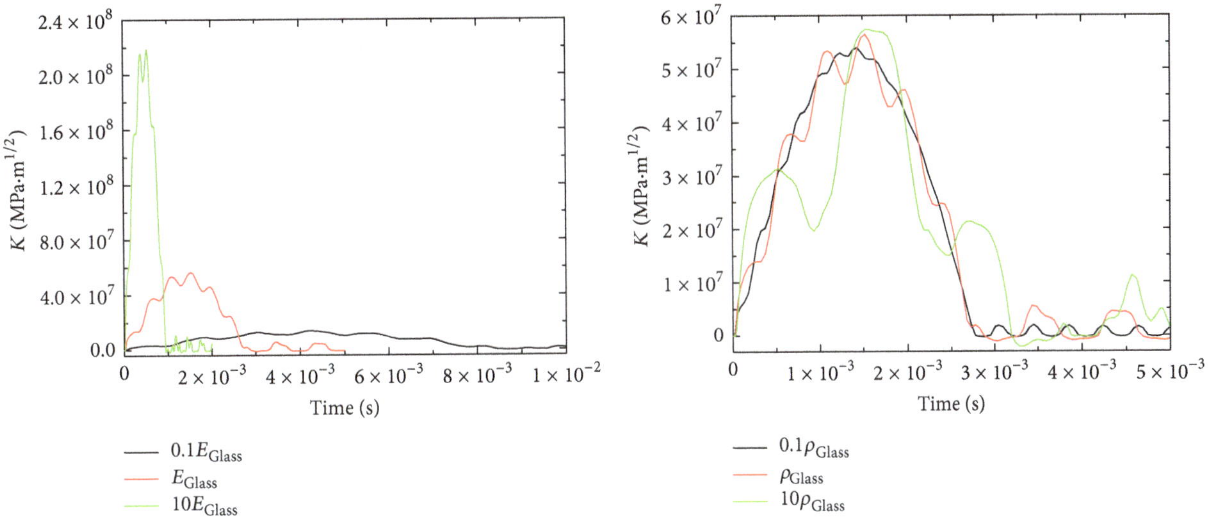

(a) Glass Young's modulus

(b) Glass density

FIGURE 13: Mode I DSIF at node 1 of different glass properties with the impact velocity of 5 m/s and crack length of 0.02 m.

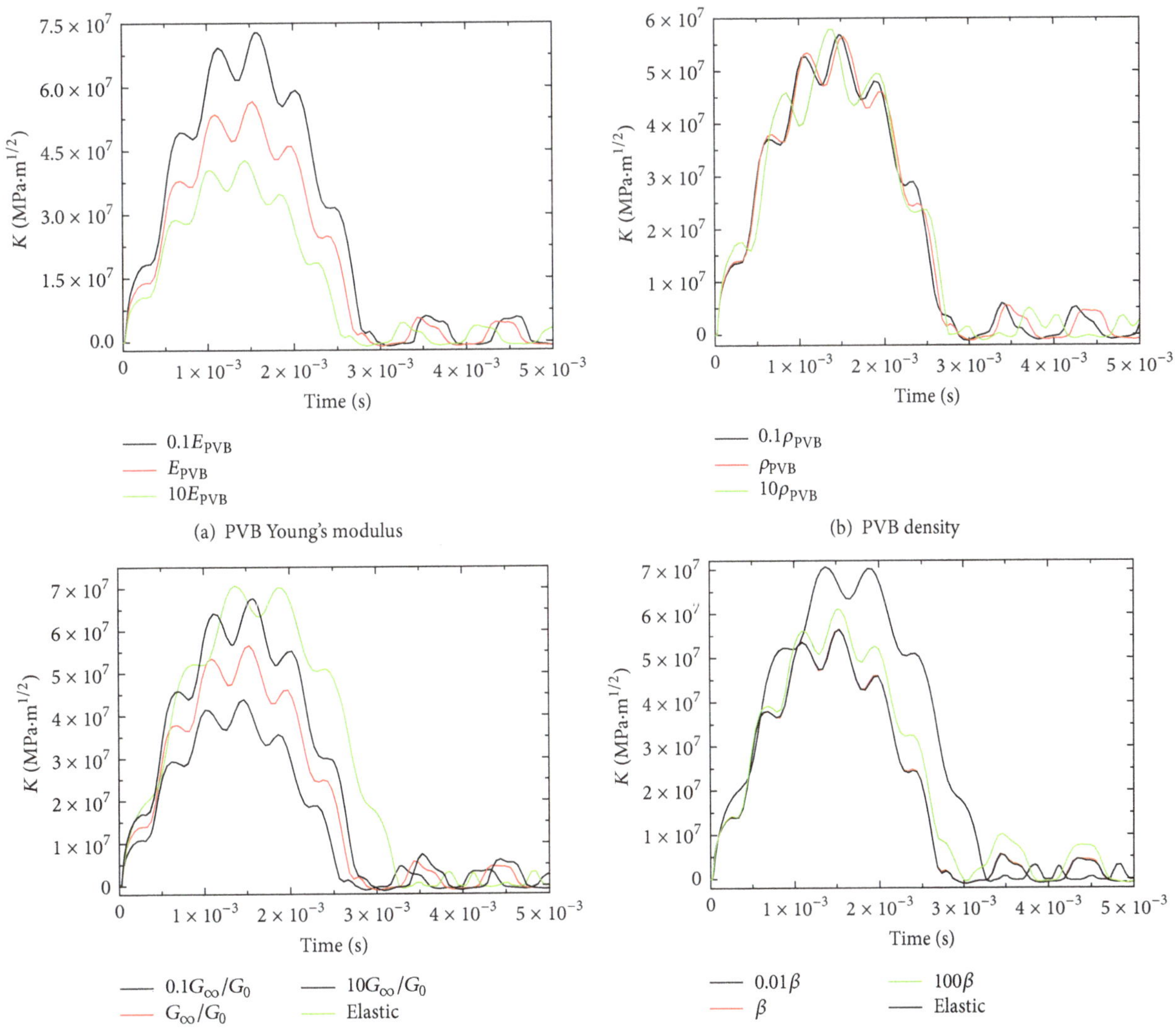

(a) PVB Young's modulus

(b) PVB density

(c) Shear modulus ratio

(d) Delay factor

FIGURE 14: Mode I DSIF at node 1 of different PVB properties with the impact velocity of 5 m/s and crack length of 0.02 m.

layers, PVB interlayer has an important influence on DSIFs of glass layer.

This research may provide basic crack propagation laws for researches on cracking mechanism of PVB laminated glass under impact loading conditions in the future.

Competing Interests

The authors declare that they have no competing interests.

Acknowledgments

The work is supported by the National Natural Science Foundation of China (11372164 and 11102099) and Tsinghua University under Grants nos. 20121080050 and 20132001016.

References

[1] J. Belis, J. Depauw, D. Callewaert, D. Delincé, and R. Van Impe, "Failure mechanisms and residual capacity of annealed glass/SGP laminated beams at room temperature," *Engineering Failure Analysis*, vol. 16, no. 6, pp. 1866–1875, 2009.

[2] C. V. G. Vallabhan, Y. C. Das, M. Magdi, M. Asik, and J. R. Bailey, "Analysis of laminated glass units," *Journal of Structural Engineering*, vol. 119, no. 5, pp. 1572–1585, 1993.

[3] J. Xu, Y. Li, B. Liu, M. Zhu, and D. Ge, "Experimental study on mechanical behavior of PVB laminated glass under quasi-static and dynamic loadings," *Composites Part B: Engineering*, vol. 42, no. 2, pp. 302–308, 2011.

[4] G. Herndon, K. Allen, A. Roberts, D. Phillips, and S. A. Batzer, "Automotive side glazing failure due to simulated human interaction," *Engineering Failure Analysis*, vol. 14, no. 8, pp. 1701–1710, 2007.

[5] M. Larcher, G. Solomos, F. Casadei, and N. Gebbeken, "Experimental and numerical investigations of laminated glass subjected to blast loading," *International Journal of Impact Engineering*, vol. 39, no. 1, pp. 42–50, 2012.

[6] P. A. Hooper, R. A. M. Sukhram, B. R. K. Blackman, and J. P. Dear, "On the blast resistance of laminated glass," *International Journal of Solids and Structures*, vol. 49, no. 6, pp. 899–918, 2012.

[7] K. Fischer and I. Häring, "SDOF response model parameters from dynamic blast loading experiments," *Engineering Structures*, vol. 31, no. 8, pp. 1677–1686, 2009.

[8] X. Zhang, H. Hao, and G. Ma, "Laboratory test and numerical simulation of laminated glass window vulnerability to debris impact," *International Journal of Impact Engineering*, vol. 55, pp. 49–62, 2013.

[9] X. Zhang, H. Hao, and Z. Wang, "Experimental study of laminated glass window responses under impulsive and blast loading," *International Journal of Impact Engineering*, vol. 78, pp. 1–19, 2015.

[10] X. Zhang and H. Hao, "Experimental and numerical study of boundary and anchorage effect on laminated glass windows under blast loading," *Engineering Structures*, vol. 90, pp. 96–116, 2015.

[11] Y. Liu and B. Liaw, "Drop-weight impact tests and finite element modeling of cast acrylic/aluminum plates," *Polymer Testing*, vol. 28, no. 8, pp. 808–823, 2009.

[12] J. Stenzler and N. Goulbourne, "Impact mechanics of transparent multi-layered polymer composites," in *Proceedings of the Society for Experimental Mechanics-Sem Annual Conference and Exposition on Experimental and Applied Mechanics*, 2009.

[13] H. Chai and G. Ravichandran, "On the mechanics of fracture in monoliths and multilayers from low-velocity impact by sharp or blunt-tip projectiles," *International Journal of Impact Engineering*, vol. 36, no. 3, pp. 375–385, 2009.

[14] H. Park and W. W. Chen, "Experimental investigation on dynamic crack propagating perpendicularly through interface in glass," *Journal of Applied Mechanics*, vol. 78, no. 5, Article ID 051013, 2011.

[15] A. J. Hsieh and J. W. Song, "Measurements of ballistic impact response of novel coextruded PC/PMMA multilayered-composites," *Journal of Reinforced Plastics and Composites*, vol. 20, no. 3, pp. 239–254, 2001.

[16] M. Hebert, C.-E. Rousseau, and A. Shukla, "Shock loading and drop weight impact response of glass reinforced polymer composites," *Composite Structures*, vol. 84, no. 3, pp. 199–208, 2008.

[17] J. J.-W. Lee, I. K. Lloyd, H. Chai, Y.-G. Jung, and B. R. Lawn, "Arrest, deflection, penetration and reinitiation of cracks in brittle layers across adhesive interlayers," *Acta Materialia*, vol. 55, no. 17, pp. 5859–5866, 2007.

[18] W. Zhang, S. Tekalur, and L. Huynh, "Impact behavior and dynamic failure of PMMA and PC plates," *Dynamic Behavior of Materials*, vol. 1, pp. 93–104, 2011.

[19] W. Hu, Y. Wang, J. Yu, C.-F. Yen, and F. Bobaru, "Impact damage on a thin glass plate with a thin polycarbonate backing," *International Journal of Impact Engineering*, vol. 62, pp. 152–165, 2013.

[20] N. Vandenberghe, R. Vermorel, and E. Villermaux, "Star-shaped crack pattern of broken windows," *Physical Review Letters*, vol. 110, no. 17, Article ID 174302, 2013.

[21] J. Xu, Y. Sun, B. Liu et al., "Experimental and macroscopic investigation of dynamic crack patterns in PVB laminated glass sheets subject to light-weight impact," *Engineering Failure Analysis*, vol. 18, no. 6, pp. 1605–1612, 2011.

[22] J. Chen, J. Xu, X. Yao et al., "Experimental investigation on the radial and circular crack propagation of PVB laminated glass subject to dynamic out-of-plane loading," *Engineering Fracture Mechanics*, vol. 112-113, pp. 26–40, 2013.

[23] J. Chen, J. Xu, X. Yao, X. Xu, B. Liu, and Y. Li, "Different driving mechanisms of in-plane cracking on two brittle layers of laminated glass," *International Journal of Impact Engineering*, vol. 69, pp. 80–85, 2014.

[24] J. Chen, J. Xu, B. Liu, X. Yao, and Y. Li, "Quantity effect of radial cracks on the cracking propagation behavior and the crack morphology," *PLoS ONE*, vol. 9, no. 7, Article ID e98196, 2014.

[25] M. Timmel, S. Kolling, P. Osterrieder, and P. A. Du Bois, "A finite element model for impact simulation with laminated glass," *International Journal of Impact Engineering*, vol. 34, no. 8, pp. 1465–1478, 2007.

[26] H. Hao and X. Zhang, "Analysis and design of glass windows against blast and impact loads," in *Proceedings of the 11th International Conference on Shock & Impact Loads on Structures*, Ottawa, Canada, 2015.

[27] T. Pyttel, H. Liebertz, and J. Cai, "Failure criterion for laminated glass under impact loading and its application in finite element simulation," *International Journal of Impact Engineering*, vol. 38, no. 4, pp. 252–263, 2011.

[28] Y. Peng, J. Yang, C. Deck, and R. Willinger, "Finite element modeling of crash test behavior for windshield laminated glass," *International Journal of Impact Engineering*, vol. 57, pp. 27–35, 2013.

[29] F. W. Flocker and L. R. Dharani, "Modelling fracture in laminated architectural glass subject to low velocity impact," *Journal of Materials Science*, vol. 32, no. 10, pp. 2587–2594, 1997.

[30] D. Sun, F. Andreiux, and A. Ockewitz, "Modeling of the failure behavior of windscreens and component tests," in *Proceedings of the 4th LS-DYNA Users Conference*, Bamberg, Germany, 2005.

[31] X. Zhang, H. Hao, and G. Ma, "Parametric study of laminated glass window response to blast loads," *Engineering Structures*, vol. 56, pp. 1707–1717, 2013.

[32] M. Larcher, M. Teich, N. Gebbeken et al., "Simulation of laminated glass loaded by air blast waves," *Applied Mechanics and Materials*, vol. 82, pp. 69–74, 2011.

[33] S. J. Bennison, A. Jagota, and C. A. Smith, "Fracture of glass/poly(vinyl butyral) (Butacite®) laminates in biaxial flexure," *Journal of the American Ceramic Society*, vol. 82, no. 7, pp. 1761–1770, 1999.

[34] J. Xu and Y. Li, "Crack analysis in PVB laminated windshield impacted by pedestrian head in traffic accident," *International Journal of Crashworthiness*, vol. 14, no. 1, pp. 63–71, 2009.

[35] W. Xu and M. Zang, "Four-point combined DE/FE algorithm for brittle fracture analysis of laminated glass," *International Journal of Solids and Structures*, vol. 51, no. 10, pp. 1890–1900, 2014.

[36] M. Y. Zang, Z. Lei, and S. F. Wang, "Investigation of impact fracture behavior of automobile laminated glass by 3D discrete element method," *Computational Mechanics*, vol. 41, no. 1, pp. 73–83, 2007.

[37] W. Gao and M. Zang, "The simulation of laminated glass beam impact problem by developing fracture model of spherical DEM," *Engineering Analysis with Boundary Elements*, vol. 42, pp. 2–7, 2014.

[38] S. Chen, M. Zang, D. Wang, Z. Zheng, and C. Zhao, "Finite element modelling of impact damage in polyvinyl butyral laminated glass," *Composite Structures*, vol. 138, pp. 1–11, 2016.

[39] X. Xu, J. Chen, J. Xu, Y. Li, and X. Yao, "Modeling and experimental studies of crack propagation in laminated glass sheets," *SAE International Journal of Materials & Manufacturing*, vol. 7, no. 2, pp. 328–336, 2014.

[40] J. Xu, Y. Li, X. Chen et al., "Characteristics of windshield cracking upon low-speed impact: numerical simulation based on the extended finite element method," *Computational Materials Science*, vol. 48, no. 3, pp. 582–588, 2010.

[41] J. Xu, Y. Li, X. Chen et al., "Numerical study of PVB laminated windshield cracking upon human head impact," *Computers, Materials & Continua*, vol. 18, no. 2, pp. 183–211, 2010.

[42] S. Zhao, L. R. Dharani, L. Chai, and S. D. Barbat, "Analysis of damage in laminated automotive glazing subjected to simulated head impact," *Engineering Failure Analysis*, vol. 13, no. 4, pp. 582–597, 2006.

[43] J. Dolbow and T. Belytschko, "A finite element method for crack growth without remeshing," *International Journal for Numerical Methods in Engineering*, vol. 46, pp. 131–150, 1999.

[44] T. Belytschko and T. Black, "Elastic crack growth in finite elements with minimal remeshing," *International Journal for Numerical Methods in Engineering*, vol. 45, no. 5, pp. 601–620, 1999.

[45] T. Belytschko, N. Moës, S. Usui, and C. Parimi, "Arbitrary discontinuities in finite elements," *International Journal for Numerical Methods in Engineering*, vol. 50, no. 4, pp. 993–1013, 2001.

[46] ABAQUS 6.13 analysis user's guide, Dassault Systems, 2013.

[47] J. Lambros and A. J. Rosakis, "Dynamic crack initiation and growth in thick unidirectional graphite/epoxy plates," *Composites Science and Technology*, vol. 57, no. 1, pp. 55–65, 1997.

[48] X. Zhang, Y. Zou, H. Hao, X. Li, G. Ma, and K. Liu, "Laboratory test on dynamic material properties of annealed float glass," *International Journal of Protective Structures*, vol. 3, no. 4, pp. 407–430, 2012.

[49] M. Peroni, G. Solomos, V. Pizzinato, and M. Larcher, "Experimental investigation of high strain-rate behaviour of glass," *Applied Mechanics and Materials*, vol. 82, pp. 63–68, 2011.

[50] L. Freund, *Dynamic Fracture Mechanics*, Cambridge University Press, Cambridge, UK, 1998.

2

Interfacial Effects on the Spherulitic Morphology of Isotactic Polystyrene Thin Films on Liquid Substrates

Takashi Sasaki,[1] Masaaki Nakagiri,[1] and Satoshi Irie[2]

[1]*Department of Materials Science and Engineering, University of Fukui, 3-9-1 Bunkyo, Fukui 910 8507, Japan*
[2]*Department of Frontier Fiber Technology and Science, University of Fukui, 3-9-1 Bunkyo, Fukui 910 8507, Japan*

Correspondence should be addressed to Takashi Sasaki; sasaki@matse.u-fukui.ac.jp

Academic Editor: Luigi Nicolais

The influence of interfaces on the morphology of flat spherulites of isotactic polystyrene (iPS) grown in thin films on liquid substrates was investigated. Amorphous iPS thin films spin-cast from a solution were annealed for cold crystallization on glycerol and silicone oil (nonsolvents for iPS). The number density of grown spherulites was revealed to be higher on the glycerol substrate than on the silicone oil substrate. This implies that the primary nucleation rate of crystallization is greater at the iPS/glycerol interface than at the iPS/silicone oil interface. The results may be consistent with the previous findings that concern the molecular interaction between atactic polystyrene and nonsolvents at the interface. In some cases, holes were formed in the thin films during the cold crystallization due to dewetting, which also significantly affect the spherulite morphology via, for example, transcrystallization.

1. Introduction

Since anomalous dynamics of nanosized polymer materials was reported, growing interest is focused on polymer thin films of which the thickness is typically smaller than 100 nm [1–3]. The origin of the dynamical properties (such as glass transition) of nanosized polymers that are different from bulk ones is still unclear. However, there is considerable evidence that the surface (interface) of the materials plays an important role [3]. The interfacial effects are emphasized as the size of the material becomes small. This is also the case for polymer crystallization behaviors, where the nucleation event that controls the crystallization rate and resulting morphology is very susceptible to the material's interfacial feature [4]. It is most likely that crystallization in thin films mainly occurs via heterogeneous nucleation, and the interfacial effects on the crystallization behaviors are closely related to the nucleation process at the interface.

The effects of the interface on polymer crystallization may be interpreted in principle, in terms of surface energetics and rheological properties. However, it seems to still be a long road to complete elucidation of the mechanism of the interfacial influence based on detailed molecular interactions. This issue is essentially important to develop nucleating reagents for polymer crystallization to achieve desired thermal and mechanical properties.

Crystallization in thin polymer films is considered to proceed in a quasi-two-dimensional manner; in particular, at high supercoolings, flat crystallites or spherulites frequently grow, reflecting the geometry of the substrate that supports the film. For example, crystallites composed of flat-on lamellae were found to grow in ultrathin films of poly(ε-caprolactone) [5]. A branching morphology of a flat-on lamellar crystal was observed for isotactic polystyrene (iPS) in an ultrathin film with a thickness of 11 nm [6]. On the other hand, depletion-induced banding has been reported for melt-grown iPS in thin films with a thickness of <100 nm [7]. Poly(ethylene oxide) (PEO) exhibits a finger-like crystalline morphology in dewetting holes where very thin polymer layers still remain even after the dewetting [8–10]. This characteristic finger-like morphology observed for ultrathin layers is considered to occur via a diffusion-controlled growth mechanism. An apparent surface effect on the rate of crystallization was reported for porous poly(L-lactic acid) (PLLA) prepared through freeze-drying [11, 12]. The overall crystallization rate is raised as the specific surface area of the porous PLLA increases. The enhanced crystallization rate is at least partially relevant to the large surface area of the porous PLLA. It has also been reported recently that the crystallization is activated at air/PLLA interface [13].

Dewetting is frequently observed when a polymer ultrathin film undergoes crystallization on a substrate [4], which is governed by the mechanical properties of the polymer film as well as the polymer/substrate interfacial free energy. On a liquid substrate of a nonsolvent, dewetting tends to become remarkable: the polymer film on the liquid shrinks when heated above its glass transition temperature (T_g), which can provide a simple method to measure T_g of a polymer thin film [14–16]. When a crystallizable polymer thin film is annealed on a liquid substrate, both shrinkage and crystallization are expected to occur. In this case, the shrinkage can significantly affect the crystallization behaviors, which are expected to give rise to characteristic crystalline morphologies. For some cases, the shrinkage induces hole formation in thin films, which act as a nucleation promoter for crystallization.

In this study, we examine the morphology of iPS spherulites grown via cold crystallization in ultrathin films. iPS is a typical polymer that has been reported to exhibit characteristic crystalline morphologies in thin films under various conditions [6, 7, 17, 18]. Here we employ for the first time liquid substrates, that is, silicone oil and glycerol; these are nonsolvent for iPS. The iPS thin films with a thickness of 35 nm were prepared through spin-casting, and they were annealed for cold crystallization on the liquid substrates to investigate the interfacial effects on the crystalline morphology in thin films.

2. Materials and Methods

iPS was purchased from Scientific Polymer Products Inc. (M_w = 400,000, triad tacticity 90%) and was used without further purification. Silicone oil (Sigma-Aldrich Inc. #175633) was used as received. Spin-casting was performed to prepare the iPS thin films: iPS was dissolved in cyclohexanone at 150°C for 30 min with stirring to obtain a solution of 0.63 wt% iPS content. After filtration of the solution, spin-cast films were prepared at 2,000 rpm for 10 s on a clean glass substrate (SiO_2). The thickness of the obtained films was estimated to be 35 nm from transmission electron microscopy (TEM) by using a Jeol JEM-2100 operated at 200 kV (the specimens were obliquely shadowed with Au). We confirmed that the above thickness was consistent with that from the ellipsometry measurement, which was performed by using an Otsuka Electronics FE-5000. For cold crystallization experiments on liquid substrates, the iPS thin film was peeled off the glass substrate by immersing the specimen into water. The peeled film floating on water was then moved carefully onto glycerol (liquid substrate) by using a handmade wire mesh. Then the film floating on glycerol was covered with a small amount (less than 1.0 mL) of another liquid (silicone oil or glycerol), or with no liquid (air). Schematics of the film on these liquid substrates are shown in Figure 1. As the sample has an iPS/SiO_2 interface, we used spin-cast films without pealing.

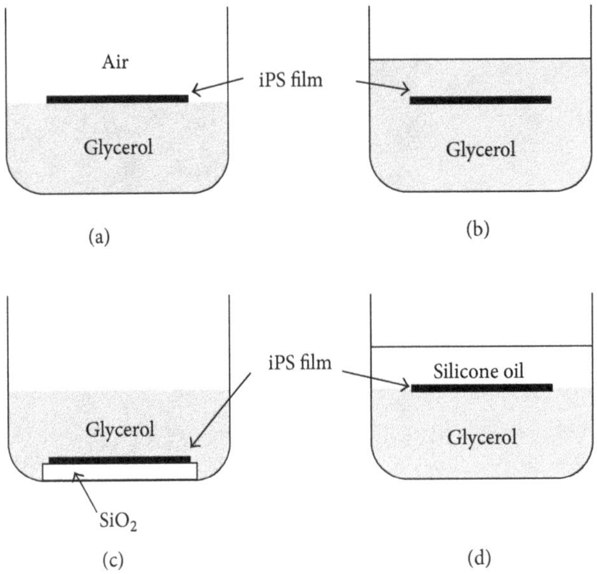

FIGURE 1: Schematics of the iPS film possessing polymer/liquid interfaces; (a) air/iPS/glycerol, (b) glycerol/iPS/glycerol, (c) glycerol/iPS/SiO_2, and (d) silicone oil/iPS/glycerol.

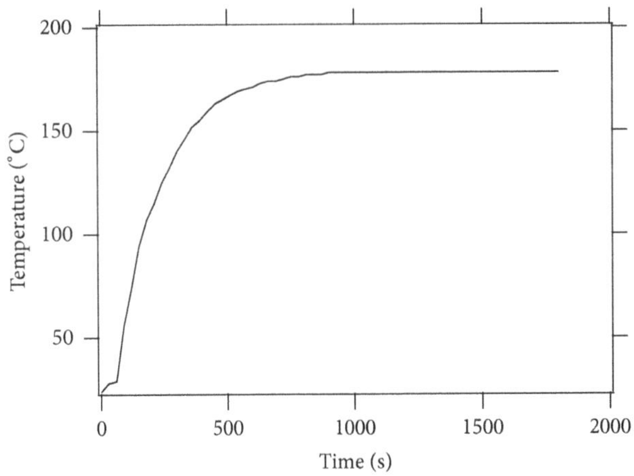

FIGURE 2: Typical profile of sample temperature versus time during the heating for cold crystallization. Here the final crystallization temperature $T_{c,f}$ was found to be 178°C.

The thin iPS films on various substrates thus prepared were subject to heating up to a final crystallization temperature $T_{c,f}$ of 160–178°C in an oil heating bath. The time-evolution of the sample temperature during the heating for crystallization ($T_{c,f}$ = 178°C) is shown in Figure 2; there was a rather long period for temperature raising; that is, the annealing was not isothermal. Similar profiles were observed for $T_{c,f}$ = 160–178°C. The gradual temperature rising was unavoidable because of a large heat capacity of the oil bath. The typical size of the liquid substrate was 10 g. The crystallized thin film was moved onto a microgrid covered with an elastic carbon layer for TEM observation. Here, we also prepared specimens with the film being upside down for the samples possessing two different interfaces, that is, air/iPS/glycerol,

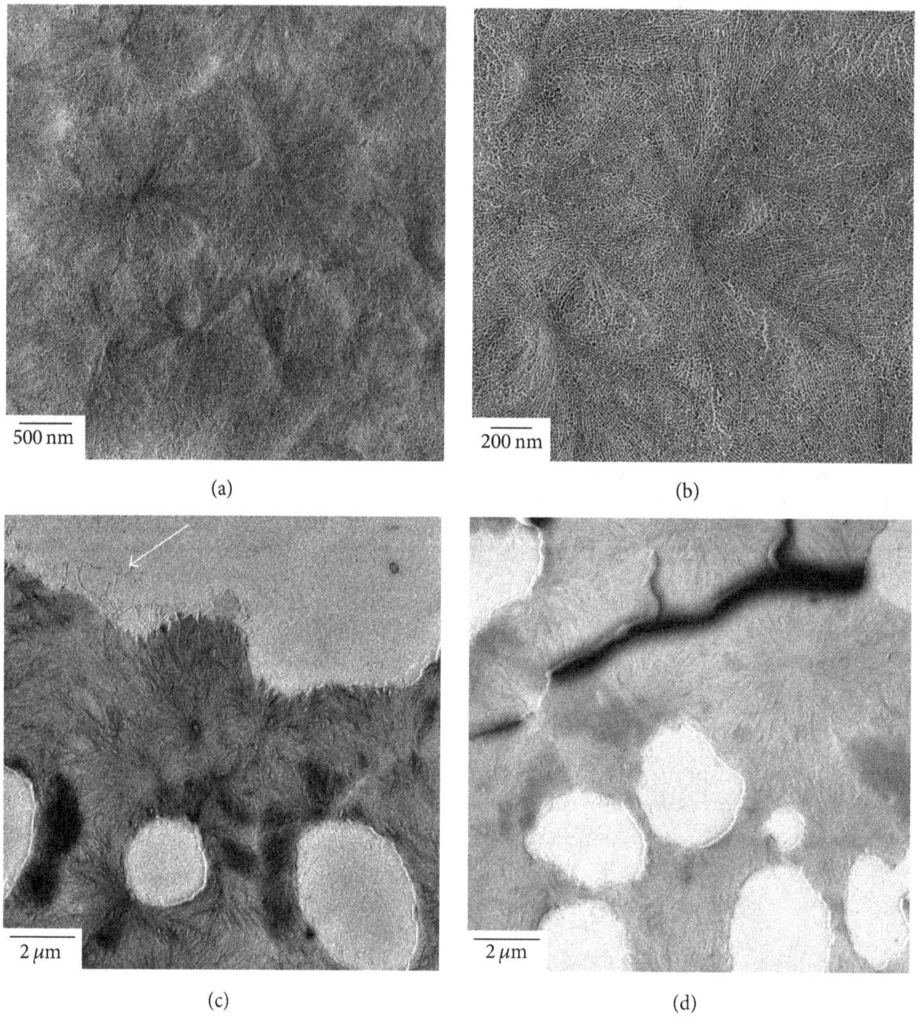

FIGURE 3: TEM images of iPS thin films crystallized via a temperature profile with $T_{c,f}$ = 160°C (a and b) and 178°C (c and d) on glycerol substrates (air/iPS/glycerol). A fibril-like texture is seen outside the edge of the film, as indicated by the arrow in (c).

glycerol/iPS/SiO_2, and silicone oil/iPS/glycerol. The specimen was dried under vacuum for several hours at room temperature. Finally, the upper surface of the specimen was shadowed with Au for the TEM observation.

3. Results and Discussion

We confirmed from the TEM observation that spin-cast iPS film without annealing is homogeneously flat and it does not exhibit any signs of crystallization nor holes due to dewetting. Therefore, the morphologies of annealed films described below were formed via cold crystallization on annealing. Figure 3 shows TEM images of iPS thin films crystallized on glycerol substrate (air/iPS/glycerol). We found that the films crystallized via a temperature profile of $T_{c,f}$ = 160°C are wholly packed with flat spherulites (Figures 3(a) and 3(b)), which have grown in an almost two-dimensional manner. In the case of $T_{c,f}$ = 178°C, on the other hand, we observed holes of several μm in size in the film, which resulted from dewetting (Figures 3(c) and 3(d)). In some of the holes, fibril-like morphology was observed as indicated by an arrow in Figure 3(c). This texture was considered to be formed on glycerol surface on annealing: after the formation of the hole, a very thin layer of iPS still remains on the dewetted surface of glycerol (pseudo-dewetting), which underwent crystallization to form thin crystallites. A similar behavior has been reported for PEO thin layers on silicone substrates, where finger-like branched structures are formed [8, 9]. The present thin texture that formed in the hole is composed of thin curved wires rather than a finger-like structure. Similar morphology of curved array that consists of lamellae has been reported to occur from the edge of a crystalline sheaf for melt-grown iPS crystallites [17].

Figure 4 shows TEM images of iPS thin films of which both sides contacted with glycerol on crystallization (glycerol/iPS/glycerol). The film exhibited a greater number density of spherulites than that of air/iPS/glycerol as shown in Table 1. This suggests that the primary nucleation is promoted at the iPS/glycerol interface compared with the

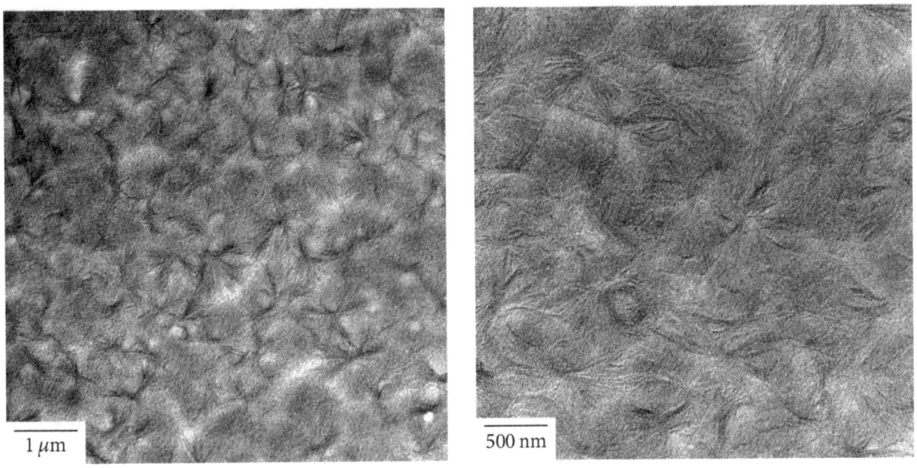

(a) (b)

FIGURE 4: TEM images of an iPS thin film crystallized with both sides contacted with glycerol (glycerol/iPS/glycerol). The final crystallization temperature $T_{c,f}$ was 178°C.

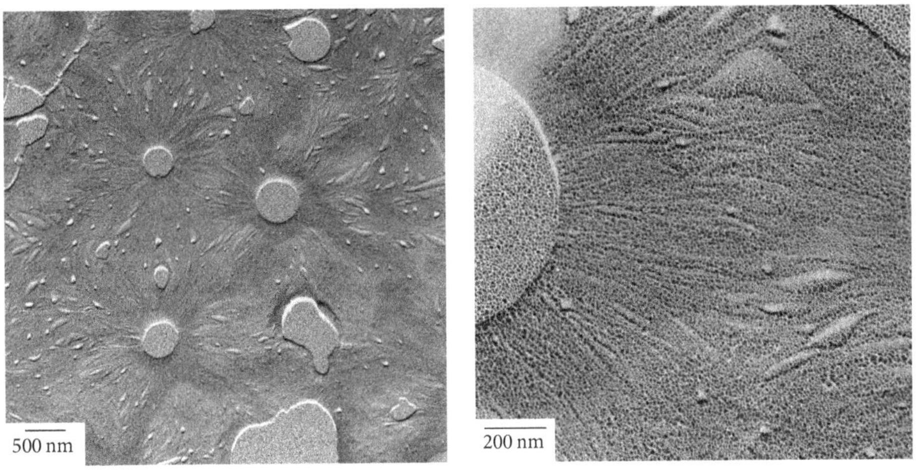

(a) (b)

FIGURE 5: TEM images of an iPS thin film crystallized between glycerol and SiO_2 substrates (glycerol/iPS/SiO_2). The final crystallization temperature $T_{c,f}$ was 178°C.

TABLE 1: Number density of spherulites n for $T_{c,f}$ = 178°C.

Interfaces	$n/\mu m^{-2}$
Air/iPS/glycerol	0.15
Glycerol/iPS/glycerol	1.8
Glycerol/iPS/SiO_2	0.38
Silicone oil/iPS/glycerol	0.024

air/iPS interface; that is, glycerol tends to promote the primary nucleation of iPS crystal. It is also noted that no hole formation occurred in this glycerol/iPS/glycerol thin film on crystallization; this suggests that shrinkage as a whole occurs rapidly in viscous glycerol rather than forming small holes that could act as nuclei of hole formation. Furthermore, it may be considered that, for the glycerol/iPS/glycerol system, the primary nucleation is activated on both sides of the iPS film by glycerol, leading to an enhanced crystallization rate. As a result, the film becomes rigid before the hole formation starts to occur via dewetting.

Figure 5 shows TEM images of the film crystallized between the glycerol and SiO_2 substrate (glycerol/iPS/SiO_2). Here again, holes due to dewetting are observed. In this system, the iPS film is attached to the solid substrate which prevents shrinking as a whole; as a result, small holes are sporadically formed. The size of the holes was less than 1 μm, which is smaller than those for the air/iPS/glycerol system. This may be due to a stronger adhesion to the iPS/SiO_2 interface, which limits the growth of the hole. In addition, it has been reported that the relaxation of polystyrene chains at an SiO_2 interface is rather hindered [19]. We should also note

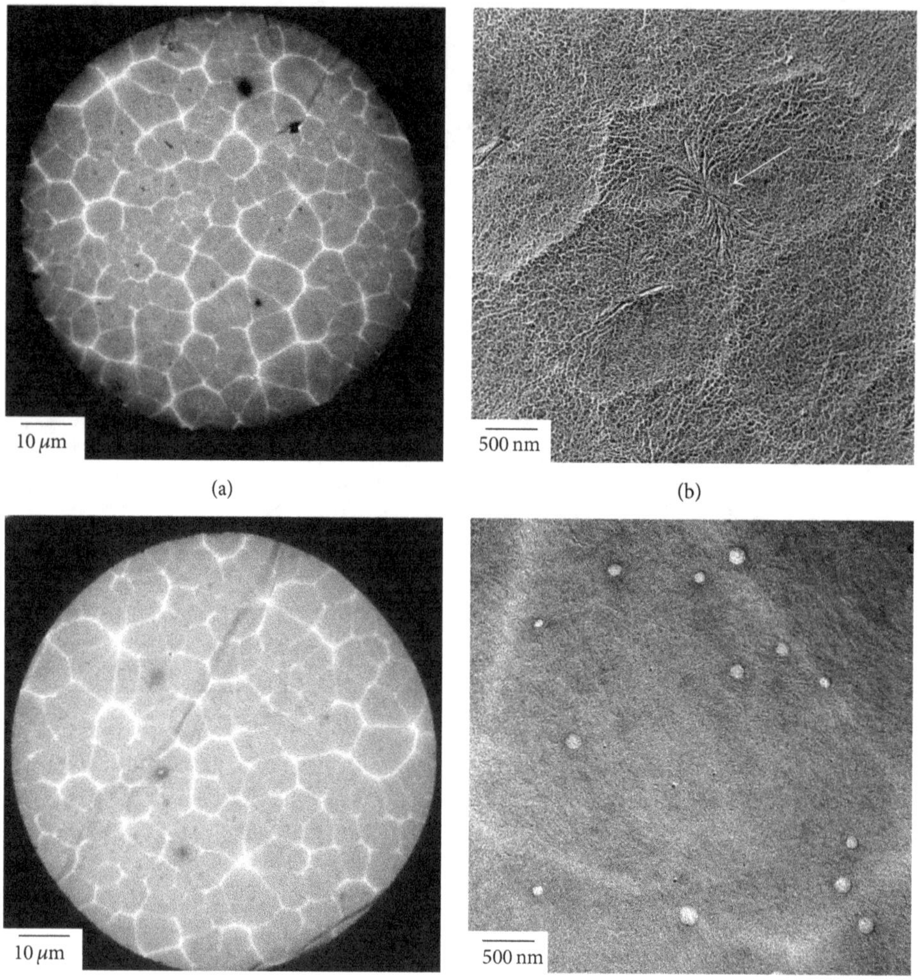

(a) (b) (c) (d)

FIGURE 6: TEM images of iPS thin films crystallized between silicone oil and glycerol (silicone oil/iPS/glycerol); Au was deposited on the side that contacted with glycerol (a and b) and with silicone oil (c and d). The final crystallization temperature $T_{c,f}$ was 178°C. The arrow in panel (b) indicates a texture of dominant lamellae.

the lamellae that have grown in the direction perpendicular to the contour of the hole (transcrystallization). This suggests that the primary nucleation preferentially occurs at the edge of the hole. This indicates that the formation of the holes due to dewetting occurs first, and then the crystallization starts. The number density of the spherulites for glycerol/iPS/SiO_2 is smaller than that for glycerol/iPS/glycerol, while it is greater than that for air/iPS/glycerol (Table 1). The former result probably indicates that SiO_2 has a weaker tendency to activate the nucleation than glycerol. The latter may be due to the holes that provide nucleation sites for transcrystallization, which leads to a greater number of spherulites. The air/iPS/glycerol system also exhibits holes, but no morphological signs of transcrystallization were observed at the hole edge.

The tendency of preferential nucleation at the iPS/glycerol interface was further confirmed by the morphology of the crystallized film between silicone oil and glycerol (silicone oil/iPS/glycerol). Figure 6 shows TEM images of the thin films of silicone oil/iPS/glycerol. The film was wholly packed with impinged spherulites, and no holes due to dewetting were observed. Au shadowing was executed on the side contacted with glycerol (Figures 6(a) and 6(b)) and on the opposite (silicone oil) side (Figures 6(c) and 6(d)). The average size of the spherulites was almost the same for the images observed on both sides, which implies that the spherulites are flat and penetrate from one side to the other. However, we found a detailed difference between the textures on the two sides: on the glycerol side, a fine texture of initial dominant lamellae is seen in the center part of the spherulite, which formed at an early stage of crystallization, as indicated by an arrow in Figure 6(b), whereas no such texture is observed on the silicone oil side (Figure 6(d)). This again indicates that the primary nucleation preferentially occurs at the glycerol interface, suggesting that glycerol has a strong tendency to activate the primary nucleation compared with silicone oil.

Dewetting behavior is mainly governed by the balance of interfacial force. Zisman's critical surface tension γ_c serves as a criterion of wettability of liquid to a substrate [20, 21]. When

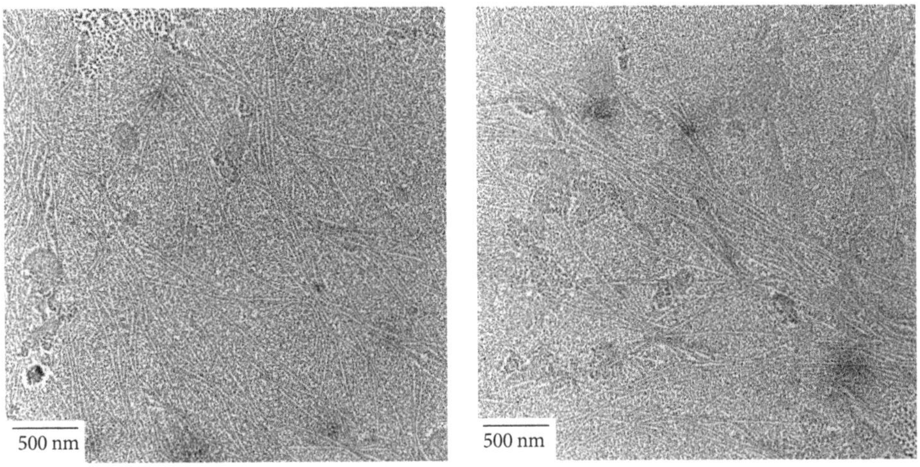

(a) (b)

FIGURE 7: TEM images of an iPS thin film crystallized on SiO_2 substrate (air/iPS/SiO_2). The final crystallization temperature $T_{c,f}$ was 178°C.

a surface tension γ of a liquid is greater than γ_c, dewetting is expected to occur. The critical surface tension of wetting for polystyrene was estimated to be 33 mN m^{-1} [21], while γ (glycerol) = 63.3 mN m^{-1} and γ (silicone oil) = 14.6 mN m^{-1} at 25°C. Thus, glycerol is considered to promote the dewetting of iPS film, while silicone oil wets polystyrene. We estimated the interfacial free energy at 25°C for glycerol/iPS and silicone oil/iPS interfaces to be 35.4 and 14.7 mN m^{-1}, respectively, while the surface tension of polystyrene is 40.7 mN m^{-1} at 25°C (this is not equal to γ_c). Hole formations due to dewetting were observed for the systems of air/iPS/glycerol and glycerol/iPS/SiO_2; in these systems, glycerol plays a role in promoting the hole formation. In contrast, no holes formed for the system of silicone oil/iPS/glycerol; in this case, dewetting itself may be suppressed by silicone oil.

As for the system of air/iPS/SiO_2, only branched lamellar textures formed on annealing at 178°C, but no mature spherulites were observed as shown in Figure 7. We found no hole formation. In this system, the iPS layer seems to be strongly attached to the glass substrate as reported in [19] so that dewetting is inhibited. As a result the polymer layer is forced to retain the original geometry of a thin film. In such a confined system, lamellar growth in the direction normal to the film surface is rather limited, which results in the incomplete formation of spherulite.

We now discuss the origin of the observed interfacial effects on the primary nucleation described above. We speculate that the primary nucleation preferentially occurs at the interface of the iPS film; that is, the substrate acts as a nucleating reagent. The size of critical nucleus depends on the surface free energy [22]. This is smaller for a nucleation at an interface (phase boundary) than that in a bulk polymer phase because the increment of interfacial area due to nucleation is smaller for the former. It has been evidenced that the contacting phase material (solid, nonsolvent liquid, or air) significantly influences the conformational local structure of atactic polystyrene at the interface [23–28]. The orientation of the phenyl groups of polystyrene has been reported to be dependent on the hydrophobicity and surface free energy of the substrate. In particular, a tilted orientation of the phenyl group has been revealed for hydrophobic liquids and air, whereas a flat orientation has been observed for hydrogen bonding liquids such as water and glycerol due to a phenyl π-hydrogen bond interaction [27]. The flat arrangement of the phenyls might lead to efficient chain packing such as 3/1 helix of iPS crystal [29], which promotes the primary nucleation.

In contrast, the present result shows that silicone oil is less active in promoting the nucleation than glycerol (Figure 6). The number density of spherulites for the silicone oil/iPS/glycerol system was found to be very small compared with the glycerol/iPS/glycerol system (Table 1). This suggests that the silicone oil/iPS interface is extremely unfavorable for the primary nucleation. It has been revealed that hydrophobic liquids, such as hexane, form a diffuse boundary on the polystyrene surface (surface swelling) [28]. In the present study, hydrophobic silicone oil may behave similarly as a surface plasticizer. The nanoscaled surface swelling may lower the equilibrium melting temperature, which leads to a reduction of the degree of supercooling ΔT. According to a standard nucleation theory [22], the rate of nucleation depends strongly on ΔT; that is, the nucleation rate is proportional to exp $[-A/(\Delta T)^2]$, where A is a constant. This means that even a small decrease in ΔT induces a noticeable reduction in the nucleation rate. Thus it is likely that even the nanoscaled surface swelling can suppress the nucleation behavior as far as the nucleation occurs at the interface. Furthermore, we obtained consistent results of a greater crystallization rate at the iPS/glycerol interface than at the iPS/silicone oil interface from chip nanocalorimetry measurements (the method has been developed for the thermal analysis of nanosized materials [30, 31]). The details of this study will be reported in a separate paper.

4. Conclusions

In this paper, we demonstrated pronounced effects of interface on the morphology of the spherulites grown in iPS ultrathin films possessing interfaces with liquids. A strong tendency of promoting the primary nucleation of iPS crystallization at the glycerol/iPS interface was found, whereas the nucleation was shown to be unfavorable at the silicone oil/iPS interface. This may be attributed to a specific interaction between the hydroxyl groups of glycerol and the phenyl groups of iPS, as proposed in [27]. Hole formation due to dewetting was observed for the thin films of air/iPS/glycerol and glycerol/iPS/SiO_2, while no holes were found for those of glycerol/iPS/glycerol and silicone oil/iPS/glycerol. In the former case, the holes significantly affect the crystalline morphology: a fibril-like texture in the hole was seen for air/iPS/glycerol, and a transcrystallization morphology near the edge of the hole was observed for glycerol/iPS/SiO_2. To elucidate further the effect of molecular interactions at the interface on polymer crystallization process, experiments are to be carried out for variety of polymers and liquid substrates that possess specific functional groups.

Competing Interests

The authors declare that there is no conflict of interests regarding the publication of this paper.

Acknowledgments

The authors are indebted to Otsuka Electronics, Co., Ltd., for the ellipsometry measurements of film thickness.

References

[1] J. L. Keddie, R. A. L. Jones, and R. A. Cory, "Size-dependent depression of the glass transition temperature in polymer films," *Europhysics Letters*, vol. 27, no. 1, pp. 59–64, 1994.

[2] M. Alcoutlabi and G. B. McKenna, "Effects of confinement on material behaviour at the nanometre size scale," *Journal of Physics: Condensed Matter*, vol. 17, no. 15, pp. R461–R524, 2005.

[3] M. D. Ediger and J. A. Forrest, "Dynamics near free surfaces and the glass transition in thin polymer films: a view to the future," *Macromolecules*, vol. 47, no. 2, pp. 471–478, 2014.

[4] J.-U. Sommer and G. Reiter, "Crystallization in ultra-thin polymer films: morphogenesis and thermodynamical aspects," *Thermochimica Acta*, vol. 432, no. 2, pp. 135–147, 2005.

[5] V. H. Mareau and R. E. Prud'homme, "In-situ hot stage atomic force microscopy study of poly(ε-caprolactone) crystal growth in ultrathin films," *Macromolecules*, vol. 38, no. 2, pp. 398–408, 2005.

[6] K. Taguchi, H. Miyaji, K. Izumi, A. Hoshino, Y. Miyamoto, and R. Kokawa, "Growth shape of isotactic polystyrene crystals in thin films," *Polymer*, vol. 42, no. 17, pp. 7443–7447, 2001.

[7] Y. Duan, Y. Jiang, S. Jiang, L. Li, S. Yan, and J. M. Schultz, "Depletion-induced nonbirefringent banding in thin isotactic polystyrene thin films," *Macromolecules*, vol. 37, no. 24, pp. 9283–9286, 2004.

[8] G. Reiter and J.-U. Sommer, "Crystallization of adsorbed polymer monolayers," *Physical Review Letters*, vol. 80, no. 17, pp. 3771–3774, 1998.

[9] G. Reiter and J.-U. Sommer, "Polymer crystallization in quasi-two dimensions. I. Experimental results," *The Journal of Chemical Physics*, vol. 112, no. 9, pp. 4376–4383, 2000.

[10] G. Dorenbos, J.-U. Sommer, and G. Reiter, "Polymer crystallization on pre-patterned substrates," *Journal of Chemical Physics*, vol. 118, no. 2, pp. 784–791, 2003.

[11] T. Sasaki, N. Yamauchi, S. Irie, and K. Sakurai, "Differential scanning calorimetry study on thermal behaviors of freeze-dried poly(L-lactide) from dilute solutions," *Journal of Polymer Science Part B: Polymer Physics*, vol. 43, no. 2, pp. 115–124, 2005.

[12] T. Sasaki, D. Morino, and N. Tabata, "Origin of enhanced cold crystallization rate for freeze-dried poly(L-lactide) from solutions," *Polymer Engineering and Science*, vol. 51, no. 9, pp. 1858–1865, 2011.

[13] A. Udagawa, T. Fujie, Y. Kawamoto, A. Saito, S. Takeoka, and T. Asahi, "Interfacial effects on the crystallization and surface properties of poly(l-lactic acid) ultrathin films," *Polymer Journal*, vol. 48, no. 2, pp. 157–161, 2016.

[14] H. Bodiguel and C. Fretigny, "Viscoelastic dewetting of a polymer film on a liquid substrate," *The European Physical Journal E*, vol. 19, no. 2, pp. 185–193, 2006.

[15] H. Bodiguel and C. Fretigny, "Viscoelastic properties of ultrathin polystyrene films," *Macromolecules*, vol. 40, no. 20, pp. 7291–7298, 2007.

[16] J. Wang and G. B. McKenna, "Viscoelastic and glass transition properties of ultrathin polystyrene films by dewetting from liquid glycerol," *Macromolecules*, vol. 46, no. 6, pp. 2485–2495, 2013.

[17] D. C. Bassett and A. S. Vaughan, "On the lamellar morphology of melt-crystallized isotactic polystyrene," *Polymer*, vol. 26, no. 5, pp. 717–725, 1985.

[18] K. L. Beers, J. F. Douglas, E. J. Amis, and A. Karim, "Combinatorial measurements of crystallization growth rate and morphology in thin films of isotactic polystyrene," *Langmuir*, vol. 19, no. 9, pp. 3935–3940, 2003.

[19] H. Tsuruta, Y. Fujii, N. Kai et al., "Local conformation and relaxation of polystyrene at substrate interface," *Macromolecules*, vol. 45, no. 11, pp. 4643–4649, 2012.

[20] W. A. Zisman, "Relation of the equilibrium contact angle to liquid and solid constitution," *Advances in Chemistry*, vol. 43, pp. 1–51, 1964.

[21] M. K. Chaudhury, "Interfacial interaction between low-energy surfaces," *Materials Science and Engineering: R: Reports*, vol. 16, no. 3, pp. 97–159, 1996.

[22] K. Armitstead and G. Goldbeck-Wood, "Polymer crystallization theories," *Advances in Polymer Science*, vol. 100, pp. 218–311, 1991.

[23] K. S. Gautam, A. D. Schwab, A. Dhinojwala, D. Zhang, S. M. Dougal, and M. S. Yeganeh, "Molecular structure of polystyrene at air/polymer and solid/polymer interfaces," *Physical Review Letters*, vol. 85, no. 18, pp. 3854–3857, 2000.

[24] K. A. Briggman, J. C. Stephenson, W. E. Wallace, and L. J. Richter, "Absolute molecular orientational distribution of the polystyrene surface," *The Journal of Physical Chemistry B*, vol. 105, no. 14, pp. 2785–2791, 2001.

[25] P. T. Wilson, K. A. Briggman, W. E. Wallace, J. C. Stephenson, and L. J. Richter, "Selective study of polymer/dielectric interfaces with vibrationally resonant sum frequency generation via

thin-film interference," *Applied Physics Letters*, vol. 80, no. 17, pp. 3084–3086, 2002.

[26] P. T. Wilson, L. J. Richter, W. E. Wallace, K. A. Briggman, and J. C. Stephenson, "Correlation of molecular orientation with adhesion at polystyrene/solid interfaces," *Chemical Physics Letters*, vol. 363, no. 1-2, pp. 161–168, 2002.

[27] C. S.-C. Yang, P. T. Wilson, and L. J. Richter, "Structure of polystyrene at the interface with various liquids," *Macromolecules*, vol. 37, no. 20, pp. 7742–7746, 2004.

[28] A. Horinouchi, N. L. Yamada, and K. Tanaka, "Aggregation states of polystyrene at nonsolvent interfaces," *Langmuir*, vol. 30, no. 22, pp. 6565–6570, 2014.

[29] G. Natta, P. Corradini, and I. W. Bassi, "Crystal structure of isotactic polystyrene," *Il Nuovo Cimento*, vol. 15, supplement 1, pp. 68–82, 1960.

[30] H. Huth, A. A. Minakov, and C. Schick, "Differential AC-chip calorimeter for glass transition measurements in ultrathin films," *Journal of Polymer Science Part B: Polymer Physics*, vol. 44, no. 20, pp. 2996–3005, 2006.

[31] M. Ahrenberg, E. Shoifet, K. R. Whitaker, H. Huth, M. D. Ediger, and C. Schick, "Differential alternating current chip calorimeter for in situ investigation of vapor-deposited thin films," *Review of Scientific Instruments*, vol. 83, no. 3, Article ID 033902, 2012.

3

Numerical Research on the Pullout Failure of GFRP Bolt

Yanqing Wang,[1] Beicheng Wu,[1] Yu Guo,[2] and Haifeng Yang[3]

[1]*School of Materials Science & Engineering, China University of Mining & Technology, Xuzhou 221116, China*
[2]*School of Mines, China University of Mining & Technology, Xuzhou 221116, China*
[3]*School of Mechatronic Engineering, China University of Mining & Technology, Xuzhou 221116, China*

Correspondence should be addressed to Yu Guo; guoyu@cumt.edu.cn and Haifeng Yang; yhf002@163.com

Academic Editor: Carlo Santulli

The fiber pullout is a main failure after the fiber is broken in the tension of glass fiber reinforced polymer (GFRP) bolt. In this paper, the numerical analysis is done on the distribution of both fiber normal force and interface shear stress. The results show that, on the ideal interface, the fiber pullout occurs from the lower end to the upper end of the matrix gradually, and both the normal stress of the fiber and the shear stress of the ideal interface gradually increase from the lower end to the upper end. With the increase of the interface layer thickness, the shear stress concentration area on the interface is enlarged while the stress applied is reduced, and the displacement of GFRP deformation is increasing sharply. This means that the capacity of GFRP deformation is enhanced. As a soft elastic body, the interface layer with a smaller elastic modulus can make the fiber stress and the interface shear stress sharply small and well dispersed. In addition, the load can be effectively transferred to the reinforced phase fibers in a bigger interfacial layer elastic modulus with a certain strength.

1. Introduction

The glass fiber reinforced polymer (GFRP) is widely used as structural composite material in the varied fields of engineering construction which is composed of the adhesive solution of resins (unsaturated polyester resin, epoxy resin, phenolic resin, etc.) as a matrix phase and the glass fiber as an enhancement phase. Compared with the traditional bolt, GFRP bolt has such characteristics as excellent corrosion resistance, high strength, low cost, and easy installation.

This paper focuses on the pullout failure at the fiber broken point which is one of main failure patterns of GFRP bolt [1]. However, as the GFRP bolt is mainly used in underground tunnel and construction support, the mechanical research on GFRP bolt support is mainly concentrated in the field of macromechanics of bolt distribution and installation by the early literatures, and the researches on the microscopic damage of GFRP bolt are not enough.

Recently, there are two kinds of important theoretical research development in the microscopic mechanical field of GFRP composite. One is application and development of mesomechanics [2–4], and the other one is the proposed property study of interface phase structure [5–7]. The fiber-matrix interface phase behavior plays an important role in determining the mechanical properties of GFRP composites such as elastic modulus, tensile strength, and fracture toughness, and it has received considerable attention and extensive investigations. Various interfacial interaction models based on classical shear-lag theory have been adopted in analysis of the stress transfer between enhancement phase and matrix phase, especially, it was a mature way to combine shear-lag model and Monte-Carlo simulation method [8]. First of all, the van der Waals interfacial interaction between the carbon nanotubes (CNTs) and polymer matrix and the interfacial chemical bonding have been modelled with the Lennard-Jones potential [9], and the many-body bond-order potential [10, 11], respectively, in analysis of the shear strength of CNTs-polymer interfaces. Modified classical shear-lag theory with introducing nonlinear interface laws for CNTs-polymer interfaces has been established based on the van der Waals interfacial interaction [12, 13] and chemical bonding [14], respectively. And cohesive zone finite element models with a nonlinear interface cohesive law have been established to model the pullout response of CNTs-coated fibers [15, 16]. However, linear interface law was also proposed for

TABLE 1: Physical and mechanical property parameters of the fiber and the matrix.

	Density kg/mm^{-3}	Elastic modulus/GPa	Poisson's ratio	Yield strength GPa	Yield strain%
Fiber	2.40×10^{-6}	72	0.3	1.6	2.3
Matrix	1.20×10^{-6}	2.2	0.25	0.1	3.12

CNTs pullout from a polymer matrix to investigate fracture toughness enhancement and bond breaking [17]. Moreover, shear-lag behaviors widely exist in the following composite fields such as cementitious composites, ceramic matrix composites, metal matrix composites, and nanocomposites. A type of steel-concrete composite structure with double-box (the DBSCCS model) was proposed to illustrate and predict shear-lag characteristics by obtaining its strain functions and shear-lag coefficients. It provides more reference value for engineering design and structure optimization in some extent for the steel-concrete composite structures [18]. Damage and failure process of the unidirectional fiber reinforced ceramic matrix composites (FRCMCs) under tensile stress was studied with consideration of evolving compliant interface (ECI) which can respond to the failure of debonding and sliding of interface efficiently [19]. The strengthening behavior of SiC reinforced Al matrix composites was well explained by the modified shear-lag model [20]. Short-fiber reinforced metal matrix composites were modelled by modified shear-lag theory based fatigue crack growth life prediction. It shows good agreement with the experimental data for the low cycle and high cycle fatigue applications [21]. An analytical approach to predict initiation and growth of interfacial delamination in the through-silicon via structure was developed by combining a cohesive zone model with a shear-lag model [22]. Interfacial shear stress transfer of a monolayer graphene on top of a polymer substrate subjected to uniaxial tension was investigated by a cohesive zone model integrated with a shear-lag model. It provides valuable insight and design guidelines for enhancing interfacial shear stress transfer in nanocomposites, stretchable electronics, and other applications based on graphene and other 2D nanomaterials [23].

The abundant applications of modified classical shear-lag theory with varying degrees of complexity in analyzing interfacial properties of composites above inspire us to establish an analytical model to discuss stress distribution and the pullout evolution for GFRP bolt. Of course, they can be determined accurately by the single fiber pullout experiment with less experimental material, and it is not affected by the strain performance of the matrix [24–26]. However, there are a large number of experimental samples needed in the experiments which are complexly prepared in a relatively small size (microns). The critical length of the fiber embedded in the matrix can only be obtained by constant experiments (the length of the fiber embedded in each sample is changed based on the result of the previous experiment). So, these samples cannot be prepared at the same time in advance. In addition, no common interface model or standard specification is presently formulated for single fiber pullout experiment; there are too many complexes influencing factors in the experiment, and the experiment is not repeated and multiple [27, 28]. We emphatically derive the governing constitutive equations for a single fiber pullout model with a cohesive fiber-matrix interface by introducing an interfacial bilinear cohesive law into the classical shear-lag model in Section 2.6. Then, we discuss shear stress in the interface, axial stress in the fiber, and the evolution of pullout with finite element calculations.

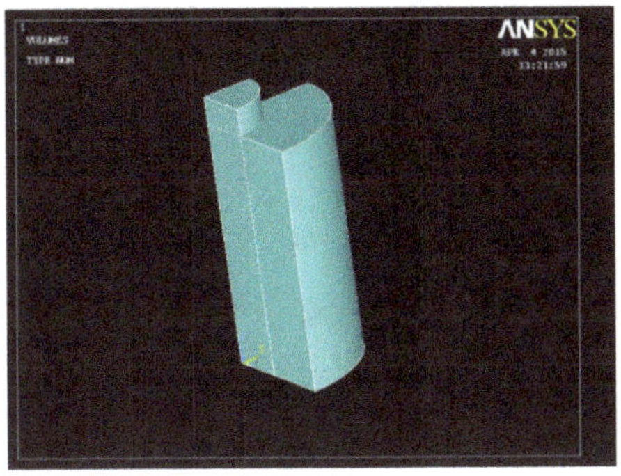

FIGURE 1: A quarter of the geometric model.

2. Establishment of Finite Element Model

2.1. Geometric Model. In the body of the GFRP bolt for the numerical research in this paper, the long and evenly distributed glass fiber and the bolt were isotropic and perpendicular to the cross-section, and the matrix was made of unsaturated polyester resin. The abstract model built for the calculations could be described as one fiber at a diameter of 8 μm that was embedded in a cylindrical matrix at a diameter of 18 μm, and each fiber had the same length of 40 μm. The lower end face of each fiber was fixed, and its upper one was pulled with a gradual increase of tensile. As the model was axisymmetric, only a quarter of the whole geometric model was used in order to reduce amount of calculation, as shown in Figure 1.

2.2. Material Properties. In the numerical analysis on GFRP bolt in this paper, the fiber and the matrix were homogeneous, that is, the impact of inhomogeneity on the fiber pullout was ignored. The elastic modulus and Poisson's ratio of the fiber and the matrix especially presented a negative change with temperature. Table 1 shows the physical and mechanical property parameters of the fiber and the matrix.

In the initial state, it was assumed that the interface between the fiber and the matrix was fine without debonding or sliding, and there were no such defects as bubble, impurity, and residual stress in the fiber and the matrix.

2.3. Definition of Boundary Conditions and Load. In the GFRP bolt numerical analysis in this paper, the upper

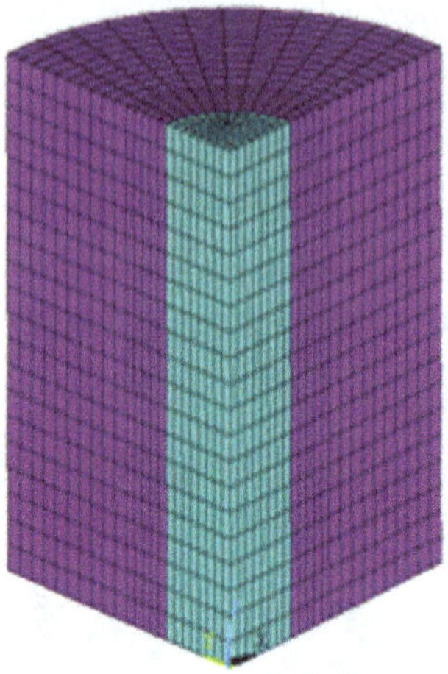

FIGURE 2: Mesh generation of the model.

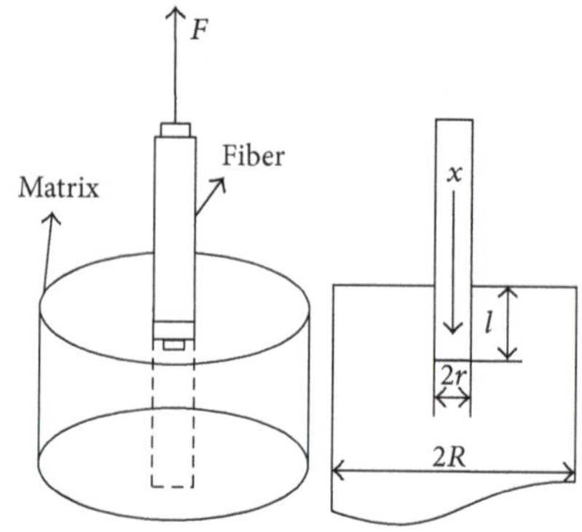

FIGURE 3: Analysis of stress concentration in the model of single fiber pullout.

direction parallel to the fiber axis was the positive direction of Z-axis, and the right direction perpendicular to the fiber axis was the positive direction of Y-axis; then the direction of X-axis was determined by Descartes coordinate system. The lower end face of the XOY plane was fixed. Only unidirectional stretching was loaded in the positive direction of Z-axis at a loading rate of 0.04 mm/s, and no compression, shear failure, or rotation occurred.

2.4. Unit Selection and Meshing. 163 explicit structural shell units were selected as the calculating nits, and they supported all nonlinear characteristics in the explicit dynamics numerical analysis [29, 30]. In addition, the fastest Belytschko-Tsay algorithm was applied [31]. The regular hexahedral element was easily used to divide the regular geometry model [31]. Therefore, the mapped mesh method was used which could reduce the total number of elements to improve the computational efficiency. The model after meshing is shown in Figure 2.

2.5. Material Model. In the constitutive relation of elastic plastic material, the bilinear isotropic (BISO) was represented to build a stress-strain curve [32]. In the BISO model, the stress-strain relationship curve was linear according to the bigger specific elastic modulus before the material yielding. However, it was linear according to the smaller specific elastic modulus after the material yielding. Moreover, the yield strength of the material increased in one direction, and so did the yield strength in the other direction accordingly.

2.6. Mechanical Model and Constitutive Equation. The mechanical model numerical built in this paper is shown in Figure 3.

The average shear strength of the interface τ_i is

$$\tau_i = \frac{F}{\pi d l}, \quad (1)$$

where F denotes load at a complete pullout of the fiber, l stands for the length of the fiber embedded in the matrix, and d is the diameter of the fiber.

Assuming the length of the fiber embedded in the matrix is X, and the stress of the fiber before pullout is δf, then, the following equation is satisfied [33]:

$$\frac{d^2\delta_f}{dx^2} = -\frac{2u_m\delta_F}{\ln(R/r)\,E_f r^2}, \quad (2)$$

where R is the radius of the matrix; r is the radius of the fiber; u_m is elastic modulus of matrix; and E_f is elastic modulus of the fiber.

Formula (2) can be integrated as follows:

$$\sigma_f = \frac{\sigma \sin[n(l-x)/r]}{\sin(nl/r)}. \quad (3)$$

When $x = 0$,

$$\sigma_f = \delta = \frac{P}{(\pi r^2)}. \quad (4)$$

When $x = l$,

$$\sigma_f = 0. \quad (5)$$

The equilibrium equation of shear-lag model is

$$\frac{d\sigma_f}{dx} = \frac{2\tau_i}{r}. \quad (6)$$

Formula (3) is substituted into Formula (6); τ_i is got as follows:

$$\tau_i = -\frac{n\sigma \mathrm{con}[n(l-x)/r]}{4\sin(nl/r)}. \quad (7)$$

By Formula (7), the shear stresses of the upper and lower ends of the glass fiber embedded in the matrix are τ_1 and τ_2, respectively:

$$\tau_1 = -\frac{n\sigma}{4\sin(nl/r)} \quad (8)$$

$$\tau_2 = -\frac{n\sigma\cot(nl/r)}{4}. \quad (9)$$

According to Formulas (8) and (9), τ_1 is smaller than τ_2 in the ideal interface.

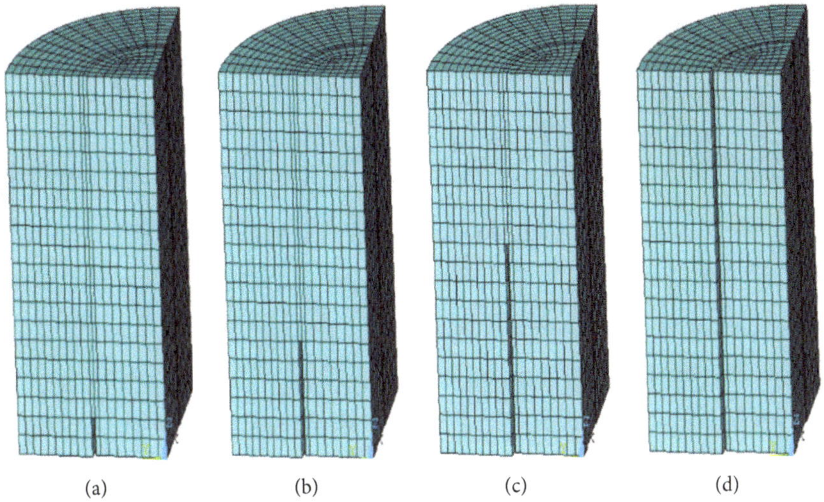

FIGURE 4: Development of the single fiber pullout on the ideal interface at (a) 10%, (b) 30%, (c) 50%, and (d) 100% of the whole fiber.

2.7. Failure Criterion. The numerical simulation for the pullout of single fiber was carried out in two steps. The first step was to judge if the fibers were broken or not. If not, the pullout of single fiber likely occurred; if yes, the length of fiber embedded in the matrix was reduced and the process above was repeated until the fiber was not broken. The second step was to repeat the numerical simulation according the new adjusted length of the fiber embedded in the matrix.

For the first step the max tensile stress criterion was applied; that is, the axial tensile stress at any point on the fiber was greater than or equal to the yield strength of the fiber, as expressed by the following Formula (10); the fiber was determined to be broken.

$$\sigma_f \geq \left[\sigma_f^s\right], \tag{10}$$

where σ_f is the axial tensile stress at any point on the fiber and $[\sigma_f^s]$ is the yield strength of the fiber.

For the second step the max shear stress criterion was applied; that is, the shear stress at any point on the interface was greater than or equal to the yield shear strength of the interface, as expressed by the following Formula (11); the fiber pullout occurred.

$$\tau_i \geq \left[\tau_i^s\right], \tag{11}$$

where τ_i is the shear stress at any point on the interface and $[\tau_i^s]$ is the yield shear strength of the interface.

Based on the above said, the numerical research on the pullout failure of GFRP bolt was done finally in the ANSYS14.0 LS-DYNA secondary development platform, using the numerical model built in the APDL-An ANSYS parametric design language, the extension program written by Visual C ++ language, and the calculating units chosen as 163 explicit structural shell units with the fastest Belytschko-Tsay algorithm and element life & death technology.

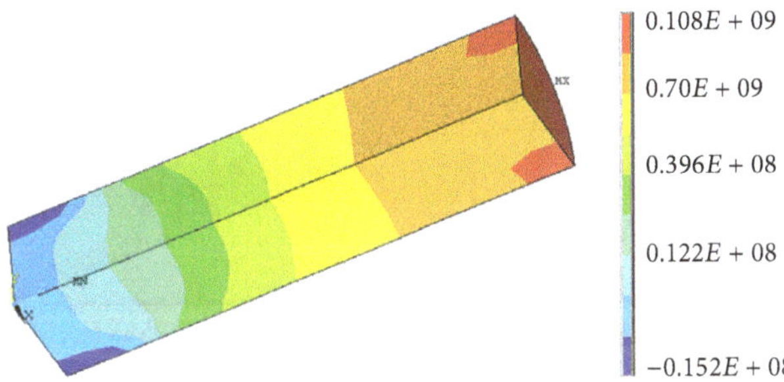

FIGURE 5: The axial stress distribution of the fiber with the ideal interface.

3. Result and Analysis

3.1. The Fiber Pullout on the Ideal Interface. Figure 4 shows the development of the single fiber pullout on the ideal interface. As can be seen from the figure, the fiber pullout occurs from the lower end of the matrix, gradually extending to the upper end, and the complete pullout of a single fiber can be realized finally. This is because there is a maximum shear stress at the lower end of the fiber embedded in the matrix. This agrees with the previous theoretical analyses of stress concentration at the end of the fiber and also presents the characteristics of unidirectional fiber reinforced composite [34–36].

The axial stress distribution of the fiber and the shear stress distribution of the ideal interface are shown in Figures 5 and 6, respectively. The axial stress of the fiber gradually increases from the lower end to the upper end, and the axial stress at the upper end is an order of magnitude compared with that of the lower end. It is consistent with the experimental result that the fiber is always broken at the upper end in the single fiber pullout experiments. However, the maximum shear stress of the ideal interface appears at the lower end of the matrix. Once the fiber pullout comes into being, the shear stress of the ideal interface gradually becomes

FIGURE 6: The shear stress distribution of the ideal interface.

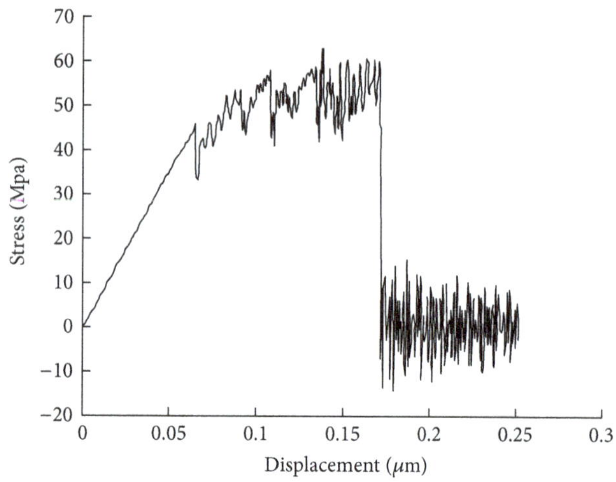

FIGURE 7: The axial stress-strain curve of the single fiber pullout under the condition of ideal interface.

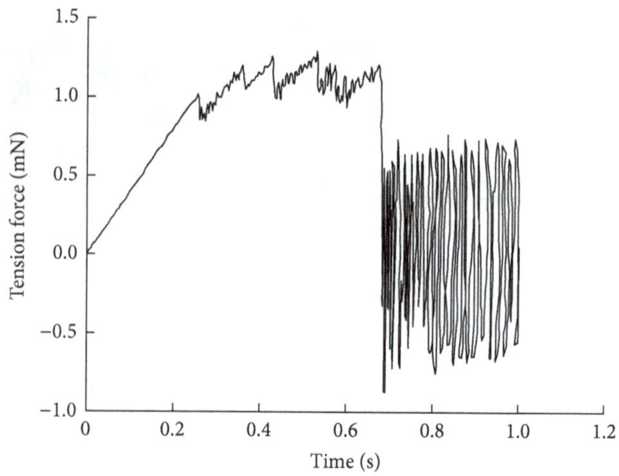

FIGURE 8: The tension force-time curve of the fiber under the condition of ideal interface.

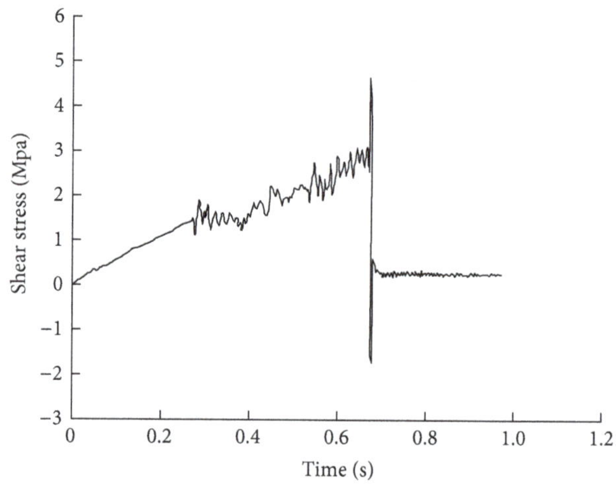

FIGURE 9: The shear stress-time curve of the ideal interface.

bigger and bigger from the broken point to the upper end of the matrix because of the energy release and shear stress reduction at every broken point.

Figure 7 shows the stress-strain curves of ideal interface during the pullout of single fiber. It can be divided into the four sections of linear-elastic deformation, partial pullout, complete pullout, and interface friction. This agrees with the conclusions drawn in literature [37, 38]. However, there is an obvious stress oscillation in the interface friction section in our study in a full consideration of the friction between the fiber and the matrix and the use of material model BISO.

The tension force-time curve of the fiber and the shear stress-time curve of the ideal interface are, respectively, shown in Figures 8 and 9. While $t = 0.6$ s, the tension force of the fiber and the shear stress of the ideal interface are 0.97 mN and 3.8 MPa, respectively. According to Formula (1) and considering calculation of a quarter of the whole model, the tension of the fiber force (F) is available as a theoretical value, that is, $F = 0.954$ mN, which is close to the numerical simulation value of 0.97 mN. Therefore, the accuracy of the numerical simulation is proved.

3.2. Effect of the Interface Layer Thickness on Single Fiber Pullout. The elastic modulus of the interface is set as 50 GPa, while the interface layer thickness values are set as 1/8, 2/8, and 3/8 of the fiber radius, and then the corresponding values are, respectively, 0.5 μm, 1 μm, and 1.5 μm. It should be noted that the interface is a separate layer and should be divided into fine grids to ensure the accuracy of the numerical simulation.

Figure 10 shows the development of single fiber pullout with the interface thickness of 3/8 of fiber radius. It is apparent that the fiber pullout starts from the lower end and then gradually extends to the upper end, and the complete pullout of a single fiber is available finally. The single fiber pullout with the other two interface thickness values develops in the same process. However, a bigger gap of pullout can be obviously observed with thickening of the bigger interface layer.

Figures 11 and 12 show, respectively, both the axial stress distribution of the fiber and the shear stress distribution of the interface at a moment of $t = 0.38$ s at different interface layer thicknesses of 1/8, 2/8, and 3/8 of the fiber radius. Without obvious effect from the interface layer thickness, no matter what the interface layer thickness value is, the axial normal

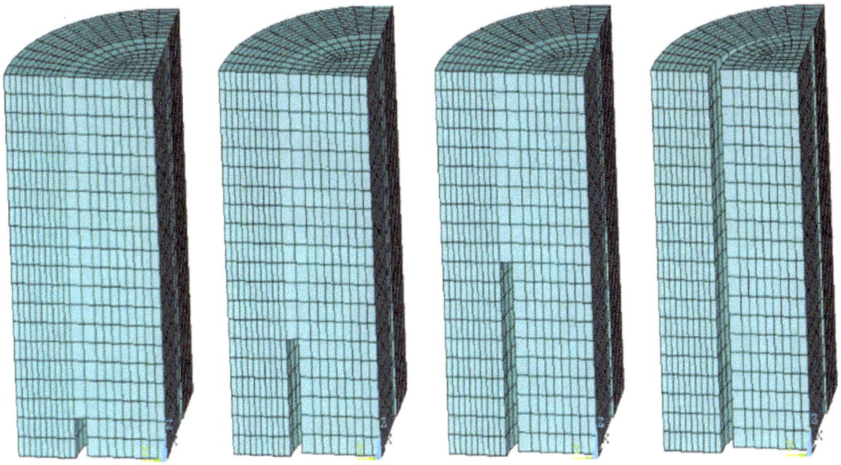

(a) (b) (c) (d)

FIGURE 10: Development of the single fiber pullout with the interface layer thickness of 3/8 of the fiber radius at (a) 10%, (b) 30%, (c) 50%, and (d) 100% of the whole interface layer.

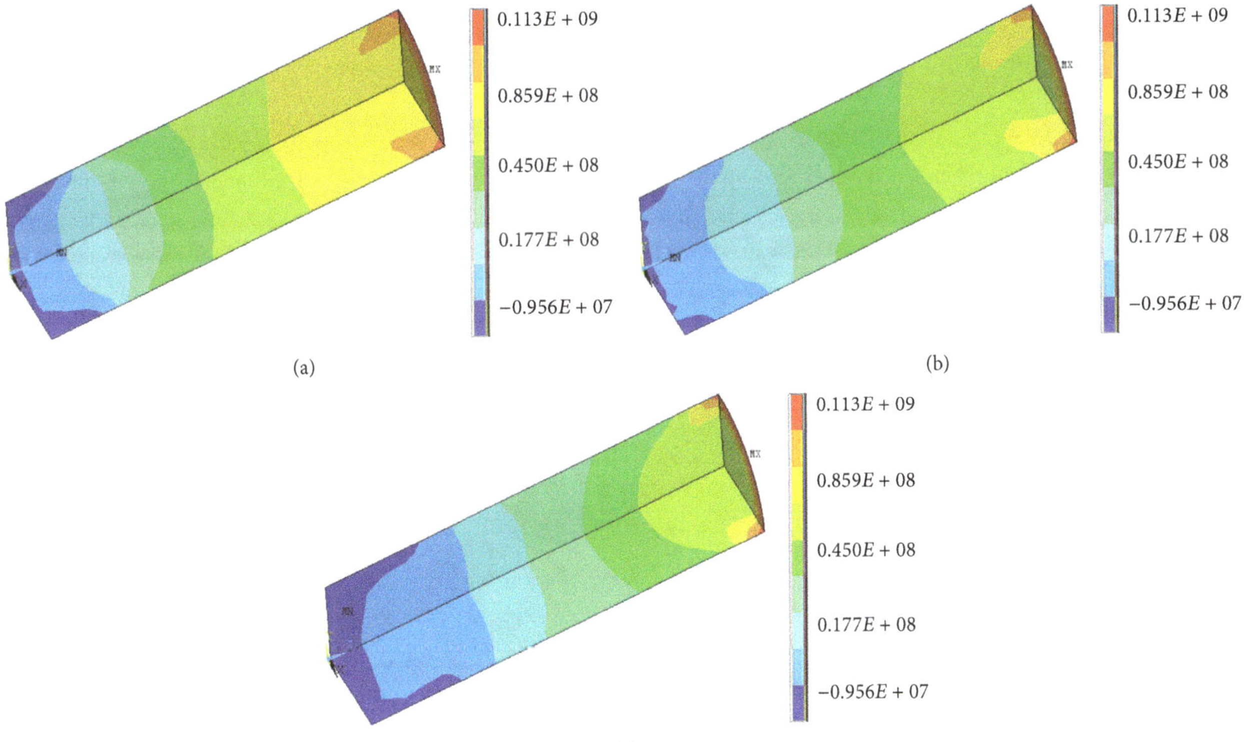

FIGURE 11: Axial stress distribution of the fiber with different interface layer thickness at(a) 1/8, (b) 2/8, and (c) 3/8 of fiber radius.

stress concentration of the fiber has the nearly same distribution with the bigger concentration at the upper end, as shown in Figure 11. However, with thickening of the interface layer, the shear stress concentration area of the interface is enlarged but the degree of shear stress concentration is reduced. This means that the interface layer thickness can effectively reduce the shear stress concentration, as shown in Figure 12.

Figures 13 and 14 show, respectively, the axial stress-strain curve of the fiber and the shear stress-strain curve of the interface with different interface layer thicknesses of 1/8, 2/8, and 3/8 of the fiber radius. For each stress-strain curve and shear stress-strain curve based on thickness values of interface layers, there is nearly the same varying curve which includes three stages: linear-elastic deformation, plastic deformation, and pullout. Because the elastic modulus (50 GPa) of interface is unchanged, the GFRP bolt has nearly the same linear-elastic deformation and is not affected by the interface layer thickness. However, both the axial stress

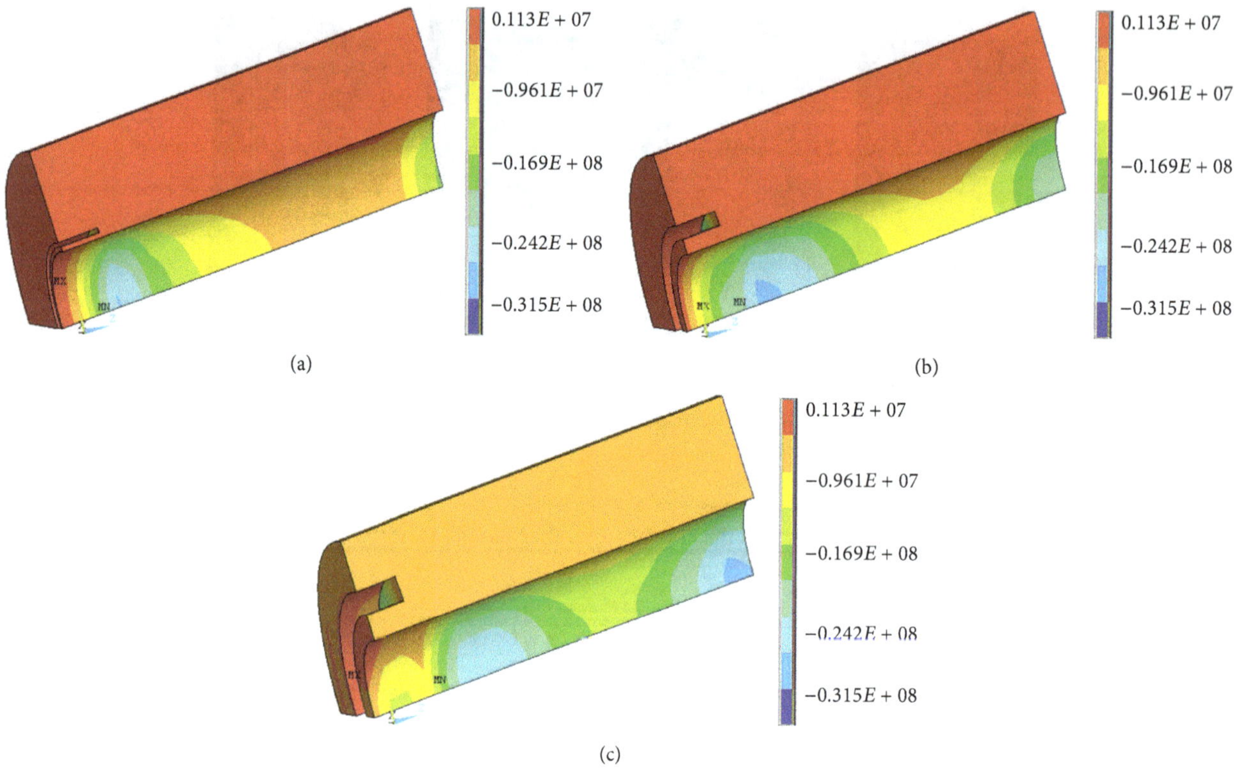

FIGURE 12: The shear stress distribution of the interface with different interface layer thickness at (a) 1/8, (b) 2/8, and(c) 3/8 of the fiber radius.

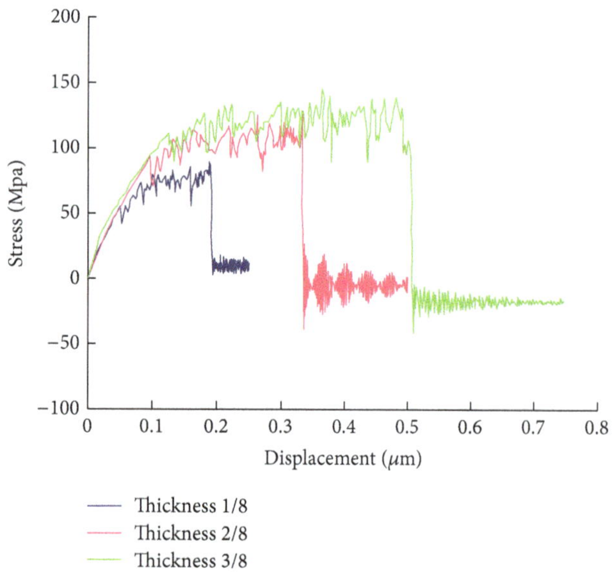

FIGURE 13: The axial stress-strain curve of the fiber in the single fiber pullout with the different thickness of interface layer.

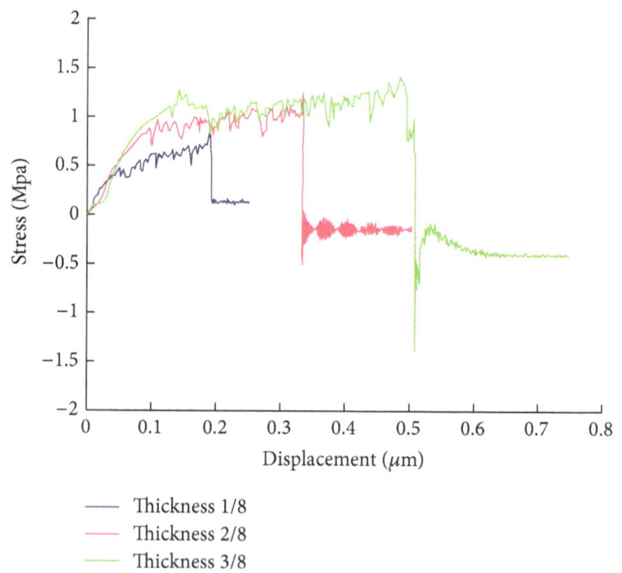

FIGURE 14: The shear stress-strain curve of the interface in the single fiber pullout with the different thickness of interface layer.

of the fiber and the shear stress of the interface increase a little bit and the displacement in the plastic deformation increases sharply, and the plastic deformation is enhanced as the interface layer thickness increases. There is obvious concussion observed in the pullout stage because of the friction of interface.

3.3. Effect of the Interface Layer Elastic Modulus on Single Fiber Pullout. The thickness of the interface layer is set as 1/8 of the fiber radius, that is, 0.5 μm, while the interface layer elastic moduli are set as 70 GPa, 50 GPa, and 2.5 GPa, respectively. Figures 15 and 16 show, respectively, both axial normal stress distribution of the fiber and shear stress distribution of the

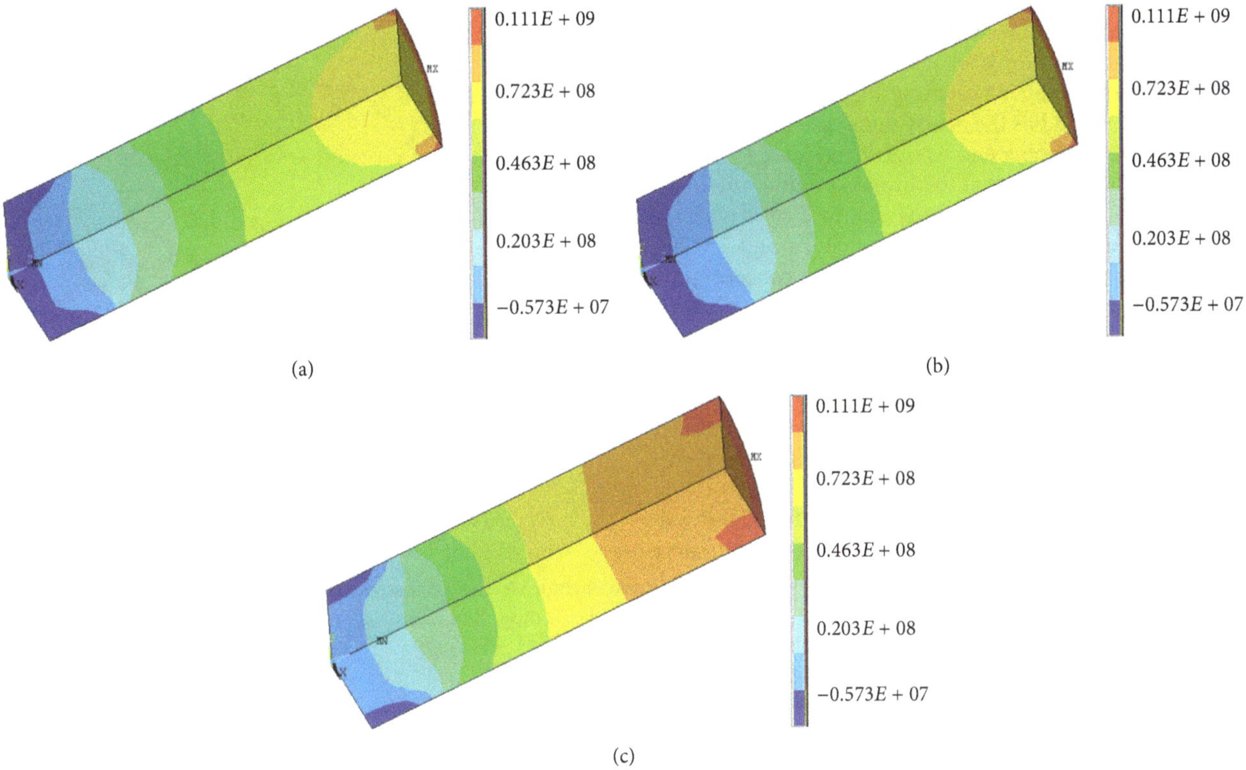

FIGURE 15: The axial stress distribution of the fiber at $t = 0.38$ s with different interface layer elastic modulus of (a) 70 GPa, (b) 50 GPa, and (c) 2.5 GPa.

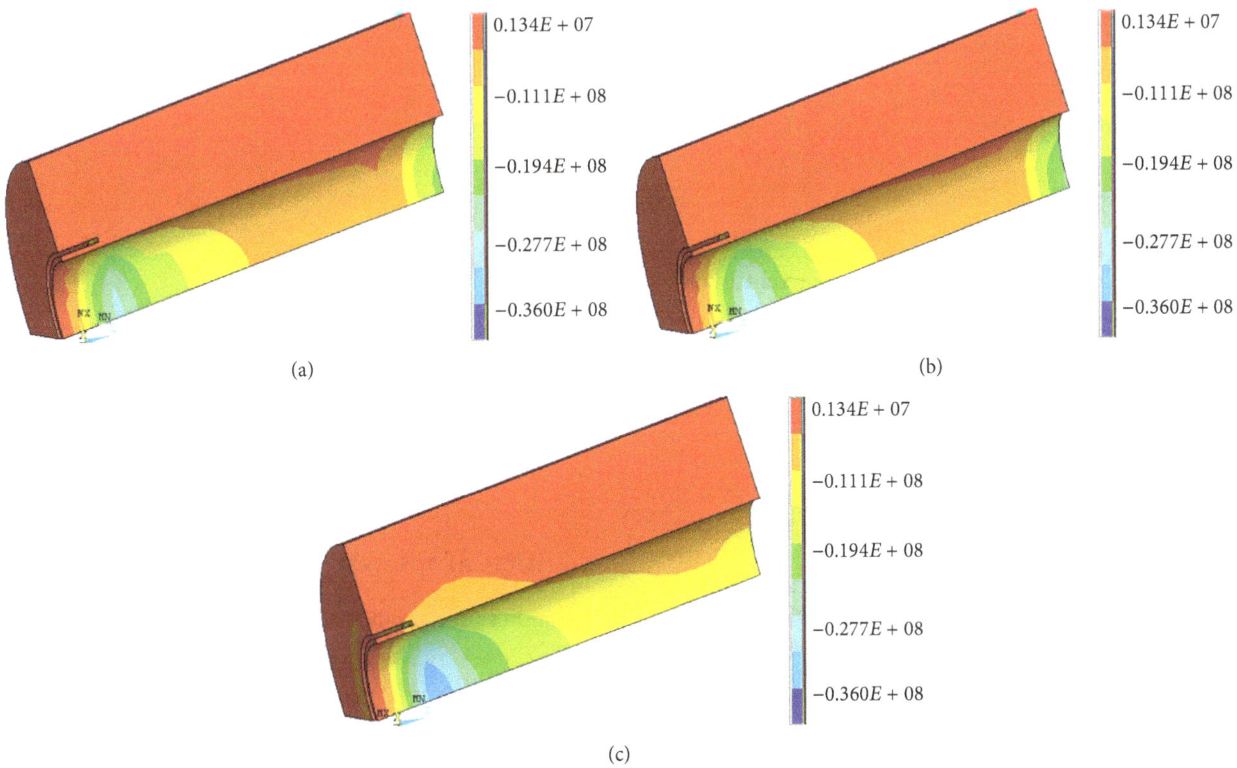

FIGURE 16: The shear stress distribution of the interface at $t = 0.38$ s with different interface layer elastic moduli of (a) 70 GPa, (b) 50 GPa, and (c) 2.5 GPa.

interface at the moment (t) of 0.38 s with different interface layer elastic moduli of 70 GPa, 50 GPa, and 2.5 GPa. No matter what the interface layer elastic modulus is, the axial normal stress of the fiber becoming bigger from the lower end to the upper end has the nearly same distribution with a concentration at the upper end, as shown in Figure 15. For the shear stress of the interface at the broken point of the lower end, there is a concentration on the interface layer with the three different elastic moduli of 70 GPa, 50 GPa, and 2.5 GPa. However, at the upper end, the shear stress concentration cannot be observed on the interface layer with an elastic modulus of 2.5 GPa but on the interface layer with bigger elastic moduli of 70 GPa and 50 GPa, as shown in Figure 16. This is because the shear stress of the interface can be well dispersed by the bigger plastic deformation on the interface layer with a smaller elastic modulus of 2.5 GPa. It corresponds with the viewpoints from literatures [39].

Figures 17 and 18 show, respectively, both the axial stress-strain curve of the fiber and the shear stress-strain curve of the interface with different interface layer elastic moduli of 70 GPa, 50 GPa, and 2.5 GPa. They have three stages of linear-elastic deformation, plastic deformation, and pullout and present nearly the same distribution with relative bigger interface layer elastic moduli of 70 GPa and 50 GPa. In contrast, the stress of the fiber and shear stress of the interface become less sharp and are obtained, respectively, with the smaller interface layer elastic modulus of 2.5 GPa. To carry the stress, the interface with the smaller elastic modulus, as a soft elastic body, can transfer the stress evenly on the entire interface to avoid the stress concentration. This is conducive to improving the macroductility of GFRP. Of course, the interface should have a certain elastic modulus to ensure that the load can be effectively transferred to the reinforce phase fibers at a certain strength. This is conducive to improving the macrostrength of GFRP [40]. Therefore, there is an optimal interfacial layer elastic modulus worthy of exploring.

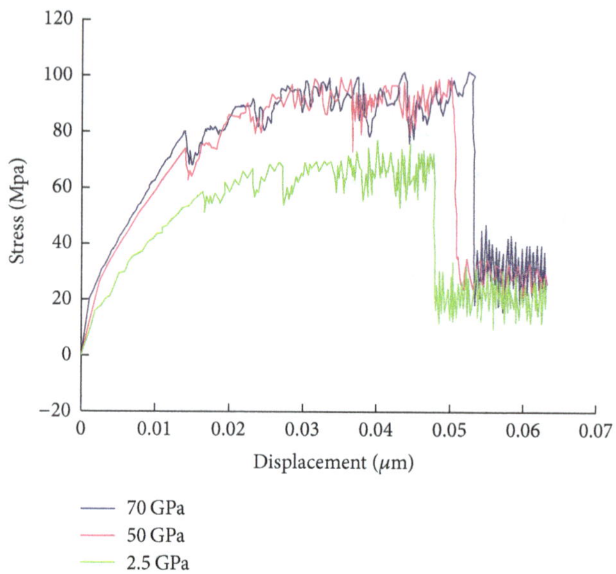

FIGURE 17: The axial stress-strain curve of the fiber with different elastic moduli of interface layer.

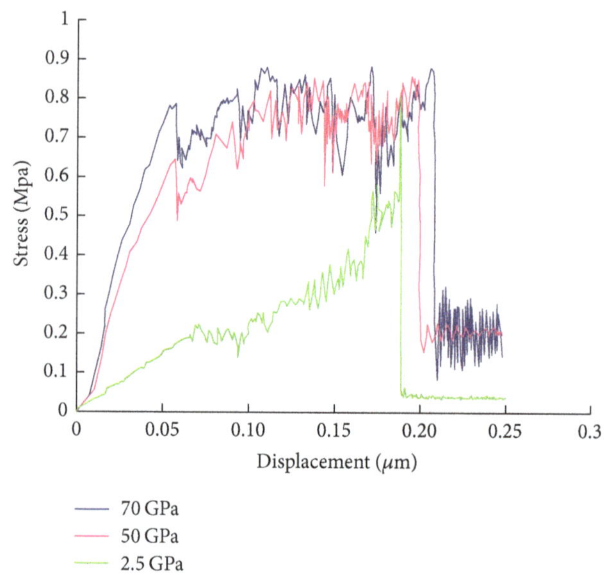

FIGURE 18: The shear stress-strain curve of the interface with different elastic moduli of interface layer.

4. Conclusions

The numerical mechanical research has been carried out for the pullout failure of the GFRP bolt in different interface conditions. There are three deformation stages included in the pullout process: linear-elastic deformation, plastic deformation, and pullout. The interface structure plays a key role for the mechanical behavior in the pullout of GFRP bolt. Firstly, there is a stress concentration at the fiber broken point at the lower and the upper ends of the matrix and the concentration at the lower end is distributed more widely and rigidly. Secondly, the interface with bigger layer thickness but less elastic modulus can make the interface shear stress distributed more evenly with less concentration. Moreover, the plastic deformation capacity of the interface increases, which means that GFRP bolt should present properties of high strength and ductility with the nonoccurrence of brittleness failure. It should be noted that it is not a fully predictive model because compressive stresses on the fiber due to matrix shrinkage, stress transfer in the interface, matrix cracking, and fiber breaking are not considered. Consequently, the limitations of the simplified models are unavoidable, and it can only essentially add to the knowledge base of numerical investigation for single fiber pullout.

Conflicts of Interest

The authors declare that they have no conflicts of interest.

Acknowledgments

The authors would like to express their thanks for the support of the Fundamental Research Funds for the Central Universities of China University of Mining and Technology (2017XKQY008).

References

[1] H. L. Han, "Test and analysis on main performances of bolt body for resin fiber-glass bolt," *Coal Science Technology*, 2005.

[2] G. L. Shen and G. K. Hu, *Mechanics of composite materials*, Tsinghua University Press, 2006.

[3] S. F. Luan, *The Study on the Destruction of Fiber Reinforced Matrix composites in mesomechanics*, Changchun Institute of Applied Chemistry Chinese Academy of Sciences, 2006.

[4] H. Li, Y. Jia, G. Mamtimin, X. Wang, W. Jiang, and L. An, "Numerical Simulation of Mesoscopic Mechanical Behaviors of Gradual Multi-Fiber-Reinforced Polymer Matrix Composites," *Macromolecular Materials and Engineering*, vol. 21, no. 8, pp. 57–60, 2006.

[5] P. I. Gonzalez-Chi and R. J. Young, "Deformation micromechanics of a thermoplastic-thermoset interphase of epoxy composites reinforced with poliethylene fiber," *Journal of Materials Science*, vol. 39, no. 23, pp. 7049–7059, 2004.

[6] Q. D. Zeng, *Qing-dun.Mesoscopic Failure Mechanism and Strength Composite Material*, Beijing Science Press, Beijing, China, 2002.

[7] Y. Guo, Y. Wang, Z. Wang, and C. Shen, "Study on the preparation and characterization of high-dispersibility nanosilica," *Science Engineering of Composite Materials*, vol. 23, no. 4, pp. 401–406, 2016.

[8] Y. Fan, "The Monte-Carlo Simulation of Composite Micromechanics Based on Shear-lag Model," *School of Aeronautics and Astronautics Shanghai Jiao Tong University Shanghai*, 2012.

[9] Y. L. Chen, B. Liu, X. Q. He, Y. Huang, and K. C. Hwang, "Failure analysis and the optimal toughness design of carbon nanotube-reinforced composites," *Composites Science Technology*, vol. 70, no. 9, pp. 1360–1367, 2010.

[10] S. J. V. Frankland, A. Caglar, D. W. Brenner, and M. Griebel, "Molecular Simulation of the Influence of Chemical Cross-Links on the Shear Strength of Carbon Nanotube-Polymer Interfaces," *Accounts of Chemical Research*, vol. 106, no. 12, pp. 592–600, 2002.

[11] S. J. V. Frankland and V. M. Harik, "Analysis of carbon nanotube pull-out from a polymer matrix," *Surface Science*, vol. 525, no. 1-3, pp. L103–L108, 2003.

[12] L. Y. Jiang, Y. Huang, H. Jiang et al., "A cohesive law for carbon nanotube/polymer interfaces based on the van der Waals force," *Journal of the Mechanics and Physics of Solids*, vol. 54, no. 11, pp. 2436–2452, 2006.

[13] W. B. Lu, J. Wu, J. Song, K. C. Hwang, L. Y. Jiang, and Y. Huang, "A cohesive law for interfaces between multi-wall carbon nanotubes and polymers due to the van der Waals interactions," *Computer Methods in Applied Mechanics & Engineering*, vol. 197, no. no, pp. 3261–3267, 2008.

[14] L. Y. Jiang, "A cohesive law for carbon nanotube/polymer interface accounting for chemical covalent bonds," *Mathematics and Mechanics of Solids*, vol. 15, no. 3, pp. 718–732, 2010.

[15] P. K. Agnihotri, K. K. Kar, and S. Basu, "Cohesive zone model of carbon nanotube-coated carbon fiber/polyester composites," *Modelling and Simulation in Materials Science and Engineering*, vol. 20, no. 3, Article ID 035014, pp. 523–530, 2012.

[16] W. T. Vestrand, J. A. Wren, A. Panaitescu et al., "A numerical study on carbon nanotube–hybridized carbon fibre pullout," *Composites Science & Technology*, vol. 91, no. 2, pp. 38–44, 2014.

[17] S. J. V. Frankland, V. M. Harik, G. M. Odegard, D. W. Brenner, and T. S. Gates, "The stressstrain behavior of polymernanotube composites from molecular dynamics simulation," *Composites Science & Technology*, vol. 63, no. 11, pp. 1655–1661, 2003.

[18] S. Hu, J. Yu, C. Wei, and Z. Zhang, "Shear lag behavior and parametric sensitivity analysis of steel–concrete composite structure with double-box," *Archive of Applied Mechanics*, pp. 1–15, 2017.

[19] Z. Sun, H. Shao, X. Niu, and Y. Song, "Failure simulation of unidirectional fiber-reinforced ceramic matrix composites based on evolving compliant interfacial debonding model," *Materials Science and Engineering A*, vol. 663, pp. 78–85, 2016.

[20] L. Xin, W. Yang, Q. Zhao et al., "Effect of extrusion treatment on the microstructure and mechanical behavior of SiC nanowires reinforced Al matrix composites," *Materials Science and Engineering A*, vol. 682, pp. 38–44, 2017.

[21] A. Tevatia and S. K. Srivastava, "Modified shear lag theory based fatigue crack growth life prediction model for short-fiber reinforced metal matrix composites," *International Journal of Fatigue*, vol. 70, pp. 123–129, 2015.

[22] S.-K. Ryu, T. Jiang, J. Im, P. S. Ho, and R. Huang, "Thermomechanical failure analysis of through-silicon via interface using a shear-lag model with cohesive zone," *IEEE Transactions on Device and Materials Reliability*, vol. 14, no. 1, pp. 318–326, 2014.

[23] G. Guo and Y. Zhu, "Cohesive-Shear-Lag Modeling of Interfacial Stress Transfer between a Monolayer Graphene and a Polymer Substrate," *Journal of Applied Mechanics, Transactions ASME*, vol. 82, no. 3, Article ID 031005, 2015.

[24] Y. Wang, Z. Wang, C. Shen, and Y. Wu, "Research on enhancement of GFRP-anchors torsional strength," *Science Engineering of Composite Materials*, vol. 19, no. 4, pp. 423–429, 2012.

[25] Y. Wang, Y. Guo, R. Cui, Z. Wang, and Y. Wu, "Preparation and mechanical properties of nano-silica/UPR polymer composite," *Science Engineering of Composite Materials*, vol. 21, no. 4, pp. 471–477, 2014.

[26] Z. Xu, Y. Huang, C. Zhang, and G. Chen, "Influence of rare earth treatment on interfacial properties of carbon fiber/epoxy composites," *Materials Science Engineering A*, vol. 444, no. 1-2, pp. 170–177, 2007.

[27] S. W. Wen, J. C. Zeng, J. Y. Xiao, F. B. Yang, Z. Wang, and C. T. Zhang, "Study on the Interfacial Shear Strength of Bf/Epoxy Resin Composites by Single Fiber Pull-Out Test," *Journal of Hunan University*, vol. 34, no. 5, pp. 53–57, 2007.

[28] X. Ji, Y. Dai, B.-L. Zheng, L. Ye, and Y.-W. Mai, "Interface end theory and re-evaluation in interfacial strength test methods," *Composite Interfaces*, vol. 10, no. 6, pp. 567–580, 2003.

[29] G. L. Ling and Z. F. Li, *Mastering ANSYS 14.0 Premium Edition*, Tsinghua University Press, BeiJing, China, 2013.

[30] H. Fang, H. Shi, Y. Wang, Y. Qi, and W. Liu, "Experimental and theoretical study of sandwich panels with steel facesheets and GFRP core," *Advances in Materials Science and Engineering*, vol. 2016, Article ID 7159205, 12 pages, 2016.

[31] H. S. Zhang and R. X. Hu, *ANSYS14.5/LS-DYNA Nonlinear finite element analysis of examples of the tutorial*, China Machine Press, BeiJing, China, 2013.

[32] H. C. Zhang, *ANSYS 14.0 The engineering examples, parsing and FAQ*, China Machine Press, BeiJing, China, 2013.

[33] X. Z. Wen, "Method for improving interfacial shear strength of sprayed fiber epoxy composite," *Mechanical Engineering Materials*, vol. 31, pp. 38–41, 2007.

[34] C.-H. Hsueh, "Interfacial debonding and fiber pull-out stresses of fiber-reinforced composites. VIII: The energy-based debonding criterion," *Materials Science and Engineering A*, vol. 159, no. 1, pp. 65–72, 1992.

[35] Q. Meng and Z. Wang, "Theoretical analysis of interfacial debonding and fiber pull-out in fiber-reinforced polymer-matrix composites," *Archive of Applied Mechanics*, vol. 85, no. 6, pp. 745–759, 2015.

[36] X. Chen, I. J. Beyerlein, and L. C. Brinson, "Curved-fiber pull-out model for nanocomposites. Part 2: Interfacial debonding and sliding," *Mechanics of Materials*, vol. 41, no. 3, pp. 293–307, 2009.

[37] C.-H. Hsueh, "Interfacial debonding and fiber pull-out stresses of fiber-reinforced composites," *Materials Science and Engineering A*, vol. 123, no. 1, pp. 1–11, 1990.

[38] M. S. Camara, Y. Ducq, and R. Dupas, "A methodology for the evaluation of interoperability improvements in inter-enterprises collaboration based on causal performance measurement models," *International Journal of Computer Integrated Manufacturing*, vol. 27, no. 2, pp. 103–119, 2014.

[39] J. G. Goree and R. S. Gross, "Analysis of a unidirectional composite containing broken fibers and matrix damage," *Engineering Fracture Mechanics*, vol. 13, no. 3, pp. 563–578, 1980.

[40] Q. Shan, *Interphase Microstructure Optimization of Carbon Fiber Reinforced Composite Materials*, Harbin Engineering University, 2009.

4

Mode II Fracture of GFRP Laminates Bonded Interfaces under 4-ENF Test

Zhi-peng Zhong[1,2] and Hong Liu[1]

[1]*Department of Civil Engineering, Shanxi University, Taiyuan, Shanxi 030013, China*
[2]*Key Laboratory of C&PC Structures of the Ministry of Education, Southeast University, Nanjing 210096, China*

Correspondence should be addressed to Hong Liu; liuhong1966@sxu.edu.cn

Academic Editor: Katsuyuki Kida

This experiment studies the mode II fracture behavior of an adhesively bonded joint composed of GFRP laminates. A new beam model is presented to calculate the mode II ERR for GFRP bonded 4-ENF specimens. In this model, the deformation of 4-ENF specimens caused by the relative deflection angle between the upper and lower layers and by the bending deformation of the upper and lower layers, respectively, is introduced; the effect of the adhesive layer deformation is presented. The closed-form analytical solutions of compliance and energy release rate based on the crack compliance method are obtained. The high accuracy of present analytical solutions are verified by finite element analysis through bonded GFRP 4-ENF specimens and compared to the rigid joint model and the CBT model. The interfacial crack propagation is numerically simulated using shear fracture toughness determined in this experiment, from which the predicted critical load results are in good agreement with the experimental results. The conclusion indicates that the compliance and ERR can accurately be predicted using the new bonded 4-ENF beam model.

1. Introduction

To combat problems associated with concrete/steel deck fatigue and corrosion, glass fiber-reinforced polymer (GFRP) bridge deck structures have been widely used in the bridge engineering field in recent years. Advantages of using GFRP include its light weight, high strength, and corrosion resistance [1, 2]. A GFRP bridge panel system is comprised of modular tube form units integrated via adhesive bonding. The adhesively bonded interface usually transfers shear forces and allows the GFRP modular tubes to bear loads together. However, the adhesive bonding joints are the weak links in the entire structure. They can easily delaminate and crack due to overloading or fatigue. Therefore, the reliability of the adhesive bonding joints plays a key role in ensuring the integrity, and influences the structural behavior of GFRP bridge decks [3, 4]. To design effective, reliable bonding joints, the accurate determination of shear fracture toughness of GFRP adhesively bonded joints under mode II loading is necessary.

Fracture toughness is the characterization of the bonding interface properties of GFRP. In the literature, there has been much study of fracture toughness [5–13]. However, it is the stable cracking problem that should be investigated in the interfacial debonding test. The end-notched flexure (ENF) test and the four-point bend end-notched flexure (4-ENF) test have been commonly employed to determine the mode II bonding fracture toughness. The original ENF test is essentially unstable and requires to measure the crack sliding displacement of the specimen during the experimental process [5]. This test procedure is too complex to achieve accurate results. To avoid the disadvantages of the ENF test, a 4-ENF test was proposed by Martin and Davidson [6, 7]. The 4-ENF test is stable and does not require measuring crack length during the experiment. What is more, there is only pure moment, and no shear force exists within the crack tip of the 4-ENF test, which reduces the frictional effect across the delaminated interface. Other alternatives for mode II fracture research are the end-loaded split (ELS) test [8–10] and the over notched flexure (ONF) test [11, 12]. The ELS test raises stable crack propagation, but it contains large vertical deflection for thin specimens. The ONF test seems to be simple and also results in stable crack propagation, similar to the 4-ENF test [13].

To get accurate values for the mode II fracture toughness of materials, both theoretical and experimental fracture characteristic research have been conducted in past decades.

Russell and Street [14] performed a simple beam theory that neglected the crack tip singularity and the transverse shear deformation of the specimen, so the energy release rate (ERR) is significantly underestimated in their solution. Carlsson et al. [15] suggested Timoshenko's beam theory to obtain the flexibility and ERR of 4-ENF beams. However, there was some deviation from the result obtained by using finite element analysis. To get an accurate solution, several modifications have been provided in recent years. Whitney and Sun [16] proposed a higher order beam theory based on Reissner's Principle to analyze an ENF specimen. The shear deformation is considered and the shear singularity at the crack tip is omitted and regarded as the surface tensile stress in the higher order beam theory. Wang and Williams [17] derived an additional length of v_h to the original crack length by introducing a correction factor v. Then the ERR was evaluated by the formulation of classical beam theory (CBT) using the adjusted crack length. Corleto and Hogan [5] employed two-parameter semi-infinite elastic foundation to the Timoshenko beam model to obtain the compliance and ERR of ENF specimens, in which the crack tip deformation was considered in their model. Ding and Kortschot [18] adapted a simplified analysis of a beam supported on a shear spring without accounting for the transverse shear deformation in order to find a simple and precise analytical solution. By using the superposition principle, Wang and Qiao [19] decomposed ENF into a three-point uncracked bending beam and an asymmetric cracked beam under shear action on the crack surface. The compliance and ERR of the ENF are obtained based on the compatibility of deformation.

Currently, most of the literature is confined to GFRP laminates delamination, and a method is developed that can predict delamination of GFRP laminates through analysis. In practice, however, in fiber-reinforced polymer (FRP) engineering structures, mode II crack growth is the common failure form of adhesive bonding joints for the bonding structure (e.g., GFRP bridge decks and FRP strengthening concrete structures) [20]. As there are some deviations (e.g., joint dimensions and failure modes) from the available results, it is necessary to study interfacial debonding behavior of adhesively bonded composite joints [21]. However, there is little research into the behavior of adhesively bonded joints composed of pultruded FRP laminates. It is necessary to study the mode II bond interface fracture toughness of adhesive joints. Therefore, research into the validity and practical formulas for determining mode II fracture toughness of FRP adhesive bonding joints is necessary.

This research is to understand the mode II fracture behavior of adhesive-bonded composite joints using pultruded GFRP 4-ENF specimens. To better represent the effect of adhesive layer stiffness on the deflection and ERR on GFRP 4-ENF specimens, a two-dimensional model based on adhesive layer deformable beam theory is used in this paper. The adherends are treated as a distinct deformation Euler-Bernoulli beam, and the adhesive layer is treated as Goodman's elastic interlayer. The longitudinal deformation of adherends is caused by the relative deflection angle between the upper and lower layers, the longitudinal displacement of the adherends, and the vertical displacement caused by the bending deformation of the upper and lower layers. The compatibility of deformation at the mid-plane of the bonding specimen is satisfied by incorporating deformation of the adhesive layer and bonding interface. Using a combination of the beam force balance equation and the adhesive constitutive equation, the theoretical calculation formula of deflection deformation of 4-ENF beams is derived. In addition, mode II ERR G_{II} is calculated using the compliance method, which is compared with finite element analysis and existing literature results. Further, a 4-ENF test on GFRP bonded specimens was conducted to determine the critical energy release rate G_{IIc} based on load versus displacement curves. Finally, an interfacial crack propagation process is simulated by the virtual crack closure technique (VCCT) applying existing G_{IIc}. The critical load for interfacial crack initiation also predicted by finite element analysis is validated by comparing the experimental results.

2. Analysis of 4-ENF Specimens with an Adhesive Layer

A typical 4-ENF specimen with an adhesive layer under loading of P at the points of $x = L_0$ and $x = a + L_1$ by the left rollers can be modelled as a simply supported beam with a span of L and an end traction free crack of length a (Figure 1). In this paper, the 4-ENF test takes place under the symmetric loading, so the length $a + L_1 = L - L_0$. The top beam and bottom beam have thicknesses of h_1 and h_2, respectively. The thickness of the adhesive layer is h_a and the beam width is b. Considering that the span of L is greater than the bonded beam thickness, transverse shear deformation is not incorporated.

Next, considering the uncracked region ($a < x < L$) in Figure 1, the internal forces are expressed as a beam theory (Figure 2). N_i, M_i, and Q_i ($i = 1, 2$) are defined as the internal axial forces, bending moments, and transverse shear forces in sublayers 1 and 2, respectively.

According to the internal forces and stresses shown in Figure 2, the equilibrium equations for an infinitesimal isolated body of two adherends are given by

$$\frac{dN_1(x)}{dx} = b\tau(x), \qquad \frac{dQ_1(x)}{dx} = b\sigma(x) \tag{1}$$

$$\frac{dN_2(x)}{dx} = -b\tau(x), \qquad \frac{dQ_2(x)}{dx} = -b\sigma(x) \tag{2}$$

$$\frac{dM_1(x)}{dx} = Q_1(x) - b\tau(x)\frac{h_1}{2}, \qquad \frac{dM_2(x)}{dx} = Q_2(x) - b\tau(x)\frac{h_2}{2}, \tag{3}$$

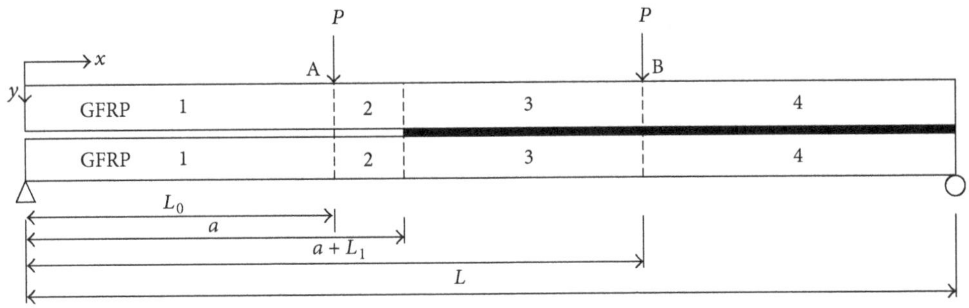

FIGURE 1: Configuration of the 4-ENF beam.

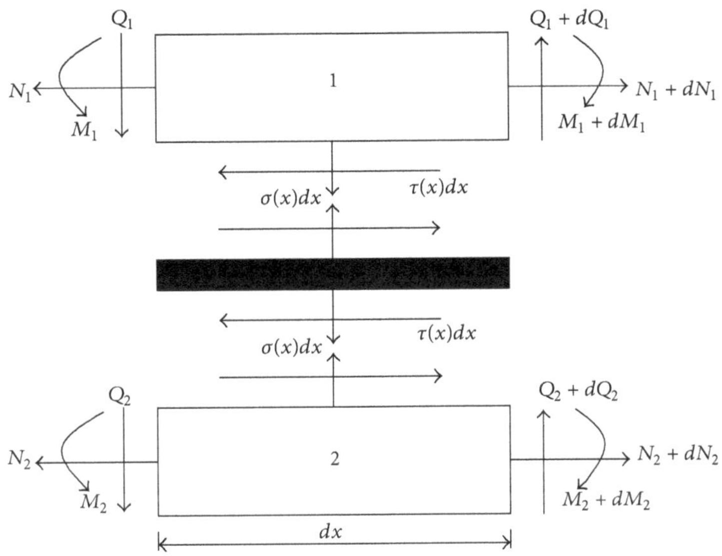

FIGURE 2: Infinitesimal isolated body diagram of sublayers.

where $\sigma(x)$ and $\tau(x)$ are the peel and shear stresses of the adhesive layer, respectively.

According to GFRP laminated beam theory, the relationship between the internal forces and displacement for two adherends can be described as follows:

$$N_i = A_{ii}\frac{du_i}{dx} + B_{ii}\frac{d\phi_i}{dx},$$
$$M_i = B_{ii}\frac{du_i}{dx} + D_{ii}\frac{d\phi_i}{dx}, \tag{4}$$

where u_i and ϕ_i are the ith sublayer's longitudinal and rotation angle, respectively; A_{ii}, B_{ii}, and D_{ii} are the ith sublayer's tension stiffness, coupling stiffness, and flexural stiffness, respectively; $(A_{ii}, B_{ii}, D_{ii}) = b\int_{-h_i/2}^{h_i/2} Q_{ii}(1, z, z^2)dz$, where Q_{ii} is the off-axis stiffness coefficient, and $i = 1, 2$ denotes sublayers 1 and 2.

For the 4-ENF beam, the axial force on the bonded cross section is zero. The equilibrium conditions can be shown as follows:

$$N_T(x) = N_1(x) + N_2(x) = 0, \tag{5}$$

where $N_T(x)$ is the resultant axial internal force. From (5), we can obtain

$$N_1(x) = -N_2(x). \tag{6}$$

The shear force on the cross section is described as

$$Q_1(x) + Q_2(x) + Q_a(x) = Q_T(x), \tag{7}$$

where $Q_T(x)$ and $Q_a(x)$ are the resultant transverse shear force and the shear force of the adhesive.

The integral bending behavior of the 4-ENF beam can be considered as the superposition of the overall bending action and the bending action for the upper and lower layers, respectively.

$$M_{tc}(x) = -N_2\frac{h_2 + h_a}{2} + N_1\frac{h_1 + h_a}{2}$$
$$M_{12}(x) = M_1(x) + M_2(x) \tag{8}$$
$$M_T(x) = M_{12}(x) + M_{tc}(x),$$

where $M_T(x)$ is the resultant transverse moment of the cross section for the bonded beam.

Taking the derivative of the 1st equation and 2nd equation for (3) and adding the two equations, then combining with (4), the deflection differential equation is derived:

$$E_1 I_1 \frac{d^4 w_1}{dx^4} + E_2 I_2 \frac{d^4 w_2}{dx^4} = b d_3 \frac{d\tau(x)}{dx}, \quad (9)$$

where $d_3 = (h_1 + h_2 + 2h_a)/2$.

When the 4-ENF beam is in linear elastic range, the longitudinal displacement is caused by the relative deflection angle between the upper and lower layers, the longitudinal displacement of the adherends, and the vertical displacement caused by the bending deformation of the upper and lower layers.

Based on the linear superposition principle, the lower surface displacement of the upper layer is

$$u_u = u_1 - \frac{h_1 + h_a}{2}\phi + \frac{h_1}{2}\phi_i. \quad (10)$$

The upper surface displacement of the lower layer is

$$u_l = u_2 + \frac{h_1 + h_a}{2}\phi + \frac{h_2}{2}\phi_i, \quad (11)$$

where ϕ is the relative deflection angle between the upper and lower layers.

Combined with (10) and (11), the constitutive relation of the adhesive layer is expressed as follows:

$$\tau(x) = \frac{G(u_u - u_l)}{h_a} = \frac{G}{h_a}\left(u_1 - \frac{h_1 + h_a}{2}\phi + \frac{h_1}{2}\phi_i - u_2 - \frac{h_1 + h_a}{2}\phi - \frac{h_2}{2}\phi_i\right). \quad (12)$$

Equation (12) by second derivative is reduced to

$$\frac{d^2\tau(x)}{dx^2} = \frac{G}{h_a}\left(\left(\frac{1}{E_1A_1} + \frac{1}{E_2A_2}\right)\frac{dN_1}{dx} + \frac{h_1}{2}\frac{d^3 w_1}{dx^3} + \frac{h_2}{2}\frac{d^3 w_2}{dx^3}\right). \quad (13)$$

For the configuration of the 4-ENF specimen, the vertical extrusion can be ignored; that is, $w_i = w$, $d\phi/dx = d\phi_i/dx = -d^2 w_i/dx^2$. By differentiating (13) and combining this with (12), the deflection differential equation of the 4-ENF specimen is obtained as

$$\frac{d^6 w}{dx^6} = \frac{G_a}{h_a}\left(\frac{b}{E_1A_0} + \frac{b d_3 (h_1/2 + h_2/2)}{(E_1I_1 + E_2I_2)}\right)\frac{d^4 w}{dx^4}, \quad (14)$$

where $1/E_1A_0 = 1/E_1A_1 + 1/E_2A_2$.

Solving (14), the deflection of the 4-ENF specimen is obtained as

$$w = B_1 e^{sx} + B_2 e^{-sx} + B_3 x^3 + B_4 x^2 + B_5 x + B_6, \quad (15)$$

where

$$s = \sqrt{\frac{G_a}{h_a}\left(\frac{b}{E_1A_0} + \frac{b d_3 h_1 + b d_3 h_2}{2(E_1I_1 + E_2I_2)}\right)}. \quad (16)$$

For crack area 1 ($0 \le x \le L_0$), the bending moment of the upper layer is

$$M_1(x) = \alpha P x, \quad (17)$$

where $\alpha = D_1/(D_1 + D_2)$, D_1 and D_2 are the bending stiffness on upper and lower layer, respectively.

Substituting (15) and (17) into (4) yields

$$w(x) = -\frac{\alpha P x^3}{6D_1} + C_1 x + C_2. \quad (18)$$

For crack area 2 ($L_0 \le x \le a$), the bending moment of upper layer is

$$M_1(x) = \alpha P L_0. \quad (19)$$

Substituting (18) and (19) into (4) yields

$$w(x) = -\frac{\alpha P L_0}{2D_1}x^2 + C_3 x + C_4. \quad (20)$$

The boundary conditions and displacement compatibility for the 4-ENF specimen can be expressed as

$$\begin{gathered} w|_{x=0} = 0, \\ w|_{x=L} = 0, \\ w|_{x=L_0^-} = w|_{x=L_0^+}, \\ w'|_{x=L_0^-} = w'|_{x=L_0^+}, \\ w|_{x=a^-} = w|_{x=a^+}, \\ w'|_{x=a^-} = w'|_{x=a^+}, \\ w''|_{x=a^+} = -\frac{PL_0}{D}, \\ w'''|_{x=a^+} = 0, \\ w''|_{x=L^-} = 0, \\ w'''|_{x=L^-} = \frac{\alpha P}{D_1}, \end{gathered} \quad (21)$$

where $D = b(h_1 + h_2 + h_a)^3/12$. From (21), the ten unknown constants B_1, B_2, B_3, B_4, B_5, B_6, C_1, C_2, C_3, and C_4 can be obtained as detailed in Appendix, from which the vertical displacement w of the 4-ENF beam can be determined.

From [19], the ERR from the 4-ENF test data is accurately calculated based on the compliance method:

$$G_{II} = \frac{P^2}{2b}\frac{dC}{da}. \quad (22)$$

As mentioned in [22], the compliance calibration method is very suitable for the linear elastic loading configuration characterized by a single applied load and a single induced displacement. To realize the linear elastic loading configuration, the single applied load is deemed as the total load applied to the bonded beam, and the single induced displacement is deemed as the displacement at the center of the span [6]. Therefore, the compliance C of the 4-ENF beam is defined as

$$C = \frac{\overline{w}}{P_T} = \frac{w_L + w_R}{2\left(P_L + P_R\right)} = \frac{w_L + w_R}{4P}, \tag{23}$$

where $P_T = 2P$ is the applied load, $\overline{w}$ is the displacement at the center of the span, w_L and w_R are the displacements at the left and right loading rollers, respectively, which can be obtained by the above-mentioned deflection formula as follows:

$$\begin{aligned} w_L &= -\frac{\alpha P L_0^3}{6D_1} + C_1 L_0 \\ w_R &= B_1 e^{s(L-L_0)} + B_2 e^{-s(L-L_0)} + B_3\left(L - L_0\right)^3 \\ &\quad + B_4\left(L - L_0\right)^2 + B_5\left(L - L_0\right) + B_6. \end{aligned} \tag{24}$$

By combining with (23) and (24), the compliance C is determined as

$$\begin{aligned} C &= \frac{F_0}{24DD_1 s^3\left(sL - 2\right)} \\ &\quad + \frac{2\alpha D L_0^2 s - D_1 L_0^2 s - \alpha D L_0}{8DD_1 s} a, \end{aligned} \tag{25}$$

where F_0 is given in Appendix.

Then, by combining with (25), the energy release rate, G_{II}, is obtained as follows:

$$G_{\text{II}} = \frac{P_T^2}{2b}\frac{dC}{da} = \frac{2P^2}{b}\left(\frac{2\alpha D L_0^2 s - D_1 L_0^2 s - \alpha D L_0}{8DD_1 s}\right). \tag{26}$$

From (26), we can see that the fracture toughness can be directly obtained for each testing process by determining the critical load, from which the length of crack propagation is not measured. The critical load is determined by examining the typical load-displacement curve during the fracture experiment.

3. Result Verification and Discussion

To verify the accuracy of this model, the analytical results for G_{II} of the GFRP bonded interface 4-ENF specimen are examined by comparison with a rigid joint model, conventional beam theory, and two-dimensional finite element analysis (FEA). The geometry of the designed 4-ENF bonded GFRP specimen [shown in Figure 1] is set as follows: the section size of the GFRP beam is $b \times h = 40 \times 4$ mm; the span of the GFRP beam is $L = 260$ mm. The two GFRP beams are bonded together using 1 mm thick epoxy resin, and the prefabricated crack of length a is located at the end of the bonded beam. The material properties of the GFRP laminates are as follows: $E_x = 34.9$ GPa; $E_y = 7.29$ GPa; $\nu_{xy} = 0.258$; $G_{xy} = 3.92$ GPa. The material properties of the epoxy resin are $E_a = 1.5$ GPa; $\nu_a = 0.3$.

In numerical analysis, a typical two-dimensional finite element model of the 4-ENF test was undertaken using finite element software ANSYS. The layered Plane42 element was selected to model the GFRP laminates plate. The prefabricated crack is modelled as a discontinuity element in the middle of the GFRP 4-ENF specimen, setting Targe169 and Conta172 contact elements on the upper and lower contact surfaces of the prefabricated crack zone. The bonded section of the upper and lower beams is set as a Combin14 spring element. Considering that there exists contact phenomena between the upper and lower surfaces of the prefabricated crack, the elements near the crack tip are divided into $0.2h$, as suggested in [4], where h is the thickness of beam. Since the number of finite elements is so large that the refined meshing sizes are only seen with difficulty in the GFRP 4-ENF specimen numerical model, only a typical FE mesh characteristic near the crack tip is shown in Figure 3, in which the red line denotes the crack.

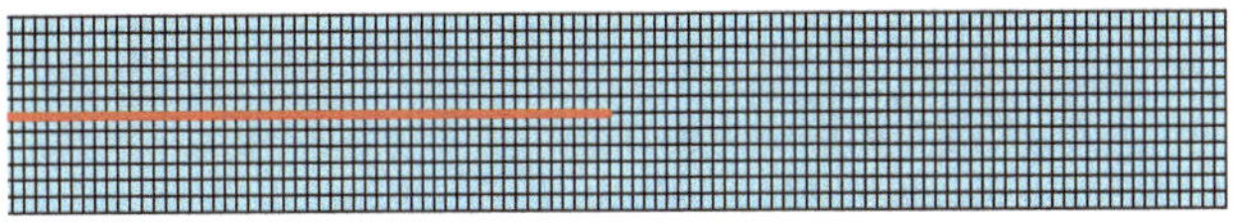

FIGURE 3: Interface mesh of the crack tip area in 4-ENF model.

The relational value of the compliances for the 4-ENF specimens with the crack length calculated by different methods (i.e., CBT, rigid joint model, and FEA) is shown in Figure 4(a). The linear curve of the compliance crack length can be observed for the bonded 4-ENF specimen. From Figure 4(a), we can see that the compliance is larger than the rigid joint model and CBT. Moreover, the rigid joint model undervalues the compliance because it ignores the deformation of the adhesive layer. By taking account into the adhesive layer deformation, the present model gives a compliance value closer to the FEA results, which indicates that the adhesive layer deformation is another important factor in contributing to the compliance of 4-ENF specimen. The energy release rate G_{II} of the adhesive joint crack front can be calculated using the VCCT [23] in FEA. The variation of ERR with regard to the crack length adopted by CBT, the rigid joint model, and FEA is shown in Figure 4(b). As shown in Figure 4(b), the ERR based on all models is a constant value, which is independent of crack length. What is more, the calculated ERR is in good agreement with the FEA result, with a difference of 3.3%, which shows that the current theoretical model is accurate and reasonable.

4. The GFRP-GFRP Bonded Interface Fracture Toughness by Four-Point Bending Test

While many studies on the interfacial delamination fracture toughness of FRP material have been conducted, there are few studies on the interfacial fracture toughness for FRP bonded specimens with an adhesive layer. However, the related research shows that the bond performance has an influence on the whole performance and on local failure behavior of the adhesive structure [20, 21]. As bond interface crack propagation is different from delamination, the crack propagation of the bonded specimens is unstable. To measure

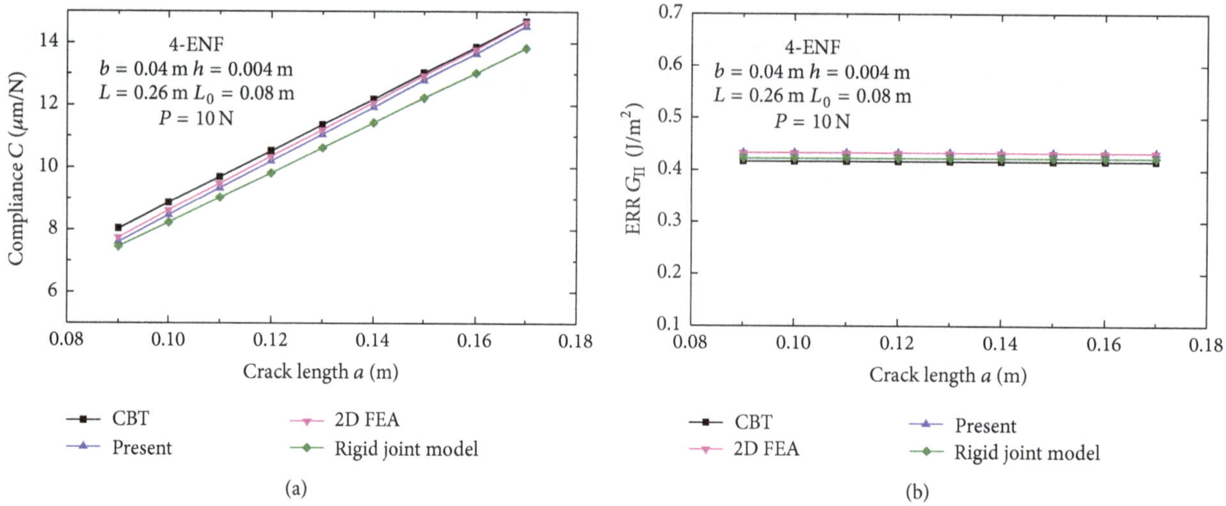

FIGURE 4: Typical compliance and ERR evaluated as a function of crack length for 4-ENF: (a) compliance; (b) ERR.

TABLE 1: Mechanical properties of pultruded GFRP composite and adhesive layer.

Material name	E_x (GPa)	E_y (GPa)	v_{xy}	G_{xy} (GPa)
GFRP composite	34.9	7.29	0.29	3.92
Adhesive layer	1.5	1.5	0.30	0.58

the bonded interfacial fracture toughness accurately, the fabrication of the 4-ENF specimens mentioned in the previous section is used to study the GFRP-GFRP bonded interface. The 4-ENF specimens are made of pultruded GFRP laminate bonded by a 1 mm thick adhesive layer with a crack length a at the end of the bonded beam. Equal amounts of epoxy resin and curing agent were used in the adhesive layer. The material mechanical properties of pultruded GFRP and adhesive are obtained from ASTM [24–26], as shown in Table 1. The geometry of the designed 4-ENF bonded GFRP specimen is $L = 260$ mm; $b \times h = 40 \times 4$ mm. The length of the precrack is 150 mm. The fracture test is performed on a material testing system (MTS) servohydraulic testing machine [see Figure 5(a)]. The loading schematic diagram is shown in Figure 5(b). The experiment was performed adopting the displacement control with a loading speed 0.5 mm/min.

During the experiment, the load-displacement information is continuously and automatically saved by the MTS machine. A typical load-displacement curve for the 4-ENF bonded specimen interfacial fracture is shown in Figure 6. Initially, the displacement increases linearly as the applied load increases. Meanwhile, the elastic strain energy is accumulating in the bonded specimen. When the strain energy increases to the critical value, the crack extension occurs and the elastic strain energy is released. After that, the applied load is decreased, and the propagation of the crack is arrested.

Based on the load-deflection curve, the compliances for the 4-ENF specimens were obtained, and the critical ERR were calculated by applying the critical loads and the compliance rate change dC/da accordingly. Table 2 summarizes a group of experimental data. For each specimen the value of the critical energy release rate G_{IIc} was calculated by (26) or VCCT [23] using the crack initiation critical load. An average value of 365.46 J/m^2 was obtained with the coefficient of variation being 19.8%, which is reasonable for bonded composite material. Therefore, it can be observed that G_{IIc} for the GFRP 4-ENF laminates specimens is determined to be 365.46 J/m^2.

To validate the fracture energy of the interfacial crack determined in the experiment, a numerical simulation of the whole bonded interfacial crack growth process is carried out. Based on the determined average value of G_{IIc} (G_{IIc} = 365.46 J/m^2) shown in Table 2, the whole interfacial crack propagation process is simulated using the VCCT [23, 27]. When the working ERR in the crack tip satisfies the fracture criterion $G_{II} \geq G_{IIc}$, the interfacial crack starts to grow. The comparison between the experimental result and numerical prediction for both load-deflection curves (P versus w) is shown in Figure 7. It is clear that the displacement for the load point in the experiment is consistent with the finite element predicted value, which indicated that the previously determined fracture toughness by (26) is true.

5. Conclusions

In this study, a beam model including the effect of the adhesive layer deformation is presented to calculate the mode II energy release rate G_{II} of GFRP bonded four-point bending end-notched flexure (4-ENF) specimens for fracture.

(1) The analytical formula of compliance and ERR based on the crack compliance method for 4-ENF are obtained. According to the model, a linear relation between the average compliance and the crack length is found for the bonded 4-ENF specimens.

(2) The accuracy of the analytical solutions for both the compliance and ERR are verified by finite element analysis through the GFRP bonded interface

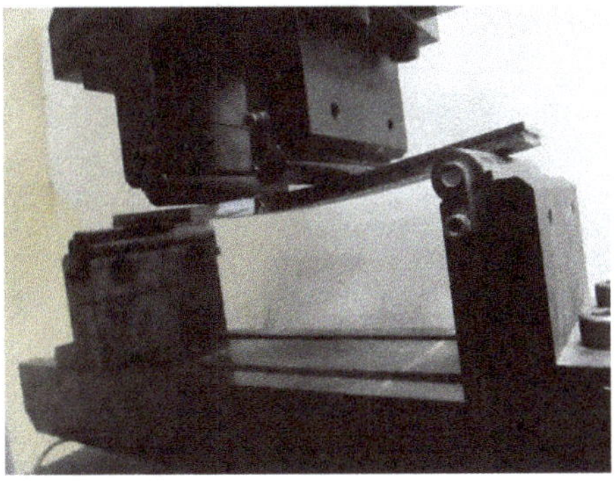

(a)

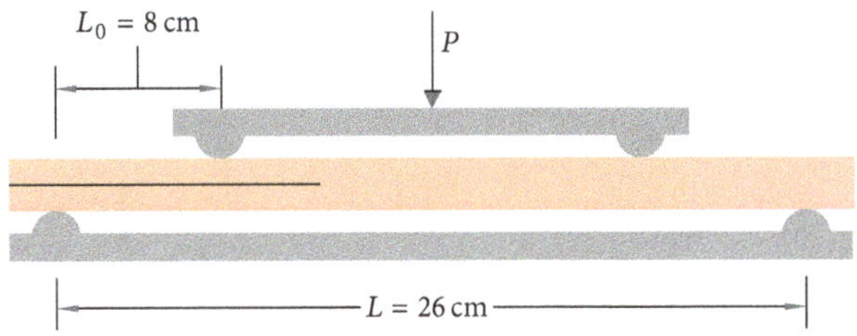

(b)

FIGURE 5: Testing configuration: (a) MTS apparatus; (b) schematic diagram for 4-ENF test.

TABLE 2: Results of critical load P_c and fracture toughness G_{IIc}.

Specimen	Critical load P_c (N)	G_{IIc} calculated by Eq. (26) (J/m^2)	G_{IIc} obtained from VCCT (J/m^2)	Error (%)
1	570.00	337.90	349.43	3.30
2	612.00	389.53	402.65	3.26
3	593.00	365.71	378.11	3.28
4	507.00	267.33	276.63	3.36
5	670.00	466.86	482.30	3.20
Average	590.40	365.46	377.82	3.38
COV (%)	10.09	19.94	19.89	

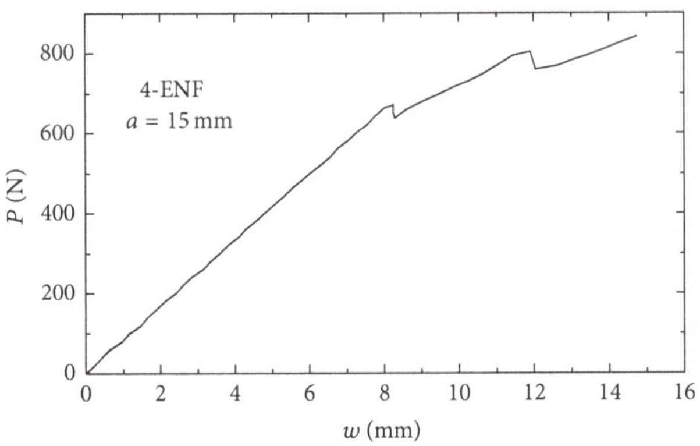

FIGURE 6: Typical load-displacement curve for 4-ENF GFRP bonded specimens.

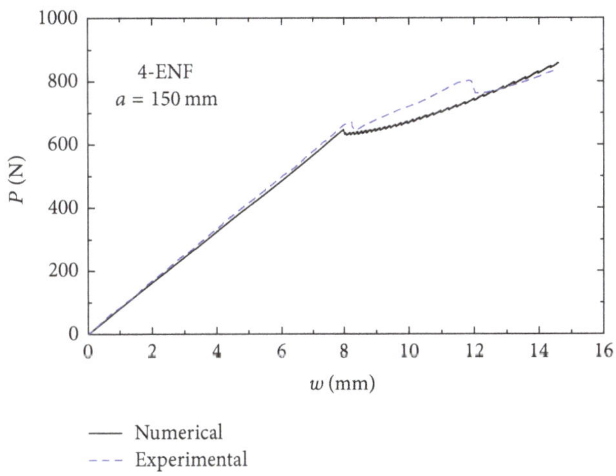

FIGURE 7: Comparison of numerical solution and experimental result for crack growth.

specimen when compared to the rigid joint model and the CBT model. The compliance and ERR of the present model are closely related to the results obtained by the finite element analysis, demonstrating that the present model is accurately evaluating the compliance and ERR of the four-point bending fracture test.

(3) The interfacial crack propagation is numerically simulated using fracture toughness G_{IIc} determined in the experiment, by which the predicted P_c is in good agreement with the experimental results.

The results obtained from both the theoretical model method and corresponding finite element analysis indicate that the compliance is linearly correlated with crack length and that the calculation of the ERR does not require measuring crack length propagation in the experiment for the bonded 4-ENF beam. By considering adhesive deformation, the present model can calculate the compliance and estimate ERR accurately without knowing the crack length for the bonded 4-ENF beam.

Appendix

Coefficient of Adhesively Bonded GFRP 4-ENF Model

$$\begin{aligned}B_1 = {} & P\left[\left(D_1SL_0 + D\alpha + DSa\alpha - DS\alpha L\right)e^{-Sa}\right.\\
& \left. - \left(D_1SL_0 + D\alpha\right)e^{-SL}\right]\\
& \cdot\left(DD_1S^3\left[2\left(e^{S(a-L)} + e^{-S(a-L)}\right)\right.\right.\\
& \left.\left. + \left(e^{-S(a-L)} - e^{S(a-L)}\right)(Sa - SL) - 4\right]\right)^{-1}\end{aligned}$$

$$\begin{aligned}B_2 = {} & P\left[\left(D_1SL_0 - D\alpha + DSa\alpha - DS\alpha L\right)e^{Sa}\right.\\
& \left. - \left(D_1SL_0 - D\alpha\right)e^{SL}\right]\\
& \cdot\left(DD_1S^3\left[2\left(e^{S(a-L)} + e^{-S(a-L)}\right)\right.\right.\\
& \left.\left. + (SL - Sa)\left(e^{S(a-L)} - e^{-S(a-L)}\right) - 4\right]\right)^{-1}\end{aligned}$$

$$\begin{aligned}B_3 = {} & -P\left[\left(D_1SL_0 + D\alpha\right)e^{S(a-L)} + \left(D\alpha - D_1SL_0\right)\right.\\
& \left. \cdot e^{-S(a-L)} - 2D\alpha\right]\\
& \cdot\left(6DD_1\left[(Sa - SL)\left(e^{S(a-L)} - e^{-S(a-L)}\right)\right.\right.\\
& \left.\left. - 2\left(e^{S(a-L)} + e^{-S(a-L)}\right) + 4\right]\right)^{-1}\end{aligned}$$

$$\begin{aligned}B_4 = {} & -P\left[\left(D_1LL_0S^2 + D_1SL_0 - D\alpha + Da\alpha S\right)e^{S(a-L)}\right.\\
& + \left(D_1SL_0 + D\alpha + DSa\alpha - D_1LL_0S^2\right)e^{-S(a-L)}\\
& \left. - 2D_1SL_0 - 2DS\alpha L\right]\\
& \cdot\left(2DD_1S\left[2\left(e^{S(a-L)} + e^{-S(a-L)}\right)\right.\right.\\
& \left.\left. + (SL - Sa)\left(e^{S(a-L)} - e^{-S(a-L)}\right) - 4\right]\right)^{-1}\end{aligned}$$

$$\begin{aligned}B_5 = {} & -P\left[\left(e^{S(a-L)} + e^{-S(a-L)}\right)\left(6DS^2a^2\alpha L_0\right.\right.\\
& - 3D_1S^2a^2L_0 - 3D_1S^2L^2L_0 - 3DS^2a\alpha L^2\\
& - 2DS^2\alpha L_0^3 + 12D_1L_0 - DS^2a^3\alpha - 6D\alpha L\\
& \left. + DS^2\alpha L^3\right) + \left(e^{S(a-L)} - e^{-S(a-L)}\right)\left(DS^3a\alpha L_0^3\right.\\
& - 3D_1S^3a^2LL_0 - 3DS^3a^3\alpha L_0 - 2D_1S^3L^3L_0\\
& + 2D_1S^3a^3L_0 - DS^3\alpha LL_0^3 + 3DS^3a^2\alpha LL_0\\
& \left. + 3DS\alpha L^2 - 6D_1SaL_0 + 3DSa^2\alpha\right) + 6D_1S^2a^2L_0\\
& + 6DS^2a^2\alpha L + 6D_1S^2L^2L_0 + 4DS^2\alpha L^3\\
& + 4DS^2\alpha L_0^3 - 4DS^2a^3\alpha + 12D\alpha L - 12DS^2a^2\alpha L_0\\
& \left. - 24D_1L_0\right]\\
& \cdot\left(6DD_1S^2L\left[(SL - Sa)\left(e^{S(a-L)} - e^{-S(a-L)}\right)\right.\right.\\
& \left.\left. + 2\left(e^{S(a-L)} + e^{-S(a-L)}\right) - 4\right]\right)^{-1}\end{aligned}$$

$$\begin{aligned}B_6 = {} & P\left[\left(e^{S(a-L)} + e^{-S(a-L)}\right)\left(6D_1SL_0 - 6DSa\alpha\right.\right.\\
& + 6DS^3a^2\alpha L_0 - 3D_1S^3a^2L_0 - DS^3a^3\alpha\\
& \left. - 2DS^3\alpha L_0^3\right) + \left(e^{S(a-L)} - e^{-S(a-L)}\right)\left(3DS^2a^2\alpha\right.\\
& + 6D\alpha - DS^4\alpha LL_0^3 + 3DS^4a^2\alpha LL_0 - 3DS^4a^3\alpha L_0\\
& - 3D_1S^4a^2LL_0 + DS^4a\alpha L_0^3 + 2D_1S^4a^3L_0\\
& \left. - 6D_1S^2aL_0\right) + 12DS\alpha L - 12DS^3a^2\alpha L_0\\
& + 4DS^3\alpha L_0^3 - 4DS^3a^3\alpha - 12D_1SL_0 + 6DS^3a^2\alpha L\\
& \left. + 6D_1S^3a^2L_0\right]\\
& \cdot\left(6DD_1S^3\left[(SL - Sa)\left(e^{S(a-L)} - e^{-S(a-L)}\right)\right.\right.\\
& \left.\left. + 2\left(e^{S(a-L)} + e^{-S(a-L)}\right) - 4\right]\right)^{-1}\end{aligned}$$

$$\begin{aligned}C_1 = {} & -P\left[\left(e^{S(a-L)} + e^{-S(a-L)}\right)\left(-12DS^2a\alpha LL_0\right.\right.\\
& - 3D_1S^2L^2L_0 + 12D_1L_0 + 6DS^2\alpha LL_0^2\\
& + 6DS^2a^2\alpha L_0 - 3D_1S^2a^2L_0 - DS^2a^3\alpha\\
& + 3DS^2a^2\alpha L + DS^2\alpha L^3 - 2DS^2\alpha L_0^3 - 3DS^2a\alpha L^2\\
& \left. + 6D_1S^2aLL_0\right) + \left(e^{S(a-L)} - e^{-S(a-L)}\right)\\
& \cdot\left(9DS^3a^2\alpha LL_0 - 6D_1SaL_0 - 6D_1S^3a^2LL_0\right.\\
& + 3DS\alpha L^2 + DS^3a\alpha L_0^3 + 6D_1SLL_0\\
& + 6D_1S^3aL^2L_0 - 6DSa\alpha L - 6DS^3a\alpha L^2L_0\\
& - 3DS^3a^3\alpha L_0 + 2D_1S^3a^3L_0 + 3DSa^2\alpha\\
& - 3DS^3a\alpha LL_0^2 + 3DS^3\alpha L^2L_0^2 - 2D_1S^3L^3L_0\\
& \left. - DS^3\alpha LL_0^3\right) - 12DS^2a\alpha L^2 - 24D_1L_0\\
& + 24DS^2a\alpha LL_0 + 6D_1S^2a^2L_0 + 4DS^2\alpha L_0^3\\
& - 12DS^2\alpha LL_0^2 + 12DS^2a^2\alpha L - 12D_1S^2aLL_0\\
& + 6D_1S^2L^2L_0 - 4DS^2a^3\alpha - 12S^2a^2D\alpha L_0\\
& \left. + 4DS^2\alpha L^3\right]\\
& \cdot\left(6DD_1S^2L\left[(SL - Sa)\left(e^{S(a-L)} - e^{-S(a-L)}\right)\right.\right.\\
& \left.\left. + 2\left(e^{S(a-L)} + e^{-S(a-L)}\right) - 4\right]\right)^{-1}\end{aligned}$$

$$C_2 = 0$$

$$\begin{aligned}
C_3 = {} & -P\Big[\left(e^{S(a-L)}+e^{-S(a-L)}\right)\left(12D_1L_0-3D_1S^2a^2L_0\right.\\
& -12DS^2a\alpha LL_0-3DS^2a\alpha L^2+6DS^2a^2\alpha L_0\\
& -2DS^2\alpha L_0^3+3DS^2a^2\alpha L+6D_1S^2aLL_0-DS^2a^3\alpha\\
& \left.-3D_1S^2L^2L_0+DS^2\alpha L^3\right)+\left(e^{S(a-L)}-e^{-S(a-L)}\right)\\
& \cdot\left(6D_1SLL_0+2D_1S^3a^3L_0-6DSa\alpha L\right.\\
& +9DS^3a^2\alpha LL_0-2D_1S^3L^3L_0+3DS\alpha L^2\\
& +6D_1S^3aL^2L_0-3DS^3a^3\alpha L_0-6DS^3a\alpha L^2L_0\\
& -6D_1S^3a^2LL_0-6D_1SaL_0+3DSa^2\alpha\\
& \left.+DS^3a\alpha L_0^3-DS^3\alpha LL_0^3\right)+6D_1S^2L^2L_0\\
& +6D_1S^2a^2L_0-12DS^2a\alpha L^2-24D_1L_0\\
& +4DS^2\alpha L_0^3+4DS^2\alpha L^3-4DS^2a^3\alpha\\
& -12DS^2a^2\alpha L_0-12D_1S^2aLL_0+24DS^2a\alpha LL_0\\
& +12DS^2a^2\alpha L\Big]\\
& \cdot\Big(6DD_1S^2L\left[(SL-Sa)\left(e^{S(a-L)}-e^{-S(a-L)}\right)\right.\\
& \left.+2\left(e^{S(a-L)}+e^{-S(a-L)}\right)-4\right]\Big)^{-1}\\
C_4 = {} & -\frac{\alpha PL_0^3}{6D_1}\\
F_0 = {} & \left(6DS\alpha L_0-6D_1S^3LL_0^2-6DS\alpha L+3DS^3\alpha L^2L_0\right.\\
& \left.-3DS^3\alpha LL_0^2+3D_1S^3L_0^3+6D_1SL_0+9DS^3\alpha L_0^3\right)\\
& -\left(D_1S^4L_0^4-3D_1S^4L^2L_0^2+6DS^2\alpha LL_0\right.\\
& \left.-3DS^2\alpha L_0^2-6D\alpha+6D_1S^2L_0^2+4DS^4\alpha LL_0^3\right)
\end{aligned} \quad (A.1)$$

Conflicts of Interest

The authors declare that they have no conflicts of interest.

Acknowledgments

The work is supported by the National Basic Research Program (973) of China (no. 2012CB026200), the Fundamental Research Funds for the Central Universities (126545010), and the Southeast University, Key Laboratory of Concrete and Prestressed Concrete Structure of Ministry of Education (CPCSME2015-03).

References

[1] C. E. Bakis, L. C. Bank, V. L. Brown et al., "Fibre-reinforced polymer composites for construction, state—of-the-Art Review," *Journal of Composites for Construction*, vol. 6, no. 2, pp. 73–87, 2002.

[2] R. M. W. Reising, B. M. Shahrooz, V. J. Hunt, A. R. Neumann, A. J. Helmicki, and M. Hastak, "Close look at construction issues and performance of four fiber-reinforced polymer composite bridge decks," *Journal of Composites for Construction*, vol. 8, no. 1, pp. 33–42, 2004.

[3] A. Zhou and T. Keller, "Joining techniques for fiber reinforced polymer composite bridge deck systems," *Composite Structures*, vol. 69, no. 3, pp. 336–345, 2005.

[4] Z. Jiang, S. Wan, Z. Zhong, M. Li, and K. Shen, "Determination of mode-I fracture toughness and non-uniformity for GFRP double cantilever beam specimens with an adhesive layer," *Engineering Fracture Mechanics*, vol. 128, pp. 139–156, 2014.

[5] C. R. Corleto and H. A. Hogan, "Energy Release Rates for the ENF Specimen Using a Beam on an Elastic Foundation," *Journal of Composite Materials*, vol. 29, no. 11, pp. 1420–1436, 1995.

[6] R. H. Martin and B. D. Davidson, "Mode II fracture toughness evaluation using four point bend, end notched flexure test," *Plastics, Rubber and Composites*, vol. 28, no. 8, pp. 401–406, 1999.

[7] C. Schuecker and B. D. Davidson, "Evaluation of the accuracy of the four-point bend end-notched flexure test for mode II delamination toughness determination," *Composites Science and Technology*, vol. 60, no. 11, pp. 2137–2146, 2000.

[8] S. Hashemi, A. J. Kinloch, and J. G. Williams, "Mechanics and mechanisms of delamination in a poly(ether sulphone)—fibre composite," *Composites Science and Technology*, vol. 37, no. 4, pp. 429–462, 1990.

[9] H. Wang and T. Vu-Khanh, "Use of end-loaded-split (ELS) test to study stable fracture behaviour of composites under mode II loading," *Composite Structures*, vol. 36, no. 1-2, pp. 71–79, 1996.

[10] A. Szekrényes and J. Uj, "Beam and finite element analysis of quasi-unidirectional composite SLB and ELS specimens," *Composites Science and Technology*, vol. 64, no. 15, pp. 2393–2406, 2004.

[11] A. Szekrényes and J. Uj, "Mode-II fracture in E-glass-polyester composite," *Journal of Composite Materials*, vol. 39, no. 19, pp. 1747–1768, 2005.

[12] K. Tanaka, T. Yuasa, and K. Katsura, "Continuous mode II interlaminar fracture toughness measurement by over notched flexure test," in *Proceedings of the 4th European Conference on Composites: Testing and Standardisation*, pp. 171–179, Lisbon, Portugal, 1998.

[13] A. Szekrényes, "Crack stability of fracture specimens used to test unidirectional fiber reinforced material," *Experimental Mechanics*, vol. 50, no. 4, pp. 473–482, 2010.

[14] A. J. Russell and K. N. Street, "Factors affecting the interlaminar fracture energy of graphite/epoxy laminates," *Progress in Science and Engineering of Composites*, pp. 279–286, 1982.

[15] L. A. Carlsson, J. W. Gillespie Jr., and R. B. Pipes, "On the analysis and design of the end notched flexure (ENF) specimen for mode II testing," *Journal of Composite Materials*, vol. 20, no. 6, pp. 594–604, 1986.

[16] J. M. Whitney and C. T. Sun, "A higher order theory for extensional motion of laminated composites," *Journal of Sound and Vibration*, vol. 30, no. 1, pp. 85–97, 1973.

[17] Y. Wang and J. G. Williams, "Corrections for mode II fracture toughness specimens of composites materials," *Composites Science and Technology*, vol. 43, no. 3, pp. 251–256, 1992.

[18] W. Ding and M. T. Kortschot, "A simplified beam analysis of the end notched flexure mode II delamination specimen," *Composite Structures*, vol. 45, no. 4, pp. 271–278, 1999.

[19] J. Wang and P. Qiao, "Novel beam analysis of end notched flexure specimen for mode-II fracture," *Engineering Fracture Mechanics*, vol. 71, no. 2, pp. 219–231, 2004.

[20] T. Keller and H. Gürtler, "Composite action and adhesive bond between fiber-reinforced polymer bridge decks and main girders," *Journal of Composites for Construction*, vol. 9, no. 4, pp. 360–368, 2005.

[21] Y. Zhang, A. P. Vassilopoulos, and T. Keller, "Mode I and II fracture behavior of adhesively-bonded pultruded composite joints," *Engineering Fracture Mechanics*, vol. 77, no. 1, pp. 128–143, 2010.

[22] K. Kageyama, "Composite Materials: Fatigue and Fracture (Third Volume)," Tech. Rep. ASTM STP 1110, ASTM International, West Conshohocken, PA, USA, 1991.

[23] D. Xie and S. B. Biggers Jr., "Progressive crack growth analysis using interface element based on the virtual crack closure technique," *Finite Elements in Analysis and Design*, vol. 42, no. 11, pp. 977–984, 2006.

[24] ASTM D3039/D3039M-14, "Standard test method for tensile properties of polymer matrix composite materials," Tech. Rep., American Society for Testing Materials, West Conshohocken, PA, USA, 2014.

[25] ASTM, "Standard test method for flexural properties of unreinforced and reinforced plastics and electrical insulating materials by four-point bending," ASTM Standard D6272, American Society for Testing Materials, West Conshohocken, PA, USA, 2010.

[26] China NSOT, *GB/T 2567-2008 Test Methods for Properties of Resin Casting Body*, China Standard Press, Beijing, China, 2008.

[27] D. Xie and S. B. Biggers Jr., "Strain energy release rate calculation for a moving delamination front of arbitrary shape based on the virtual crack closure technique. Part I: formulation and validation," *Engineering Fracture Mechanics*, vol. 73, no. 6, pp. 771–785, 2006.

5

Effects of Thermal and Humidity Aging on the Interfacial Adhesion of Polyketone Fiber Reinforced Natural Rubber Composites

Han Ki Lee,[1] Dae Sik Kim,[1] Jong Sung Won,[2] Da Young Jin,[2] Hyun Jae Lee,[2] and Seung Goo Lee[2]

[1]*New Technology Development Team, Research & Development Division, Hyundai-ngv, Seoul 151-742, Republic of Korea*
[2]*Department of Organic Materials Engineering, Chungnam National University, Daejeon 34134, Republic of Korea*

Correspondence should be addressed to Seung Goo Lee; lsgoo@cnu.ac.kr

Academic Editor: Belal F. Yousif

Polyketone fiber is considered as a reinforcement of the mechanical rubber goods (MRG) such as tires, automobile hoses, and belts because of its high strength and modulus. In order to apply it to those purposes, the high adhesion of fiber/rubber interface and good sustainability to aging conditions are very important. In this study, polyketone fiber reinforced natural rubber composites were prepared and they were subjected to thermal and humidity aging, to assess the changes of the interfacial adhesion and material properties. Also, the effect of adhesive primer treatment, based on the resorcinol formaldehyde resin and latex (RFL), of polyketone fiber for high interfacial adhesion was evaluated. Morphological and property changes of the rubber composites were analyzed by using various instrumental analyses. As a result, the rubber composite was aged largely by thermal aging at high temperature rather than humidity aging condition. Interfacial adhesion of the polyketone/NR composites was improved by the primer treatment and its effect was maintained in aging conditions.

1. Introduction

As general reinforcements of rubber composites, rayon, nylon, and polyester fibers have been used for a long time due to their reasonable performance and usefulness. Recently, application of the para-aramid fiber which has high tenacity and modulus is gradually increasing in order to satisfy high performance requirements including specialty tire and tire weight reduction. However, in working with high performance fibers like para-aramid, it is normally difficult to obtain an excellent adhesion with simple processing and their fatigue resistance is not good [1, 2].

Polyketone fiber, recently developed, is available by copolymerization of carbon monoxide and unsaturated hydrocarbon monomer, ethylene. Hence, the cost of its raw material is cheap and the spinning process is environmentally friendly. Also, it has the additional benefits of similar tenacity and modulus to para-aramid and has higher breaking elongation and good adhesion property with a rubber matrix. Therefore, it is expected to be the reinforcement of mechanical rubber goods (MRG) such as tires, hoses, and protective gloves [3].

Adhesion strength between textile cord and rubber matrix can affect the entire performance of a tire. To improve interfacial adhesion between polar cord and nonpolar rubber matrix, an adhesive agent must be used. Since 1935, resorcinol-formaldehyde-latex (RFL) has been used to promote the cord/rubber adhesion. Generally, textile cord must be coated in RFL dip followed by precuring. Low viscosity and good wetting properties have advantages of these aqueous dips. After curing, they are transformed to an insoluble system. The latex gives required flexibility to adhesive agent and reactivity to rubber. Moreover, RF resin provides a desirable heat and fatigue resistance by forming a three-dimensional network. At fiber cord/rubber interface, the functional groups of polar cord and unsaturated zones of rubber undergo reactions with methylol and vinyl groups of RFL, respectively. Physical forces originated from interpenetration and interlocking of the RF resin with the fiber can be a major factor of adhesion, as well [4–9].

On the other hand, the rubber composites are aged by surrounding temperature, humidity, or other factors in actual use environments and, therefore, if aging is conducted, physical properties of the rubber itself are degraded and also the interfacial adhesion between reinforcement and rubber are changed more or less. In particular, the temperature of rubber in the tire increases according to the car speed during driving and this can cause the aging of rubber and interfaces. Under normal driving conditions, a surface temperature of tire is up to 125°C and the faster driving raises it up to 180°C. Also, moisture is known as a factor of the aging of the tire rubber. Therefore, the temperature and humid condition has a major effect on the aging of the tire rubber. Such aging of rubber and interface may cause the failure of the tire, and, thus, the research of the aging effects by heat and humidity of the tire rubber is very important [10–15].

In this study, the effects of aging condition on the interfacial properties of the polyketone fiber reinforced natural rubber composite were investigated. Thermal and humidity aging of the polyketone/NR composites were conducted in the limited experimental range for the tire application. The effect of a primer coating for the polyketone fiber on an interfacial adhesion of the polyketone/NR composites was especially evaluated with the aging conditions. Variations of the rubber hardness, interfacial strength, chemical composition of the rubber matrix, and so forth according to the aging conditions were measured and analyzed as well as their morphological changes.

2. Experimentals

2.1. Materials. Composition of the natural rubber (NR) used in this study is given in Table 1. Same formulation was used in all experiments. Materials were weighed and mixed in an internal mixer and rolled in a roll mill. Finally, a NR sheet with thickness of 6 mm was obtained. Polyketone fiber was supplied from "H" company in Korea. Table 2 shows the properties of the fiber used in this study. Polyketone cord was prepared through the two-for-one twister from polyketone fiber supplied. Twist was in direction S-TPM 868 and direction Z-TPM 434 according to each order.

2.2. Surface Treatment. Polyketone cords were washed with acetone by using ultrasonic agitation for an hour. Then, they were washed with distilled water repeatedly. After purifying polyketone cord, their surfaces were treated with a primer consisting of RFL and epoxy. Typical formulation of the RFL used in this study is given in Table 3. Primer dipping process of the polyketone cord is schematically shown in Figure 1.

2.3. Accelerated Thermal and Humidity Aging. Thermal aging was carried out in an air circulating environmental chamber at temperatures of 105, 125, and 180°C. Because, under normal driving conditions, a surface temperature of tire is up to 125°C and the faster driving raises it up to 180°C. Aging time was set to 8 days. Data on this phenomenon was presented in the results. For humidity aging, the samples were placed in a humidity chamber for 500 h at 25, 60, 85, and 99 RH%, respectively.

Table 1: Composition of rubber compound.

Component	Weight (g)
NR	170
Topping #1	200
ZnO	35
DP-805	1.4
Stearic acid	7
SRF	105
Zeosil #155	21
CaCO3	400
A #2	56
Anti. Ble-N	7
Rosin	21
Acc. DM	8.4
Acc. CZ	2.1

Table 2: Properties of polyketone fiber used in this study.

Fineness (denier)	Strength (g/denier)	Elongation (%)
747	18	6.0

Table 3: Formulation of the RFL primer used in this study.

	Wet part (g)
RF solution	
Resorcinol	16
Formaldehyde	7
Sodium hydroxide	1
Water	200
Maturation: 25°C, 4 h	224
Final dip solution	
RF solution	224
Latex	160
Ammonium hydroxide	5
Water	120
Maturation: 25°C, 12 h	509

2.4. Surface Characterization. Surface chemical composition of polyketone fiber was analyzed by a XPS (Multilab 2000, Thermo). The XPS spectra were obtained using AlKa ($h\nu$ = 1400 eV) monochromated X-ray source with a voltage of 12 kV. Also, morphological changes of polyketone fibers were observed with a scanning electron microscope (SEM, S4700, HITACHI) with surface treatment condition. The hardness of the polyketone cord/NR composites was measured by the Shore type A durometer according to ASTM 2240. All tests were conducted at room temperature.

2.5. Thermal Properties. Using a TGA (TA Instrument), changes in thermal stability of the polyketone fiber/NR composite having thermal and humidity aging were analyzed.

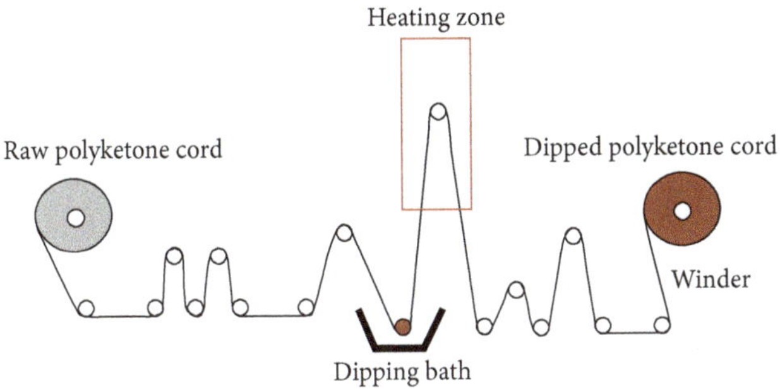

FIGURE 1: Schematics of the primer dipping process.

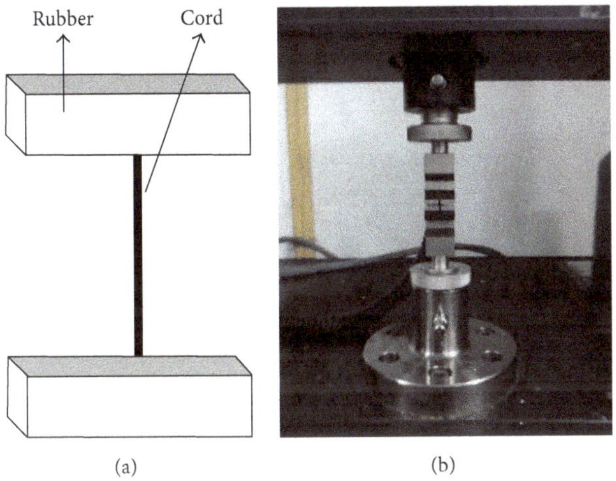

FIGURE 2: Preparation of the H-test: (a) the H-test sample; (b) photo of test set-up (ASTM D4776).

Heating rate was 10°C/min from 25°C to 800°C under nitrogen atmosphere.

2.6. Interfacial Adhesion Test. The interfacial adhesion properties were measured with the H-test method according to ASTM D4776. To prepare the H-test samples, rubber sheets were placed in the channels of a stainless steel mold. Then, the surface treated cord was embedded in rubber sheets. The shapes of the sample and test set-up are shown in Figure 2. Samples were vulcanized at 160°C, for 30 min at pressure of 48 MPa. Then the products were cut into H-shape samples. Test was performed by using an Instron 4467 tester under room temperature and the cross-head speed was fixed at 125 mm/min. Adhesion strength was calculated from the maximum load.

3. Results and Discussion

3.1. Surface Morphology and Hardness of the Polyketone/NR Composite. To investigate the morphological changes due to aging, surface structure of the polyketone/NR composites was analyzed by the SEM. Figure 3 shows surface morphologies with SEM photograph of the polyketone/NR composite aged under each condition. It is shown that the surface was roughened by thermal aging, and change of the surface occurs more severely at the higher temperature. This phenomenon is due to oxidation and hardening of NR by thermal aging. In general, the cross-linking reaction of the rubber is not the completed reaction. Therefore, a new cross-linking reaction is caused by aging and then the matrix cracks are caused from surface with the progress of the reaction [16]. On the other hand, the change in the surface structure did not occur in the humidity aged sample. Through this, it is expected that oxidation and cross-linking reaction are not caused by humidity aging.

In addition, these characteristics can be quantified as the hardness of the rubber. The hardness of rubber is defined as resistivity of surface for the pressure. Hardness of the rubber is controlled according to the usage of the product and represented as Shore A, D, and OO and number 0~100. In the case of a tire, generally Shore 60 A is the appropriate value. Increase in hardness by aging affects the performance of the product. Figure 4 shows the surface hardness of the polyketone/NR composites thermal aged under 105°C and 125°C. It is shown that hardness increased with aging time. In the case of sample aged at 125°C, compared to sample aged at 105°C, the hardness is rapidly increased in the beginning but finally is similar. However, crumbling of surface during the measurement occurred more frequently in the case of sample aged at 125°C. This phenomenon is because increase of hardness by oxidation occurred more than by cross-linking in the case of sample aged at 125°C. In addition, the effect of primer on the hardness is not observed. Figure 5 shows the surface hardness of the polyketone/NR composites thermal aged under 180°C. The increase of hardness was greater than in the other case because of much more oxidation [17]. In addition, in the case of aging temperature of 180°C, there is no significant difference between sample aged for 2 days and 8 days. It is due to the fast degradation at an early stage of 180°C thermal aging followed by the very slow degradation [18]. Also, as with other cases, the effect of primer on the hardness is not observed. Therefore, the processing of the primer has no adverse effects on the physical properties of the NR. However, in the case of the humidity aging, results are different. Figure 6 shows the surface hardness of the polyketone/NR composites aged under each humidity condition. Unlike the results of the thermal aging, change in the hardness by the humidity aging is not observed. Therefore, from the results of SEM observation, it can be said that oxidation and cross-linking reaction are not caused by humidity aging in our experimental range.

3.2. Surface Chemical Composition of the Polyketone/NR Composite. To further investigate the above-mentioned results, the surface chemical composition was analyzed. Table 4 shows functional groups in surface of the polyketone/NR composites. It is shown that C-C bond was reduced and C=O, COOH, C-OH, and O1s increased in the case of thermal aged specimens. This result is due to oxidation by thermal aging. When C-C bonds are broken by thermal energy during thermal aging, carbon radicals are formed. In the presence of atmospheric oxygen, the functional groups containing oxygen such as hydroperoxide are formed from reaction of

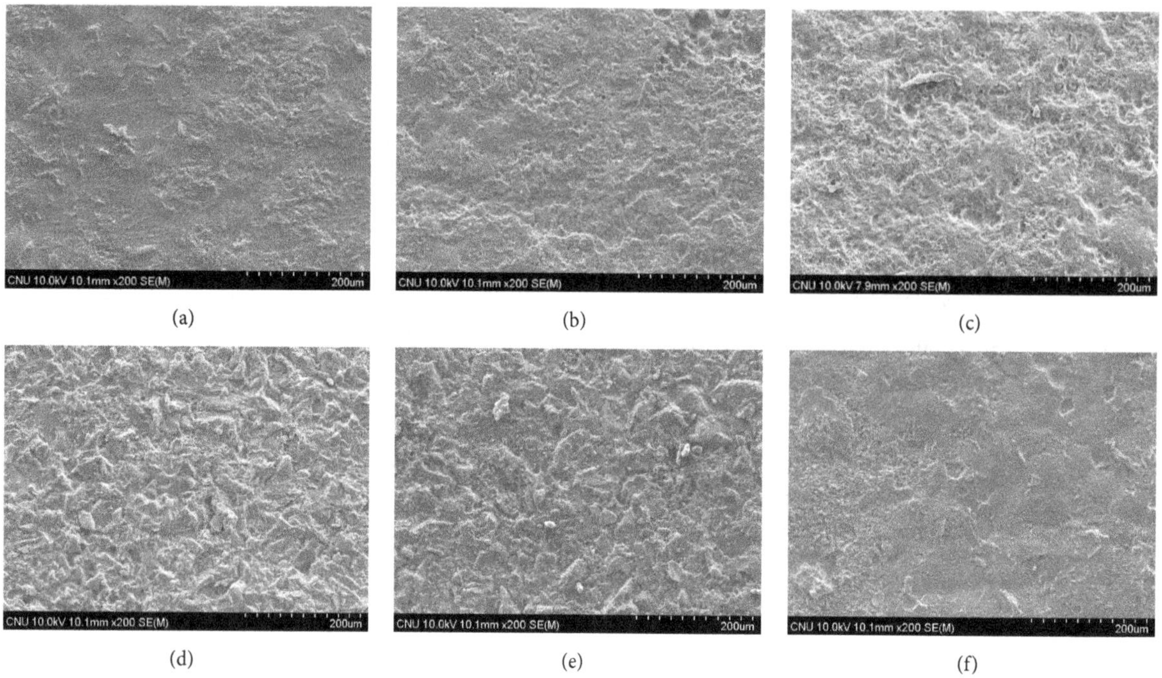

FIGURE 3: SEM photographs of polyketone/NR composite surface; (a) nonaged, (b) thermal aged at 105°C (8 days), (c) 125°C (8 days), (d) 180°C (2 days), and (e) 180°C (8 days), and (f) humidity aged at 99% (500 h).

TABLE 4: Surface functional groups analyzed by XPS of the polyketone/NR composite as thermal aging and humidity aging.

Surface treatment		Binding energy (eV)				
		285	287.6	289.2	286.1	532
		The concentration of correlative functional groups (%)				
		C-C	C=O	COOH	C-OH	Ols
Raw		72.81	—	3.52	6.44	17.23
Thermal aging	105°C (8 days)	59.34	11.6	8.25	—	20.81
	125°C (8 days)	60.07	12.66	5.01	—	22.26
	180°C (2 days)	47.89	5.9	5.92	15.16	25.13
Humidity aging	99% (500 h)	77.86	—	2.93	3.07	16.14

carbon radical [19]. In all conditions of the thermal aging, these phenomena appear to be identical to each other; the oxidation occurs more severely with an increase in aging temperature. However, in the case of humidity aging, this phenomenon does not occur since there is no oxidation. Therefore, the analysis result of humidity aged sample is substantially the same as the raw sample.

Table 5 shows the results of XPS surface composition analysis. Similar conclusion can be obtained through this result. It can be seen that the oxygen content increased due to oxidation. But, in the case of humidity aging, the oxidation does not occur, because the decomposition of the main-chain does not occur since thermal energy is not applied. In addition, in the case of aging temperature of 180°C, there is no significant difference between sample aged for 2 days and for 8 days. This is because most of the oxidation is completed at an early stage by high temperature and subsequent aging proceeds very slowly.

TABLE 5: Chemical composition of the polyketone/NR composite as thermal aging and humidity aging.

Surface treatment		Chemical composition (%)		Atomic ratio
		C	O	O/C
Raw		82.77	17.23	0.208
Thermal aging	105°C (8 days)	79.19	20.81	0.262
	125°C (8 days)	77.74	22.26	0.286
	180°C (2 days)	74.87	25.13	0.335
	180°C (8 days)	74.44	25.56	0.343
Humidity aging	99% (500 h)	83.86	16.14	0.192

3.3. Thermal Properties of the Polyketone/NR Composite. To investigate the changes in thermal properties of polyketone/NR composite by aging, TGA analysis was carried out.

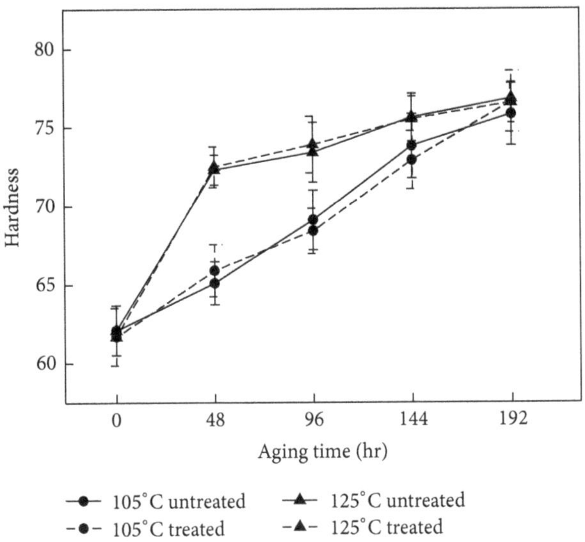

FIGURE 4: Hardness of the polyketone/NR composite after thermal aging at 105 and 125°C.

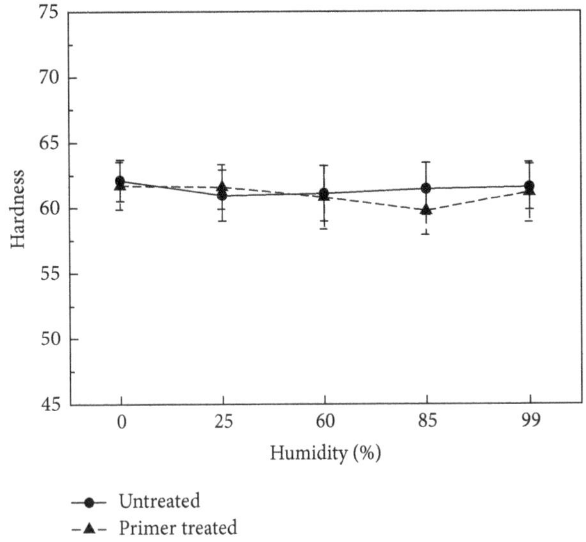

FIGURE 6: Hardness of the polyketone/NR composite after humidity aging.

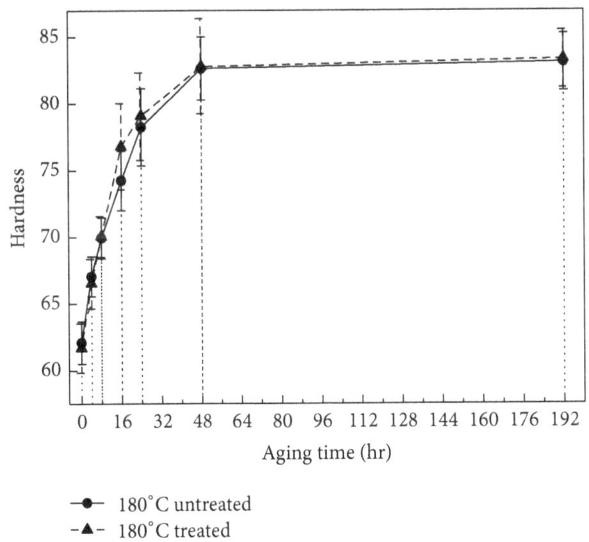

FIGURE 5: Hardness of the polyketone/NR composite after thermal aging at 180°C.

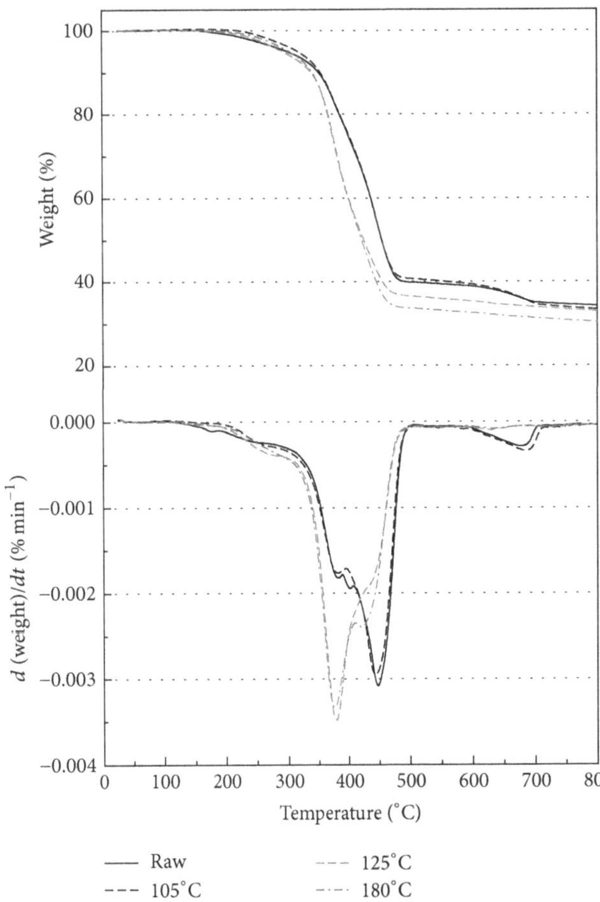

FIGURE 7: TGA and DTG curve of thermal aged polyketone/NR composites.

Figure 7 shows TGA and DTG curves of the polyketone/NR composites aged under each temperature. Thermal decomposition peak appears prominently at 370~450°C. But in the case of raw sample and aged sample at 105, thermal decomposition peak at 450°C appears most significantly, while in the case of sample aged at 125°C and 180°C, peak at 370°C appears most significantly. This difference in decomposition behavior is due to the difference in aging mechanism. In general, the mechanism of thermal aging is classified into two types. One is due to cross-linking and the other is due to chain degradation. If thermal aging is conducted at temperature higher than 120°C, aging caused by chain degradation becomes more active by oxidation [19]. Therefore, in the case of sample aged at 125°C and 180°C, peak is at 450°C which means the main-chain decomposition appears smaller than the case of sample aged at 105°C. This difference in aging mechanism appears to be the difference in increase rate of the initial hardness in the previous analysis results of the hardness.

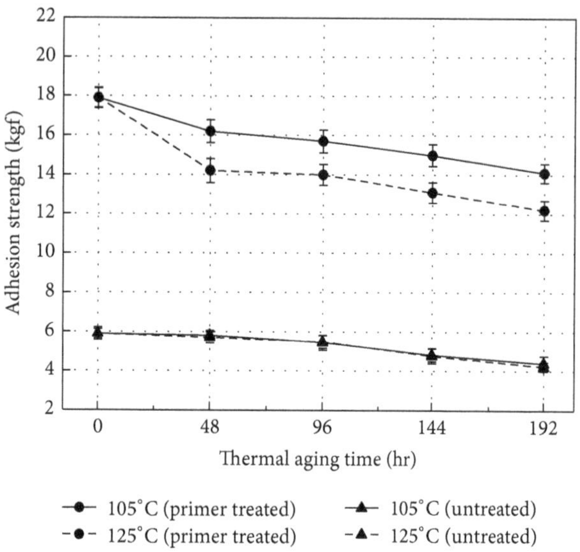

FIGURE 8: Adhesion properties of thermal aging at 105 and 125°C with and without primer treatment.

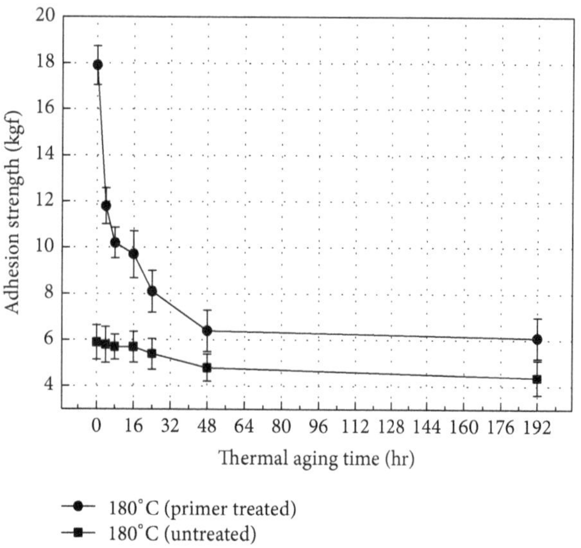

FIGURE 9: Adhesion properties of thermal aging at 180°C with and without primer treatment.

3.4. Interfacial Properties of the Polyketone/NR Composite. To investigate the change in interfacial adhesion strength of the polyketone/NR composite by primer treatment and aging, H-test was carried out. Figures 8 and 9 show the effect of thermal aging on interfacial adhesion strength of the polyketone/NR composites with a primer treatment. The adhesion strength increased largely, almost three times, by a primer treatment compared to untreated case. Therefore, primer coating is very effective in high surface adhesion between polyketone cord and rubber matrix. Also, adhesion strength in all samples decreased by thermal aging. However, adhesion strength of the primer coated cases decreased more than noncoated cases by thermal aging. This result is due to oxidation of NR during thermal aging and the reason of this phenomenon can be verified through observing adhesion fracture surfaces. Figure 10 shows SEM photographs in adhesion fracture surface of the primer treated and untreated polyketone/NR composite aged under each temperature. In the case of untreated composite, the NR is not properly impregnated into the fiber due to a lack of compatibility. Therefore, the interfacial adhesive strength is weak and the rubber residues are not left on the fiber surface. However, in the case of treated composites, the NR is well impregnated into the fiber and there remain rubber residues on the fracture surface. Rubber residues are observed to be thicker at higher aging temperatures. These results mean that most of the fracture is generated from the rubber region weakened by oxidation. Thus it can be considered that although rubber matrix degraded by oxidation at thermal aging, its interfacial region, caused by a primer coating, was not degraded largely by a protection effect of primer. Nevertheless, finally at highest temperature and long aging time, interfacial strength of the composite decreased largely due to the matrix degradation. Also, in the case of aging temperature of 180°C, there is no significant difference between sample aged for 2 days and 8 days as in the other analysis results. It is due to the fast degradation at an early stage of 180°C thermal aging followed by the very slow degradation [20].

Figure 11 shows the effect of humidity aging on the interfacial strength of the polyketone/NR composites with a primer treatment. However, change in the interfacial strength under the humidity aging is not observed, because the decomposition of the main-chain does not occur when high thermal energy is not applied, and, thus, it means that oxidation is not caused by humidity aging. And infiltration of moisture into the matrix and interface does not occur in our experimental humidity-time range [21].

4. Conclusions

The effects of thermal and humidity aging on the characteristics of the polyketone fiber reinforced natural rubber composite were investigated and analyzed in accordance with the primer coating of fiber. The rubber hardness of the polyketone/NR composite increased with increasing thermal aging time because of the oxidation and the cross-linking reaction of the rubber matrix. Oxygen content of the NR matrix increased with the thermal aging temperature and time due to the surface oxidation at an elevated temperature. Lots of oxygen containing groups were increased by thermal aging. Oxidation of the NR matrix by thermal aging caused cracked and roughened morphologies and it brought the decrease of the interfacial strength of the composites. In addition, the oxidation occurred more actively in the aging temperature above 125°C. In the cases of using the primer treated polyketone fiber, the composites have higher adhesion strength than the untreated one. Strong adhesion between the primer treated polyketone fiber and NR matrix protected the degradation of the interfacial strength against thermal aging. On the other hand, aging of the polyketone/NR composites under various humidity conditions did not have a significant impact on the interfacial adhesion and hardness. Therefore, it can be considered that moisture condition has almost no influence on the polyketone/NR composite interface in our experimental humidity-time range.

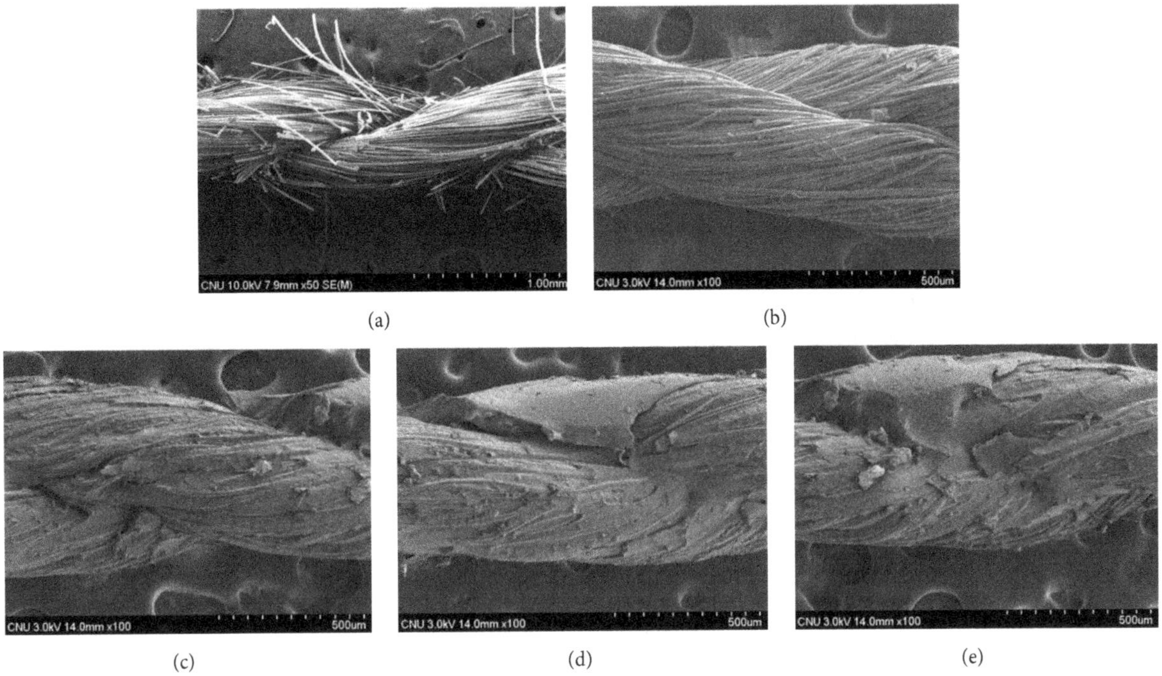

FIGURE 10: SEM photographs of adhesion fracture surface of polyketone/NR composites; (a) primer untreated and nonaged, (b) primer treated and nonaged, and (c) primer treated and thermal aged at 105°C, (d) 125°C, and (e) 180°C.

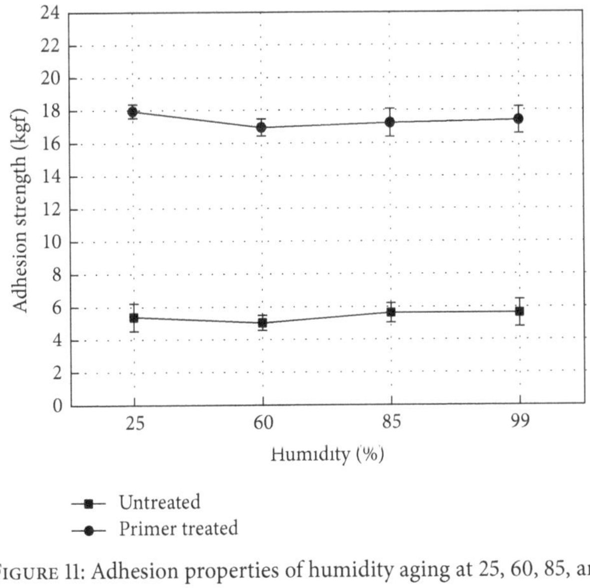

FIGURE 11: Adhesion properties of humidity aging at 25, 60, 85, and 99.

Competing Interests

The authors declare that they have no competing interests.

Acknowledgments

This research was financially supported by the "Fundamental R&D Program for Technology of the Graduate Student Education Program for Research of Hybrid and Super Fiber Materials" through the Ministry of Trade, Industry & Energy (MOTIE) and the Korea Institute for Advancement of Technology (KIAT) (N0000993).

References

[1] D. K. Setua and S. K. De, "Short silk fiber reinforced natural rubber composites," *Rubber Chemistry and Technology*, vol. 56, no. 4, pp. 808–826, 1983.

[2] L. A. Goettler and K. S. Shen, "Short fiber reinforced elastomers," *Rubber Chemistry and Technology*, vol. 56, no. 3, pp. 619–638, 1983.

[3] Y. Zuigyo and M. Yamamoto, "Basic study of PK fiber tire cord," *Tire Science and Technology*, vol. 35, no. 4, pp. 317–325, 2007.

[4] M. Jamshidi, F. Afshar, N. Mohammadi, and S. Pourmahdian, "Study on cord/rubber interface at elevated temperatures by H-pull test method," *Applied Surface Science*, vol. 249, no. 1–4, pp. 208–215, 2005.

[5] A. K. Naskar, A. K. Mukherjee, and R. Mukhopadhyay, "Studies on tyre cords: degradation of polyester due to fatigue," *Polymer Degradation and Stability*, vol. 83, no. 1, pp. 173–180, 2004.

[6] T. S. Solomon, "Systems for tire cord-rubber adhesion," *Rubber Chemistry and Technology*, vol. 58, no. 3, pp. 561–576, 1985.

[7] M. Jamshidi and F. A. Taromi, "Estimation of dynamic adhesion from static test results in the model cord-RFL-rubber system," *Journal of Adhesion Science and Technology*, vol. 20, no. 15, pp. 1693–1703, 2006.

[8] R. B. Durairaj, "Resorcinol formaldehyde latex (RFL) adhesives and applications," in *Resorcinol: Chemistry, Technology and Applications*, pp. 263–339, Springer, 2005.

[9] W. B. Wennekes, *Adhesion of RFL-Treated Cords to Rubber: New Insights into Interfacial Phenomena*, University of Twente, 2008.

[10] W. J. van Ooij, "Mechanism and theories of rubber adhesion to steel tire cords—an overview," *Rubber Chemistry and Technology*, vol. 57, no. 3, pp. 421–456, 1984.

[11] N. Rattanasom, A. Poonsuk, and T. Makmoon, "Effect of curing system on the mechanical properties and heat aging resistance of natural rubber/tire tread reclaimed rubber blends," *Polymer Testing*, vol. 24, no. 6, pp. 728–732, 2005.

[12] A. K. Naskar, S. K. De, A. K. Bhowmick, P. K. Pramanik, and R. Mukhopadhyay, "Characterization of ground rubber tire and its effect on natural rubber compound," *Rubber Chemistry and Technology*, vol. 73, no. 5, pp. 902–911, 2000.

[13] J. M. Baldwin and D. R. Bauer, "Rubber oxidation and tire aging— a review," *Rubber Chemistry and Technology*, vol. 81, no. 2, pp. 338–358, 2008.

[14] Y. Iyengar, "Factors in rubber compounds affecting the adhesion of polyester tire cords," *Journal of Applied Polymer Science*, vol. 15, no. 2, pp. 267–276, 1971.

[15] M. Forrest, *Long-Term and Accelerated Ageing Tests on Rubbers*, vol. 110, iSmithers Rapra Publishing, 2000.

[16] D. Huang, B. J. LaCount, J. M. Castro, and F. Ignatz-Hoover, "Development of a service-simulating, accelerated aging test method for exterior tire rubber compounds I. Cyclic aging," *Polymer Degradation and Stability*, vol. 74, no. 2, pp. 353–362, 2001.

[17] A. Tóth,, I. Bertóti,, M. Blazsó, G. Bánhegyi, A. Bognar, and P. Szaplonczay, "Oxidative damage and recovery of silicone rubber surfaces. I. X-ray photoelectron spectroscopic study," *Journal of Applied Polymer Science*, vol. 52, no. 9, pp. 1293–1307, 1994.

[18] E. J. Blackman and E. B. McCall, "Relationships between the structures of natural rubber vulcanizates and their thermal and oxidative aging," *Rubber Chemistry and Technology*, vol. 43, no. 3, pp. 651–663, 1970.

[19] J. I. Cunneen, "Oxidative aging of natural rubber," *Rubber Chemistry and Technology*, vol. 41, no. 1, pp. 182–208, 1968.

[20] V. G. Geethamma, G. Kalaprasad, G. Groeninckx, and S. Thomas, "Dynamic mechanical behavior of short coir fiber reinforced natural rubber composites," *Composites Part A: Applied Science and Manufacturing*, vol. 36, no. 11, pp. 1499–1506, 2005.

[21] A. Hodzic, J. K. Kim, A. E. Lowe, and Z. H. Stachurski, "The effects of water aging on the interphase region and interlaminar fracture toughness in polymer-glass composites," *Composites Science and Technology*, vol. 64, no. 13-14, pp. 2185–2195, 2004.

6

Photon Conversion and Radiation Synergism in Eu/Tb Complexes Incorporated Poly Methyl Methacrylate

P. Xiao, J. J. Zhang, Z. Q. Wang, and H. Lin

School of Textile and Material Engineering, Dalian Polytechnic University, Dalian 116034, China

Correspondence should be addressed to J. J. Zhang; zhangjj@dlpu.edu.cn and H. Lin; wjs@dlpu.edu.cn

Academic Editor: Mikhael Bechelany

Green and red emissions in terbium and europium benzoic acids (TBA and EBA) incorporated poly methyl methacrylates (PMMA) are exhibited. Larger intensity parameters Ω_2 (11.13×10^{-20} cm^2) and Ω_4 (13.15×10^{-20} cm^2) of Eu^{3+} indicate a high inversion asymmetrical and strong covalent environment around lanthanide (Ln) ions and maximum emission cross section $\sigma_{em\text{-}max}$ (2.45×10^{-21} cm^2) for the dominant transition $^5D_0 \rightarrow {}^7F_2$ of Eu^{3+} reveals the effective photon conversion capacity in EBA incorporated PMMA. Color variation and fluorescence enhancement are validated to be originated from energy transfer and synergistic effect in TBA/EBA coincorporated PMMA system. The results indicate that Eu and Tb complexes incorporated PMMA with effective photon conversion hold great prospect in increment of lighting quality and enhancement of solar-cell efficiency employed in outer space.

1. Introduction

Dye-sensitized solar cells (DSSCs) have attracted extensive attention as one of the promising alternatives for silicon solar cells [1–4]. However, most of the solar radiation energy in the ultraviolet (UV) and infrared (IR) regions is not utilized since DSSCs have a maximum absorption in visible region, which limits the power conversion efficiency of DSSCs [5]. Hence, developing low-cost and high-efficient technologies evokes a great deal of interest in the past decades [6]. Recently, investigations have been reported on the efficient fluorescence generated in lanthanide (Ln) ions incorporated TiO_2 as photoelectrode, which are deemed to be a new route for high-quality conversion efficiency in DSSCs [7–9]. Ln ions can enhance light harvesting via converting UV and near infrared (NIR) radiation to visible emission, so that the conversion efficiency of solar cells will be improved effectively [10–21].

The luminescence of Ln ions originates from the intra-4f transitions, which are forbidden in principle, resulting in relatively low emission efficiency [22]. An effective approach to increase the luminescent efficiency is that chelating Ln ions with organic ligands, which possess advantages such as long emission lifetimes, narrow bandwidths, and large Stokes shifts [23–34]. Since the Ln organic complexes are poor in thermal performances, mechanical stabilities, and processing properties, incorporating these complexes into organic polymers is an ideal method to make up these shortcomings [35–43]. Poly methyl methacrylate (PMMA) exhibits high transparency in the visible spectral region, excellent weather resistance, and high mechanical strength. Moreover, PMMA contains carbonyl groups along with its carbon chain that can be interacted with Ln complexes [44–50]. These characteristics make it suitable as a host material for lanthanide complex doping.

Here in this work, terbium and europium benzoic acid (TBA and EBA) coincorporated PMMA have been prepared. Green and red fluorescence have been observed in TBA and EBA incorporated PMMA, respectively, and the variation of color coordinates originated from synergistic effect of TBA and EBA have been captured. Spontaneous emission probabilities, fluorescence branching ratios, and emission cross sections of intrinsic transitions of Eu^{3+} in PMMA have been evaluated. Thermogravimetric analysis (TGA) was carried out to give a detailed discussion of thermal property of PMMA after the addition of TBA. The analyses of lanthanide complexes incorporated PMMA reveal a promising application prospect for improving illumination quality of lighting

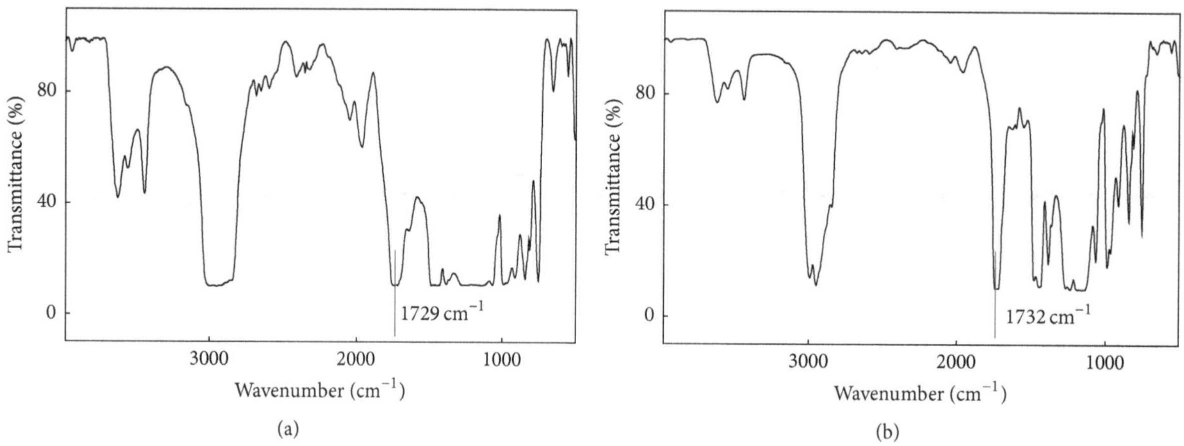

FIGURE 1: FTIR spectra of (a) PMMA and (b) 1 wt% TBA incorporated PMMA.

devices and conversion efficiency of solar cells employed in outer space.

2. Experiments

TBA and EBA were synthesized according to the following procedures. Initially, $LnCl_3$ was synthesized from the reaction between Ln oxides and hydrochloric acid. Then $LnCl_3$ ethanol solution was added dropwise into the benzoic acid (BA) ethanol solution, and the molar ratio of Ln ions to BA was 1 : 3. Subsequently, sodium hydroxide dissolved in ethanol was used to adjust the pH of the mixture to remain between 6 and 7. Instantaneously the precipitate was formed and reaction mixture was stirred adequately for 2 hours at room temperature. The precipitate was filtered off by a Büchner funnel repeatedly, washed with ethanol and water, and dried in an oven at 80°C overnight.

A certain amount of PMMA powder was dissolved in tetrahydrofuran (THF) solution followed by the addition of the lanthanide complexes. The well-mixed solutions were heated at 60°C for 1 hour using a thermostat water bath and the samples were prepared by evaporating the solvent slowly from the solutions.

The infrared spectra of the samples were recorded at room temperature using a Spectrum One-B FTIR spectrophotometer in the range of 400–4000 cm^{-1}. Thermogravimetric analysis scan was carried out by an American TA Company Q-50 thermogravimetric analysis system at a heating rate of 10°C/min under dynamic nitrogen atmosphere with a stream of 50 mL/s. Using the Metricon 2010 prism coupler, the refractive indices of the 0.2 wt% EBA incorporated PMMA were measured to be 1.4870 and 1.4777 at 635.96 and 1536.9 nm, respectively. The refractive indices at other wavelengths can be solved by Cauchy's equation $n = A + B/\lambda^2$ with $A = 1.4758$ and $B = 4526\,\text{nm}^2$. The excitation and emission spectra were recorded by a Hitachi F-7000 fluorescence spectrophotometer and the relative spectral power distributions were measured by a Jobin Yvon Fluorolog-3 spectrophotometer with the calibration of a referenced halogen lamp. The luminescence pictures of the samples were taken using a Sony α200 digital camera.

3. Results and Discussion

Infrared spectroscopy is a useful technique to reflect the supramolecular interaction between the molecules. Taking an example of TBA incorporated PMMA, Fourier transform infrared (FTIR) spectra of pure PMMA and TBA incorporated PMMA in the spectral range 400–4000 cm^{-1} are shown in Figure 1. The peaks at 989 and 1384 cm^{-1} are attributed to C–O–C symmetric stretching and O–CH_3 deformation, respectively, corresponding to fingerprint vibrations of the PMMA [51]. The intense and sharp band at 1729 cm^{-1} for the PMMA corresponds to the (–COOR), which is assigned to symmetrical stretching vibration of the carbonyl group [52]. The asymmetric stretching vibration modes of CH_3 and CH_2 show a strong broad band centered at about 2952 cm^{-1}, indicating a high hydrogen content in the carbon backbone of PMMA [53]. Moreover, new peaks at 1603 and 1554 cm^{-1} appear, corresponding to the asymmetric stretching vibration of COO–, which demonstrates the coordination between the BA and the Ln ions [54].

The IR spectra show that the positions of carbonyl vibration in PMMA and TBA incorporated PMMA are located at 1729 and 1732 cm^{-1}, respectively. The wavenumber difference reveals that the terbium complex is stabilized through chemical interactions with the oxygen atoms of the carbonyl group of PMMA [55], since the oxygen atoms of the carbonyl group in PMMA are a kind of hard base and lanthanide ions are regarded as a kind of hard acid according to HSAB theory [56]. When Tb^{3+} ion interacts with oxygen atom of the carbonyl group, the electron density changes, consequently leading to the position shift corresponding to its characteristic vibration [57].

Thermal property is of great significance to PMMA for practical applications as optical materials and it is strongly influenced by the chain lengths of the polymer, the placements of the ester, and the methyl groups along the polymer backbone [58]. Thermogravimetric analyses (TGA) of PMMA and TBA incorporated PMMA have been carried out to explore their degradation behavior, as illustrated in Figure 2. TGA thermogram of PMMA appears with a maximum value at 143°C on the weight loss rate curve and shows

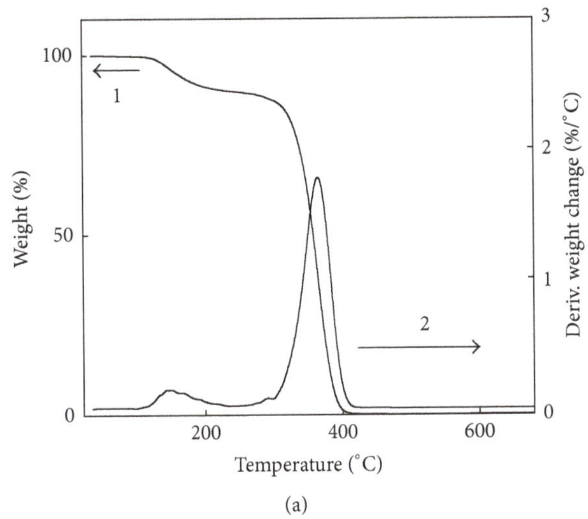

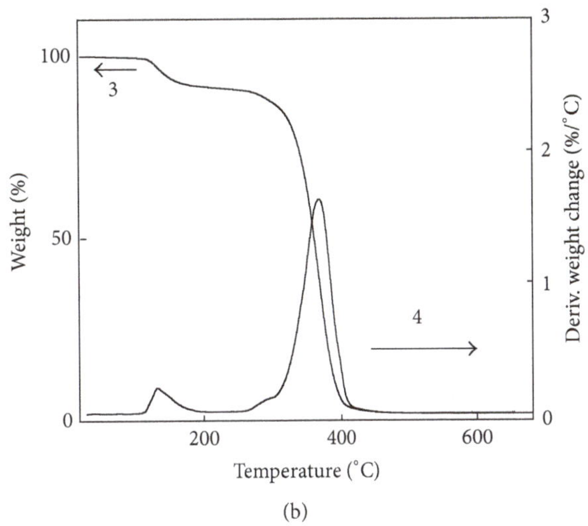

FIGURE 2: Thermogravimetric analysis of (a) PMMA and (b) 1 wt% TBA incorporated PMMA. Curves 1 and 3 are the weight loss curves, and curves 2 and 4 are the weight loss rate curves.

a 10 wt% total weight loss at 260°C, which is attributed to the solvent removal from the TBA incorporated PMMA. The figure reveals that the major weight loss events that occurred begin at the temperature of 300°C, leaving essentially no residue at 420°C, indicating the thermal decomposition behavior of pure PMMA. In addition, the TGA data show that TBA incorporated PMMA has no weight loss event within 260–330°C due to the water molecules coordinated to the Tb^{3+} ion of the hydrate precursor complex after the doping reaction, and it can be deduced that the TBA framework decomposed and turned into terbium oxide at the temperature of 450°C.

TBA incorporated PMMA emits bright green fluorescence under the radiation of UV lamp as shown in inserted photo of Figure 3. The excitation spectrum of PMMA polymer incorporated with TBA by monitoring the emission at 546 nm is shown as curve 1 in Figure 3. The excitation spectrum is dominated by an intense broad band in the 220–300 nm region, which can be assigned to the absorption of Ln organic complex. The intense band confirms that TBA incorporated PMMA can be efficiently excited under the short-wavelength UV radiation.

Curve 2 in Figure 3 presents the emission spectrum of the PMMA polymer incorporated with TBA under 254 nm excitation. From the emission spectrum, four emission transitions have been identified at 488, 546, 583, and 619 nm, respectively, assigning to the characteristic $^5D_4 \rightarrow {}^7F_J$ (J = 6, 5, 4, 3) transitions within the $4f^8$ configuration of the Tb^{3+} ions [59–61]. The most intense emission peak corresponding to $^5D_4 \rightarrow {}^7F_5$ transition of Tb^{3+} ions is located at 546 nm with a full-width at half-maximum (FWHM) of only 10 nm, and its intensity is apparently stronger than other peaks, resulting in a bright green emission under 254 nm UV radiation. Furthermore, the PMMA matrix influences the coordination environment of Tb^{3+} ions, which causes the change of energy transfer probability from BA to Tb^{3+} and the increase of electric-dipole transition intensity [62].

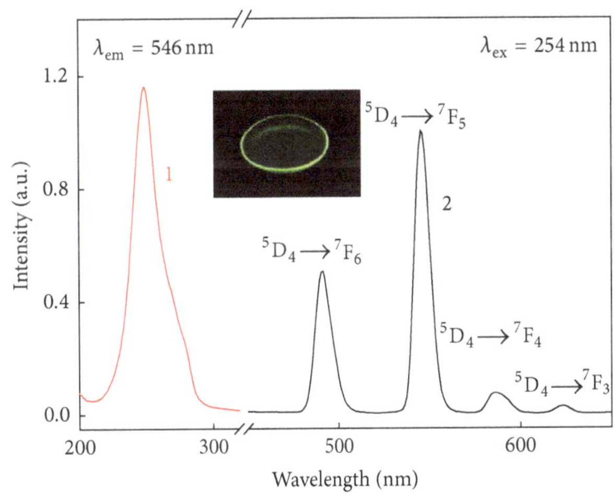

FIGURE 3: Excitation (curve 1) and emission (curve 2) spectra for 0.2 wt% TBA incorporated PMMA. Inserted photo: fluorescence from TBA incorporated PMMA under 254 nm UV radiation.

Luminescence color for the lanthanide complexes incorporated PMMA is a psychophysical property of human eye and it needs to be expressed mathematically for precision. Thus, marking color coordinates on standard chromaticity diagram is necessary. The Commission International del'Eclairage- (CIE-) 1931 color coordinates can be calculated using the following formula:

$$x = \frac{X}{X+Y+Z},$$
$$y = \frac{Y}{X+Y+Z}, \quad (1)$$
$$z = \frac{Z}{X+Y+Z},$$

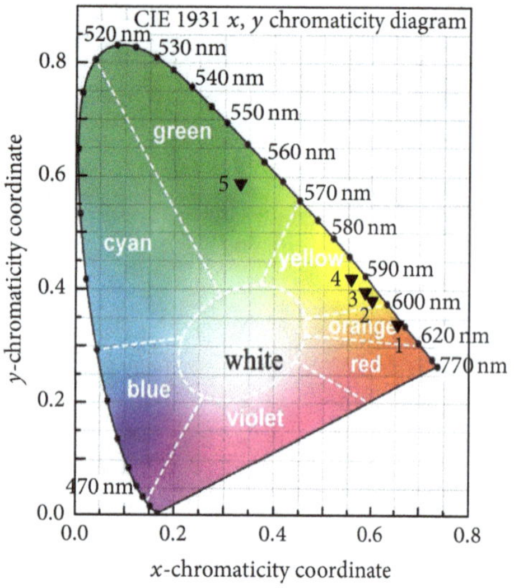

FIGURE 4: CIE (x, y) chromaticity diagram indicating the color coordinates of the fluorescence in samples, and the coordinates corresponding to points from 1 to 5 are in 0.2 wt% EBA, 0.4 wt% TBA-0.2 wt% EBA, 0.8 wt% TBA-0.2 wt% EBA, 1.4 wt% TBA-0.2 wt% EBA, and 0.2 wt% TBA incorporated PMMA.

where X, Y, and Z are the three tristimulus values. The tristimulus values for a color can be derived through the relative spectral power distribution $P(\lambda)$ by

$$X = \int P(\lambda)\,\overline{x}(\lambda)\,d\lambda,$$
$$Y = \int P(\lambda)\,\overline{y}(\lambda)\,d\lambda, \tag{2}$$
$$Z = \int P(\lambda)\,\overline{z}(\lambda)\,d\lambda,$$

where λ is the wavelength of the equivalent monochromatic light and $\overline{x}(\lambda)$, $\overline{y}(\lambda)$, and $\overline{z}(\lambda)$ are the three color-matching functions. The values of (x, y) coordinate for TBA incorporated PMMA are calculated to be (0.3327, 0.5877), as point 5 in Figure 4.

The emission spectrum of EBA incorporated PMMA under 254 nm excitation is presented in Figure 5(a), which consists of five emission bands $^5D_0 \rightarrow {}^7F_J$ $(J = 0, 1, 2, 3, 4)$ located at 578, 591, 617, 650, and 698 nm, respectively. Intensity parameters Ω_t $(t = 2, 4, 6)$, which were calculated from the emissions of Eu^{3+}, are important indicators to predict some radiative properties [63, 64]. The integrated fluorescence intensity ratio between the electric-dipole (A_{ed}) and the magnetic-dipole (A_{md}) transitions has been widely used as a criterion of the coordination status and the site asymmetry for the lanthanide ions [65–67], which is calculated based on the emission photon ratio of $^5D_0 \rightarrow {}^7F_J$ $(J = 2, 4, 6)$ transition to $^5D_0 \rightarrow {}^7F_1$ transition, following the formula

$$\frac{A_{md}}{A_{ed}} = \frac{N_J}{N_1}. \tag{3}$$

The transitions of Eu^{3+} from $^5D_0$ to ^7F_J $(J = 2, 4, 6)$ belong to electronic dipole transition and the spontaneous radiative transition probability A_{ed} from ψJ to $\psi' J'$ is described by

$$A_{ed} = \frac{64\pi^4 e^2 \nu^3}{3h(2J+1)} \frac{n(n^2+2)^2}{9} \sum_{t=2,4,6} \Omega_t \left\langle \psi J \left\| U^{(t)} \right\| \psi' J' \right\rangle^2, \tag{4}$$

where h is the Planck constant, e is the electron charge, n is the refractive index, J is the total angular momentum of the excited state, ν is the wavenumber of transition, Ω_t is the Judd-Ofelt parameters, and the term $\langle \psi J \| U^{(t)} \| \psi' J' \rangle^2$ is the square of the reduced matrix elements. The transition $^5D_0 \rightarrow {}^7F_1$ is a magnetic-dipole transition and the spontaneous transition probability A_{md} is reduced by

$$A_{md} = \frac{64\pi^4 \nu^3}{3h(2J+1)} n^3 S_{md}, \tag{5}$$

where S_{md} is the magnetic-dipole line strength, which is a constant and independent of the medium, so the value of A_{md} can be estimated by the value of fluoride glass ($A'_{md} = 60.3\ s^{-1}$). The relationship is

$$A_{md} = \left(\frac{n}{n'}\right)^3 A'_{md}, \tag{6}$$

where $n'(= 1.522)$ is the refractive index of fluoride glasses [68].

Based on the relative spectral power distribution $P(\lambda)$ of the EBA incorporated PMMA, which is shown in Figure 6(a), the relative photon distribution $N(\nu)$ has been derived as

$$N(\nu) = \frac{\lambda^3}{hc} P(\lambda), \tag{7}$$

where λ is the wavelength, h is the Planck constant, and c is the velocity of light. The relative photon distribution was presented in Figure 6(b), and the emitted photon numbers were calculated using the Gaussian multipeaks fitting and listed in Table 1.

Due to the selection rules and the characteristics of transition intensities in Eu^{3+} ions, each of the $\langle \psi J \| U^{(t)} \| \psi' J' \rangle^2$ parameters determines the intensities of the transitions since the remaining two are zero [69]. The intensity parameters Ω_t $(t = 2, 4)$ of EBA incorporated PMMA were calculated to be 11.13×10^{-20} and $13.15 \times 10^{-20}\ cm^2$, respectively. The value of Ω_6 was not obtained due to the absence of $^5D_0 \rightarrow {}^7F_6$ transition emission. The value of Ω_2 and Ω_4 reflects large chemical bond overlap polarizability to some degree, indicating a high inversion asymmetrical and strong covalent environment around Eu^{3+} in PMMA [67].

The spectroscopic parameters including spontaneous emission probabilities of electric-dipole transition A_{ed} and magnetic-dipole transition A_{md}, total transition probability A_T, fluorescence branching ratios β, and radiative lifetime τ_{rad} are calculated by Ω_t values and listed in Table 2. The

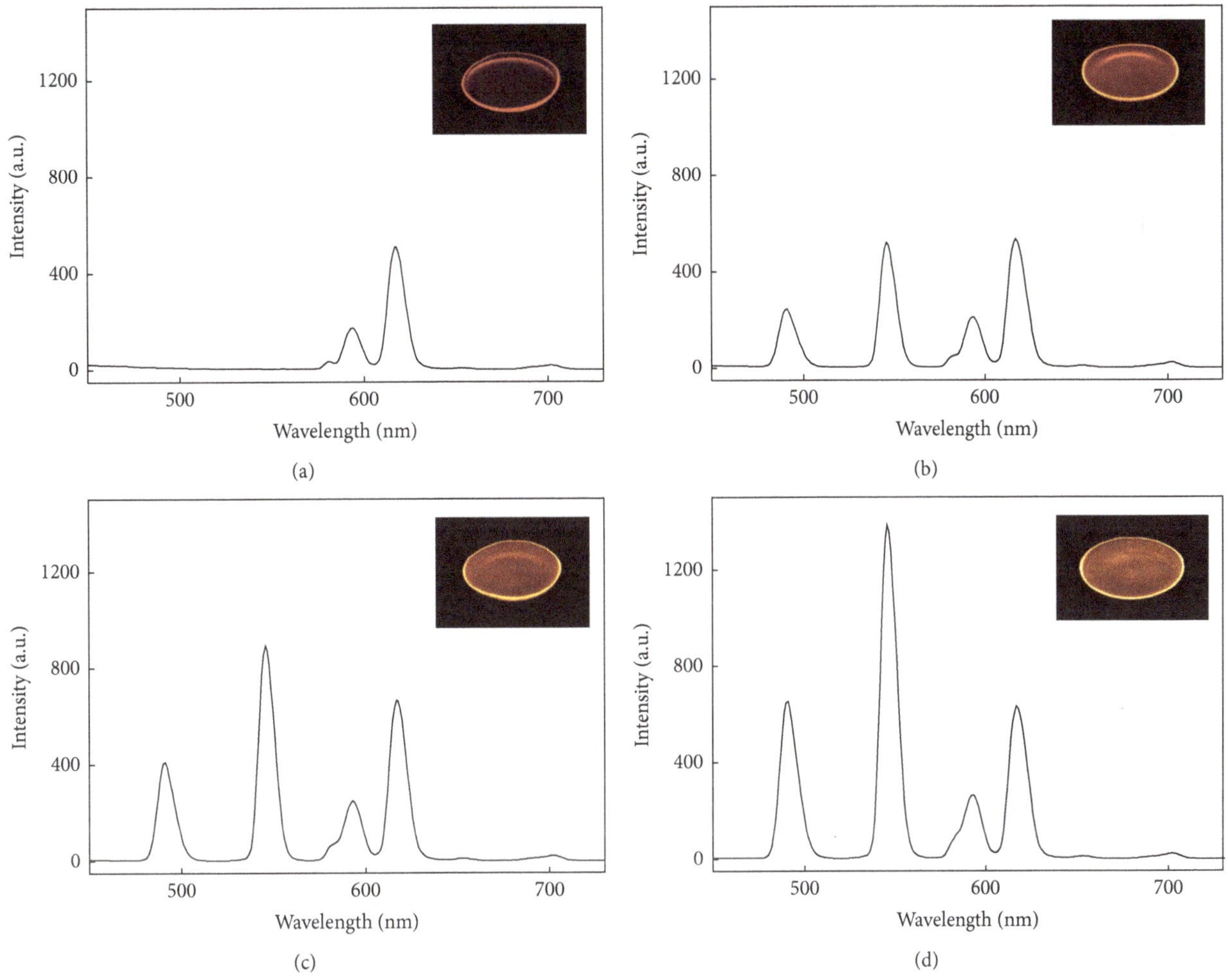

FIGURE 5: Emission spectra of (a) 0.2 wt% EBA, (b) 0.4 wt% TBA-0.2 wt% EBA, (c) 0.8 wt% TBA-0.2 wt% EBA, and (d) 1.4 wt% TBA-0.2 wt% EBA incorporated PMMA under 254 nm UV radiation. Inserted photos: fluorescence from corresponding samples.

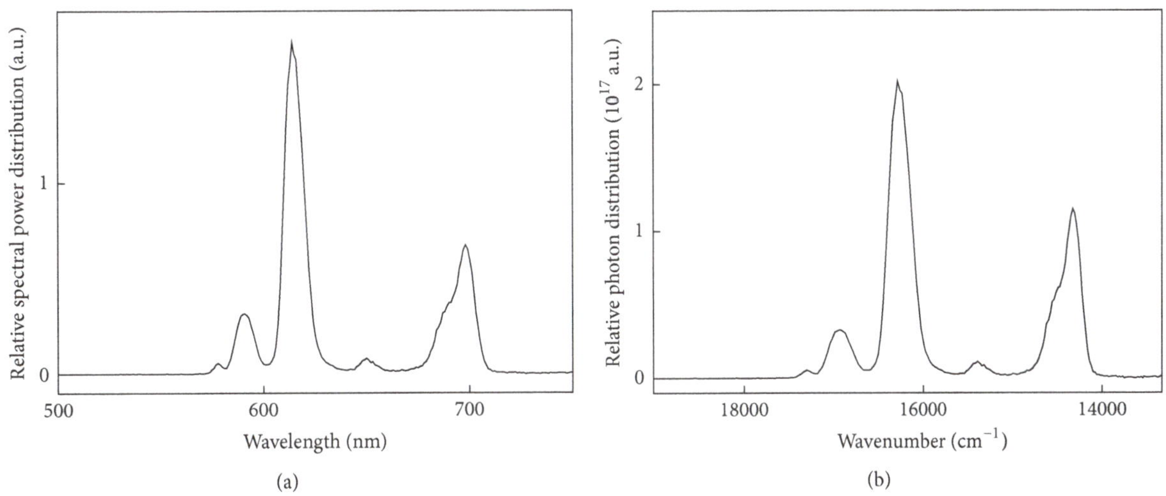

FIGURE 6: (a) Relative spectral power distribution and (b) relative photon distribution of 0.2 wt% EBA incorporated PMMA under 254 nm excitation.

Table 1: Relative photon numbers and number ratios in EBA incorporated PMMA.

Transition	Wave number (cm^{-1})	Relative photon number (10^{19} a.u.)	Relative photon number ratio
$^5D_0 \rightarrow {}^7F_1(N_1)$	16920	1.002	—
$^5D_0 \rightarrow {}^7F_2(N_2)$	16256	5.800	$N_2/N_1 = 5.79$
$^5D_0 \rightarrow {}^7F_4(N_4)$	14327	3.326	$N_4/N_1 = 3.32$

Table 2: Predicted spontaneous transition probabilities, branching ratios, and radiative lifetime of EBA incorporated PMMA.

Transition	$U^{(2)2}$	$U^{(4)2}$	$U^{(6)2}$	A_{ed} (s^{-1})	A_{md} (s^{-1})	A_T (s^{-1})	β (%)	τ_{rad} (ms)
$^5D_0 \rightarrow {}^7F_1$	0	0	0	0	56.44	570.88	9.89	1.75
$^5D_0 \rightarrow {}^7F_2$	0.0032	0	0	326.71	0		57.23	
$^5D_0 \rightarrow {}^7F_4$	0	0.0023	0	187.73	0		32.88	

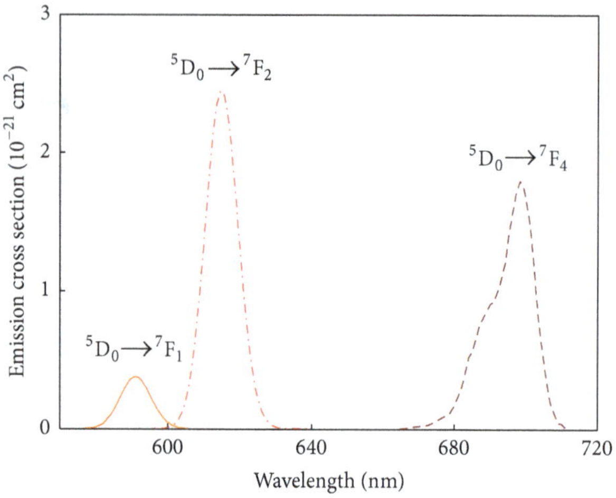

Figure 7: Emission cross section profiles of 0.2 wt% EBA incorporated PMMA.

fluorescence branching ratio β of the $^5D_0 \rightarrow {}^7F_2$ transition centered at 617 nm accounts for 57.23%, which is higher than those of other transitions, predicting that the $^5D_0 \rightarrow {}^7F_2$ transition emission is dominant among the transitions of Eu^{3+} in EBA incorporated PMMA.

To further evaluate the energy extraction efficiency for optical material, it is of great significance to discuss the emission cross section σ_e. The emission cross section for the transition emissions arising from $^5D_0$ level of Eu^{3+} reveals the radiative transition probabilities, which are determined through the Fuchtbauer-Ladenburg (FL) formula [69]:

$$\sigma_e = \frac{A_{rad}}{8\pi c n^2} \frac{\lambda^5 N(\lambda)}{\int \lambda N(\lambda)\, d\lambda}, \quad (8)$$

where n, A_{rad}, and $N(\lambda)$ represent the refractive index, spontaneous emission probability, and emission photon distribution, respectively. The σ_e profiles are shown in Figure 7 and the maximum values of σ_e for $^5D_0 \rightarrow {}^7F_J$ ($J = 1, 2, 4$) transitions are deduced to be 0.377×10^{-21}, 2.451×10^{-21}, and 1.799×10^{-21} cm^2, respectively, and the large emission cross section enables the EBA incorporated PMMA to be a promising material for active optical devices.

Table 3: CIE chromaticity coordinates of the lanthanide complexes coincorporated PMMA.

Points in CIE diagram	EBA (wt%)	TBA (wt%)	x-coordinate	y-coordinate
1	0.2	0	0.6554	0.3397
2	0.2	0.4	0.6015	0.3816
3	0.2	0.8	0.5863	0.3967
4	0.2	1.4	0.5569	0.4200
5	0	0.2	0.3327	0.5877

The visible emission spectra of TBA/EBA coincorporated PMMA under 254 nm excitation are presented in Figure 5. With the introduction of TBA into EBA incorporated PMMA, seven strong and sharp emission bands are recorded under the same excitation condition. The 491 and 546 nm emission bands are attributed to the $^5D_4 \rightarrow {}^7F_6$ and $^5D_4 \rightarrow {}^7F_5$ characteristic transitions of Tb^{3+}, and the emission bands at 582, 593, and 617 nm are ascribed to the transitions from both Tb^{3+} and Eu^{3+} ions. Other bands peaking at 653 and 702 nm are typical emission features for the $^5D_0 \rightarrow {}^7F_J$ ($J = 3, 4$) transitions of Eu^{3+} [60]. EBA single incorporated and TBA/EBA coincorporated PMMA samples exhibit various color fluorescence under the excitation of 254 nm UV radiation, as shown in the inserted photo of Figure 5. In order to exhibit the trace of color variation, the CIE color coordinates for EBA single incorporated and TBA/EBA coincorporated PMMA samples were calculated based on the relative spectral power distributions, as listed in Table 3. In the cases of a fixed concentration of EBA, the fluorescence color coordinates in CIE chromaticity diagram move along right boundary from the bottom to the top with the concentration of TBA varying from 0 to 1.4 wt%, as marked points 1, 2, 3, and 4 in Figure 4. The variation of color coordinates is originated from the superimposed effect of TBA and EBA, indicating that Ln complexes incorporated PMMA are a promising candidate for radiation conversion system and lighting device. Furthermore, intrinsic transition intensities of Eu^{3+} increase due to energy migration from Tb^{3+}, and the energy transfer process can be further exposed through the excitation behaviors [70, 71].

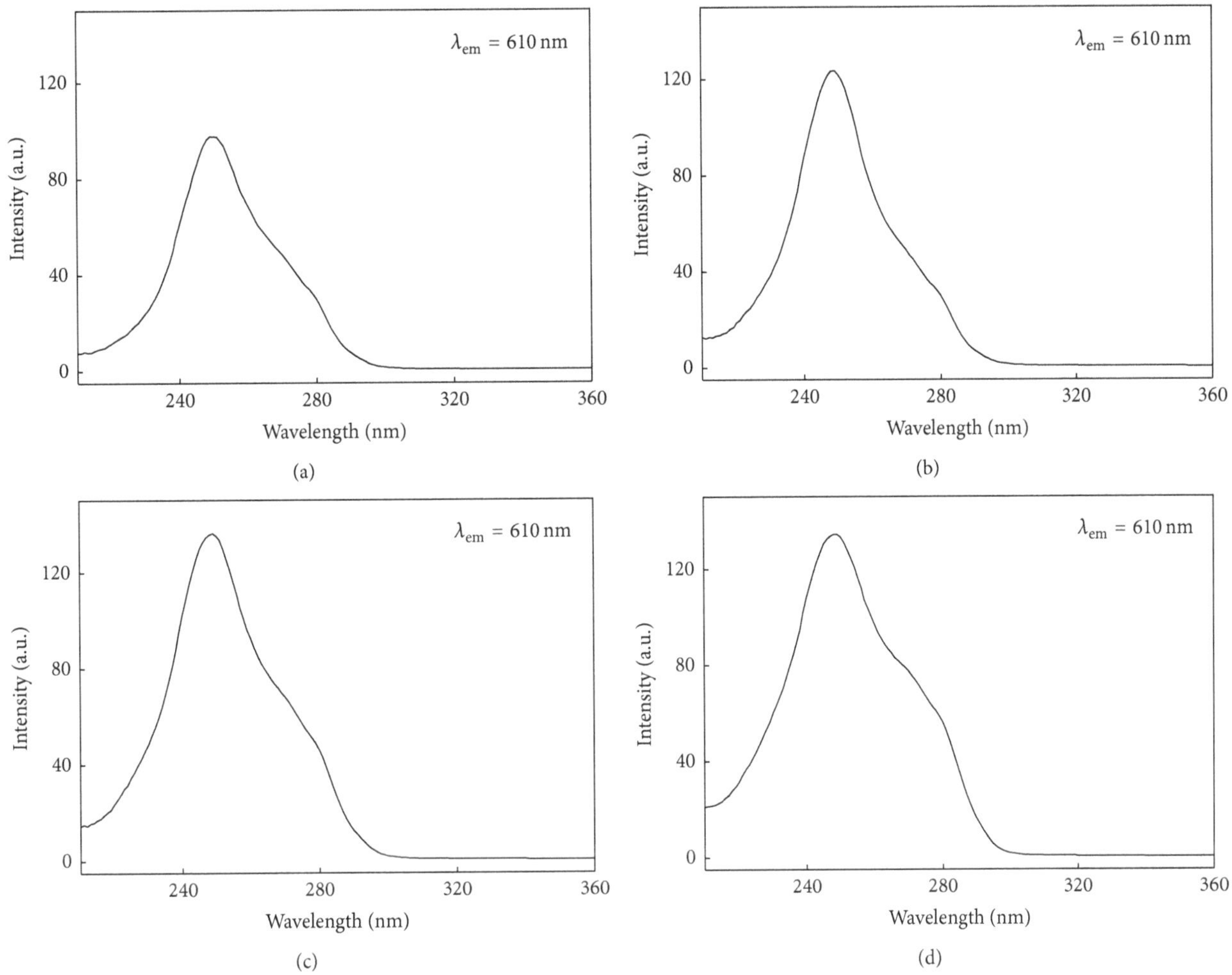

FIGURE 8: Excitation spectra of (a) 0.2 wt% EBA, (b) 0.4 wt% TBA-0.2 wt% EBA, (c) 0.8 wt% TBA-0.2 wt% EBA, and (d) 1.4 wt% TBA-0.2 wt% EBA incorporated PMMA.

In order to ascertain the occurrence of energy transfer between Tb^{3+} and Eu^{3+}, the excitation spectra are recorded by monitoring at 610 nm, which corresponds to $^5D_0 \rightarrow {}^7F_2$ emission transition of Eu^{3+}. It is comprised of a broad excitation band in the spectral region of 240–280 nm as presented in Figure 8, and the highest intensity is achieved in 0.8 wt% TBA-0.2 wt% EBA incorporated PMMA. The enhancement factor of 610 nm emissions between TBA/EBA coincorporated and EBA single incorporated PMMA under different excitation wavelengths has been derived from the excitation spectra, as shown in Figure 9. The enhancement factor is obviously larger than 1 when the excitation wavelength is located in 220–300 nm, which indicates that TBA/EBA coincorporated PMMA can be more efficiently excited by UV radiation compared with EBA single incorporated PMMA, especially in the short-wavelength UV region.

The luminescence mechanism of Ln complex incorporated PMMA is depicted in Figure 10. In EBA/TBA single incorporated PMMA, the Ln complex is excited from the singlet ground state S_0 to the singlet excited state S_1 by UV absorption first and then relaxes to the triplet state T_2 via intersystem crossing; after that an internal conversion $T_2 \rightarrow T_1$ takes place in triplet state. Subsequently, the energy is nonradiatively transferred from the triplet state of the ligands to the resonance state of the coordinated Tb^{3+} or Eu^{3+} [72–75]. Finally, the Tb^{3+} or Eu^{3+} emits visible characteristic fluorescence by the radiative transitions from the excited state to the ground states. Besides, in EBA/TBA coincorporated PMMA, a part of excitation energy in Tb^{3+} can be transferred to $^5D_3$ level of Eu^{3+} with the assistance of chemical bonds and then the excited Eu^{3+} transition radiatively to ^7F_J ($J = 0, 1, 2, 3, 4$) emitting visible light, which consequently bring about the enhanced visible fluorescence emission and improved photon conversion. Effective intramolecular and intermolecular energy transfers make TBA/EBA coincorporated PMMA promising for enhancement of lighting quality and conversion efficiency of solar cells employed in outer space.

4. Conclusions

Efficient green and red fluorescence is observed from terbium and europium benzoic acids (TBA and EBA) incorporated poly methyl methacrylate (PMMA) under ultraviolet (UV)

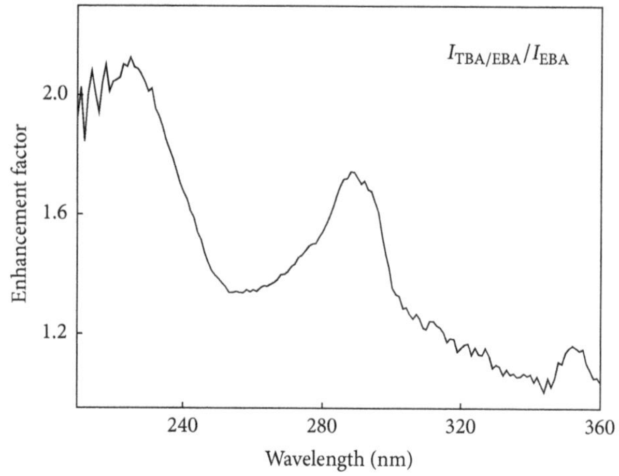

FIGURE 9: Enhancement factor of 610 nm fluorescence between 0.8 wt% TBA-0.2 wt% EBA coincorporated and 0.2 wt% EBA incorporated PMMA.

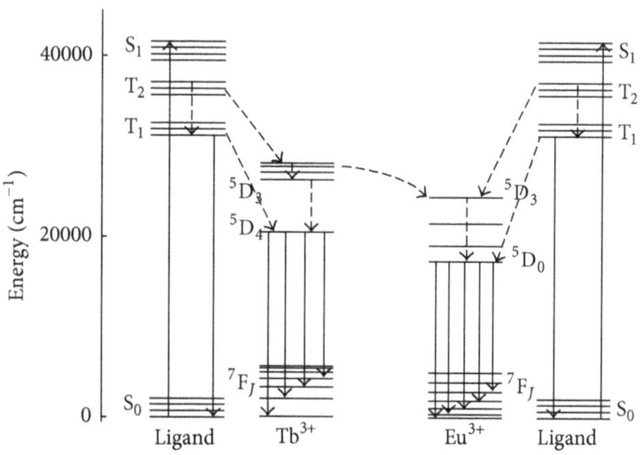

FIGURE 10: Schematic diagram of luminescence mechanism in TBA/EBA incorporated PMMA.

radiation. Judd-Ofelt intensity parameters Ω_2 and Ω_4 of EBA incorporated PMMA have been derived to be 11.13×10^{-20} and 13.15×10^{-20} cm^2, respectively, indicating a high inversion asymmetrical and strong covalent environment around Ln ions in PMMA. The maximum emission cross sections for $^5D_0 \rightarrow {}^7F_2$ transition in EBA incorporated PMMA are 2.451×10^{-21} cm^2, revealing the potential UV→Visible photons conversion capacity of Eu^{3+}. Color variation and fluorescence enhancement in TBA/EBA coincorporated PMMA are originated from superimposed effect of TBA and EBA. Furthermore, emission intensity of Eu^{3+} has been enhanced due to energy migration from Tb^{3+} with the assistant of chemical bonds. Effective photon conversion and superimposed effect in TBA/EBA coincorporated PMMA suggest that the Ln complexes incorporated PMMA is a potential candidate for lighting quality improvement and conversion efficiency increment of DSSCs.

Competing Interests

The authors declare that there are no competing interests regarding the publication of this paper and regarding the grants that they have received.

Acknowledgments

This work was supported by the National Natural Science Foundation of China (Grant no. 61275057).

References

[1] A. Yella, H.-W. Lee, H. N. Tsao et al., "Porphyrin-sensitized solar cells with cobalt (II/III)-based redox electrolyte exceed 12 percent efficiency," *Science*, vol. 334, no. 6056, pp. 629–634, 2011.

[2] S. H. Ko, D. Lee, H. W. Kang et al., "Nanoforest of hydrothermally grown hierarchical ZnO nanowires for a high efficiency dye-sensitized solar cell," *Nano Letters*, vol. 11, no. 2, pp. 666–671, 2011.

[3] J. Qi, W. Liu, C. Biswas et al., "Enhanced power conversion efficiency of CdS quantum dot sensitized solar cells with ZnO nanowire arrays as the photoanodes," *Optics Communications*, vol. 349, Article ID 20030, pp. 198–202, 2015.

[4] H. Choi, Y.-S. Chen, K. G. Stamplecoskie, and P. V. Kamat, "Boosting the photovoltage of dye-sensitized solar cells with thiolated gold nanoclusters," *The Journal of Physical Chemistry Letters*, vol. 6, no. 1, pp. 217–223, 2015.

[5] N. Yao, J. Huang, K. Fu et al., "Efficiency enhancement in dye-sensitized solar cells with down conversion material ZnO: Eu^{3+}, Dy^{3+}," *Journal of Power Sources*, vol. 267, pp. 405–410, 2014.

[6] S. Ivanova and F. Pellé, "Strong 1.53 μm to NIR-VIS-UV upconversion in Er-doped fluoride glass for high-efficiency solar cells," *Journal of the Optical Society of America B: Optical Physics*, vol. 26, no. 10, pp. 1930–1938, 2009.

[7] H. Hafez, M. Saif, and M. S. A. Abdel-Mottaleb, "Down-converting lanthanide doped TiO_2 photoelectrodes for efficiency enhancement of dye-sensitized solar cells," *Journal of Power Sources*, vol. 196, no. 13, pp. 5792–5796, 2011.

[8] L.-L. Li and E. W.-G. Diau, "Porphyrin-sensitized solar cells," *Chemical Society Reviews*, vol. 42, no. 1, pp. 291–304, 2013.

[9] T. Aguilar, J. Navas, D. M. De Los Santos et al., "TiO_2 and pyrochlore $Tm_2Ti_2O_7$ based semiconductor as a photoelectrode for dye-sensitized solar cells," *Journal of Physics D: Applied Physics*, vol. 48, no. 14, Article ID 145102, 2015.

[10] G. Katsagounos, E. Stathatos, N. B. Arabatzis, A. D. Keramidas, and P. Lianos, "Enhanced photon harvesting in silicon multicrystalline solar cells by new lanthanide complexes as light concentrators," *Journal of Luminescence*, vol. 131, no. 8, pp. 1776–1781, 2011.

[11] M. Sendova-Vassileva, "Model calculation of the effectiveness of Tb^{3+} containing glass as a wavelength converter in thin film solar cells," *Journal of Materials Science*, vol. 46, no. 22, pp. 7184–7190, 2011.

[12] V. D. Rodríguez, V. K. Tikhomirov, J. Méndez-Ramos, A. C. Yanes, and V. V. Moshchalkov, "Towards broad range and highly efficient down-conversion of solar spectrum by Er^{3+}-Yb^{3+} co-doped nano-structured glass-ceramics," *Solar Energy Materials and Solar Cells*, vol. 94, no. 10, pp. 1612–1617, 2010.

[13] V. Kumar, A. F. Khan, and S. Chawla, "Intense red-emitting multi-rare-earth doped nanoparticles of YVO_4 for spectrum

conversion towards improved energy harvesting by solar cells," *Journal of Physics D: Applied Physics*, vol. 46, no. 36, Article ID 365101, 2013.

[14] J. J. Velázquez, A. C. Yanes, J. del-Castillo, C. Guzmán-Afonso, and V. D. Rodríguez, "Dual emission in multiphase SiO_2–SnO_2–LaF_3 nanostructured glass-ceramics for simultaneous UV and NIR solar spectrum conversion," *Science of Advanced Materials*, vol. 7, no. 11, pp. 2272–2277, 2015.

[15] S. González-Pérez, J. Sanchiz, B. González-Díaz et al., "Luminescent polymeric film containing an Eu(III) complex acting as UV protector and down-converter for Si-based solar cells and modules," *Surface and Coatings Technology*, vol. 271, pp. 106–111, 2015.

[16] C.-Y. Sun, X.-J. Zheng, X.-B. Chen, L.-C. Li, and L.-P. Jin, "Assembly and upconversion luminescence of lanthanide-organic frameworks with mixed acid ligands," *Inorganica Chimica Acta*, vol. 362, no. 2, pp. 325–330, 2009.

[17] J. Lin, Q. Wang, C. Tan, and H. Chen, "Luminescence recognition behavior concerning different anions by lanthanide complex equipped with electron-withdraw groups and in PMMA matrix," *Synthetic Metals*, vol. 160, no. 15-16, pp. 1780–1786, 2010.

[18] G. Gheno, M. Bortoluzzi, R. Ganzerla, and F. Enrichi, "Inorganic pigments doped with tris(pyrazol-1-yl)borate lanthanide complexes: a photoluminescence study," *Journal of Luminescence*, vol. 145, pp. 963–969, 2014.

[19] C. K. Oliveira, V. P. de Souza, L. L. da Luz et al., "Synthesis, crystal structure and luminescent properties of lanthanide extended structure with asymmetrical dinuclear units based on 2-(methylthio)benzoic acid," *Journal of Luminescence*, vol. 170, pp. 528–537, 2016.

[20] A. M. Darwish, M. T. Sagapolutele, S. Sarkisov, D. Patel, D. Hui, and B. Koplitz, "Double beam pulsed laser deposition of composite films of poly(methyl methacrylate) and rare earth fluoride upconversion phosphors," *Composites Part B: Engineering*, vol. 55, pp. 139–146, 2013.

[21] S. Biju, Y. K. Eom, J.-C. G. Bünzli, and H. K. Kim, "A new tetrakis β-diketone ligand for NIR emitting Ln^{III} ions: luminescent doped PMMA films and flexible resins for advanced photonic applications," *The Journal of Materials Chemistry C*, vol. 1, no. 42, pp. 6935–6944, 2013.

[22] J. Zhang, W.-X. Li, B.-Y. Ao, S.-Y. Feng, and X.-D. Xin, "Fluorescence enhancement of europium(III) perchlorate by benzoic acid on bis(benzylsulfinyl)methane complex and its binding characteristics with the bovine serum albumin (BSA)," *Spectrochimica Acta Part A: Molecular and Biomolecular Spectroscopy*, vol. 118, pp. 972–980, 2014.

[23] F. Zinna, U. Giovanella, and L. D. Bari, "Highly circularly polarized electroluminescence from a chiral europium complex," *Advanced Materials*, vol. 17, pp. 1791–1795, 2015.

[24] R. Bonzanini, E. M. Girotto, M. C. Gonçalves, E. Radovanovic, E. C. Muniz, and A. F. Rubira, "Effects of europium (III) acetylacetonate doping on the miscibility and photoluminescent properties of polycarbonate and poly(methyl methacrylate) blends," *Polymer*, vol. 46, no. 1, pp. 253–259, 2005.

[25] D. S. Bodas, S. K. Mahapatra, and S. A. Gangal, "Comparative study of spin coated and sputtered PMMA as an etch mask material for silicon micromachining," *Sensors and Actuators, A: Physical*, vol. 120, no. 2, pp. 582–588, 2005.

[26] Y. A. Ustynyuk, N. E. Borisova, V. A. Babain et al., "N,N′-Dialkyl-N,N′-diaryl-1,10-phenanthroline-2,9-dicarboxamides as donor ligands for separation of rare earth elements with a high and unusual selectivity. DFT computational and experimental studies," *Chemical Communications*, vol. 51, no. 35, pp. 7466–7469, 2015.

[27] J. Zhang, L.-X. Wang, L. Zhang, Y. Chen, and Q.-T. Zhang, "Co-luminescence properties of terbium ions-benzoic acid-phen complexes doped with europium ions," *Rare Metals*, vol. 32, no. 6, pp. 599–604, 2013.

[28] Y. Wu, J. Wang, X. Kong et al., "A dinuclear europium(III) complex with thenoyltrifluoroacetonato and 1-(2-pyridylzao)-2-naphtholato ligands and its optical properties," *Inorganica Chimica Acta*, vol. 370, no. 1, pp. 346–352, 2011.

[29] B. Kokuoz, C. Kucera, J. R. DiMaio, and J. Ballato, "Organic-inorganic hybrid nanoparticles with enhanced rare-earth emissions," *Optical Materials*, vol. 31, no. 9, pp. 1327–1330, 2009.

[30] B. Yan and X.-F. Qiao, "Rare-earth/inorganic/organic polymeric hybrid materials: molecular assembly, regular microstructure and photoluminescence," *Journal of Physical Chemistry B*, vol. 111, no. 43, pp. 12362–12374, 2007.

[31] B. Gao, L. Chen, and T. Chen, "Effect of electron-donating substituent groups on aromatic ring on photoluminescence properties of complexes of benzoic acid-functionalized polysulfone with Eu(III) ions," *Physical Chemistry Chemical Physics*, vol. 17, no. 38, pp. 25322–25332, 2015.

[32] Q. Zhang, Y. Sheng, K. Zheng, X. Qin, P. Ma, and H. Zou, "Novel organic-inorganic amorphous photoactive hybrid films with rare earth (Eu^{3+}/s Tb^{3+}) covalently embedded into silicon-oxygen network via sol-gel process," *Materials Research Bulletin*, vol. 70, pp. 379–384, 2015.

[33] G. Sharma and A. K. Narula, "Effect of doping of calcium fluoride nanoparticles on the photoluminescence properties of europium complexes with benzoic acid derivatives as secondary ligands and 2-aminopyridine as primary ligand," *Optical Materials*, vol. 46, pp. 438–443, 2015.

[34] W. X. Wu, T. X. Wang, X. Wang et al., "Hybrid solar concentrator with zero self-absorption loss," *Solar Energy*, vol. 84, no. 12, pp. 2140–2145, 2010.

[35] Q. Ma, W. Yu, X. Dong, J. Wang, G. Liu, and J. Xu, "Electrospinning fabrication and properties of Fe_3O_4/Eu$(BA)_3$phen/PMMA magnetic-photoluminescent bifunctional composite nanoribbons," *Optical Materials*, vol. 35, no. 3, pp. 526–530, 2013.

[36] V. I. Gerasimova, A. A. Antoshkov, Y. S. Zavorotny, A. O. Rybaltovskii, and D. A. Lemenovskii, "Optical properties of europium(III) β-diketonate/polymer-doped systems using supercritical carbon dioxide," *Journal of Luminescence*, vol. 134, pp. 339–344, 2013.

[37] T. Monzón-Hierro, J. Sanchiz, S. González-Pérez et al., "A new cost-effective polymeric film containing an Eu(III) complex acting as UV protector and down-converter for Si-based solar cells and modules," *Solar Energy Materials and Solar Cells*, vol. 136, pp. 187–192, 2015.

[38] L. Liu, M. Yu, J. Zhang, B. Wang, W. Liu, and Y. Tang, "Facile fabrication of color-tunable and white light emitting nano-composite films based on layered rare-earth hydroxides," *Journal of Materials Chemistry C*, vol. 3, no. 10, pp. 2326–2333, 2015.

[39] S. Rodríguez, P. Elizondo, S. Bernès, N. Pérez, R. Bustos, and E. García-España, "Mechanochemical synthesis of an Eu(III) complex. Preparation and Luminescence Properties of PMMA:[$C_{42}H_{38}N_5O_{19}$Eu] Hybrid Films," *Polyhedron*, vol. 85, pp. 10–14, 2015.

[40] H. Jiu, J. Ding, Y. Sun, J. Bao, C. Gao, and Q. Zhang, "Fluorescence enhancement of europium complex co-doped

with terbium complex in a poly(methyl methacrylate) matrix," *Journal of Non-Crystalline Solids*, vol. 352, no. 3, pp. 197–202, 2006.

[41] Z. Zhang, C. Yu, L. Liu et al., "Efficient near-infrared (NIR) luminescent PMMA-supported hybrid materials doped with tris-β-diketonate Ln^{3+} complex (Ln = Nd or Yb)," *Journal of Photochemistry and Photobiology A: Chemistry*, vol. 314, pp. 104–113, 2016.

[42] H. Q. Gu, Y. J. Hou, F. Xu, and S. H. Wang, "Electrospinning preparation, thermal, and luminescence properties of $Eu_2(BTP)_3(Phen)_2$ complex doped in PMMA," *Colloid and Polymer Science*, vol. 293, no. 8, pp. 2201–2208, 2015.

[43] R. Puthiyottil, S. Varghese, U. Gopalakrishnapanicker, and J. T. Guthrie, "Characterization of europium complex-doped PMMA and PMMA-EVA polymer networks," *Polymer—Plastics Technology and Engineering*, vol. 53, no. 11, pp. 1111–1118, 2014.

[44] A. K. Singh, S. K. Singh, H. Mishra, R. Prakash, and S. B. Rai, "Structural, thermal, and fluorescence properties of $Eu(DBM)_3$Phenx complex doped in PMMA," *Journal of Physical Chemistry B*, vol. 114, no. 41, pp. 13042–13051, 2010.

[45] H. Jiu, L. Zhang, G. Liu, and T. Fan, "Fluorescence enhancement of samarium complex co-doped with terbium complex in a poly(methyl methacrylate) matrix," *Journal of Luminescence*, vol. 129, no. 3, pp. 317–319, 2009.

[46] J. Garcia-Torres, P. Bosch-Jimenez, E. Torralba-Calleja et al., "Highly efficient luminescent materials: influence of the matrix on the photophysical properties of Eu(III) complex/polymer hybrids," *Journal of Photochemistry and Photobiology A: Chemistry*, vol. 283, pp. 8–16, 2014.

[47] M. S. Attia, A. M. Othman, A. O. Youssef, and E. El-Raghi, "Excited state interaction between hydrochlorothiazide and europium ion in PMMA polymer and its application as optical sensor for hydrochlorothiazide in tablet and serum samples," *Journal of Luminescence*, vol. 132, no. 8, pp. 2049–2053, 2012.

[48] H.-G. Liu, X.-S. Feng, K. Jang et al., "Influences of compositions and ligands on photoluminescent properties of Eu(III) ions in composite europium complex/PMMA systems," *Journal of Luminescence*, vol. 127, no. 2, pp. 307–315, 2007.

[49] K. Buczko and M. Karbowiak, "Colour-tuneable double-layer polymeric films doped with lanthanide β-diketonate complexes," *Journal of Luminescence*, vol. 143, pp. 241–253, 2013.

[50] R. Puthiyottil, S. Varghese, U. Gopalakrishnapanicker, and J. T. Guthrie, "Optical properties of europium (2,2′-bipyridine-N,N-dioxide)$_2$ $(NO_3)_3$ complex-doped poly(methyl methacrylate)," *Polymer Engineering and Science*, vol. 53, no. 1, pp. 119–124, 2013.

[51] M. Bortoluzzi, G. Paolucci, M. Gatto et al., "Preparation of photoluminescent PMMA doped with tris(pyrazol-1-yl)borate lanthanide complexes," *Journal of Luminescence*, vol. 132, no. 9, pp. 2378–2384, 2012.

[52] T. Liu, G. Duan, Y. Zhang, J. Fang, and Z. Zeng, "Synthesis and characterization of the luminescent lanthanide complexes with two similar benzoic acids," *Spectrochimica Acta—Part A: Molecular and Biomolecular Spectroscopy*, vol. 74, no. 4, pp. 843–848, 2009.

[53] E. B. Gibelli, J. Kai, E. E. S. Teotonio, O. L. Malta, M. C. F. C. Felinto, and H. F. Brito, "Photoluminescent PMMA polymer films doped with Eu^{3+}-β-diketonate crown ether complex," *Journal of Photochemistry and Photobiology A: Chemistry*, vol. 251, pp. 154–159, 2013.

[54] B. Yue, Y.-N. Chen, H.-B. Chu, Y.-R. Qu, A.-L. Wang, and Y.-L. Zhao, "Synthesis, crystal structures and fluorescence properties of dinuclear Tb(III) and Sm(III) complexes with 2,4,6-tri(2-pyridyl)-1,3,5-triazine and halogenated benzoic acid," *Inorganica Chimica Acta*, vol. 414, pp. 39–45, 2014.

[55] H. Zhang, H. Song, B. Dong et al., "Electrospinning preparation and luminescence properties of europium complex/polymer composite fibers," *Journal of Physical Chemistry C*, vol. 112, no. 25, pp. 9155–9162, 2008.

[56] F. H. Walters, "Design of corrosion inhibitors: use of the hard and soft acid-base (HSAB) theory," *Journal of Chemical Education*, vol. 68, no. 1, pp. 29–31, 1991.

[57] H. Zhang, H. Dai, Y. Liu, J. Deng, L. Zhang, and K. Ji, "Surfactant-mediated PMMA-templating fabrication and characterization of three-dimensionally ordered macroporous Eu_2O_3 and Sm_2O_3 with mesoporous walls," *Materials Chemistry and Physics*, vol. 129, no. 1-2, pp. 586–593, 2011.

[58] Y. Hasegawa, H. Kawai, K. Nakamura, N. Yasuda, Y. Wada, and S. Yanagida, "Molecular design of luminescent Eu(III) complexes as lanthanide lasing material and their optical properties," *Journal of Alloys and Compounds*, vol. 408–412, pp. 669–674, 2006.

[59] H. Boubekri, M. Diaf, K. Labbaci, L. Guerbous, T. Duvaut, and J. P. Jouart, "Synthesis and optical properties of Tb^{3+} doped CdF_2 single crystals," *Journal of Alloys and Compounds*, vol. 575, pp. 339–343, 2013.

[60] V. Naresh and S. Buddhudu, "Energy transfer based enhanced red emission intensity from (Eu^{3+}, Tb^{3+}):LFBCd optical glasses," *Journal of Luminescence*, vol. 137, pp. 15–21, 2013.

[61] L. Mei, J. Xie, H. Liu, L. Liao, Y. Zhang, and M. Li, "Luminescence properties and energy transfer of Ce^{3+}/Tb^{3+} co-doped Ca9La(PO4)$_5$$(SiO_4)F_2$ phosphor," *Optics Communications*, vol. 335, pp. 90–93, 2015.

[62] J. Guan, B. Chen, Y. Sun, H. Liang, and Q. Zhang, "Effects of synergetic ligands on the thermal and radiative properties of $Eu(TTA)_3nL$-doped poly(methyl methacrylate)," *Journal of Non-Crystalline Solids*, vol. 351, no. 10-11, pp. 849–855, 2005.

[63] M. P. Hehlen, M. G. Brik, and K. W. Krämer, "50th anniversary of the Judd-Ofelt theory: an experimentalist's view of the formalism and its application," *Journal of Luminescence*, vol. 136, pp. 221–239, 2013.

[64] Y. Liu, D. Tu, H. Zhu, and X. Chen, "Lanthanide-doped luminescent nanoprobes: controlled synthesis, optical spectroscopy, and bioapplications," *Chemical Society Reviews*, vol. 42, no. 16, pp. 6924–6958, 2013.

[65] C. E. Secu, D. Predoi, M. Secu, M. Cernea, and G. Aldica, "Structural investigations of sol-gel derived silicate gels using Eu^{3+} ion-probe luminescence," *Optical Materials*, vol. 31, no. 11, pp. 1745–1748, 2009.

[66] Y. Gandhi, I. V. Kityk, M. G. Brik, P. R. Rao, and N. Veeraiah, "Influence of tungsten on the emission features of Nd^{3+}, Sm^{3+} and Eu^{3+} ions in ZnF_2-WO_3-TeO2 glasses," *Journal of Alloys and Compounds*, vol. 508, no. 2, pp. 278–291, 2010.

[67] R. T. Moura Jr., A. N. Carneiro Neto, R. L. Longo, and O. L. Malta, "On the calculation and interpretation of covalency in the intensity parameters of $4f$-$4f$ transitions in Eu^{3+} complexes based on the chemical bond overlap polarizability," *Journal of Luminescence*, vol. 170, part 2, pp. 420–430, 2016.

[68] M. Dejneka, E. Snitzer, and R. E. Riman, "Blue, green and red fluorescence and energy transfer of Eu^{3+} in fluoride glasses," *Journal of Luminescence*, vol. 65, no. 5, pp. 227–245, 1995.

[69] V. Naresh, B. H. Rudramadevi, and S. Buddhudu, "Cross-relaxations and non-radiative energy transfer from (^{4}G5/2)

Sm^{3+} →(5D0) Eu^{3+}: B_2O_3-ZnO glasses," *Journal of Alloys and Compounds*, vol. 632, pp. 59–67, 2015.

[70] N. T. Kalyani, S. J. Dhoble, and R. B. Pode, "Enhancement of photoluminescence in various $Eu_xRe_{(1-x)}TTA_3Phen$ (Re = Y, Tb) complexes molecularly doped in PMMA," *Indian Journal of Physics*, vol. 86, no. 7, pp. 613–618, 2012.

[71] Y. Luo, Q. Yan, S. Wu, W. Wu, and Q. Zhang, "Inter- and intra-molecular energy transfer during sensitization of $Eu(DBM)_3Phen$ luminescence by $Tb(DBM)_3Phen$ in PMMA," *Journal of Photochemistry and Photobiology A: Chemistry*, vol. 191, no. 2-3, pp. 91–96, 2007.

[72] A. de Bettencourt-Dias and S. Viswanathan, "Nitro-functionalization and luminescence quantum yield of Eu(III) and Tb(III) benzoic acid complexes," *Dalton Transactions*, vol. 34, pp. 4093–4103, 2006.

[73] Y. H. Kim, N. S. Baek, and H. K. Kim, "Sensitized emission of luminescent lanthanide complexes based on 4-naphthalen-1-yl-benzoic acid derivatives by a charge-transfer process," *ChemPhysChem*, vol. 7, no. 1, pp. 213–221, 2006.

[74] S. Lis, M. Elbanowski, B. Makowska, and Z. Hnatejko, "Energy transfer in solution of lanthanide complexes," *Journal of Photochemistry and Photobiology A: Chemistry*, vol. 150, no. 1–3, pp. 233–247, 2002.

[75] K. Lun, Q. L. Ma, M. Yang et al., "Electricity-magnetism and color-tunable trifunction simultaneously assembled into one strip of flexible microbelt via electrospinning," *Chemical Engineering Journal*, vol. 279, pp. 231–240, 2015.

7

Main Controlling Factor of Polymer-Surfactant Flooding to Improve Recovery in Heterogeneous Reservoir

Dandan Yin[1,2] and Dongfeng Zhao[1,2]

[1]*School of Petrochemical and Energy Engineering, Zhejiang Ocean University, Zhoushan 316022, China*
[2]*United National-Local Engineering Laboratory of Harbor Oil & Gas Storage and Transportation Technology, Zhejiang Ocean University, Zhoushan 316022, China*

Correspondence should be addressed to Dandan Yin; aayindan@163.com

Academic Editor: Luigi Nicolais

This study aims to analyze the influence of viscosity and interfacial tension (IFT) on the recovery in heterogeneous reservoir and determines the main controlling factors of the polymer-surfactant (SP) flooding. The influence of the salinity and shearing action on the polymer viscosity and effects of the surfactant concentration on the IFT and emulsion behavior between chemical agent and oil were studied through the static and flooding experiments. The results show that increasing the concentration of polymer GF-11 (HPAM) can reduce the influence of the salinity and GF-11 has high shear-resistance property. In the condition of the Jilin Oilfield, the oil/water IFT can reach 10^{-3} mN/m when the surfactant concentration is 0.3 wt%. The lower the IFT is, the easier the emulsion of SP and oil is formed. Seven flooding experiments are conducted with the SP system. The results show that the recovery can be improved for 5.02%–15.98% under the synergistic effect of the polymer and surfactant. In the heterogeneous reservoir, the contribution of oil recovery is less than that of the sweep volume.

1. Introduction

Polymer can increase the viscosity of injected fluid, decrease the mobility ratio of water and oil, and then expand the sweep coefficient [1, 2]. Meanwhile, as the polymer is viscoelastic, it can also improve the oil displacement efficiency [3, 4]. However, the IFT between the oil and the polymer solution is quite high; it possesses no oil solubility and emulsifiability [5]. And the recovery just can be increased by 10%, so polymer flooding is limited. The alkali-surfactant-polymer (ASP) system can reduce both the mobility ratio and the IFT, so it can effectively reduce the remaining oil saturation [6–8]. The alkali in ASP system can react with the organic acid in the oil to generate surfactant [9]. As a result, the IFT can be reduced to a larger extent due to the synergy between surfactant added and generated. What is more, the alkali can also reduce the surfactant absorption in the reservoir. However, the alkali will react with the clay mineral in the reservoir and produce the precipitate. It will result in the reservoir damage, difficulty demulsification, and the reduction of the ASP system viscosity [10]. As a new flooding method, the SP system without alkali can solve the above problems [11]. As the system is free of alkali, it can play the viscoelasticity of polymers to the maximum degree and erase the corrosion and scaling caused by alkali. While reaching an ultralow IFT, the oil displacement efficiency of SP system can be closed to the ASP system. Meanwhile, the SP system can decrease environment pollution [12].

SP flooding could enhance oil recovery because polymer can increase sweep efficiency and surfactant can improve the oil displacement efficiency. The polymer viscosity influences the sweep efficiency and the displacement efficiency depends on the IFT. The viscosity is influenced by the salinity, temperature, absorption, and shearing action [13]. And the IFT is influenced by the reservoir temperature, the composition of crude oil, and the concentration change caused by the action between the injection fluid and the rock [14]. Main controlling factor of polymer-surfactant

TABLE 1: Composition of injected water.

Ions	$K^+ + Na^+$	Mg^{2+}	Ca^{2+}	Cl^-	SO_4^{2-}	HCO^{3-}	CO_3^{2-}	TDS
Content mg/L	2043	36	39	1347	24	3126	135	6750

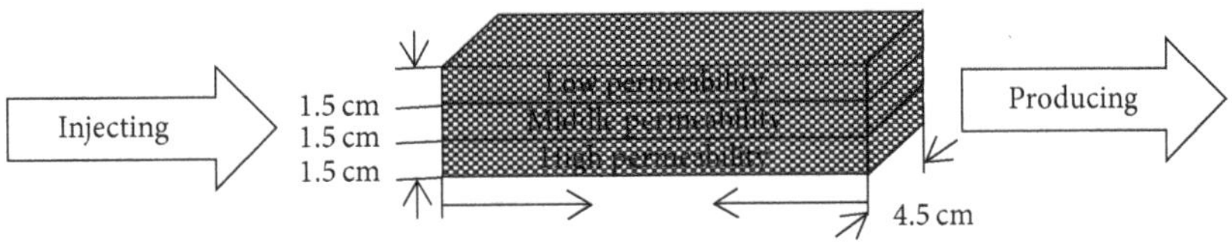

FIGURE 1: The two-dimensional vertical heterogeneous model.

flooding to improve recovery is different in different reservoir condition. According to the capillary number equation, if the injection can reduce the IFT to 10^{-3} mN/m, the highest recovery can be reached in homogeneous reservoirs [15–17], while, in the heterogeneous reservoir, the mobility ratio plays an important role in spreading to the middle-low permeability layer. The surfactant can activate the residual oil and form emulsion, the emulsion expands the swept volume. According to the different characteristics of the reservoir, the main controlling factors of SP flooding should be cleared out and provide the basis for the system optimization for SP flooding, especially for the Jilin Oilfield of high salinity and heavy heterogeneous.

According to the characteristics of Jilin Oilfield, a series of static experiments were carried out to study the influence of salinity and shearing action on the viscosity; other experiments are done to analyze the effect of surfactant concentration on the IFT and emulsion. According to the core flooding experiments, the influence of viscosity and IFT on the recovery is discussed. And the main controlling factors of SP flooding are analyzed. The results provide the basis on the optimization of the chemical and the injecting method.

2. Experimental Studies

2.1. Equipment. The core holder is 30 cm long, which hold the core with external pressure that is 1-2 MPa more than the inlet pressure. Other equipment includes a flowmeter, DV viscometer, high pressure middle vessel, automatic metering plunger pump, thermotank, pressure acquisition system, and constant flow pump. The pressures are recorded by the pressure acquisition system and the output liquid is collected to calculate the recovery.

2.2. Materials. The mother liquor of polymer and surfactant are prepared by the injection water; the salinity of injection water is shown in Table 1. The two-dimensional vertical heterogeneous positive rhythm artificial model with the size of 30 cm × 4.5 cm × 4.5 cm is used in the experiment, as shown in Figure 1. The model is cemented by quartz sand and epoxy resin, and the formation is divided into high, medium, and low permeability layers, with a permeability of 630 mD, 120 mD, and 30 mD, respectively, in order to simulate the main oil layer of Honggang Oilfield.

The polymer is GF-11 produced by Daqing Assistant Factory, and GF-11 is partially hydrolyzed polyacrylamide (HPAM). The average relative molecular weight of GF-11 is 2500×10^4, the degree of hydrolysis is 25%, and the solid content is 88%. The surfactant is anionic sulfonate, and the effective content is 40 wt%, 24 wt% is unsulfonated oil, 33 wt% is volatile content, and 3.0 wt% is inorganic salt. The average relative molecular weight is 645.

SP systems were prepared by the polymer whose relative molecular weight is 2500×10^4, the surfactant, and different composition water. The composition of crude oil is shown in Table 2. To get the same viscosity with reservoir oil, we used simulated oil in the experiment. The simulated oil is prepared by the degassed oil and kerosene with the proportion of 5 : 3. The viscosity of oil is 17 mPa·s in the reservoir temperature. of 45°C.

2.3. Measurement of Viscosity and IFT. The viscosity of SP system is measured by the HAAKE rotary rheometer, under the shearing rate of $7.31\,s^{-1}$. IFT is measured by TX500 at 1000 r/min after 120 min under 45°C.

2.4. Emulsification Experiment. The emulsion is prepared by the chemical and crude oil stirring at the rate of 1500 r/min for 1 minute at 45°C. The volume ratio of oil and water is 3 : 7. The micromorphology of emulsion is observed by the ZEISS SteREO Discovery.V20 microscope. In the emulsion stability experiment, putting the emulsion into the glass tube, the oil and water will separate from each other because of the density difference. The water separating proportion is defined as the ratio of the separated water volume and the original water volume after separating 2 hours and it is used to evaluate the stability of emulsion; the reciprocal of stability constant of Turbiscan Lab Expert (Formulaction Inc.) is used to evaluate the stability of emulsion at the reservoir temperature. The bigger the TSI^{-1} is, the more stable the emulsion is.

2.5. Core Flooding Experiments. At the temperature of 45°C and the constant flow of 0.1 mL/min, the experiments of water flooding, surfactant flooding, polymer flooding, and SP flooding are conducted. The experiment steps are as follows:

(1) Measure the dry weight, length, and diameter of the core; the permeability is measured by N_2.

Table 2: Composition of crude oil.

Crude oil	Hydrocarbon	Aromatic Hydrocarbons	Resin	Asphaltene
Content, wt%	59.9	22.4	4.3	8.0

(2) Vacuum the core, saturate it with the reservoir water, weigh the wet weight, and calculate the porosity. Then, saturate the core with oil and calculate the original oil saturation.

(3) Displace the oil in the core with the reservoir water; measure the output oil, water, and pressure until the water cut of the output liquid is 98%. After the water flooding, the chemical flooding is done until the water cut is 98%.

3. Results and Discussion

3.1. Influence Factor of SP System

3.1.1. Viscosity of SP System. The SP solution is prepared by the injection water of the block, and the salinity of injected water is less than that of the original formation water. Although after long-term water flooding, the salinity of formation water is nearly that of the injected water. But due to mobility control of the polymer, the injected chemical agents can sweep the area which is unsweep during water flooding. These areas are still high salinity, so the viscosity of polymer would reduce and then oil recovery decreases. Therefore it is necessary to evaluate the influence of salinity on the viscosity of polymer solution.

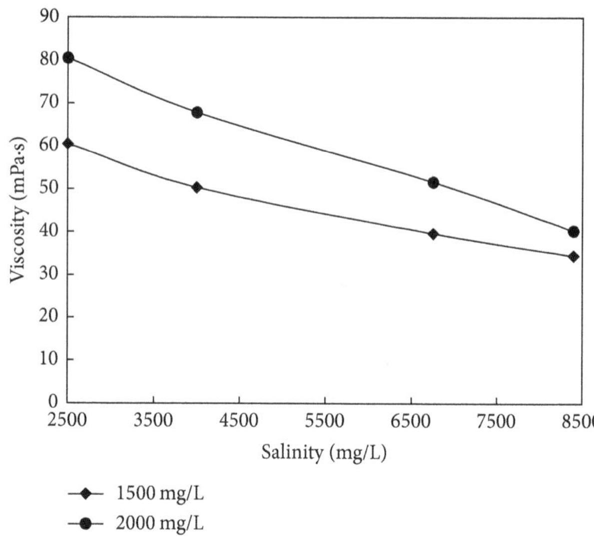

Figure 2: Viscosity of SP solution with different salinity when polymer concentrations are 1500 mg/L and 2000 mg/L.

Figure 2 is the viscosity curve along with salinity when polymer concentration is 1500 mg/L and 2000 mg/L. It shows that viscosity decreases as salinity increases. When polymer concentration is 1500 mg/L and salinity increases from 2500 mg/L to 8400 mg/L, viscosity decreases from 60.47 mPa·s to 34.51 mPa·s and viscosity retention rate is 56.01%; when polymer concentration is 2000 mg/L and salinity increases from 2500 mg/L to 8400 mg/L, viscosity decreases from 80.49 mPa·s to 40.33 mPa·s and viscosity retention rate is 50.11%. So the higher the polymer concentration is, as salinity increases, the lower the viscosity retention rate is. According to the theory of molecular motion, the more the polymer molecular chain stretches, the more serious the intermolecular winding is, the higher the viscosity is. Double electric layer on surface of polymer molecular chain is compressed and negative charge of polymer molecular chain is blocked, which will lead to weakening of the repulsion between molecular chains, and then polymer molecular chain curls up, entanglement between molecular diminishes, and viscosity reduces. Therefore polymer concentration should be appropriately increased in order to overcome the effects of salinity on polymer solution viscosity.

In order to reduce oil-water mobility ratio effectively and enhance oil recovery, polymer solution in the formation needs high viscosity. But when polymer solution is injected, it has to pass through pump, valve, pipeline, perforation, and the pore and throat of rock and the shearing action during flooding will change the size of polymer molecule. In order to simulate the effect on the viscosity when the polymer solution is affected by the mechanical shear, inject polymer solution into one-meter long cores with the concentration of 1500 mg/L, 2000 mg/L, 2500 mg/L confected by injected water and formation water. Take a sample at the exit of the core, determine the viscosity of produced fluid at 45°C, and evaluate the shear resistance of polymers.

The results are shown in Table 3. For polymer solution of 1500 mg/L, 2000 mg/L, and 2500 mg/L, the viscosity retention rate of the solution prepared with the injected water is 71.99%, 72.15%, and 74.74%, respectively, while the viscosity retention rate of the solution prepared with the formation water is, respectively, 67.33%, 68.21% and 71.32%. The results show that this polymer has strong ability to resist shear and can be used for oil displacement.

3.1.2. IFT Behavior of Crude Oil/Chemical System. As SP system has not the synergistic effect of alkali, surfactant is the key to enhance oil recovery. Excellent surfactant can keep the IFT ultralow (10^{-3} mN/m) [18]. But the activity of surfactants in the formation is affected by adsorption, retention, salinity, and dilution formation of water, so it is necessary to measure IFT between JN-1 solution and crude oil of different concentration.

The results are shown in Figure 2 that the range of surfactant concentration is 0.05–0.5 wt%, IFT of oil-surfactant-polymer solution changes with different surfactant concentrations. As shown in Figure 3 when the surfactant concentration increases from 0.05 wt% and 0.3 wt%, IFT drastically reduces from 8.4 mN/m to 4.11×10^{-3} mN/m. This is due to the fact that the surfactant molecules on the oil-water

TABLE 3: The viscosity of the polymer solution under different conditions.

Polymer concentration mg/L	Injected water		Formation water	
	Viscosity before shearing, mPa·s	Viscosity after shearing, mPa·s	Viscosity before shearing, mPa·s	Viscosity after shearing, mPa·s
1500	39.67	28.96	30.14	20.29
2000	51.64	37.26	42.31	28.86
2500	72.46	54.16	59.47	42.41

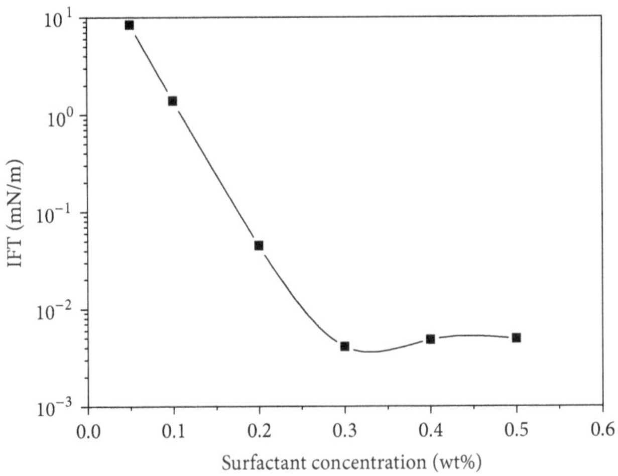

FIGURE 3: Minimum IFT between surfactant solution and oil with different surfactant concentrations.

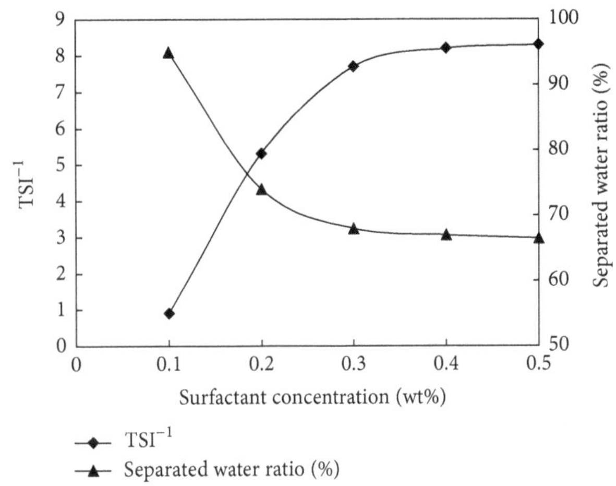

FIGURE 4: Relationship of TSI^{-1} and separated water ratio when surfactant concentration varies. The polymer concentration is 2000 mg/L.

interface increase as the surfactant concentration increases. When the surfactant concentration is 0.3 wt%, the adsorption of the surfactant molecules is saturated on the oil-water interface and IFT reach a minimum of about .11 × 10^{-3} mN/m, so 0.3 wt% is CMC of the surfactants and IFT slowly increase with the increase of surfactant concentration. According to the value of IFT and economic factors, choose the surfactant concentration of 0.3 wt%.

3.1.3. The Emulsification Behavior of SP Solution and Oil. Emulsification is a common phenomenon in the SP flooding and also an important mechanism for EOR [9, 14, 19]. In order to evaluate the efficiency of chemical emulsifying the oil, a series of emulsification experiments are conducted, according to the description of the experimental method in Section 2.4. In each experiment, emulsion is prepared by the oil and different chemical agents.

In Figure 4, the range of surfactant concentration is 0.1–0.5 wt%, and the polymer concentration is 2000 mg/L. The figure shows the relationship between the surfactant concentration and the stability of the emulsion. From the figure, it can be concluded that when the surfactant concentration increases from 0.1 wt% to 0.3 wt%, TSI^{-1} raises rapidly and the separated water ratio decreases rapidly. With the further increase of the surfactant concentration, the TSI^{-1} curves and the separated water ratio curves become relatively gentle. It indicates that the emulsion stability is stronger when the surfactant concentration increases. From Figure 2, it can be concluded that the change trend of the IFT curves and the TSI^{-1} curves is similar. That is to say, when the surfactant concentration exceeds 0.3%, the IFT and the TSI^{-1} significantly decrease and finally reach a stable value. A phenomenon appears in the forming process of emulsion: the lower the IFT is, the easier the emulsion forms.

Figure 5 is the microscopic photos of the emulsion layer which contains the three kinds of emulsion. Figure 6 shows distribution of emulsion droplet with different chemical agent. Figure 5 shows that three kinds of SP system with the water all form the O/W emulsion. The quantity of emulsion in Figure 5(c) is more than that of Figures 5(a) and 5(b) and the diameter of emulsion in Figure 5(c) is fewer and the distribution is more uniform (as shown in Figure 6), so that the emulsion in the Figure 5(c) is the most stable. It is because that the IFT in Figure 5(c) is the lowest. And it can be inferred that the low IFT is quite important in the formation of emulsion, which coincides with the above phenomenon and analysis.

3.2. Core Flooding Experiment. Under the reservoir temperature of 45°C, the heterogeneous cores of three layers are used for the SP flooding experiments. The chemical agents were prepared by the injection water. The results of polymer flooding, surfactant flooding, and SP flooding are shown in Table 4.

TABLE 4: Experiment data of oil displacement efficiency.

Number	Polymer concentration	Surfactant concentration	Viscosity, mPa·s	IFT mN/m	Water flooding Recovery, %	chemical flooding Recovery, %	Incremental Recovery, %
(1)	1500	0	39.67	—	39.98	49.21	9.23
(2)	0	0.3	0.67	3.53×10^{-3}	40.42	43.35	2.93
(3)	1500	0.3	40	4.12×10^{-3}	40.15	53.55	13.40
(4)	2000	0.3	51.64	6.33×10^{-3}	41.11	55.09	13.98
(5)	1500	0.2	39.47	4.51×10^{-2}	38.97	50.08	11.11
(6)	1500	0.4	38.79	4.6×10^{-3}	40.34	53.86	13.52
(7)	1500	0.3	28.97	4.92×10^{-3}	38.92	43.97	5.02

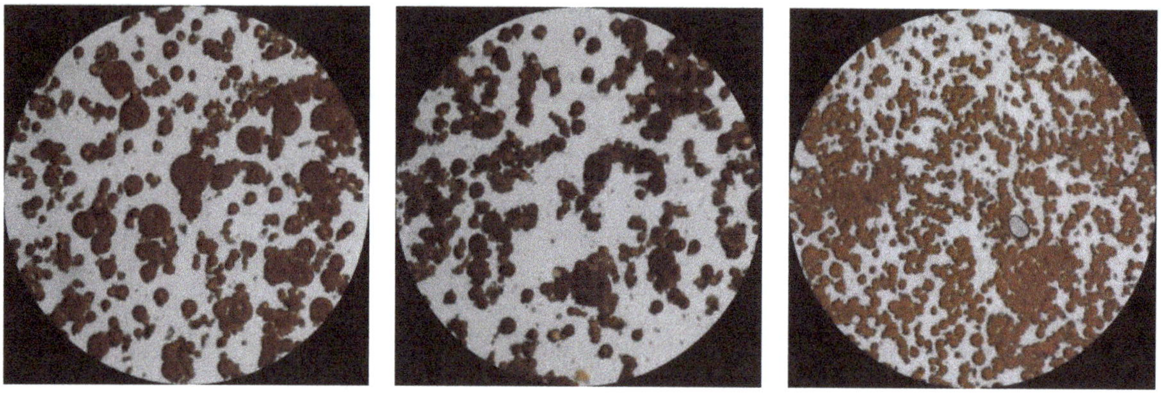

(a) (b) (c)

FIGURE 5: Micrographs of emulsions with different chemical agent: (a) 0.1 wt% surfactant + 2000 mg/L polymer; (b) 0.2 wt% surfactant + 2000 mg/L polymer; (c) 0.3 wt% surfactant + 2000 mg/L polymer.

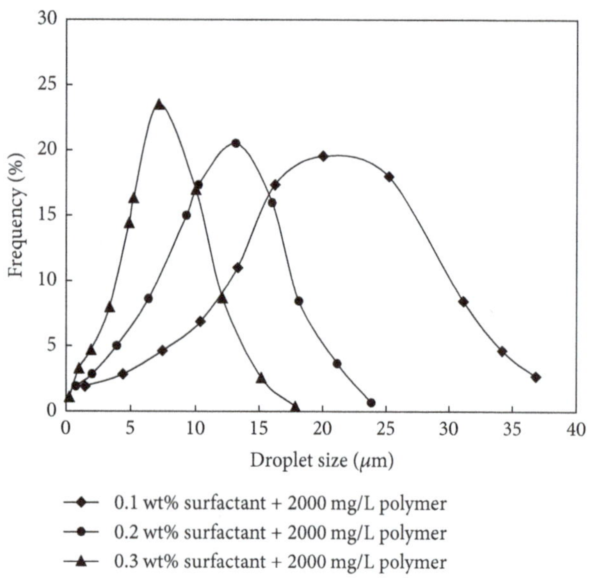

FIGURE 6: Distribution of emulsion droplet with different chemical agent.

3.2.1. *The Influence of the Viscosity on EOR.* The oil displacement experiments results of different chemical are shown in Table 3. Comparing the results of experiments (2), (3), and (4), it can be concluded that, under the conditions of a certain surfactant concentration, the recovery increases with the increase of the viscosity of polymer solution after water flooding. From the water cut and the pressure curves, it is indicated that the higher the viscosity is, the bigger the flow resistance of the injection fluid in the high permeability is and the higher the injection pressure is. And the injection fluid will turn into the middle-low permeability layers with high oil saturation and cut down the water cut. When the injection viscosity increases from 0.67 mPa·s to 40 mPa·s, the recovery improves 10.47%. And when the injection viscosity increases from 40 mPa·s to 51.64 mPa·s, the recovery improves 0.58%. This indicates that, with the increase of the injection viscosity, the recovery improves, but the degree of increment becomes smaller. Mainly because the possibility of winding between polymer molecular threads is increasing and the radius of the polymer molecular thread becomes large due to the increase of polymer concentration (see Figure 7). According to the literature [20], when the ratio of the radius of pore in the core and the radius of polymer molecular threads is greater than 5.5, the compatibility of pore and throat of the core is bad. The quantity of the polymer which can enter the middle and small pores with high residual oil saturations after water flooding is smaller. When the incremental recovery becomes obviously

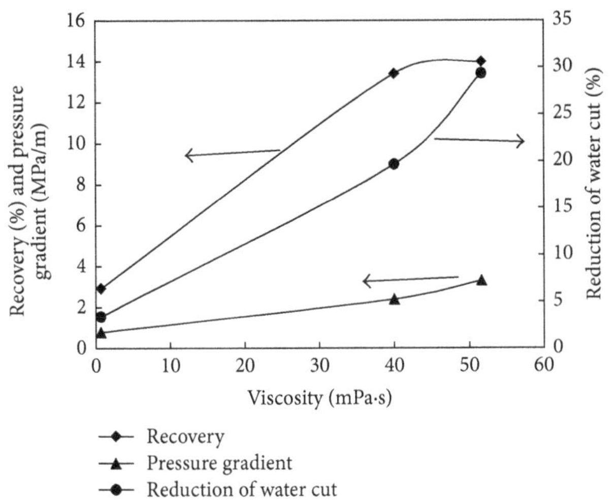

FIGURE 7: Relationship between enhanced recovery, maximum pressure gradient, reduction of water cut, and viscosity.

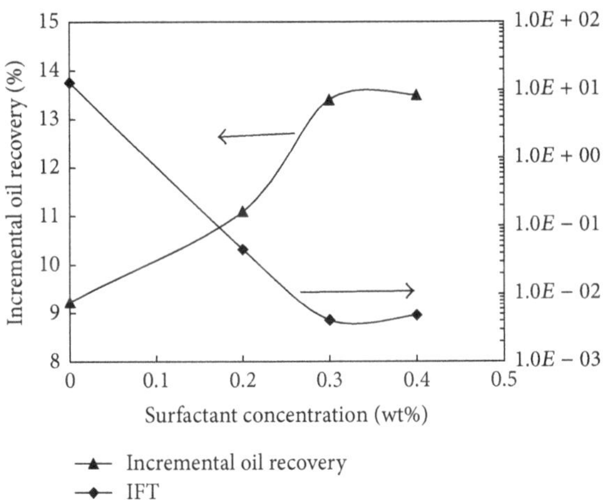

FIGURE 8: Relationship of IFT and incremental oil recovery when surfactant concentration varies. The polymer concentration is 1500 mg/L.

little, the injection viscosity is called the critical viscosity. At this time, the bigger viscosity, namely, the bigger polymer concentration, means less economic performance.

The SP system with polymer concentration 1500 mg/L and the surfactant concentration 0.3 wt% was injected into the core to simulate shearing during the injecting process in oilfield and then the SP system shared was used to displace oil. Comparing experiments (3) and (7), it can be concluded that the viscosity changes a lot after shearing and the recovery has an obvious decrease from 13.4% to 5.02%. This indicates that the system has a weaker mobility.

3.2.2. The Influence of the IFT on the EOR. In order to find the influence of the surfactant concentration of SP system on the EOR, 4 experiments were conducted. The range of surfactant concentration is 0–0.4 wt% and the slug size of injection chemical is 0.6 PV. It can be concluded from Figure 8 that EOR depends on the surfactant concentration at constant polymer concentration. When the range of surfactant concentration is 0–0.3 wt%, the values of EOR increase rapidly from 9.23% to 13.40%. When the surfactant concentration has a further increase from 0.3 to 0.4 wt%, the trend of the EOR value becomes gentle. This is because the surfactant molecules are adsorbed on the oil-water contact surface. It will reduce the IFT to a rather low degree and the emulsion is easy to form. So the residual oil can be flooded more easily. Comparing the curves in Figures 8 and 9, it can be concluded that the lower the IFT is, the more stable the emulsion is and the higher the incremental oil recovery is.

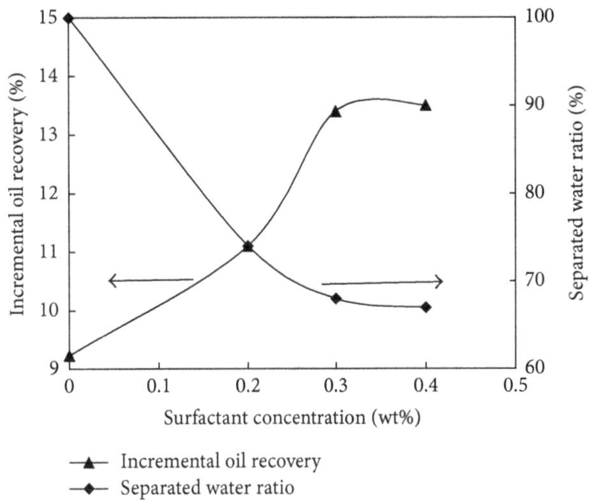

FIGURE 9: Relationship of separated water ratio and incremental oil recovery when surfactant concentration varies. The polymer concentration is 1500 mg/L.

The micromorphology of the output liquid in the process of chemical flooding is shown in Figure 10. Figures 10(a) and 10(b) prove that the O/W emulsion is formed in the process of injecting. This illustrates that the oil increment in the SP flooding is mainly in the form of O/W emulsion. At the same time, the results show that emulsion plays quite an important role for EOR in SP flooding.

3.2.3. Main Controlling Factor of SP Flooding to Improve Recovery. According to the traditional theory of chemical flooding, the basic methods of chemical flooding for EOR are enlarging the sweep volume and improving cleaning efficiency [21, 22]. But the two methods have different contribution to enhance oil recovery. Thus, it is of great importance to clear the main factors of SP flooding, which provides theoretical guidance for system optimization.

Comparing experiments number (1), (2), and (3), it can be concluded that different kinds of chemical have different effects on oil recovery. The SP flooding has the largest oil recovery increases, followed by the polymer flooding, and the last is surfactant flooding. Comparing with the polymer flooding and SP flooding, the surfactant flooding has better oil displacement efficiency, but as surfactant solution viscosity is low, the flow resistance does not rise and the injection pressure is not higher than the water flooding (Figure 11),

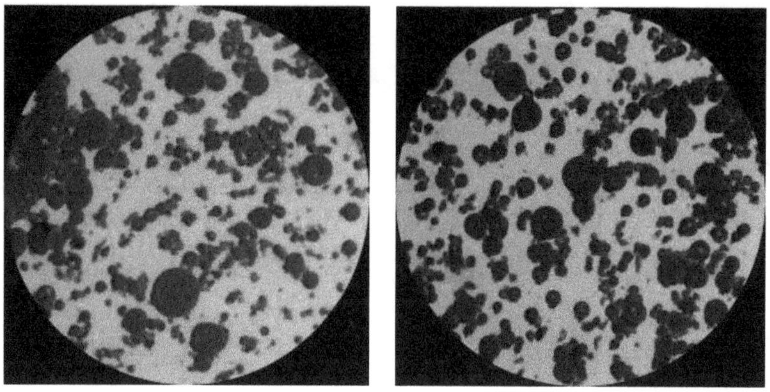

FIGURE 10: Micrographs of produced fluids in different production periods when chemical solution was injected: (a) in the earlier production period; (b) in the later production period.

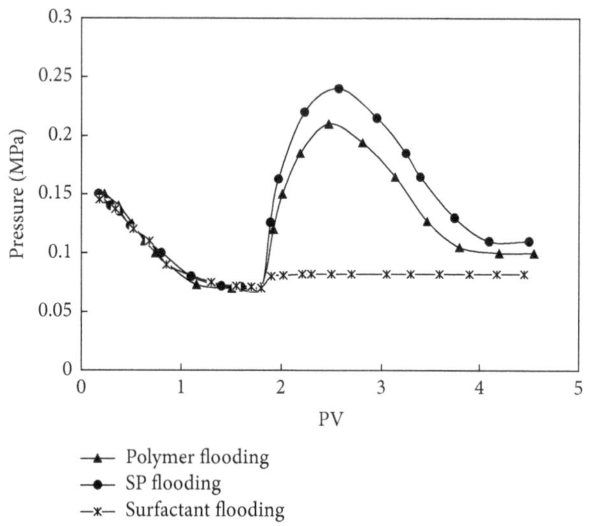

FIGURE 11: Relationship between inject pressure of different agent and pore volume.

which indicates that the surfactant solution is still flowing in the water channel and does not expand the swept volume. The incremental recovery is mainly from the action of washing oil. Further analysis showed that the incremental recovery of SP flooding is higher than the sum of incremental recovery of the polymer flooding and surfactant flooding, which indicates that the polymer carries surfactant into more pore volume and the surfactant will wash oil well with synergistic effect of polymer. Thus, it is quite important to expand the swept volume in the process of chemical flooding. The premise that the surfactant improves the oil displacement efficiency must be that the polymer can expand the swept volume.

SP flooding not just improves oil displacement recovery but also emulsifies oil into different sizes of drops which play the role of deep steering, improve the sweep factors of the following injection liquid, and reach the highest recovery. The oil drops are relatively large and hard to go through the throat. This is called the resistance effect (Jamin effect). The accumulation of resistance effect will result in improving the resistance factor that the liquid passes through the high permeability throat and raises the sweep factor of the following liquid. Meanwhile, under the stable effect of polymer, the emulsion can play the role of deep steering for a long time, compensating for the lack of polymer profile control ability.

4. Conclusions

(1) Salinity and shearing have a negative effect on both viscosity of polymer solution and oil recovery, so increasing the concentration of the polymer to maintain high viscosity is necessary.

(2) Different from the polymer flooding and SP flooding, the surfactant solution can only flow in the water channel and cannot improve the recovery effectively. The improving recovery of SP flooding is higher than the sum of improving recovery of the polymer and surfactant. It shows that the polymer carries surfactant into more pore volume and the oil displacing action gets full play, forming synergistic effect. It is a further explanation that the SP flooding is an effective method after the polymer flooding.

(3) The main factor of SP flooding in the heterogeneous reservoir is the mobility control action. As a result, when choosing the SP system, the viscosity should be considered first and the IFT then comes second. When the system reaches the ultralow IFT, the emulsion is easy to form and it is advantageous to the start of residual oil.

Conflicts of Interest

The authors declare that they have no conflicts of interest.

Acknowledgments

The authors would like to express their appreciation for the financial support received from the National Natural Science Foundation of China (51604243) and School of Petrochemical and Energy Engineering of Zhejiang Ocean University for permission to publish this paper.

References

[1] E. Allen and D. V. Boger, *The Influence of Rheological Properties on Mobility Control in Polymer-Augmented Waterflooding*, Society of Petroleum Engineers, 1988.

[2] X. Han, W. Wang, and Y. Xu, *The Viscoelastic Behavior of HPAM Solutions in Porous Media And It's Effects on Displacement Efficiency*, Society of Petroleum Engineers, 1995.

[3] D. Wang et al., *Viscous-Elastic Polymer Can Increase Microscale Displacement Efficiency in Cores*, Society of Petroleum Engineers, 2000.

[4] W. Demin et al., "Viscous-elastic polymer fluids rheology and its effect upon production equipment," *SPE Production & Facilities*, vol. 19, no. 04, pp. 209–216, 2004.

[5] N. Jin-gang, "Practices and understanding of polymer flooding enhanced oil recovery technique in Daqing oilfield," *Petroleum Geology & Oilfield Development in Daqing*, vol. 5, 024 pages, 2004.

[6] S. Gao, H. Li, and H. Li, "Laboratory investigation of combination of alkaline-surfactant-polymer for Daqing EOR," *SPE Reservoir Engineering*, vol. 10, no. 03, pp. 194–197, 1995.

[7] W. Demin et al., "Pilot test of alkaline surfactant polymer flooding in Daqing oil field," *SPE Reservoir Engineering*, vol. 12, no. 04, pp. 229–233, 1997.

[8] W. Demin et al., *An Alkaline Bio-Surfactant Polymer Flooding Pilots in Daqing Oil Field*, Society of Petroleum Engineers, 1999.

[9] H. S. Al-Hashim et al., *Alkaline Surfactant Polymer Formulation for Saudi Arabian Carbonate Reservoirs*, Society of Petroleum Engineers, 1996.

[10] R. A. Wattenbarger and D. H. Thurnau, *Application of SSOR to Three-Dimensional Reservoir Problems*, Society of Petroleum Engineers, 1976.

[11] Y. Wang, F. Zhao, and B. Bai, *Optimized Surfactant IFT and Polymer Viscosity for Surfactant-Polymer Flooding in Heterogeneous Formations*, Society of Petroleum Engineers, 2010.

[12] A. Feng et al., "Study of surfactant-polymer flooding in heavy oil reservoirs," *Journal of Dispersion Science and Technology*, (just-accepted), 2012.

[13] W. Kang, L. Meng, and J. Niu, "Mechanism of the effect of salinity on HPAM solution viscosity," *Polymer Materials Science and Engineering*, vol. 22, no. 5, p. 175, 2006.

[14] K. Wanli et al., "Interactions between alkali/surfactant/polymer and their effects on emulsion stability," *Colloids and Surfaces A: Physicochemical and Engineering Aspects*, vol. 175, no. 1, pp. 243–247, 2000.

[15] P. Shen et al., *The Influence of Interfacial Tension on Water-Oil Two-Phase Relative Permeability*, Society of Petroleum Engineers, 2006.

[16] D. O. Shah, *Improved Oil Recovery by Surfactant and Polymer Flooding*, Elsevier, 2012.

[17] W. R. Foster, "Low-tension waterflooding process," *Journal of Petroleum Technology*, vol. 25, no. 02, pp. 205–210, 1973.

[18] J. C. Melrose, in *Improved Oil Recovery by Surfactant and Polymer Flooding*, D. O. Shah and R. S. Schechter, Eds., 1978, p. 590, Academic Press, New York, USA, 1977.

[19] S. N. Ashrafizadeh, E. Motaee, and V. Hoshyargar, "Emulsification of heavy crude oil in water by natural surfactants," *Journal of Petroleum Science and Engineering*, vol. 86-87, pp. 137–143, 2012.

[20] Y. Q. Li et al., "Study on matching relationship of polymer hydrodynamic size and pore throat size for stratum in sand reservoir," in *Proceedings of the Offshore Technology Conference-Asia*, Offshore Technology Conference, Kuala Lumpur, Malaysia, 2014.

[21] H. Jirui et al., *Study of the Effect of ASP Solution Viscoelasticity on Displacement Efficiency*, 2001.

[22] J. Hou, Z. Liu, and H. Xia, *Viscoelasticity of ASP Solution Is A More Important Factor of Enhancing Displacement Efficiency than Ultra-Low Interfacial Tension In ASP Flooding*, Society of Petroleum Engineers, 2001.

8

DC Electrical Ageing of XLPE under Hydrostatic Pressure

Fadila Benlizidia Lalam and Faouzi Djemmal

Quantum Chemistry and Physics Laboratory, Mouloud Mammeri University, Tizi-Ouzou, Algeria

Correspondence should be addressed to Fadila Benlizidia Lalam; flalam@yahoo.fr

Academic Editor: Antonio Riveiro

The experimental electrical ageing, of cross-linked polyethylene films 100 μm thick, was investigated under high hydrostatic pressure of 300 bar and at atmospheric pressure. The tests are conducted on direct current (dc) for up to 1000 h ageing and at temperature of 70°C. The use of the Weibull statistic, with the estimation of confidence bounds at 90%, has shown that the hydrostatic pressure has a real effect on the lifetime. These lifetime data are qualitatively analyzed with the inverse power model. It was found that thermally activated process is able to describe the pressure effect on the electrical ageing of XLPE.

1. Introduction

Since the advent of synthetic polymer development, cross-linked polyethylene (XLPE) has become the commonly used insulating material for modern high-voltage extruded cables, due to a combination of low material and processing costs, reliability, and appropriate electrical and mechanical properties [1, 2]. This widespread use of XLPE may be attributed also to a maximum operation temperature of 90°C, emergency temperature of 130°C, and a short circuit maximum temperature of 250°C [3, 4].

However, it is well known that, under service conditions, these insulators will likely undergo a graduation deterioration in performance as a dielectric material that can ultimately lead to failure. Consequently, the study of this phenomenon, namely, electric ageing, or a long-term dielectric breakdown, of these materials, is necessary. The interest of such a study is double: on the one hand, it allows the evaluation of the time to breakdown or lifetime of these insulators under the effect of the considered stresses; on the other hand, it looks for a correlation between the ageing process and the stresses causing it.

The main purpose of our work has therefore been to examine the effect of hydrostatic pressure, on the electrical ageing, by a dc voltage, on insulators used in submarine cables. We have already investigated this effect on other insulators polymers [5]. This paper gives the long-term performance of XLPE films at a temperature of 70°C and two pressures: 1 and 300 bar. These parameter values (300 bar and 70°C) can be considered as the real use conditions of the submarine cables.

2. Experimental Procedure

2.1. Materials under Study. The samples, which were 100 μm thick films of XLPE, were cut from unaged cables. The insulation thickness was precisely controlled by micrometer. The principal characteristics of these XLPE specimens are given in Table 1. Commercial confidentiality denied us knowledge of the manufacturing procedure.

2.2. Experimental Setup. The purpose of our work is to show the effect of hydrostatic pressure on the electrical ageing, by a dc voltage, on XLPE and that is why the experimental study has been carried out at two pressures: 1 bar and 300 bar and the temperature of 70°C. The latter can be considered as the temperature of XLPE insulator in service condition.

The experimental setup is classically composed of a dc generator of 100 kV, a measuring cell, and a detection system.

The samples were placed into the measuring cell composed of five high-voltage relays and five pairs of electrodes. Electrodes are made with stainless steel following Rogowsky profile. The same dc voltage (ranging from 14 to 50 kV) was applied to each specimen (they were tested in parallel). When a dielectric breakdown occurred in a sample, a positive

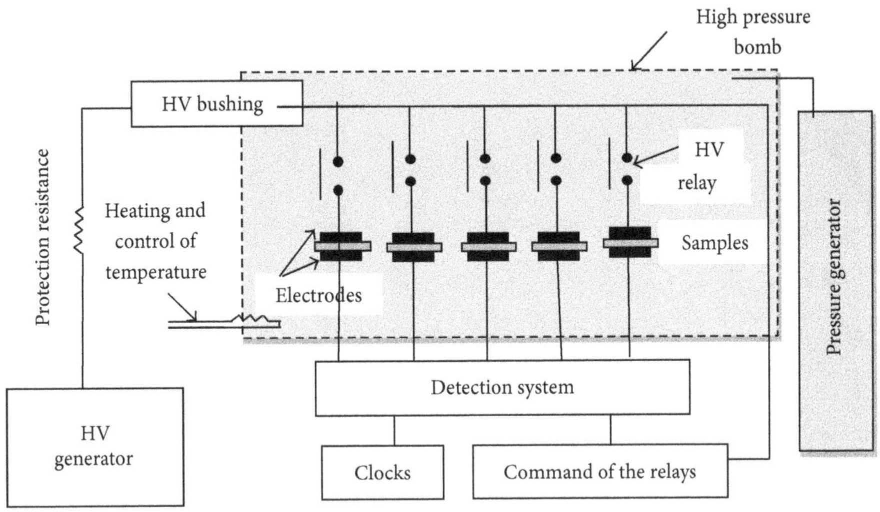

FIGURE 1: Electrical ageing measurement setup under hydrostatic pressure.

TABLE 1: Main characteristics of XLPE under study.

Density	0.923
Melting point (°C)	110
Rate of crystallinity	40
Tensile strength (MPa)	17
Elongation (%)	450
Water content (ppm)	100
Dielectric constant ε_r (1 MHz)	2.3
Loss factor	$5 \cdot 10^{-4}$

pulse was generated across the detection system by which the applied voltage was immediately removed for this sample and the breakdown time value, t_i, was recorded.

For high-pressure measurements (300 bar), the measuring cell was placed into a high-pressure bomb with a working volume of 9500 cm^3, where the pressure was obtained by a diaphragm compressor using gaseous nitrogen as the transmitting fluid. At atmospheric pressure (1 bar), the measuring cell was immersed into a chamber filled with silicone oil to avoid flashover and partial discharge effects.

For all measurements, the desired temperature was obtained, with stability of 1°C, by the use of heating resistance of 1600 W.

A schematic drawing of the experimental setup, under hydrostatic pressure, is shown in Figure 1.

3. Results and Discussion

Experimental results were obtained at 1 and 300 bar, for a temperature of 70°C in the range of an electrical field of 140–450 kV·mm^{-1} and for up to 1000 h of ageing. For each value of field, only five specimens were aged since it takes a relatively long time to perform one measurement.

3.1. Statistical Analysis. To assess the results for the breakdown of polymeric insulation, the use of statistical method is often required. Weibull distribution is now widely accepted which is based on the fact that the failure of the whole system depends on the failure of the weakest point in the system [6, 7]. The two-parameter Weibull distribution is given by

$$F(t) = 1 - \exp\left[-\left(\frac{t}{\alpha}\right)^{\beta}\right], \quad (1)$$

where t is the measured breakdown time; $F(t)$ is the probability of failure at time less than or equal to t; α is the scale parameter ($\alpha > 0$) and represents the time for which the failure probability is 63.2%; and β is a shape parameter and a measure of the spread of the data.

The time to breakdown data and tolerance bounds on percentiles, of the two-parameter Weibull distribution, were then fitted using an exact approach based on the maximum likelihood method [8].

Figures 2 and 3 give examples of the application of the Weibull distribution to our experimental data. The same types of graph were obtained for the other values of electrical fields. They show that the 90% confidence bounds (for 1 and 300 bar) do not overlap, indicating that the two sets of data are different significantly.

Even though the distribution of Weibull does not offer any physical mechanism of electric ageing, it remains essential for the analysis of the experimental data and the description of the effect of a parameter (pressure in our case). To explain the effect of pressure on the electrical ageing of XLPE, we try to model the experimental lifetime data. The models suggested in the literature are of two types: macroscopic and microscopic. The first one or phenomenological one can be considered as global description of ageing, while the second one is based on a physical approach.

3.2. The Inverse Power Law Model. The inverse power law model is an empirical model, but it is commonly used for a

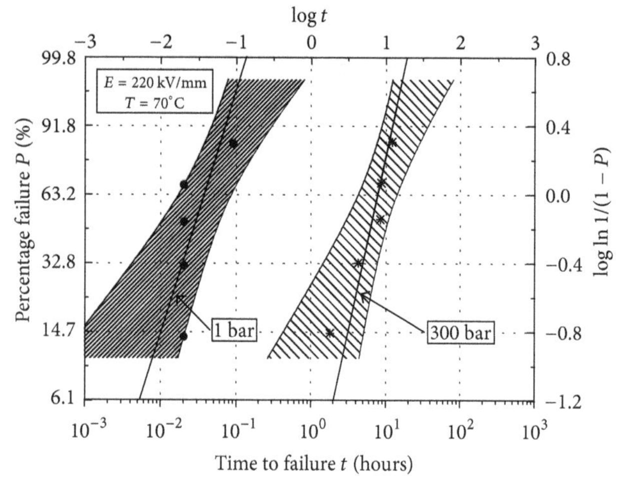

FIGURE 2: Weibull plot of the electrical ageing data of XLPE at 220 kV/mm and 70°C.

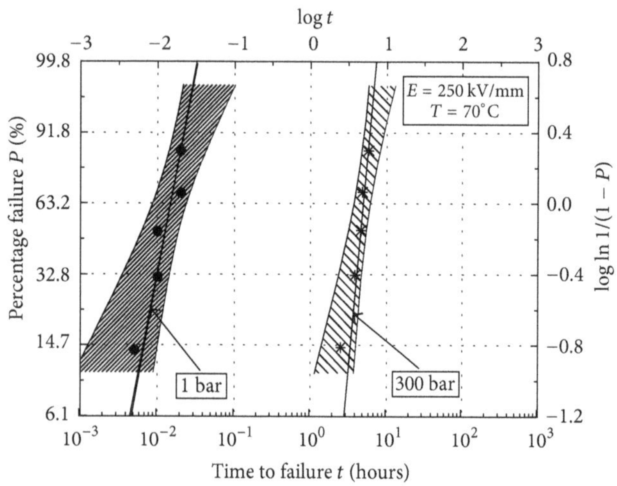

FIGURE 3: Weibull plot of the electrical ageing data of XLPE at 250 kV/mm and 70°C.

qualitative analysis [9, 10]. It is given by the following relationship:

$$t = kE^{-n}, \tag{2}$$

where t is the time to breakdown (usually it is a Weibull scale at 63.2% probability); E is the applied electrical field; k is a constant; the exponent n is the voltage endurance coefficient whose value depends on the compound, additives, and manufacturing procedures [11].

The inverse power law is considered valid, if the data being plotted on log-log graph fits a straight line.

Figure 4, which gives the fit of our experimental data, for the two considered pressures, shows the following:

(i) The reduction of the life exponent n, with pressure, is from n (1 bar) = 18 to n (300 bar) = 12. The recommendation, for testing high-voltage dc extruded cables systems, gives lower limit of ageing parameter n equal to 10 [12].

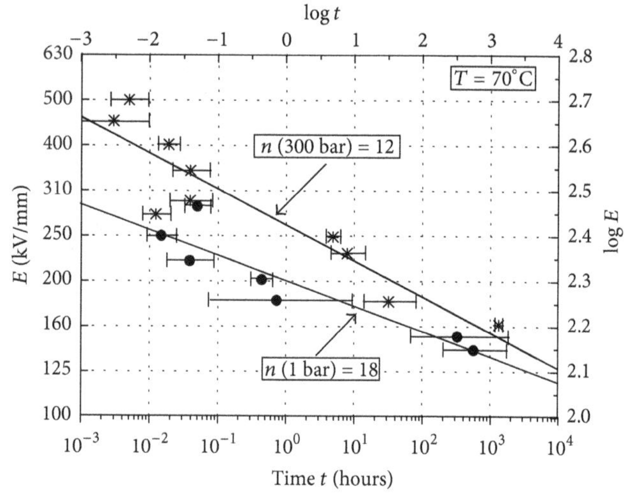

FIGURE 4: Representation of the ageing data of XLPE, according to the inverse power model, at 1 and 300 bar and 70°C. The bars to each line in this plot are 90% confidence limits.

(ii) The life curve at 300 bar is shifted towards the high fields.

3.3. Application of Crine's Model. The inverse power law model has no clear physical basis; it allows a qualitative analysis but does not give information on the origin of the breakdown processes [13]. Among the physical theories studied in our bibliography, it seems that Crine's model is adapted the most to describe our results.

This electrical life model, presented in the last three decades and modified later, assumes that electrical ageing is thermally activated process [13–17]. It was the first model to use the thermodynamic concept associated with the theory rate to describe ageing process [18]. Indeed, the energy parameter controlling the rate of ageing is the free energy change $\Delta G = \Delta H - T\Delta S$, where ΔH and ΔS are the activation enthalpy and entropy [19]. Thus, a free energy barrier ΔG shall be crossed for the system to pass from the unaged state to the aged state. Under the action of electrical field, the barrier ΔG decreases with the square of the field due to an electromechanical modification of the intrinsic free energy barrier. Then, the insulation life under dc high field E is given by

$$t \approx \left[\frac{h}{2kT}\right] \exp\left[\frac{\Delta G - (1/2)\,\varepsilon_0 \varepsilon_r \Delta V E^2}{kT}\right], \tag{3}$$

where h and k are the Planck and Boltzmann constant, ΔV is the activation volume, ε_0 and ε_r are the permittivity of free space and the dielectric constant of polymer, respectively.

This model suggested that the main ageing factor is the Maxwell stress induced by the field [17, 20]. Under the action of low and moderate electrical field, the nanocavities (the free volume of polymers) will be compressed by this compressive Maxwell stress. In this case only some nanocavities will be deformed and those located near chain ends and in the amorphous phase are more likely to be the first to be deformed. When a given number of nanocavities have been entirely

TABLE 2: Parameters of Crine's model for the ageing of XLPE at 70°C.

Pressure (bar)	ΔG (J)	ΔV (m^3)
1	$2{,}3 \cdot 10^{-19}$	$1{,}2 \cdot 10^{-25}$
300	$2{,}35 \cdot 10^{-19}$	$0{,}837 \cdot 10^{-25}$

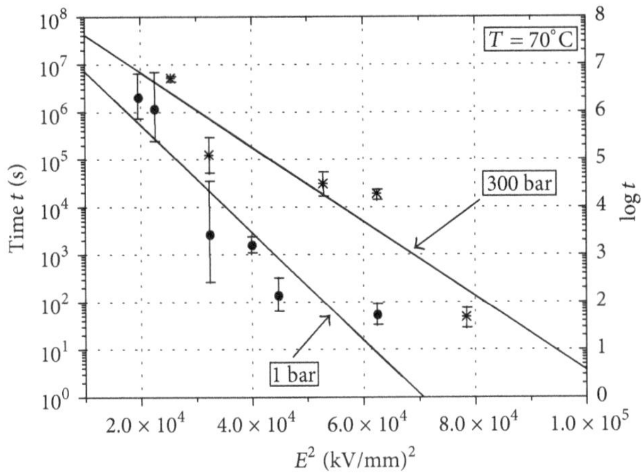

FIGURE 5: Representation of the experimental data of the XLPE at 70°C, according to Crine's model.

deformed, the amorphous phase is then significantly deformed and weak Van der Waals attraction bonds are broken; then the ageing process will considerably accelerate. The activation volume is then the inverse of the concentration of completely deformed nanocavities.

This model envisages a linear relation between E^2 and the logarithm of time (3), which we obtained with our results on Figure 5. Table 2 gives the value of ΔV and ΔG deduced from the slope and intercept, respectively, in these plots.

The free energy and the activation volume values deduced from experimental results (Table 2) seem in good agreement with existing data [17, 20].

The increase of ΔG, with pressure, can be attributed to the term $P\Delta V$ which appears in the relation of ΔG ($\Delta G = \Delta H - T\Delta S = \Delta E - T\Delta S + P\Delta V$) and which is generally negligible at atmospheric pressure [21]. The deformation of the barrier has been also reported by Crine and David [22]. They found that under a hydrostatic pressure P the barrier is deformed symmetrically (the original and final states are equally deformed by the energy term $P\Delta V$).

The decrease of ΔV with the pressure can be explained by the fact that the hydrostatic pressure induces a compression which reduces the free volume of the polymer; it has been reported that the hydrostatic pressure can change the free volume and the morphology of polymers [23].

4. Conclusion

The pressure effect on the electrical ageing of XLPE films has been investigated. The experimental measurements were carried out at pressures of 1 and 300 bar, at temperature of 70°C, in the range of an electrical field of 140–450 kV/cm, and for up to 1000 hours of ageing. The results were analyzed by Weibull's statistic with a confidence bounds of a 90% level. The inverse power law model gives a qualitative analysis: the hydrostatic pressure delays the electrical ageing of this XLPE in the time domain considered. The ageing process of this polymer could be explained by Crine's model based on the concept of thermally activated degradation and the existence of nanocavities. We assume that both of them can be influenced by an increase of hydrostatic pressure.

Conflicts of Interest

The authors declare that there are no conflicts of interest regarding the publication of this paper.

References

[1] X. R. Chen, L. B. Hu, Y. Xu, X. L. Cao, and S. M. Gubanski, "Investigation of temperature effect on electrical trees in XLPE cable insulation," in *Proceedings of the Annual Report Conference on Electrical Insulation and Dielectric Phenomena (CEIDP '12)*, pp. 612–615, Montreal, Canada, October 2012.

[2] T. L. Hanley, R. P. Burford, R. J. Fleming, and K. W. Barber, "A general review of polymeric insulation for use in HVDC cables," *IEEE Electrical Insulation Magazine*, vol. 19, no. 1, pp. 13–24, 2003.

[3] V. Vahedy, "Polymer insulated high voltage cables," *IEEE Electrical Insulation Magazine*, vol. 22, no. 3, pp. 13–18, 2006.

[4] A. M. Nobrega, M. L. B. Martinez, and A. A. A. De Queiroz, "Investigation and analysis of electrical aging of XLPE insulation for medium voltage covered conductors manufactured in Brazil," *IEEE Transactions on Dielectrics and Electrical Insulation*, vol. 20, no. 2, pp. 628–640, 2013.

[5] F. Benlizidia Lalam and T. G. Hoang, "Pressure effect on the electrical ageing of polyethylene," *Journal of Physics D: Applied Physics*, vol. 33, no. 22, pp. L133–L136, 2000.

[6] W. Weibull, "A statistical distribution function of wide applicability," *Journal of Applied Mechanics*, vol. 18, pp. 293–297, 1951.

[7] IEEE, "IEEE guide for the statistical analysis of electrical insulation voltage endurance data," IEEE Standard 930-1987, 1998.

[8] G. C. Stone and J. F. Lawless, "The application of weibull statistics to insulation aging tests," *IEEE Transactions on Electrical Insulation*, vol. EI-14, no. 5, pp. 233–239, 1979.

[9] P. Cygan and J. R. Laghari, "Models for insulation aging under electrical and thermal multistress," *IEEE Transactions on Electrical Insulation*, vol. 25, no. 5, pp. 923–934, 1990.

[10] L. Simoni, "A general approach to the endurance of electrical insulation under temperature and voltage," *IEEE Transactions on Electrical Insulation*, vol. 16, no. 4, pp. 277–289, 1981.

[11] G. Mazzanti and M. Marzinotto, *Extruded Cables for High-Voltage Direct-Current Transmission*, John Wiley & Sons, Hoboken, NJ, USA', 2013.

[12] T. Liu, Z. Lv, Y. Wang et al., "A new method of estimating the inverse power law ageing parameter of XLPE based on step-stress tests," in *Proceedings of the Annual Report Conference on Electrical Insulation and Dielectric Phenomena*, pp. 69–72, Shenzhen, China, October 2013.

[13] J. Crine, J. Parpal, and G. Lessard, "A model of aging of dielectric extruded cables," in *Proceedings of the 3rd International Conference on Conduction and Breakdown in Solid Dielectrics*, pp. 347–351, Trondheim, Norway, July 1989.

[14] C. Dang, J.-L. Parpal, and J.-P. Crine, "Electrical aging of extruded dielectric cables: review of existing theories and data," *IEEE Transactions on Dielectrics and Electrical Insulation*, vol. 3, no. 2, pp. 237–247, 1996.

[15] J. P. Crine, "A model of solid dielectric aging," in *Proceedings of the Conference Record of the 1990 IEEE International Symposium on Electrical Insulation*, pp. 25–26, Toronto, Canada, 1990.

[16] J. P. Crine, "The compensation law and polymer aging relaxations," in *Proceedings of the 3rd International Conference on Conduction and Breakdown in Solid Dielectrics*, pp. 321–325, 1989.

[17] J.-P. Crine, "On the interpretation of some electrical aging and relaxation phenomena in solid dielectrics," *IEEE Transactions on Dielectrics and Electrical Insulation*, vol. 12, no. 6, pp. 1089–1107, 2005.

[18] H. A. Alghamdi, G. Chen, and A. S. Vaughan, "Comparison of two schools of polymeric ageing models: causes and effects of space charge," in *Proceedings of the IEEE Conference on Electrical Insulation and Dielectric Phenomena (CEIDP '12)*, pp. 827–830, Montreal, Canada, October 2012.

[19] J. P. Crine, "On the use of rate theory to understand the aging of dielectrics," in *Proceedings of the International Conference on Properties and Applications of Dielectric Materials*, pp. 214–217, 1985.

[20] J. P. Crine, "A molecular model for the electrical aging of XLPE," in *Proceedings of the Conference on Electrical Insulation and Dielectric Phenomena (CEIDP '07)*, pp. 608–610, October 2007.

[21] P. Destruel, *Contribution à l'étude des propriétés diélectriques des polymères en fonction de leur histoire thermodynamique [Ph.D. thesis]*, Paul Sabatier University, Toulouse, France, 1980.

[22] J. P. Crine and E. David, "Influence of mechanical stresses on some electrical properties of polymers," in *Proceedings of the Annual Report Conference on Electrical Insulation and Dielectric Phenomena*, pp. 71–74, 2005.

[23] S. Li, W. Wang, S. Yu, and H. Sun, "Influence of hydrostatic pressure on dielectric properties of polyethylene/aluminum oxide nanocomposites," *IEEE Transactions on Dielectrics and Electrical Insulation*, vol. 21, no. 2, pp. 519–528, 2014.

9

Fabrication of Electrospun Chitosan/Nylon 6 Nanofibrous Membrane toward Metal Ions Removal and Antibacterial Effect

Akram R. Jabur, Laith K. Abbas, and Saja A. Moosa

Department of Materials Engineering, University of Technology, Baghdad, Iraq

Correspondence should be addressed to Akram R. Jabur; akram.jabut@gmail.com

Academic Editor: Luiz F. de Moura

Nylon 6/Chitosan membranes were fabricated by electrospinning onto a Millipore glass fiber filter to produce nanofibrous filter. Scanning electron microscopy (SEM) and water contact angle (WCA) were done to characterize the produced filter. Filter removal capability (adsorption) for metal ions was investigated for lead nitrate ($Pb(NO_3)_2$) and sodium chloride (NaCl). The antibacterial effect of the Nylon 6/Chitosan was investigated for *Escherichia coli*. Removal optimum values for $Pb(NO_3)_2$ and NaCl reached 87% and 75%, respectively. This research demonstrated that Nylon 6/Chitosan nanofibrous membrane has an enormous applicable potential removal for metal ions from aqueous solutions, antibacterial activity reaching 96%, and reasonable inhibition zone against *Escherichia coli* (*E. coli*).

1. Introduction

Removal of pathogens, chemicals, and heavy metals to produce clean drinking water is the most important factor that water purification focuses on. Polymer gels and several polymer solutions were electrospun to produce filters that were able to remove these contaminates. Most common used electrospun polymers are chitin [1], Chitosan derivative [2], Poly(ethylene-vinyl alcohol) [3], Poly(glycolic acid) and chitin [4], and Chitosan/PVA [5]. Recently, Chitosan and its derivatives had gained a lot of interest because of the wide range of applications in biomedicine, bioseparation, and food science. Chitosan, which is deacetylate's product of chitin (produced from the crust of crustacean shells), has desirable properties such as biofriendly, biodegradable, and antibacterial compound [6]. Also, Chitosan and chitin are able to remove metals and dyes so that they can be used to clean water; this is due to the adsorption of the metals and dyes to the cationic amine functional groups on chitosan. Various cross-linked polysaccharide materials with Chitosan can remove different metal and dye pollutants [7]. Chitosan and chitin also had the ability to remove polycyclic aromatic hydrocarbons more effectively [8]. Removal of particles like As^{+5} (arsenic) can be effected by the degree of deacetylation (DD). Adsorption and removal of particles in water can be influenced by crystallinity. Crystallinity increased by the amine group which forms a hydrogen bond to other chitosan monomers, and this affects the fiber morphology and adsorption [9]. Negatively charged dye like reactive Black 5 can be removed by Chitosan. It can positively charge in an acid environment and electrostatic forces are responsible for dye removal. It had been mentioned in the results that changing the molecular weight (MW) of chitosan from 80,100 to 308,300 reduces chitosan's ability to adsorb dye [10, 11]. This is because the changing in the internal structure of the chitosan chains and hydrogen bonding between hydroxyl and amine groups which reduced the possibility of dye binding. It had been found that the number of amine groups, determined by the DD, was affected by the removal of metal ions in water. High DD value (97%) produced higher removal efficiency, comparing to chitosan with 52% of DD value [11, 12]. Finally, nanofibers and in particular Chitosan nanofibers are better in removing metals, chemicals, and bacteria from water [11].

2. Experimental Part

2.1. Materials. The materials used in this research are Nylon 6 and formic acid, which were purchased from Sigma-Aldrich Co. (USA), molecular weight for repeat unit was 113.16 g/mol, and the total molecular weight was 25,000 g/mol. Chitosan was purchased from Cheng Du Micxy Chemical Co. Ltd. with the degree of deacetylation ((DD) ≥ 90%), molecular weight for repeat unit was 161.16 g/mol, and the total molecular weight was Mw 338,000. Lead nitrate ($Pb(NO_3)_2$) was purchased from Fluka Chemika, Germany (molecular weight 331.2 g/mol), sodium chloride (NaCl) was purchased from Edutek Chemicals, India (molecular weight 58.44 g/mol), and all agents were used without further purification. Micropore glass filter with binder (technical microfilter) has pore size of 2 μm, porosity of 90%, thickness of 1.2 mm, and fiber diameter of 124 mm.

2.2. Preparation of Nylon 6/Chitosan Blend Electrospinning Solution. Nylon 6 pellets and Chitosan powder at a concentration of 25% wt were blended in formic acid solution (25% wt/v) at 30°C using a laboratory magnetic stirrer for three hours to ensure complete dissolution of solutes and obtain a homogenous solution. Various ratios of Nylon 6/Chitosan (100/0, 90/10, 80/20, and 70/30) were utilized to prepare the blended nanofibrous membranes.

2.3. Electrospinning. The prepared electrospun solution was collected in 10 mL syringe equipped with a 22-gauge stainless steel needle tip. The electrospinning process was carried at 25 kV voltage, 0.5 mL/hr flow rate, 0.7 mm needle diameter tip, and 15 cm distance between electrodes. The electrospinning process took four hours' time and these parameters will be fixed for all the following solutions. The electrospinning process was carried out at room temperature of 25 ± 1°C and relative humidity of 25–30%. The prepared electrospun nanofiber sheet has an area of 153.86 cm^2 and diameter of 14 cm, whereas the weight of the prepared electrospun nanofiber sheet equals 0.0035 g for 1 cm^2. The electrospinning process was carried out by a bioelectrospinning/electrospray system (ESB-200), provided by (Nano NC, South Korea), which is shown in Figure 1.

2.4. Characterization. Scanning electron microscopy (Model: VEGA3 LM-TESCAN) was used in studying the surface morphology of nanofibers. Low variable pressure was used in all SEM tests for all specimens, the pressure was enough, and there was no need for any more of sample gold ion sputtering coating to chive conductivity on the surface of the specimens in SEM tests. Specimen conductivity was measured by a model (C and 7110 Inolab). Solution viscosity was measured by a viscometer of type DV-II-pro at room temperature (25 ± 1°C). Solution surface tension was measured by surface tensiometer model (JYW-200A-Laryee Technology Co.) using a ring of platinum. Solution electrical conductivity was measured using the electrical conductivity device of model (C and 7110 Inolab).

2.5. Contact Angle Measurement. The wettability of the electrospun mats was measured with deionized water contact angle measurements. Contact angle meter of the type CAM 110, Germany, was used. Deionized water was automatically dropped onto the membrane. Measurement was carried out in three seconds.

2.6. Permeability Test. Experiments were carried out at a temperature of 25 ± 1°C. Tests were done and conducted by using pressure cross flow filtration system. The diameter of membrane filter sample was 30 mm. Measuring pure water flux and salt rejection was the used method that characterized Nylon 6 with various additives membranes. Membranes were exposed to a pressure with the value of 1 bar for 50 min. They were placed with a shim and a mesh structured spacer to eliminate pressure polarization. They were pressurized with a mechanical pump controlled by pressure regulators and then the pressure was controlled to the operating pressure (1–6 bar). Figure 2 shows pressure cross flow filtration system which is designed especially to conduct the test. The permeable flux was calculated by

$$J = \frac{V}{A \times T}, \tag{1}$$

where J is the permeable flux (L/m^2× h), V is the volume of permeate (liter), A is the effective membrane surface area (m^2), and T is the time (hour).

The salt rejection was determined using atomic absorption device (AA-7000 atomic absorption spectrophotometer, Shimadzu). The rejection of salts was obtained by

$$R = \left(1 - \frac{C_p}{C_f}\right) \times 100\%, \tag{2}$$

where C_p and C_f are ion concentration in permeate and feed, respectively, and (R) is rejection as a percentage. Figure 2 shows the system used in this process. During the rejection test three cells of membranes (three layers) on one layer of microfiber were used to increase the relation ratio.

2.7. Mechanical Strength Measurement. Mechanical properties of the electrospun membrane were measured by a tensile mechanical tester of the type Tinus Olsen, H50 KT. A (5 N) load cell was used in that device. Specimen thicknesses were measured by an optical microscope. Specimens with 10 mm dimensions width and 100 mm length were tested and the extension rate at the room temperature was 0.5 mm/min. The experiments were carried out three times to calculate the average value of the results.

2.8. Antimicrobial Test

2.8.1. Antibacterial Activity Test (Disc Diffusion Method). Bacteria were grown aerobically in nutrient broth at 37°C for 12 hours [13]. Cells were washed and suspended in distilled water until reaching the final concentration of 10^6 CFU/mL. The antimicrobial susceptibility of Nylon 6 nanofibers/Chitosan was evaluated using the disc diffusion method. "Muller-Hinton agar" was prepared from a commercially available dehydrated medium according to manufacturer's instructions. The dried surface of "Muller-Hinton agar" plate was inoculated with *E. coli* by swabbing over entire the sterile agar surface. Two forms of nanofiber sterilized membrane samples were cut into small standard circles (6 mm in diameter) for each circle and placed on the surface of the inoculated media.

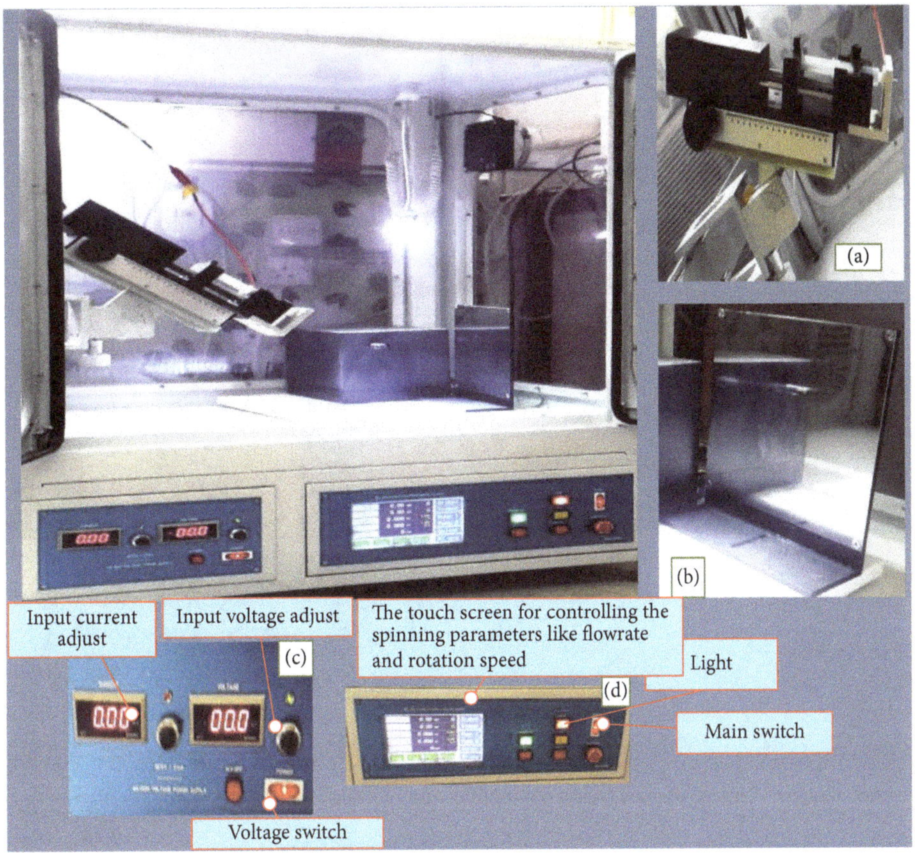

FIGURE 1: Electrospinning device: (a) syringe pump instruction, (b) collector plate, (c) power supply, and (d) touch control board.

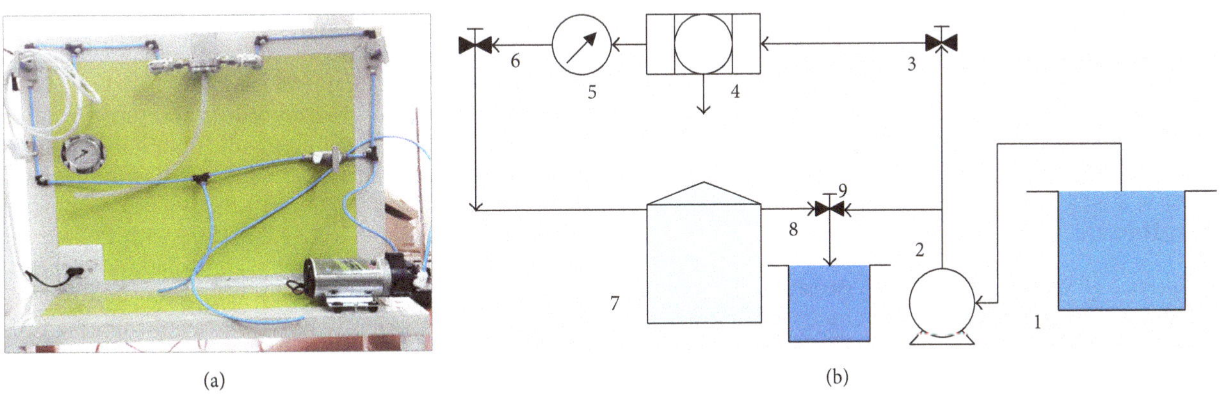

FIGURE 2: (a) Pressure cross flow image of filtration system. (b) Schematic diagram of the filtration system process. 1: tank for water sample test. 2: pump. 3: entry valve. 4: filter/module. 5: pressure gauge. 6: outlet valve. 7: open tank for permeate water sample. 8: tank of rejected water sample. 9: recycle valve.

The first form of nanofiber membrane contains additive and the other one without additive was used as a control. The plates were incubated at 37°C for 24 hours. Incubation plates then were examined in order to identify zones of on growth characteristic for antibacterial activity (halos around the fragments).

2.8.2. Antibacterial Activity Test (Optical Density Method). The antibacterial activity of electrospun nanofibers membrane was also tested by immobilizing nanofiber onto filters of a Millipore under vacuum filtration. A test water sample was prepared by inoculating 1×10^8 cells/mL *E. coli* into 250 mL of sterile normal saline water sample (0.85% NaCl in 100 mL distilled water). Water samples then were filtered through the membrane. After this step the optical density of the solution was measured by UV spectrophotometer (UV-1800 spectrophotometer, Shimadzu) at 450 nm wavelength. The number of bacteria was indirectly measured by optical density in an ultraviolet (UV) visible spectrometer and the

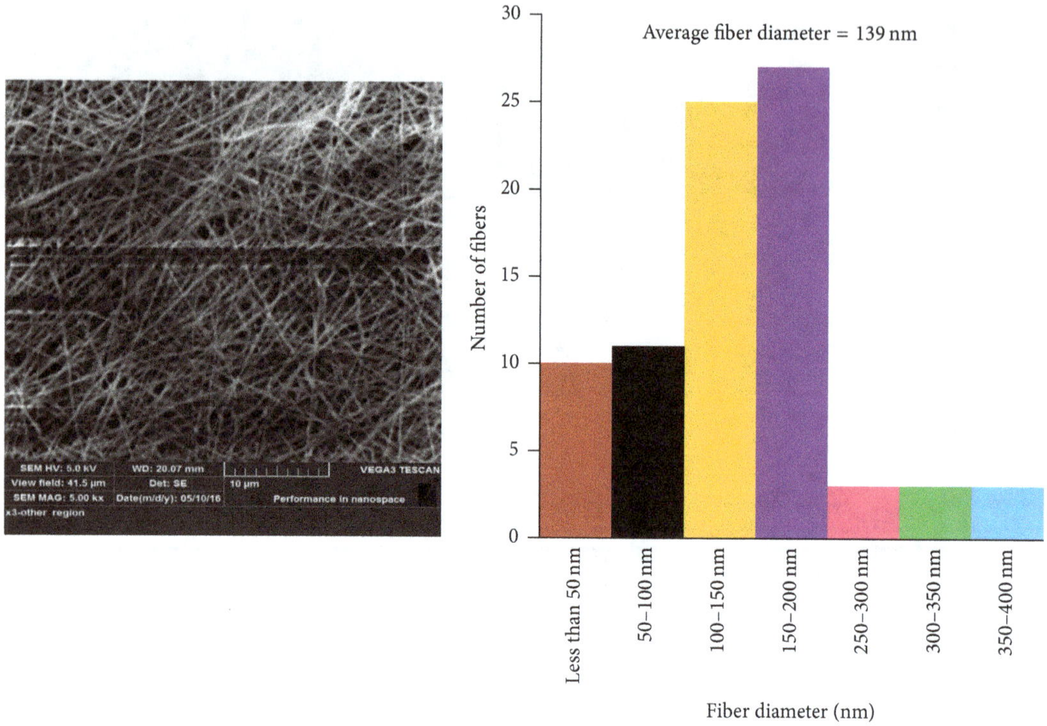

(a) (b)

FIGURE 3: (a) SEM image of pure Nylon 6 nanofibers membrane and (b) fiber diameter distribution.

antibacterial activity was evaluated quantitatively with the following equation:

$$\text{Antibacterial activity (Efficiency)} = \frac{A - B}{A} \times 100\%, \quad (3)$$

where A and B are the numbers of surviving cells in the control and test samples, respectively [14].

3. Results and Discussion

3.1. Properties of Spinning Solutions. Viscosities of Nylon 6 and Nylon 6/Chitosan solutions at various concentrations were shown in Table 1. Adding Chitosan in Nylon 6 solution led to increase the viscosity with increasing Chitosan content. Chitosan molecules seem to be very large when compared with Nylon molecules [15].

Table 1 shows the effect of Chitosan on electrical conductivity which had increased because of the increase of the Chitosan; this result agreed with "Bizarria et al.," whose result shows improving in electrical conductivity with increase in Chitosan concentration when using Chitosan blended with PEO [16]. It is obvious that the surface tension increased with increasing Chitosan concentration due to high solution viscosity, but there is no significant change in surface tension.

3.2. Electrospun Membrane Characterization. Nanofibers (SEM) images of nanofiber diameter histogram distribution are shown in Figure 3. Figure 3(a) represents the Nylon 6/Chitosan (100/0) nanofibers with an average diameter of 139 nm and the desirable morphology had been obtained. The nanofiber diameter increased with the increasing of Chitosan content in the blended solution and this is because of the increasing in the viscosity of the solution. Figure 4(a) shows an increase in fiber diameter and this had been happing because of very high viscosity values and there is difficulty in the ejection of jets from polymer solution and it results in a larger fiber diameter. This result had been matched with "Deitzel et al." [17]. The desirable morphology had been converted to a defective structure with the increasing of the Chitosan content in the blend solution. Experiments showed that increasing of Chitosan ratio above 30% will rise the solution viscosity in which electrospinning became impossible. Usually fouling decreases with an increase in hydrophilicity of the polymeric material. In fact, a decrease in contact angle leads to increase in the flux ratio, which means decreased fouling [18]; therefore, evaluating the membrane characteristics was followed by contact angle measurements. Figure 5 shows the contact angle values of droplet with nanofibrous membranes after 3-second contact.

Water contact angle (WAC) for pure Nylon 6 nanofiber reaches the value 70 while the WAC of Nylon 6/Chitosan at a weight ratio of 70/30 reaches the value of 12.8. The results indicate a significant improvement in hydrophilicity of a membrane which is increased by increasing the Chitosan content. This can be attributed to the presence of a large number of functional groups like acetamide, primary amino, and/or hydroxyl groups in Chitosan structure. These outcomes agreed with the work reported by "Zhang et al." [19].

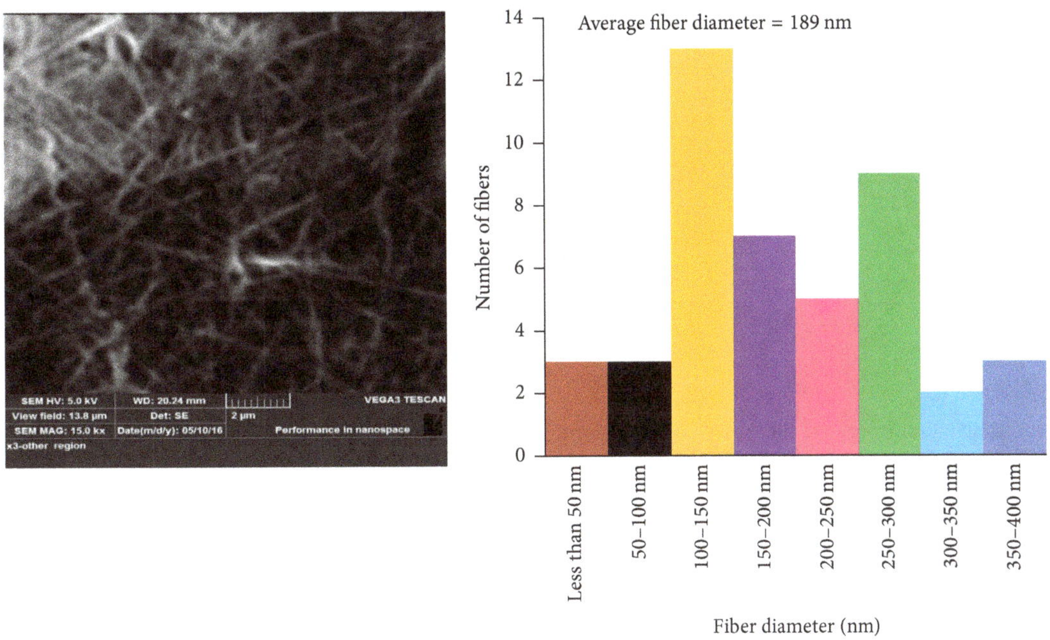

FIGURE 4: (a) SEM image of (30/70) Chitosan/Nylon 6 nanofibers membrane and (b) fiber diameter distribution.

TABLE 1: The electrospinning solution properties (these average values after three-time test).

Chitosan/Nylon ratio	Electrical conductivity (σ: mS/cm)	Viscosity cP	Surface tension (mN/m)
0/100	4.00	692	34.430
10/90	6.02	6467	36.610
20/80	7.40	11350	38.532
30/70	8.33	17300	39.452

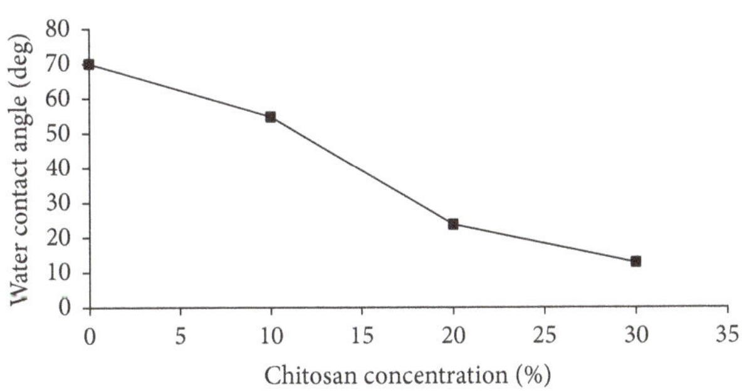

FIGURE 5: Water contact angle variations of various Nylon/Chitosan electrospun nanofibrous membranes.

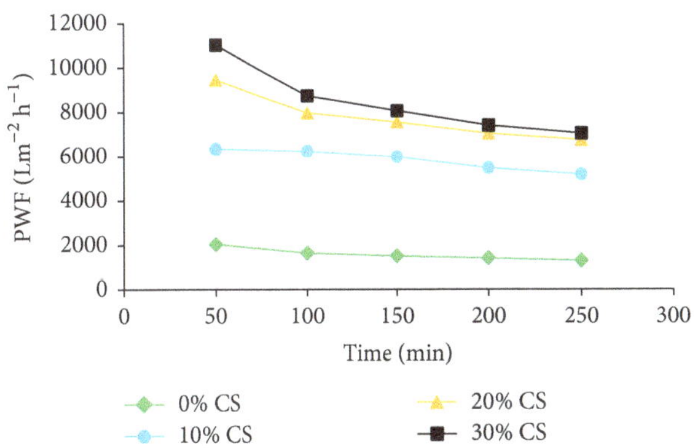

FIGURE 6: Pure water permeation flux with time of Nylon 6/Chitosan nanofiber membrane at room temperature and 1-bar pressure.

Figure 6 shows the behavior of pure water permeation flux with time at room temperature and 1-bar membrane pressure. It was found that the pure water permeation flux of Nylon 6/Chitosan nanofiber fiber membrane decreases with time from 2037.063 to 1293.53 (l/m^2·hr) for 0/100 Nylon 6/Chitosan. Figure 6 shows that flux has been improved as the concentration of Chitosan increased because Chitosan increases the hydrophilicity of the membrane, whereas the relationship between flux and pressure is explained in Figure 7 which represents that when the pressure increased the flux permeates increased, because flux is directly proportional to the pressure drop across the membrane as shown in Figure 7, the same behavior for all membranes as the pressure increased.

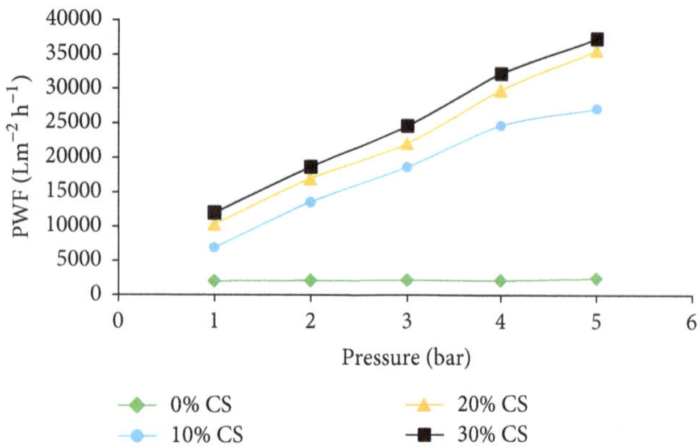

FIGURE 7: Effect of membrane pressure on the pure water permeation flux of the Nylon 6/Chitosan nanofiber membrane.

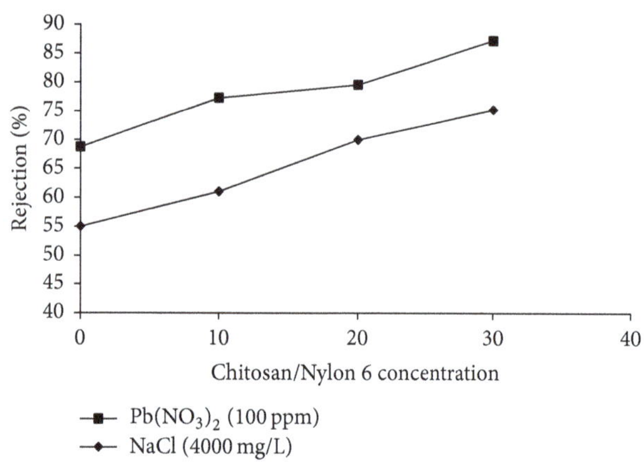

FIGURE 8: Effect of Nylon 6/Chitosan ratio on rejection of $Pb(NO_3)_2$ salt and NaCl Salt aqueous solutions with initial concentration 100 ppm and 4000 mg/L, respectively, at room temperature and 1 bar.

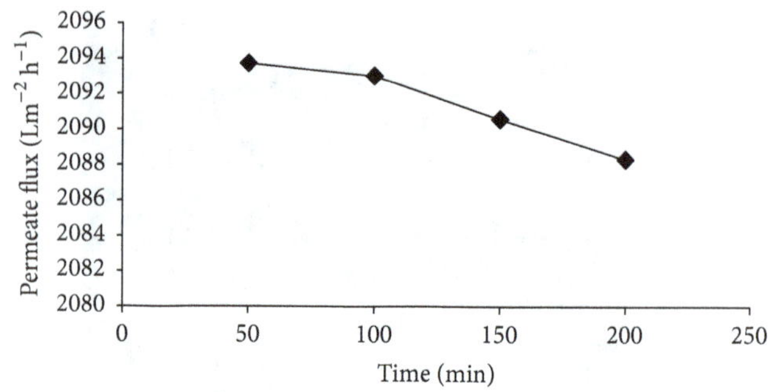

FIGURE 9: Effect of time on flux of water containing 100 ppm ($Pb(NO_3)_2$) for Nylon/10% Chitosan membrane at room temperature and 1 bar.

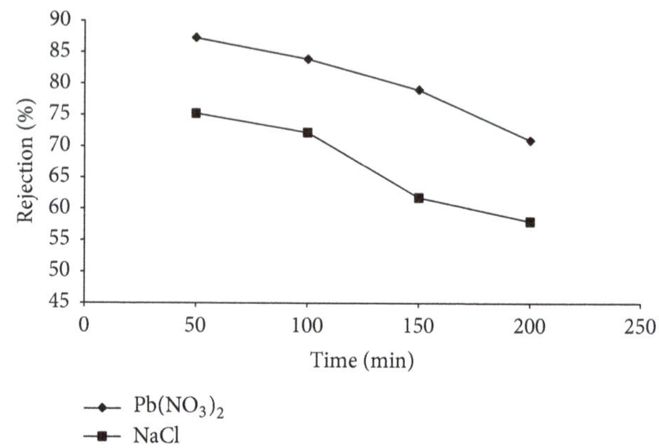

FIGURE 10: Effect of time on rejection (%) of $Pb(NO_3)_2$ and NaCl for Nylon/30% Chitosan membrane at room temperature and 1 bar.

The transport mechanism within membrane can be explained on the basis of the solution diffusion model. According to this model, transport process within the membrane involves three steps: (1) sorption at the surface of the membrane, (2) diffusion into the membrane, and (3) desorption. The hydrophilic nature of Chitosan acts as driving force for sorption of water into membrane [20].

The presence of an amine and hydroxyl groups on Chitosan developed hydrogen bonding and Van der Wall's forces which caused a change in hydrophilicity of blended membranes [20]. Due to increase electrostatic interactions, extensive cross-linkage was developed which helped in the exclusion of salt ions [20]. Chitosan and chitin had been shown to remove metals and dyes so that they can be used to clean water [7]. Various cross-linked polysaccharide materials with Chitosan can remove different metal and dye pollutants [7]. Figure 8 shows the effect of Chitosan on the rejection percentage for $Pb(NO_3)_2$ and NaCl. When the Chitosan concentration had been increased, rejection increases functional groups and electrostatic interaction. In general, increasing Chitosan concentration will increase the expulsion of metal ions due to increase in functional groups and the electrostatic force which kept within electrospun nanofiber during the electrospinning process. Another important factor of electrospun nanofiber membrane high surface area provides excessive adsorption sites for metal ions and dye adsorption and the higher porosity leads to smaller driving forces to push the water through the membrane which makes the process less energy intensive and facile [21]. Chitosan has high contents of amino and hydroxyl functional groups. Owing to these properties, Chitosan is widely used in removal of contamination from wastewater [21].

Among the salts potassium sulphate showed more flux and rejection followed by sodium chloride. This can make it clear with the assistance of ionic size and charge density. The high rejection of K^+ ion reaches the value of 80% for the membrane (PEI) and this is because of the higher ionic radius of K^+ ion, where its value is 157 pm compared with the value of 116 pm for Na^+ [22]. This clarified why the high rejection ratio of Pb^{+2} ion and rejection ratio proportion came up to 87% and Na^{+1} ion up to 75%. Nanofiber adsorption methods show that nanofibers are the best to remove metal ions while having low pressure drops and high water fluxes. Figure 9 shows the effect of time on the flux which decreases as the time increases due to membrane fouling. Figure 10 shows the efficiency of salt ions removal which also decreases with time which indicates the effect of Chitosan decrease with time.

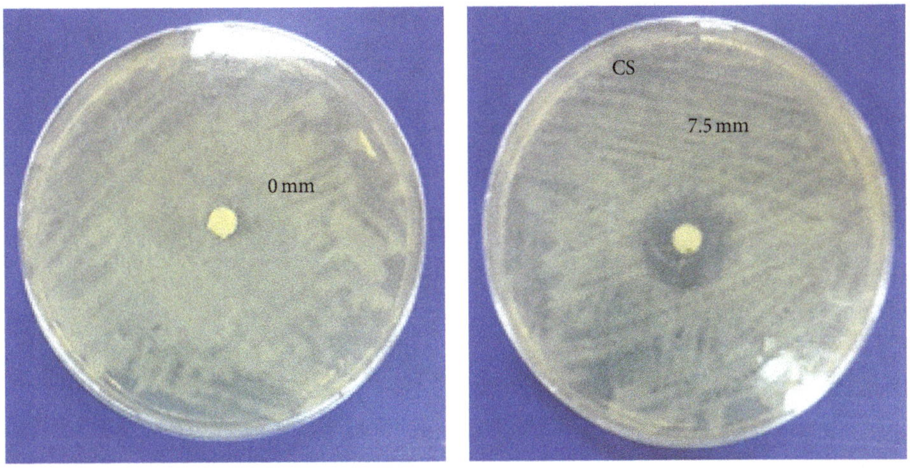

FIGURE 11: The formation inhibition zone around the Chitosan/Nylon 6 membrane after 24 hours and 37°C.

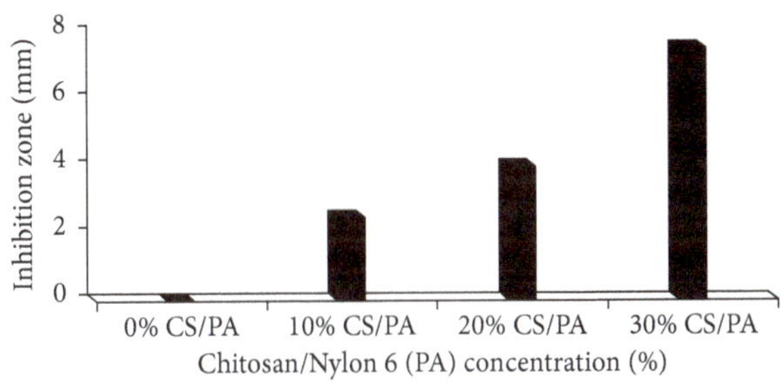

FIGURE 12: Effect of Chitosan/Nylon 6 ratio on inhibition zone against *E. coli.* bacteria.

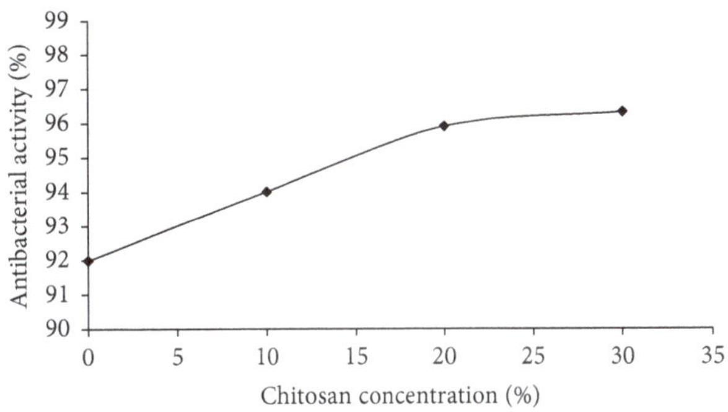

FIGURE 13: Effect of Chitosan addition on the antibacterial activity.

3.3. Antibacterial Activity Results. The Chitosan is known for its antimicrobial activity. Generally, it is accepted to say that the amine group of Chitosan can react with the anionic groups on the bacterial cell surface. Such interaction brings extensive change to the cell surface and the cell permeability [23–25]. Leakages of intracellular substances can be caused by cell permeability leading to cell death. This mechanism has been demonstrated by electron microscopy [26]. When amine groups increase the charge of the Chitosan will increase and that will cause a stronger interaction between Chitosan and cells. Figure 12 shows the effect of Chitosan on *E. coli* bacteria and the measure in the inhibition zone of 30/70-Chitosan/Nylon 6 which reached up to 8 mm and increasing the Chitosan percentage increases the inhibition zone. Figure 11 shows the formation of inhibition zone around the Chitosan/Nylon membrane.

Figure 13 shows the antibacterial activity and when the Chitosan concentration increases the antibacterial activity increases to 96% at 30/70-Chitosan/Nylon ratio. SEM image shows how the electrospun nanofiber captures and prevents the bacteria from penetrating into water, as in Figure 14.

3.4. Mechanical Test Result. The tensile strength of prepared membranes was given in Figures 15 and 16. In this research the mean tensile strength of Nylon membranes was 1.45 MPa. The tensile strength of 10/90-Chitosan/Nylon 6 membrane (2.844 MPa) was slightly higher than that of pure Nylon membrane, the tensile strength of 20/80-Chitosan/Nylon 6 membrane was 4.1 MPa, and 30/70-Chitosan/Nylon 6 membrane was 4.5 MPa. Clearly, Chitosan/Nylon 6 membrane had the highest tensile strength. There is an increase in the tensile strength of Chitosan/Nylon 6 electrospun membrane with the increase in Chitosan percentage.

3.5. Results Statistical Analysis. The standard deviation was calculated from the results obtained in Figures 5, 8, 10, and 11, which can be seen in Table 2.

As can be seen, the results of standard deviation calculations showed that increasing the wt.% of Chitosan led to making water contact angle the most affected result, which made it the most significant result, while the antibacterial activity (%) was the lowest significant result. Standard deviation calculation of rejection (%) for NaCl salts and $Pb(NO_3)_2$ salts gives a different indication as is seen, it gives an opposite indication for rejection (%) for NaCl salts, and this cannot be considered as a final result and the results obtained in Figure 8 are more reasonable because of the difficulty of capturing the ions of NaCl salts comparing with the ions of $Pb(NO_3)_2$ salts. Table 3 illustrates the standard deviation calculation for a

FIGURE 14: SEM of membrane after being exposed to bacteria, where (a) and (b) show the bacteria cluster on the membrane surface as a dendritic shape colony and (c) shows how fiber captures the bacteria.

TABLE 2: Standard deviation calculation of water contact angle, rejection %, inhibition zone against *E. coli.* bacteria, and antibacterial activity (%) results.

Results	Water contact angle (deg.)	Rejection % of NaCl salts (4000 mg/L)	Rejection % of $Pb(NO_3)_2$ salts (100 ppm)	Inhibition zone against *E. coli* bacteria (mm)	Antibacterial activity (%)
Standard deviation value	26.60	9.03	7.61	3.13	1.98

strain of the nanofiber membrane and its strength with the increase of the wt.% of Chitosan.

As it can be seen, the increase of the wt.% of Chitosan to 30% increases the strain of the membrane and also increases its tensile strength which is a desirable efficient factor.

4. Conclusions

Chitosan/Nylon 6 nanofiber membranes were prepared via electrospinning by mixing different amounts of Chitosan with Nylon 6. The addition of Chitosan improves the hydrophilicity and mechanical strength of Nylon 6 nanofiber membrane. The enhanced hydrophilicity of the nanofiber membrane can decrease the antifouling effect to make the mat a potential candidate for water filtration and improve the performance of membrane at low pressures. The Chitosan blended with Nylon 6 membranes showed better bactericidal ability as compared to the neat Nylon 6 membranes. Water filter media with a high rejection ratio against heavy metals ions reached 87% at 30/70-Chitosan/Nylon membrane.

TABLE 3: Standard deviation calculation for tensile strength and strain of the nanofiber membrane.

Results	Membrane strain (%)				Membrane tensile strength (MPa)
	0% CS	10% CS	20% CS	30% CS	
Standard deviation value	0.44	0.84	1.26	1.50	1.37

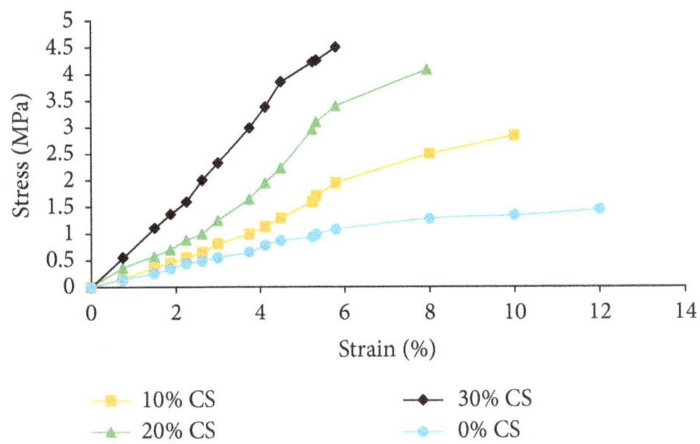

FIGURE 15: Stress-strain curve of Chitosan/Nylon 6 with various ratios.

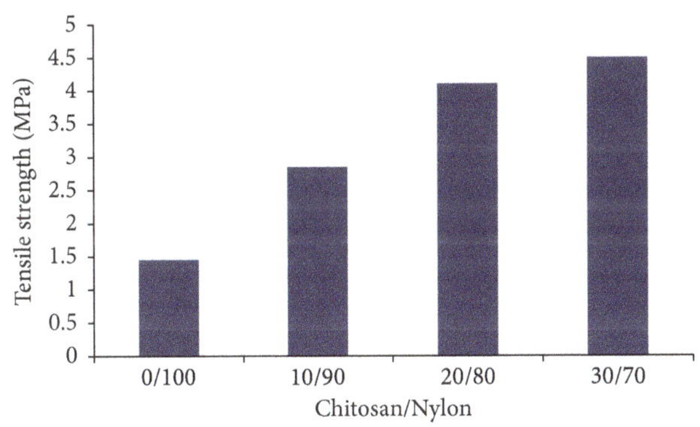

FIGURE 16: Chitosan/Nylon 6 ratio effect in mechanical strength.

Competing Interests

The authors declare that there is no conflict of interests regarding the publication of this paper.

References

[1] M. C. McManus, E. D. Boland, D. G. Simpson, C. P. Barnes, and G. L. Bowlin, "Electrospun fibrinogen: feasibility as a tissue engineering scaffold in a rat cell culture model," *Journal of Biomedical Materials Research Part A*, vol. 81, no. 2, pp. 299–309, 2007.

[2] S. M. Alipour, M. Nouri, J. Mokhtari, and S. H. Bahrami, "Electrospinning of poly(vinyl alcohol)-water-soluble quaternized chitosan derivative blend," *Carbohydrate Research*, vol. 344, no. 18, pp. 2496–2501, 2009.

[3] S. Chuangchote and P. Supaphol, "Fabrication of aligned poly (vinyl alcohol) nanofibers by electrospinning," *Nanoscience and Nanotechnology*, vol. 1, pp. 125–129, 2006.

[4] W. H. Park, L. Jeong, D. I. Yoo, and S. Hudson, "Effect of chitosan on morphology and conformation of electrospun silk fibroin nanofibers," *Polymer*, vol. 45, no. 21, pp. 7151–7157, 2004.

[5] Y. Zhang, X. Huang, B. Duan, L. Wu, S. Li, and X. Yuan, "Preparation of electrospun chitosan/poly(vinyl alcohol) membranes," *Colloid and Polymer Science*, vol. 285, no. 8, pp. 855–863, 2007.

[6] I. Aranaz, M. Mengíbar, R. Harris et al., "Functional characterization of chitin and chitosan," *Current Chemical Biology*, vol. 3, no. 2, pp. 203–230, 2009.

[7] G. Crini, "Recent developments in polysaccharide-based materials used as adsorbents in wastewater treatment," *Progress in Polymer Science*, vol. 30, no. 1, pp. 38–70, 2005.

[8] R. Crisafully, M. A. L. Milhome, R. M. Cavalcante, E. R. Silveira, D. D. Keukeleire, and R. F. Nascimento, "Removal of some polycyclic aromatic hydrocarbons from petrochemical wastewater using low-cost adsorbents of natural origin," *Bioresource Technology*, vol. 99, no. 10, pp. 4515–4519, 2008.

[9] M. Rinaudo, "Chitin and chitosan: properties and applications," *Progress in Polymer Science*, vol. 31, no. 7, pp. 603–632, 2006.

[10] E. Guibal and J. Roussy, "Coagulation and flocculation of dye-containing solutions using a biopolymer (Chitosan)," *Reactive & Functional Polymers*, vol. 67, no. 1, pp. 33–42, 2007.

[11] E. Guibal, M. Van Vooren, B. A. Dempsey, and J. Roussy, "A review of the use of chitosan for the removal of particulate and dissolved contaminants," *Separation Science and Technology*, vol. 41, no. 11, pp. 2487–2514, 2006.

[12] E. Guibal, "Interactions of metal ions with chitosan-based sorbents: a review," *Separation and Purification Technology*, vol. 38, no. 1, pp. 43–74, 2004.

[13] S. Pal, Y. K. Tak, and J. M. Song, "Does the antibacterial activity of silver nanoparticles depend on the shape of the nanoparticle? A study of the gram-negative bacterium *Escherichia coli*," *Applied and Environmental Microbiology*, vol. 73, no. 6, pp. 1712–1720, 2007.

[14] B. Son, B.-Y. Yeom, S. H. Song, C.-S. Lee, and T. S. Hwang, "Antibacterial electrospun chitosan/poly(vinyl alcohol) nanofibers containing silver nitrate and titanium dioxide," *Journal of Applied Polymer Science*, vol. 111, no. 6, pp. 2892–2899, 2009.

[15] P. Danwanichakul, D. Danwanichakul, N. Sueviriyapan, and B. Sumruan, "Nylon 6/chitosan nanofibrous structures for filtration," in *Proceedings of the 1st Mae Fah Luang University International Conference*, pp. 1–8, Pathumthani, Thailand, 2012.

[16] M. T. M. Bizarria, M. A. d'Ávila, and L. H. I. Mei, "Non-woven nanofiber chitosan/PEO membranes obtained by electrospinning," *Brazilian Journal of Chemical Engineering*, vol. 31, no. 1, pp. 57–68, 2014.

[17] J. M. Deitzel, J. Kleinmeyer, D. Harris, and N. C. Beck Tan, "The effect of processing variables on the morphology of electrospun nanofibers and textiles," *Polymer*, vol. 42, no. 1, pp. 261–272, 2001.

[18] D. Rana and T. Matsuura, "Surface modifications for antifouling membranes," *Chemical Reviews*, vol. 110, no. 4, pp. 2448–2471, 2010.

[19] H. Zhang, S. Li, C. J. Branford White, X. Ning, H. Nie, and L. Zhu, "Studies on electrospun nylon-6/chitosan complex nanofiber interactions," *Electrochimica Acta*, vol. 54, no. 24, pp. 5739–5745, 2009.

[20] S. Waheed, A. Ahmad, S. M. Khan et al., "Synthesis, characterization, permeation and antibacterial properties of cellulose acetate/polyethylene glycol membranes modified with chitosan," *Desalination*, vol. 351, pp. 59–69, 2014.

[21] M. Ghani, A. A. Gharehaghaji, M. Arami, N. Takhtkuse, and B. Rezaei, "Fabrication of electrospun polyamide-6/chitosan nanofibrous membrane toward anionic dyes removal," *Journal of Nanotechnology*, vol. 2014, Article ID 278418, 12 pages, 2014.

[22] A. Sumisha, G. Arthanareeswaran, A. F. Ismail, D. P. Kumar, and M. V. Shankar, "Functionalized titanate nanotube-polyetherimide nanocomposite membrane for improved salt rejection under low pressure nanofiltration," *RSC Advances*, vol. 5, no. 49, pp. 39464–39473, 2015.

[23] N. R. Sudarshan, D. G. Hoover, and D. Knorr, "Antibacterial action of chitosan," *Food Biotechnology*, vol. 6, no. 3, pp. 257–272, 1992.

[24] S. W. Fang, C. F. Li, and D. Shih, "Antifungal activity of chitosan and its preservative effect on low-sugar candied kumquat," *Food Protect*, vol. 56, no. 2, pp. 136–140, 1994.

[25] J. K. Hwang, H. J. Kim, S. J. Yoon, and Y. R. Pyun, "Bactericidal activity of chitosan on *E.coli*," *Advances in Chitin Science*, vol. 62, no. 3, pp. 239–243, 1999.

[26] I. M. Helander, E.-L. Nurmiaho-Lassila, R. Ahvenainen, J. Rhoades, and S. Roller, "Chitosan disrupts the barrier properties of the outer membrane of Gram-negative bacteria," *International Journal of Food Microbiology*, vol. 71, no. 2-3, pp. 235–244, 2001.

10

Ionothermal Synthesis of a Novel 3D Cobalt Coordination Polymer with a Uniquely Reported Framework: $[BMI]_2[Co_2(BTC)_2(H_2O)_2]$

Il-Ju Ko,[1,2] Hyun-Chang Oh,[1,2] Yong-Jun Cha,[1,2] Chae Hyeok Han,[1,2] and Eun-Young Choi[1]

[1]*Department of Chemistry and Biology, Korea Science Academy of KAIST, Busan, Republic of Korea*
[2]*Department of Chemistry, KAIST, Daejeon, Republic of Korea*

Correspondence should be addressed to Eun-Young Choi; faujasite@kaist.ac.kr

Academic Editor: Peter Majewski

The framework of $[RMI]_2[Co_2(BTC)_2(H_2O)_2]$ (RMI = 1-alkyl-3-methylimidazolium, alkyl; ethyl (EMI); propyl (PMI); butyl (BMI)), which has uniquely occurred in ionothermal reactions of metal salts and H_3BTC (1,3,5-benzenetricarboxylic acid), an organic ligand, reappeared in this work. Ionothermal reaction of cobalt acetate and H_3BTC with [BMI]Br ionic liquid as the reaction medium yielded the novel coordination polymer $[BMI]_2[Co_2(BTC)_2(H_2O)_2]$ (compound **B2**). Similar ionothermal reactions with different [EMI]Br and [PMI]Br as the reaction media have been previously reported to produce $[EMI]_2[Co_3(BTC)_2(OAc)_2]$ (compound **A1**) and $[PMI]_2[Co_2(BTC)_2(H_2O)_2]$ (compound **B1**), respectively. In contrast with the trinuclear secondary building unit of **A1**, the framework structure of **B1** and **B2** consists of dinuclear secondary building units in common, but with subtle distinction posed by the different size of the incorporated cations. These structural differences amidst the frameworks showed interesting aspects, including guest and void volume, and were used to explain the chemical trend observed in the system. Moreover, the physicochemical properties of the newly synthesized compound have been briefly discussed.

1. Introduction

The rising demands for new materials with befitted properties for applications in numerous fields of science and industry [1–3] have led to development and advancements of novel synthesis methods [4–10]. Ionothermal synthesis, where ionic liquids are used both as solvents and templates [11], has been recently deployed for preparation of novel structures [12, 13] amidst a great attention attributed to its unique and pragmatic physicochemical properties [11], namely, those discussed hereafter and many more [14–16]: high thermal stability [17], negligible vapor pressure [18], environmentally friendliness [19], and, most interestingly, its ability to finely tune the reaction environment by changing the cation and anion of the ionic liquid and ultimately enabling systematic approach to the final product [20, 21]. Despite such attractiveness of ionothermal synthesis and its expandability that has already been demonstrated with novel structures of various classes including zeolites [12, 13], zeotype frameworks [12, 22–24], and metal-organic frameworks or MOFs [20, 21, 25], the methodology has not been extensively employed [21], leaving a vast number of different reaction conditions for future research. The unique advantages of ionothermal synthesis may be fully practiced only when enough experiments have been carried out with various circumstances to produce a sizable number of frameworks.

We herein report a novel 3D cobalt coordination polymer $[BMI]_2[Co_2(BTC)_2(H_2O)_2]$ (compound **B2**), which has been ionothermally prepared with [BMI]Br as the solvent. Its framework is very rare in literature, previously being only uniquely exemplified by $[PMI]_2[Co_2(BTC)_2(H_2O)_2]$ (compound **B1**), a polymer obtained from a reaction analogous to ours, but with [PMI]Br as the solvent [26]. On the contrary, another analogous reaction with [EMI]Br as the solvent yielded a polymer with a notably different framework, $[EMI]_2[Co_3(BTC)_2(OAc)_2]$ (compound **A1**) [27]. Moreover, attempts have been made to synthesize novel polymers with [PEMI]Br and [HMI]Br and acquired

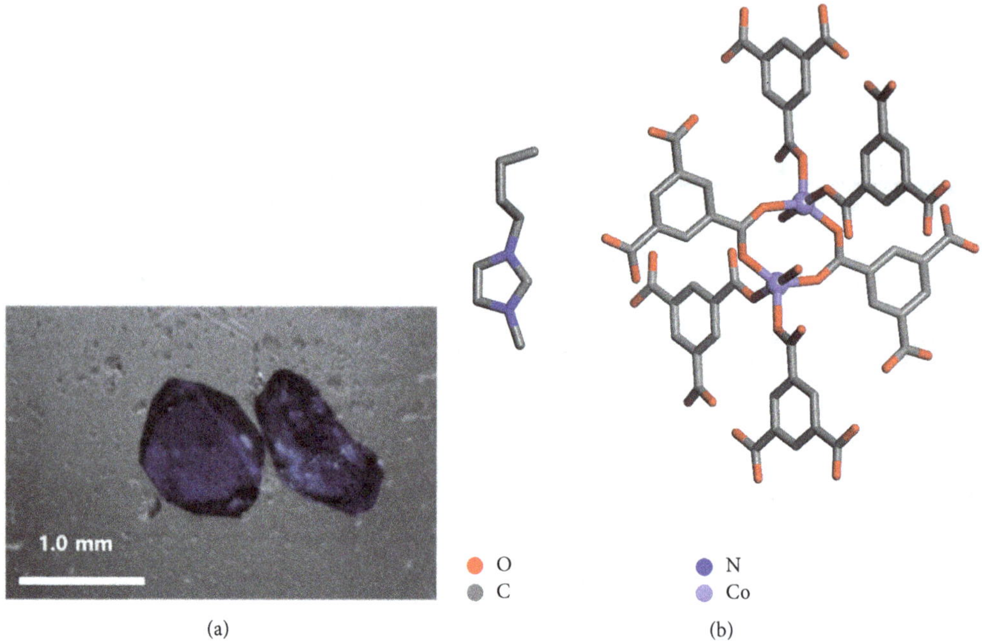

Figure 1: (a) Crystal image of synthesized $[BMI]_2[Co_2(BTC)_2(H_2O)_2]$ crystals. (b) Molecular models depicting the incorporated cations and the secondary building units (SBUs) of $[BMI]_2[Co_2(BTC)_2(H_2O)_2]$.

pink amorphous solids. In order to provide explanations to the trend introduced above, the framework of the novel compound $[BMI]_2[Co_2(BTC)_2(H_2O)_2]$ (compound **B2**) will be thenceforth examined from various aspects in relation to that of $[EMI]_2[Co_3(BTC)_2(OAc)_2]$ (compound **A1**) and $[PMI]_2[Co_2(BTC)_2(H_2O)_2]$ (compound **B1**).

2. Experimental

2.1. Crystal Preparation. For the preparation of $[BMI]_2[Co_2(BTC)_2(H_2O)_2]$ (compound **B2**), $Co(OAc)_2 \cdot 4H_2O$ (3.0 mmol, 0.7472 g) and H_3BTC (2.0 mmol, 0.4203 g) were placed altogether in a 23 mL Teflon-lined stainless-steel autoclave with [BMI]Br (20 mmol, 4.3824 g). The mixture was gradually heated in a furnace for 3 hours to reach the reaction temperature of 150°C. The temperature was held still for 3 days and slowly cooled to room temperature for 5 hours. The obtained deep blue crystals (Figure 1(a)) were filtered, washed with ethanol, and naturally dried for purification. All chemicals including ionic liquids (ILs) were commercially purchased from Chem. Tech. Research Incorporation (C-TRI) and Sigma-Aldrich Chemical Company and used without further purification.

2.2. Crystal Structure Analysis. Crystallographic data of obtained crystals were collected at −100°C on Bruker Smart Breeze diffractometer in different locations including Pusan National University. Summary of crystallographic data for compound **B2** are given in Table 1 and further details of the crystal structure have been deposited at the Cambridge Crystallographic Data Center as supplementary publication (CCDC-1525180). For crystallographic comparison, summarized data for compound **A1** and compound **B1** are also given in Table 1.

Powder X-ray diffraction (PXRD) data were recorded on a Rigaku Miniflex 600 diffractometer at 40 kV, 15 mA for Cu-$K\alpha$ radiation ($\lambda = 1.5406$ Å) with a scan speed of 5°/min, and a step size of 0.01° in 2θ.

3. Result and Discussion

The framework of compound **B2** imputes its understructure to the dinuclear unit built by connecting two cobalt atoms with six BTC^{3-} ligands as illustrated in Figure 1(b). The fundamental asymmetric unit of compound **B2** contains one cobalt (II) atom, one BTC^{3-} ligand, one [BMI] cation, and a water molecule (Table 1). Each cobalt (II) atom is five-coordinated with four oxygen atoms from carboxylate groups of BTC^{3-} ligands and one from a water molecule in a distorted trigonal bipyramid geometry. BTC^{3-} ligands exhibit μ_4 coordination modes: two of the carboxylates exhibit monodentate coordination fashion and the remaining bidentate coordination fashion. Two symmetric cobalt (II) atoms are linked by two carboxylate groups exhibiting bidentate coordination mode, each from different BTC^{3-} ligands and form the $[Co_2(\mu_2\text{-}COO)_2]$ core as in compound **B1** [26] (Figure 1(b)). Nevertheless, the distance between two neighboring cobalt (II) atoms is 3.53 Å, which is slightly shorter than that of compound **B1**, 3.56 Å [26], but such an arrangement with a long distance of 3.53 Å still eradicates the possibility of direct interaction between them. The range of Co-O distance was measured to span from 1.9830(13) Å to 2.1314(14) Å (see Table 2), exhibiting the similar range of compound **B1** [26]. The 3D framework of compound **B2**,

TABLE 1: Crystallographic data for Co-BTC metal organic frameworks synthesized with [RMI]Br ionic liquids (RMI = EMI, PMI, BMI).

	$[EMI]_2[Co_3(BTC)_2(OAc)_2]$	$[PMI]_2[Co_2(BTC)_2(H_2O)_2]$	$[BMI]_2[Co_2(BTC)_2(H_2O)_2]$
Empirical formula	$C_{22}H_{12}Co_3O_{16}$	$C_{16}H_{18}CoN_2O_7$	$C_{17}H_{20}CoN_2O_7$
Fw	709.11	409.26	423.28
Space group	Pbca	P2(1)/n	P2(1)/n
Crystal system	Orthorhombic	Monoclinic	Monoclinic
A (Å)	14.2616(4)	11.279(10)	11.1946(5)
B (Å)	16.1525(5)	10.679(10)	10.7911(5)
C (Å)	16.4867(5)	15.428(14)	15.3026(8)
α (deg)	90.00	90.00	90.00
β (deg)	90.00	95.276(14)	95.3550(10)
γ (deg)	90.00	90.00	90.00
Cell volume	3797.9(2)	1850.406	1840.52(15)
Z	4	4	4
T (K)	90(2)	273	173(2)
R factor	0.0323	0.0522	0.0183
Reference	[27]*	[26]*	This work

*The reference in the article.

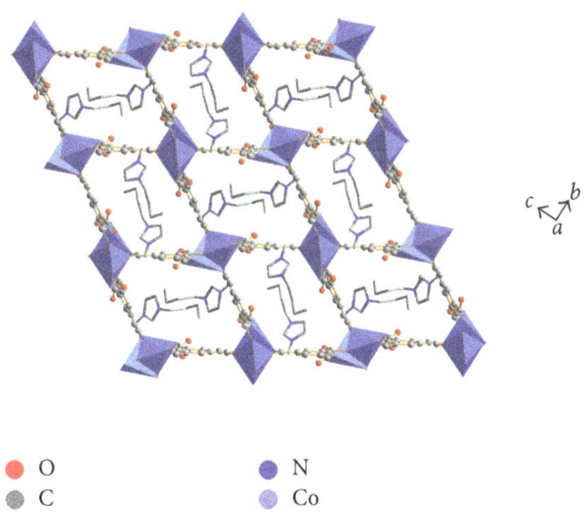

FIGURE 2: The 3D framework of $[BMI]_2[Co_2(BTC)_2(H_2O)_2]$.

given by connecting cobalt (II) core with six BTC^{3-} ligands, has rhombic channels parallel to the crystallographic *a* axe (Figure 2). [BMI] cations of ionic liquids residing in the open channels as shown in Figure 2 behave as charge balancing species for the anionic framework.

The difference between type A framework, a trinuclear system, and type B frameworks, dinuclear system, becomes evident at their secondary building units, visually depicted at Figure S1 in Supplementary Material available online at https://doi.org/10.1155/2017/3237247, along with the subtleness in distinction between compound **B1** and compound **B2** [26, 27]. However, the [RMI] molecules were present as guests in all frameworks, suggesting a role played by the alkyl chains of differing lengths in determination of framework types.

To further account for the effect of guest molecules, Connolly surfaces and volumes have been calculated for their in situ conformations and for the cation pairs in the form of which the guests occur [28] (Figure 3). Note that the [EMI] cations are so severely disordered in compound **A1** that reliable conformation data could not be obtained and therefore, data for [EMI] have been imported from isoreticular correspondent prepared with nickel in place of cobalt (CCDC number 719778) for better depiction of the trend. It has been discovered that the butyl chain of [BMI] molecule is considerably bent, making the volume of the molecule only slightly larger than [PMI] molecule compared to the increase in volume from [EMI] to [PMI]; the distance from the first carbon of the alkyl chain to the final carbon was 2.918 Å in compound **B2**, which is only slightly larger than 2.567 Å of compound **B1**. This indicates that the type B framework is stable for the cobalt-BTC system strong enough to cause the severe bending in the butyl chain of the guest molecule whilst the increase of guest volume from [EMI] to [PMI] posed the boundary between framework A and B. Nevertheless, the scarcity of the framework in systems of other metal elements suggests such a favor is limited specifically to the cobalt system (see Table S1).

It is apparent that increasing guest volume will exert a pressure on the framework to the direction of expanding its free volume. Occupied volume and free volume have been calculated for the three compounds and yielded a result corresponding to the expectation, but only partially (Figure S2). The free volume ratio was determined to be nearly the same for compounds **B1** and **B2**, suggesting the rigidness of the framework with respect to the change of free volume ratio. This result accords with the bent conformation of the [BMI] cation in compound **B2** and the failure to produce crystalline frameworks in reactions using [PEMI]Br and [HMI] in place of [BMI], respectively; this framework remains invariant over a range of synthesizing environment and such invariance may at minimum be partially attributed to the absence of any other framework that could rise from this range of synthesizing environment. Figure 4 summarizes the data regarding the occupied volume, the free volume, the guest volume, and the volume of the cation pairs that the guests found in structure.

Table 2: Selected bond lengths [Å] and bond angles [deg] of $[BMI]_2[Co_2(BTC)_2(H_2O)_2]$.

Co(1)-O(2)#1	1.9830(13)	O(2)#1-Co(1)-O(5)#3	91.75(6)
Co(1)-O(1)	1.9904(14)	O(1)-Co(1)-O(5)#3	93.06(6)
Co(1)-O(3)#2	2.0116(13)	O(3)#2-Co(1)-O(5)#3	89.12(6)
Co(1)-O(5)#3	2.0524(13)	O(2)#1-Co(1)-O(7)	84.43(6)
Co(1)-O(7)	2.1314(14)	O(1)-Co(1)-O(7)	92.24(6)
O(2)#1-Co(1)-O(1)	138.72(6)	O(3)#2-Co(1)-O(7)	89.40(6)
O(2)#1-Co(1)-O(3)#2	117.73(6)	O(5)#3-Co(1)-O(7)	174.69(6)
O(1)-Co(1)-O(3)#2	103.32(6)		

Symmetry transformations: #1: $-x+2, -y+1, -z$; #2: $-x+3, y+1/2, -z+1/2$; #3: $x-1/2, -y+1/2, z-1/2$.

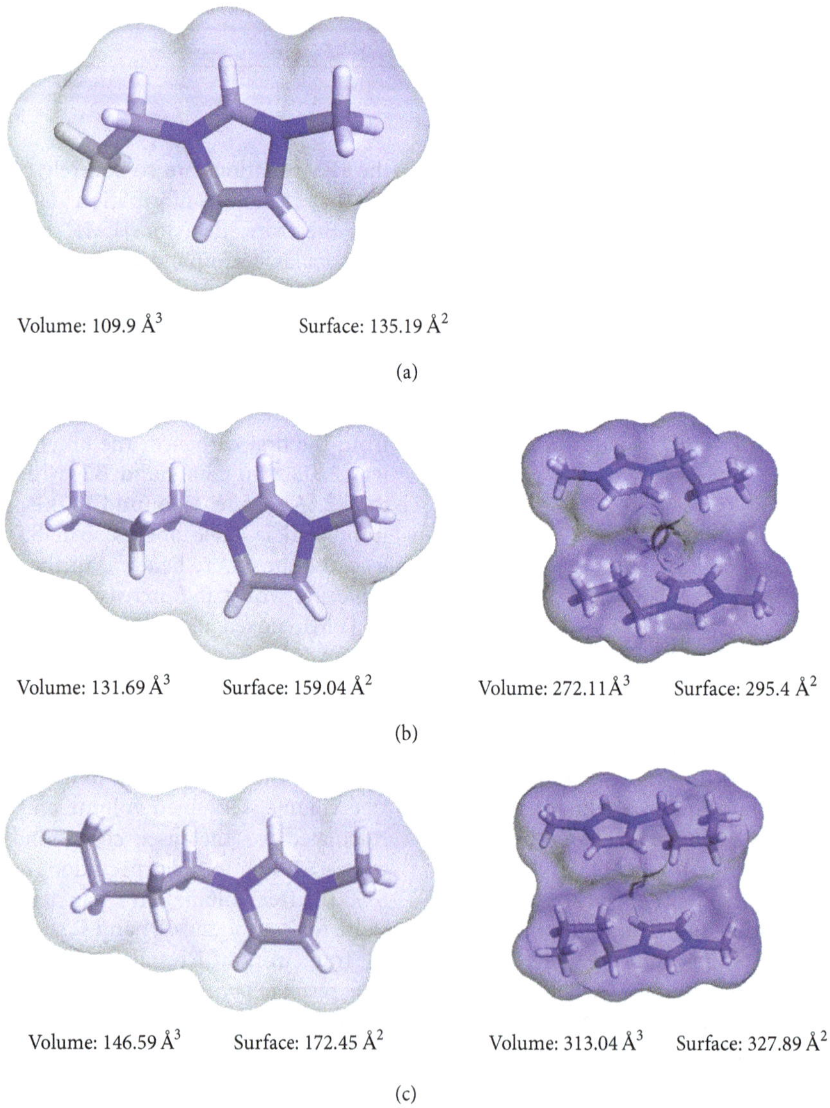

Figure 3: Connolly surfaces of the incorporated cations in their in situ conformations (left) and the cation pairs (right): (a) 1-ethyl-3-methylimidazolium, with data borrowed from the isoreticular correspondent prepared with nickel in place of cobalt, (b) 1-propyl-3-methylimidazolium, and (c) 1-butyl-3-methylimidazolium [28]. (Volume and surface area values have been calculated with the grid interval of 0.15 Å and the Connolly radius being 1.0 Å.)

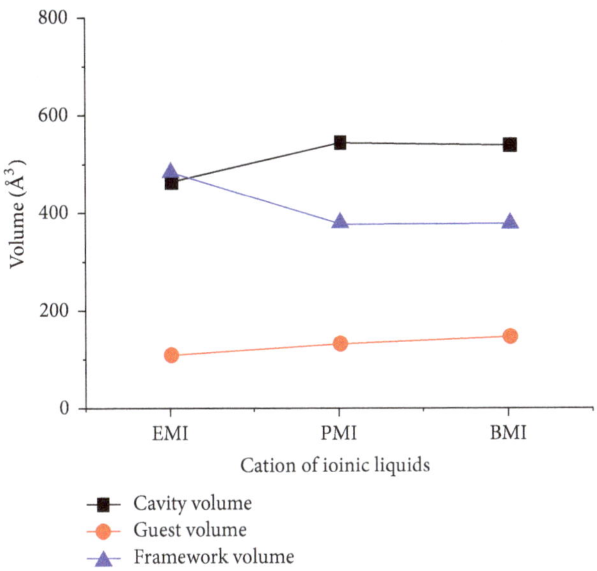

FIGURE 4: Plots of various calculated volumes per cation in each of the ionic liquids, where the data for guest and guest pair volume for EMI have been imported from the isoreticular correspondent prepared with nickel in place of cobalt for better illustration of the increasing trend of the guests.

Lastly, physicochemical properties of compound **B2** have been explored from a variety of scopes including those that have not been practiced for compound **B1** upon its synthesis. Several analyses were additionally performed to confirm the incorporation of ligands and metals. Infrared absorption spectroscopy has been practiced for the crystals and compared with H_3BTC to provide additional proof for the incorporation of BTC^{3-} ligands in the framework (Figure S3). The carboxylic C=O peak in Figure S3 is shifted from 1691 cm^{-1} of H_3BTC to 1616 cm^{-1} of compound **B2**, which supports the coordination of H_3BTC ligands to the cobalt (II) atoms. Powder X-ray diffraction pattern of the remnants of the crystals after thermogravimetric analysis (TGA) exactly matches that of cobalt oxide (Co_3O_4), proving that the incorporated metal is indeed cobalt (Figure S4). Moreover, the crystals of compound **B2** have been applied to various solvents including water, DMF, ethanol, and acetone, to further test its properties (Figure S5). The crystals remained stable under all solvents, except for water, in which a framework change was observed along with immediate color change to pink (Figure S6). The SEM image confirmed the complete change in framework structure of compound **B2** upon treatment with water (Figure S7). Additionally, powder X-ray diffraction pattern of the crystals after application of water has been compared with that of the prior state, which was discovered to be considerably different and strongly supported the change of framework in compound **B2** under water (Figure S8). To trace the direct cause of the framework change, energy dispersive spectroscopy has been performed for the crystals treated with water and the nitrogen atoms was revealed to be absent after addition of water. Since the nitrogen atoms are only included in the [BMI] cations, the framework change has been suggested to be accompanied by the evacuation of [BMI] guest molecules (Figure S9). Conceding further research taken, presumably along with the relative stability observed under other common solvents, this phenomenon, the evacuation of cation molecules out of the framework structure, may open a possibility for the ionothermal methodology in preparing coordination polymers with a cation exchange property. Lastly, thermogravimetric analysis (TGA) of compound **B2** has confirmed the framework stable up to 352°C as shown in Figure S4(a), raising the potential of the compound for industrial applications.

4. Conclusions

We have reported a novel coordination polymer $[BMI]_2[Co_2(BTC)_2(H_2O)_2]$ (compound **B2**) and compared it within the system of polymers arising from ionothermal reaction of $Co(OAc)_2$ and BTC mediated by the ionic solvent [RMI]Br. Resident inside the cavities, the cations have been expected to directly relate their sizes to the volume of the cavities. However, the accuracy of this assumption was curtailed by structural analysis of compound **B2** along with the discovery of significant bending in the alkyl chain of the BMI cations. Moreover, thermal stability and removability of the incorporated cations exhibited by compound **B2** further open the feasibility of implementing the ionothermal methodology for production of new materials with demanded properties.

Conflicts of Interest

There are no conflicts of interest regarding the publication of this paper.

Acknowledgments

This work has been supported by the Korea Science Academy of KAIST from the Ministry of Science, ICT and Future Planning. The authors also acknowledge Busan Metropolitan Office of Education for financial support.

References

[1] M. E. Davis, "Ordered porous materials for emerging applications," *Nature*, vol. 417, no. 6891, pp. 813–821, 2002.

[2] A. K. Cheetham, C. N. R. Rao, and R. K. Feller, "Structural diversity and chemical trends in hybrid inorganic-organic framework materials," *Chemical Communications*, no. 46, pp. 4780–4795, 2006.

[3] P. Wheatley, A. Butler, M. Crane et al., "NO-Releasing Zeolites and Their Antithrombotic Properties," *Journal of the American Chemical Society*, vol. 128, no. 2, pp. 502–509, 2006.

[4] S. T. Wilson, B. M. Lok, C. A. Messina, T. R. Cannan, and E. M. Flanigen, "Aluminophosphate molecular sieves: a new class of microporous crystalline inorganic solids," *Journal of the American Chemical Society*, vol. 104, no. 4, pp. 1146-1147, 1982.

[5] G. Férey, C. Mellot-Dranznieks, C. Serre et al., "A chromium terephthalate-based solid with unusually large pore volumes and surface area," *Science*, vol. 309, no. 5743, pp. 2040–2042, 2005.

[6] H. Li, M. Eddaoudi, M. O'Keeffe, and O. M. Yaghi, "Design and synthesis of an exceptionally stable and highly porous metal-organic framework," *Nature*, vol. 402, no. 6759, pp. 276–279, 1999.

[7] G. B. Hix, V. J. Carter, D. S. Wragg, R. E. Morris, and P. A. Wright, "The synthesis and modification of aluminium phosphonates," *Journal of Materials Chemistry*, vol. 9, no. 1, pp. 179–185, 1999.

[8] V. J. Carter, P. A. Wright, J. D. Gale, R. E. Morris, E. Sastre, and J. Perez-Pariente, "AlMePO-β: Inclusion and thermal removal of structure directing agent and the topotactic reconstructive transformation to its polymorph AlMePO-α," *Journal of Materials Chemistry*, vol. 7, no. 11, pp. 2287–2292, 1997.

[9] C. S. Cundy and P. A. Cox, "The hydrothermal synthesis of zeolites: history and development from the earliest days to the present time," *Chemical Reviews*, vol. 103, no. 3, pp. 663–701, 2003.

[10] R. E. Morris and S. J. Weigel, "The synthesis of molecular sieves from non-aqueous solvents," *Chemical Society Reviews*, vol. 26, no. 4, pp. 309–317, 1997.

[11] E. R. Parnham and R. E. Morris, "Ionothermal synthesis of zeolites, metal-organic frameworks, and inorganic-organic hybrids," *Accounts of Chemical Research*, vol. 40, no. 10, pp. 1005–1013, 2007.

[12] P. Wasserschied and T. Welton, *Ionic Liquids in Synthesis*, vol. 2, chapter 2, Wiley-VCH, Weinheim, Germany, 2003.

[13] E. R. Cooper, C. D. Andrews, P. S. Wheatley, P. B. Webb, P. Wormald, and R. E. Morris, "Ionic liquids and eutectic mixtures as solvent and template in synthesis of zeolite analogues," *Nature*, vol. 430, no. 7003, pp. 1012–1016, 2004.

[14] D. Camper, P. Scovazzo, C. Koval, and R. Novel, "Gas solubilities in room-temperature ionic liquids," *Industrial & Engineering Chemistry Research*, vol. 43, pp. 3049–3054, 2004.

[15] K. E. Gutowski, G. A. Broker, H. D. Willauer et al., "Controlling the aqueous miscibility of ionic liquids: aqueous biphasic systems of water-miscible ionic liquids and water-structuring salts for recycle, metathesis, and separations," *Journal of the American Chemical Society*, vol. 125, no. 22, pp. 6632-6633, 2003.

[16] J. Hoffmann, M. Nuchter, N. Ondruschka, and P. Wasserscheid, *Green Chemistry*, vol. 5, pp. 296–299, 2003.

[17] H. L. Ngo, K. LeCompte, L. Hargens, and A. B. McEwen, "Thermal properties of imidazolium ionic liquids," *Thermochimica Acta*, vol. 357-358, pp. 97–102, 2000.

[18] M. J. Earle, J. M. S. S. Esperança, M. A. Gilea et al., "The distillation and volatility of ionic liquids," *Nature*, vol. 439, no. 7078, pp. 831–834, 2006.

[19] R. D. Rogers and K. R. Seddon, "Ionic liquids—solvents of the future?" *Science*, vol. 302, no. 5646, pp. 792-793, 2003.

[20] L. Xu, E.-Y. Choi, and Y.-U. Kwon, "Combination effects of cation and anion of ionic liquids on the cadmium metal-organic frameworks in ionothermal systems," *Inorganic Chemistry*, vol. 47, no. 6, pp. 1907–1909, 2008.

[21] L. Xu, E.-Y. Choi, and Y.-U. Kwon, "A new 3D nickel(II) framework composed of large rings: Ionothermal synthesis and crystal structure," *Journal of Solid State Chemistry*, vol. 181, no. 11, pp. 3185–3188, 2008.

[22] E. R. Cooper, C. D. Andrews, P. S. Wheatley, P. B. Webb, P. Wormald, and R. E. Morris, "A new methodology for zeolite analogue synthesis using ionic liquids as solvent and template," *Studies in Surface Science and Catalysis*, vol. 158, pp. 247–254, 2005.

[23] E. R. Parnham, E. A. Drylie, P. S. Wheatley, A. M. Z. Slawin, and R. E. Morris, "Ionothermal materials synthesis using unstable deep-eutectic solvents as template-delivery agents," *Angewandte Chemie - International Edition*, vol. 45, no. 30, pp. 4962–4966, 2006.

[24] E. R. Parnham and R. E. Morris, "The ionothermal synthesis of cobalt aluminophosphate zeolite frameworks," *Journal of the American Chemical Society*, vol. 128, no. 7, pp. 2204-2205, 2006.

[25] L. Xu, E.-Y. Choi, and Y.-U. Kwon, "Ionothermal syntheses of six three-dimensional zinc metal-organic frameworks with 1-alkyl-3-methylimidazolium bromide ionic liquids as solvents," *Inorganic Chemistry*, vol. 46, no. 25, pp. 10670–10680, 2007.

[26] Y.-L. Wang, N. Zhang, Q.-Y. Liu et al., "Ionothermal syntheses and crystal structures of two cobalt(II)-carboxylate compounds with different topology," *Inorganic Chemistry Communications*, vol. 14, no. 2, pp. 380–383, 2011.

[27] Z. Lin, D. S. Wragg, and R. E. Morris, Chem. Comm., 2006.

[28] Program for Molecular Modeling and Analysis, Accelrys Software Inc., 2014.

11

New Aspects in the Mechanical Behavior of a Polycarbonate Found by an Experimental Study

Alfredo Alán Rey Calderón and Alberto Díaz Díaz

Centro de Investigación en Materiales Avanzados, S. C. Miguel de Cervantes 120, Complejo Industrial Chihuahua, 31136 Chihuahua, Chihuahua, Mexico

Correspondence should be addressed to Alberto Díaz Díaz; alberto.diaz@cimav.edu.mx

Academic Editor: Jose M. Cabrera

The aim of this paper was to analyze in detail the mechanical behavior of a polycarbonate by means of uniaxial tensile and compressive tests and to reveal new key aspects that must be taken into account in any predictive model. Uniaxial monotonic and creep-recovery tests were carried out at a variety of temperatures, stress levels, and load rates to get a complete description of the material response. Prior to mechanical testing, the material was subjected to a thermal rejuvenation in order to eliminate any previous aging and to obtain reliable and useful results. In every test, a complete determination of the strain state was assured by measuring axial and transverse strains with strain gauges. During the tests, significant asymmetry effects and viscous phenomena already reported by other authors were confirmed. The newest finding is that a nonlinear master transverse strain/axial strain curve matches perfectly with the experimental curves. This master curve is temperature- and rate-independent. Another originality of this paper is the disclosure of an instantaneous, hypoelastic-like behavior at high strain rates. The experimental observations presented in this study should be incorporated by a theoretical model whose aim is to accurately predict the mechanical behavior of polycarbonate subjected to any 3D stress state.

1. Introduction

A good mechanical design must offer a reliable performance which requires an extensive knowledge about how the material behaves. Many theoretical and experimental studies focused on the nonlinear behavior of polymers have been published. It is well known that mechanical behavior in polymers is complex: there is a dependency on strain rate [1–5]; temperature impacts the mechanical response [6–9]; moreover, hydrostatic stress affects mechanical behavior [10].

In most papers, the mechanical behavior of polymers has been studied under uniaxial stress states with a unique sign: tension or compression. Nevertheless, as Shaban states [11], only one type of experiment, commonly tension, is not enough to determine how the material behaves in different load settings. Several types of independent experiments, for example, compression, shear, and hydrostatic pressure, must be done in order to obtain a more complete characterization of materials. For example, McKenna and his coworkers [12, 13] studied the nonlinear viscoelastic behavior of specimens of some glassy polymers subjected to torsional tests in which the distance between the grips was fixed, in such a manner that a normal axial force was provoked. The derivatives of the strain energy density with respect to first and second invariants of the deformation tensor were obtained in an attempt to have a better understanding of the material behavior. In [14, 15], Knauss and Zhu carried out creep tests on a polycarbonate with different normal to shear stress ratios by means of an Arcan device and proved that the magnitude and sign of the hydrostatic stress have a very significant effect on creep behavior. An asymmetry effect occurs then in polycarbonate, and compressive tests are required to characterize the material behavior. Compressive tests that generate a quasiuniform uniaxial stress state may be hard to perform because of potential buckling and barreling effects, particularly for large strains. For this reason, Ravi-Chandar and Ma [16, 17] proposed original compressive tests with confined specimens in order to remove the aforementioned effects.

Another important aspect in the characterization of polymers that has been explored by several authors is the material response to multiple step loads [18–20]. When dealing with nonlinear materials, the application of a superposition principle is not valid. For polymer melts, it is well known that tests with reversing double step strain flows represent a challenge for those models willing to predict accurately the stress evolution; in these tests, "non-instantaneous chain retraction has complicated effects on the relaxation of stress" [19]. For polymer glasses, the understanding of the material response to multiple step loads is also necessary. McKenna and Zapas [12] carried out two-step relaxation tests with poly(methyl methacrylate) samples and adapted a model (the BKZ model [21]) to tests its accuracy in predicting the stress evolution. The model was able to predict the material behavior only if the material was subjected to previous mechanical conditioning; the model did not take into account the mechanical aging.

On the other hand, in some papers dealing with the modeling of the mechanical behavior of polymers, a constant Poisson's ratio is assumed. Apart from convenience, this may be due to the fact that many experimental studies dealing with mechanical tests on polymers do not include the measurement of transverse strains. In fact, for viscoelastic materials, Poisson's ratio is usually not constant and several authors have proved this for polymers [22–25]. For example, Tscharnuter et al. showed that for polypropylene, this ratio is rate- and temperature-dependent [22]. Pandini and coworkers [23, 24] proved that Poisson's ratios of poly (butylene terephthalate) and epoxy resins are time-dependent. Additionally, in [25], Tschoegl et al. consider that in cyclic tests, Poisson's ratio of polymers depends on frequency. Mallon et al. [26] performed creep tensile and compressive tests on PMMA, PVC, and PP polymers and proved that Poisson's ratios vary in time and depend on the sign of the stress applied.

In this document, mechanical response of a polycarbonate (PC) subjected to uniaxial tensile and compressive loads is studied. The mechanical tests carried out are monotonic and creep-recovery tests at different temperatures. As mentioned before, the recovery stage in these last tests is a second load step that is useful to have a more complete characterization of PC. Prior to mechanical tests, specimens are rejuvenated by means of a thermal process. In every test, axial and transverse strains were measured in order to have a complete description of the strain state. In these tests, it can be observed that the viscous Poisson's ratio increases in compression, but remains virtually constant in tension. In fact, a nonlinear master transverse strain/axial strain curve is found. This result is one originality of the present paper. Another originality of this paper is the disclosure of an instantaneous, hypoelastic-like behavior at high strain rates. The experimentation including tensile and compressive tests at different load rates and temperatures with a complete strain state evaluation may help to enhance the current models of PC or to create new ones and to test the accuracy of their predictions.

This article is divided into 4 sections. First, research methods and materials are introduced, detailing heat treatments and tests carried out. Then, results obtained are presented highlighting the asymmetry effect of PC. Afterwards, a discussion of results which details some important implications for the modeling of the mechanical behavior of PC is presented. Finally, conclusions of this work are shown.

2. Materials and Methods

2.1. Materials. The study material is polycarbonate (PC) Makrolon GP solid, transparent, with smooth surface, manufactured by Bayer. Specimens were machined directly from PC plates. Sample dimensions are based on ASTM (American Society for Testing and Materials) standards; for tensile tests, a bone-shaped specimen with a cross-sectional area of 13.5 mm × 3 mm inside a narrowed section was used [27]; for compressive tests, a rectangular specimen with a cross-sectional area of 25.4 mm × 12.7 mm and 55 mm height was used [28].

Physical aging is a characteristic of glassy state, in which there is a progressive change in mechanical properties. These changes are caused by the existence of a free volume (volume that is not occupied by material molecules) due to an amorphous structure, like that of polymers. The presence of this free volume affects material density and therefore its mechanical properties. If a polymer is processed at high temperatures, it tends to reduce its specific volume when cooling. This reduction keeps happening when exposing the material to a constant aging temperature. This phenomenon causes a time-dependent mechanical behavior. It is worth mentioning that molecular composition is not affected, as it is with chemical aging (this is the case of thermal degradation and photo-oxidation). Physical aging effects can be reverted by means of a thermal rejuvenation [29]. This process consists of heating the sample to a temperature above glass transition temperature (Tg) during a meaningful period of time; in this way, the physical aging time is reinitiated and the effects of the observed phenomenon are erased. During a cooling process starting from a high temperature, the reduction rate of the specific volume decreases significantly at Tg. This happens because above Tg (rubbery state), amorphous polymers are more flexible due to the initiation of a coordinated movement of large segments of material molecules [30]. Below Tg, the material state is known as glassy. In the present paper, the tests last less than 2 hours and for this reason, as suggested by McKenna [31], physical aging is neglected (physical aging times are usually 10 times greater than this test duration).

Before carrying out the mechanical tests, the material was subjected to a thermal treatment similar to that suggested by Struik [32] to rejuvenate it. This treatment consists of heating PC to 120°C for 24 hours and afterwards heating it to 155°C (value above the Tg) for 24 hours. In order to determine the Tg of PC, the following thermal analyses were done using a heating rate of 5°C/min: differential scanning calorimetry (DSC) analysis, thermomechanical analysis

(TMA), and a temperature scanning test in a dynamic mechanical analysis machine (DMA). All techniques provided virtually the same value of Tg, 140°C. Moreover, a thermogravimetric analysis (TGA) was done to confirm that material does not suffer any degradation at 155°C.

2.2. Uniaxial Mechanical Tests. Uniaxial tensile and compressive tests were carried out in a universal electromechanical testing machine, Instron 3318, in order to study the mechanical behavior of PC. The mechanical tests done were as follows:

(i) Monotonic tensile and compressive tests at different load rates: slow, moderate, and fast tests. In tension, three constant stress rates were applied: 1 MPa/min (slow), 10 MPa/min (moderate), and 100 MPa/min (fast). In compression, due to technical problems, the available universal testing machine was not able to control the tests with an assigned stress rate. For this reason, three strain rates were used in compression: 0.043%/min (slow), 0.43%/min (moderate), and 4.3%/min (fast). Tests were carried out at four different temperatures: 25°C, 50°C, 70°C, and 90°C.

(ii) Short-term tensile and compressive creep tests with a recovery stage. Tests consisted of reaching certain stress level at a 9%/min strain rate, afterwards sustaining such level for 30 minutes, then unloading at a 9%/min strain rate to a zero stress state and finally sustaining the zero stress level for 1 hour to observe the material recovery. At room temperature (25°C), stresses used were 30 MPa, 35 MPa, 40 MPa, 45 MPa, 50 MPa, and 55 MPa. At 50°C, 70°C, and 90°C, the stress levels tested were 30 MPa, 40 MPa, and 50 MPa. The creep-recovery tests considered herein are two-step loading tests. As mentioned in the introduction section, for polymer melts and glasses, the behavior prediction in reversing double-step strain flows represent a great challenge for theoretical models [12, 19]. Another test that would be interesting to perform is a multicreep test with a first tension step followed by a compression step. Nevertheless, this test is hard to perform because specimens for tensile tests have to be long enough to ensure a uniaxial stress state, and if these specimens were tested in compression, buckling would occur. It is for this reason that this load reversing test has not been performed herein.

In every case, strain gauges were used to measure the axial (ε_a) and transverse (ε_t) strains. Strains with a magnitude less than 5% are accurately measured with these gauges.

3. Results

3.1. Monotonic Tests. Figures 1(a), 1(b), and 1(c) show the stress (σ) versus axial strain (ε_a) curves at different temperatures for slow, moderate, and fast monotonic tests, respectively. For all the load rates, it can be seen that the behavior is not linear elastic and that it depends on temperature: when temperature is increased, a same stress level provokes a higher axial strain. A difference in material behavior when subjected to tension or compression is also observed: a higher stress is necessary to provoke the same absolute value of axial strain under compression than that under tension; this confirms asymmetry effects on PC behavior. It is also worth mentioning that Young's modulus measured at the very beginning of the tests is virtually the same in tension and compression for all the temperatures and load rates tested, 2.3 GPa. Figures 2(a), 2(b), and 2(c) show the transverse strain (ε_t) versus axial strain (ε_a) curves for slow, moderate, and fast monotonic tests, respectively. For each load rate, the curves merge into a single master curve independent of temperature. In tension, a quasi-constant slope (its absolute value is 0.38) is obtained for the master curve whilst in compression, the absolute value of the slope starts with a 0.38 value and as the magnitude of the axial strain grows, it increases and reaches a 0.5 value, approximately. In linear elastic materials, Poisson's ratio is less than 0.5; herein, the material is not linear elastic and the abovementioned slope is not related to the classical Poisson's ratio $\nu = -\varepsilon_t/\varepsilon_a$ but to the viscous Poisson's ratio $\mu = -\dot{\varepsilon}_t/\dot{\varepsilon}_a$. The difference of the previous slopes in tension and compression is another evidence of asymmetry in material's behavior.

In Figures 3(a), 3(b), 3(c), and 3(d), the stress versus axial strain curves obtained in tension and compression for the different load rates at 25, 50, 70, and 90°C are displayed, respectively. One can observe that at 25°C, the curves merge virtually into a single master curve but for higher temperatures, significant differences appear in these curves for stresses having a magnitude greater than 40 MPa. As the temperature increases, the material behavior is more sensitive to the load rates considered. The transverse versus axial strain curves are not shown because the curves merge once again into a single master curve which becomes independent of temperature and load rate (this will be discussed in Section 4.3).

3.2. Creep Tests with Recovery. In this section, results of tensile and compressive creep tests with recovery at different stress levels and temperatures are presented. At instant $t = 0$ s, a nonzero abrupt stress σ is applied and maintained constant for 30 minutes. At $t = 0$ s, the axial strain is ε_a^0 and as time increases (for $t \leq 30$min), a viscous strain $\varepsilon_a^v(t) = \varepsilon_a(t) - \varepsilon_a^0$ appears and grows (Figure 4). After this stage, the stress is removed and the strain is monitored for 1 hour. In Figure 4, the axial strain is plotted against time for the tensile test at 50 MPa and 50°C. During the nonzero stress stage, the maximum axial viscous strain ε_a^v attained is 0.23%. Similar to Mindel and Brown's observations [33], at the end of the recovery stage, a persistent nonzero strain ε_a^p =0.06% remains. It is likely that this persistent strain is due to plasticity, and its magnitude may be neglected as compared with the maximum viscous strain attained. Similar results have been obtained for the tests carried out at the other stress levels and temperatures. For all these reasons, the behavior

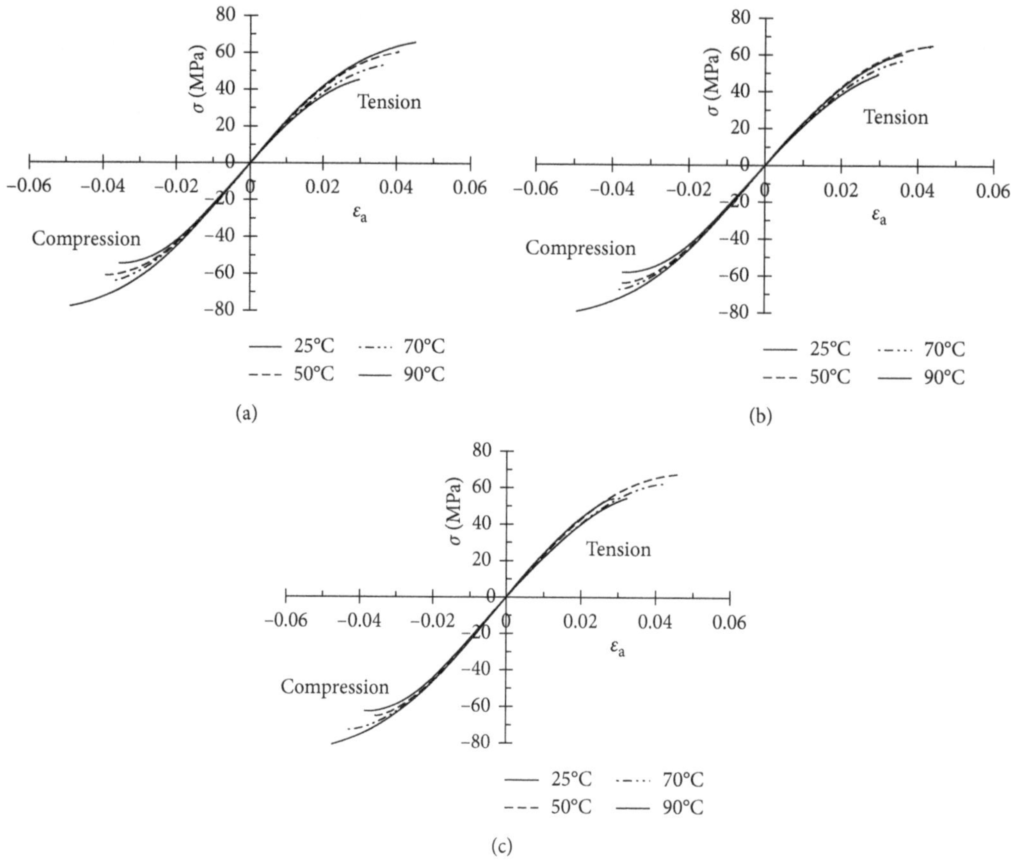

FIGURE 1: σ vs ε_a curves for slow (a), moderate (b), and fast (c) monotonic tests.

of PC is mainly viscoelastic. In what follows, only Figure 4 exhibits the evolution in time of the axial strain during the recovery stage. The plots of the transverse strain versus axial strain during the recovery stage obtained in these tests will be shown in Section 4.3.

During the application of the nonzero constant stress, it is useful to define the following axial creep compliance:

$$J_a = \frac{\varepsilon_a}{\sigma}. \tag{1}$$

Figures 5(a) and 5(b) present curves of axial creep compliances (J_a) against time at 25°C in tension and compression, respectively. The curves are significantly different: the magnitude of the creep compliance increases with the stress applied. This means that strains do not depend linearly on the applied stress σ_0 and since the material is quasi-viscoelastic, the tests reveal a nonlinear viscoelastic behavior of PC in both tension and compression. These results are consistent with the observations of Jazouli et al. [34]; they showed that polycarbonate behavior in tension is linear for a stress level below 26 MPa and nonlinear viscoelastic for a stress level above the previous value. Another important observation is that the curves in tension and compression for a same stress level are significantly different. This confirms asymmetry in the behavior of PC in creep tests.

In Figures 6(a), 6(b), and 6(c), the creep compliances are plotted against time for the different stress levels in tension at 50, 70, and 90°C, respectively. It can be seen that as temperature or stress increases, the creep compliance increases. In Figure 6(c), in the tensile test carried out at 50 MPa and 90°C, the specimen failed abruptly at the beginning of the test and for this reason, the strain could not be monitored anymore. In Figures 7(a), 7(b), and 7(c), the creep compliances versus time curves are displayed for the different stress levels in compression at 50, 70, and 90°C, respectively. Once again, a temperature or stress level increase provokes a creep compliance growth. The comparison of Figures 6 and 7 confirms once again the presence of asymmetry in the material behavior. The transverse strain versus time curves are not shown in this section because a master curve that enables deducing the transverse strain from the axial strain will be revealed in Section 4.3.

4. Discussion

In this section, the results obtained in the previous section are discussed.

4.1. Confirming Material Rejuvenation Usefulness. In some papers dealing with the mechanical characterization of PC, the authors do not specify whether they apply a rejuvenation process to the specimens or not. In order to assess the importance of the rejuvenation process, three additional monotonic tensile tests were performed using specimens in

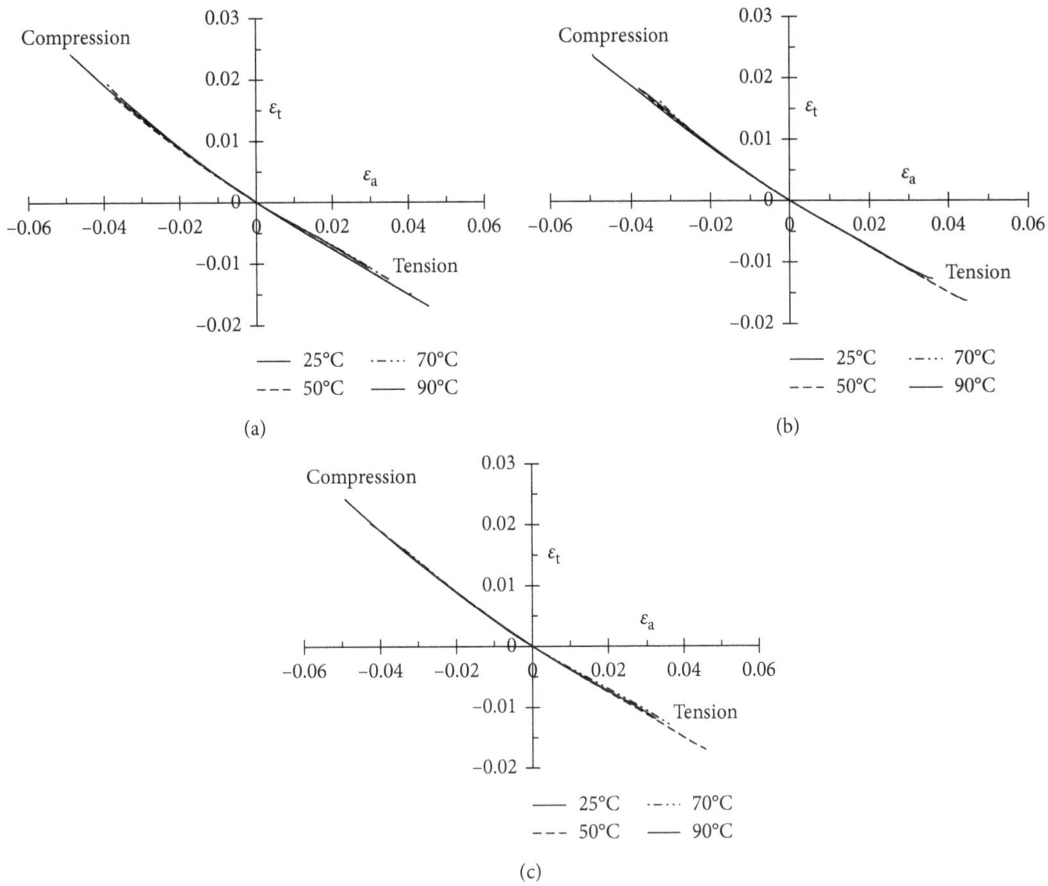

FIGURE 2: ε_t vs ε_a curves for slow (a), moderate (b), and fast (c) monotonic tests.

an as-received condition (aged specimens whose thermal history was unknown). Figure 8 shows stress versus axial strain curves in tension obtained with aged specimens tested at room temperature and three stress rates: 0.1, 1, and 10 MPa/min. A stress/strain curve obtained in a slow test (1 MPa/min) with a rejuvenated specimen was added in the previous figure so as to notice the great difference in the material behavior when the material has aged. If the aging process is not controlled, the experimental results lack usefulness because the thermal history affecting material behavior is not registered. For this reason, any mechanical characterization of PC should include a thermal rejuvenation so as to reset the thermal history of the material.

4.2. Hypoelastic-Like Behavior and Its Effects. In Figure 3(a), the slow, moderate, and fast monotonic tests carried out at 25°C yield three curves that seem to merge into a single master curve. This may be due to the fact that these tests were performed at a high load rate and that viscous strains had not enough time to evolve during the test. In order to verify the existence of the master curve and to determine the nature of the nonlinearity in this master curve, an additional 3-cycle tensile test defined by the following steps was performed:

(1) Loading at a 3%/s strain rate until a 50 MPa stress was obtained, approximately

(2) Unloading at the same strain rate

(3) Recovery for 10 minutes

(4) Loading at a 1%/s strain rate until a 50 MPa stress was obtained, approximately

(5) Unloading at a 1%/s strain rate

(6) Recovery for 10 minutes

(7) Loading at a 0.3%/s strain rate until a 50 MPa stress was obtained, approximately

(8) Unloading at a 0.3%/s strain rate

(9) Recovery for 10 minutes

The strain rates considered above yield higher stress rates (at least, 4 times greater) than those applied in Section 3.1. In Figure 9(a), the stress is plotted against the axial strain obtained in the cyclic test at 25°C. The line style corresponding for each cycle in the previous figure has been changed so as to ease their identification. Since these curves overlap, in Figure 9(b), an artificial horizontal 0.5% shift has been applied to visualize the curve obtained in each cycle. Virtually, a same curve is obtained in each cycle. Similar curves were obtained in compression and at higher temperatures: 50°C, 70°C, and 90°C (not shown in this paper).

Let us describe the stress-strain curves for each loading-unloading-recovery cycle. During the loading stage, a first linear elastic behavior stage appears for stresses less than

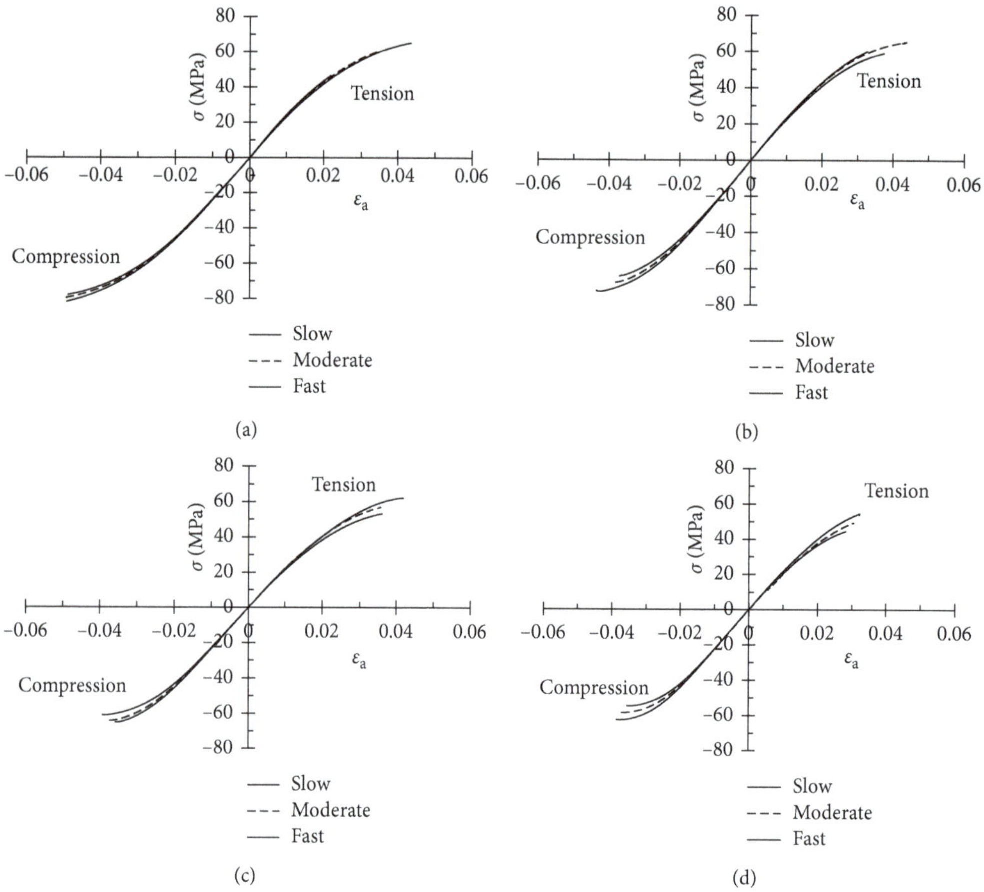

FIGURE 3: σ vs ε_a curves at 25°C (a), 50°C (b), 70°C (c), and 90°C (d) in tension and compression for different load rates.

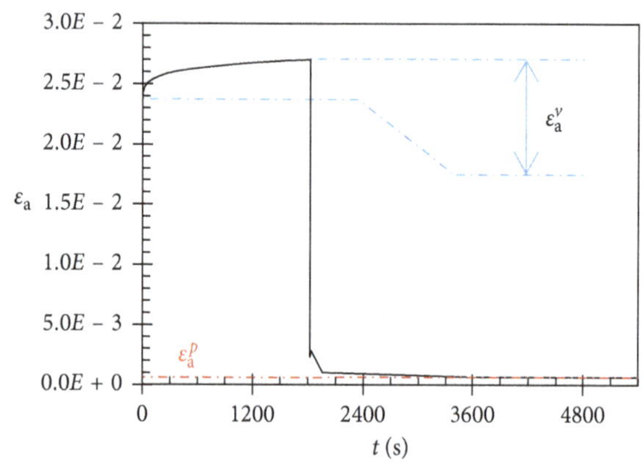

FIGURE 4: ε_a vs time curve for a tensile creep test at 50 MPa and 50°C.

30 MPa (a 2.3 GPa elastic modulus is deduced); after this, a nonlinear behavior is obtained. At 50 MPa, the axial strain is 12% greater than the expected linear elastic strain ε^l, as shown in Figure 9(a). The unloading follows a slightly different path from the loading stage and it yields also a nonlinear curve. At the end of the unloading stage, a small axial strain remains, and it marginally decreases during the recovery stage leaving a small persistent strain. The second cycle starts then from a nonzero strain and displays virtually the same stress strain-curve as the first cycle but shifted to the right due to the initial strain. The same happens for the third cycle. Let us provide an explanation to these curves. Since all cycles yield virtually the same stress-strain curve, the material behavior does not depend on the loading rates tested during loading and unloading stages, and therefore the nonlinearity is not due to viscous phenomena. Loading and unloading follow almost the same path, and therefore a hypoelastic component in the material behavior can be inferred. After loading and unloading, a very small strain remains; this strain is surely due to an instantaneous dissipative phenomenon such as plasticity. In this manner, hypoelasticity combined with a marginal plasticity can explain the shape of the curves during the loading and unloading stages. A further analysis, out of the scope of this paper, is required to make a better a characterization of the hypoelastic-like behavior of PC at high strain rates. Let us now analyze the stress-free recovery stage. This stage starts with an initial plastic strain that produces a viscous strain smaller than the initial plastic strain. In Figure 10(a), the axial strain is plotted against time during the first recovery stage. The axial strain starts with an instantaneous plastic strain, decreases during 300 seconds, and then remains virtually constant and equal to a nonzero plastic strain. It is for this reason that the curves are shifted to the right after each cycle.

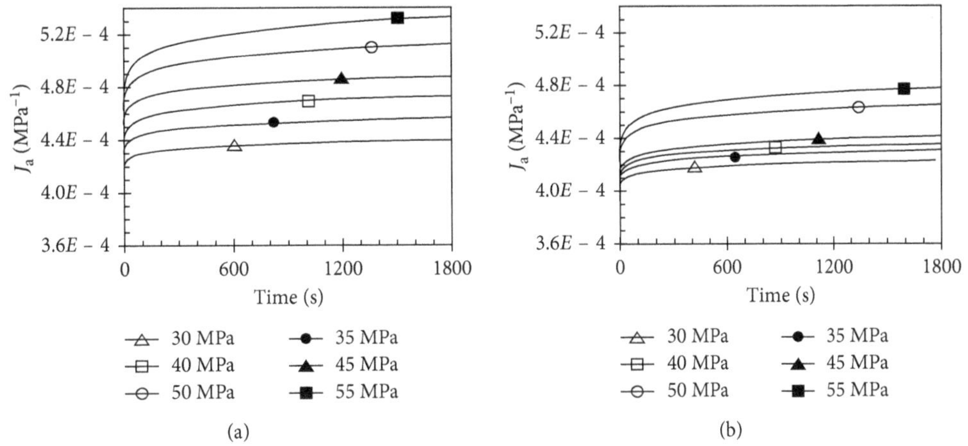

FIGURE 5: Axial creep compliance J_a vs. time curves in tension (a) and compression (b) at 25°C.

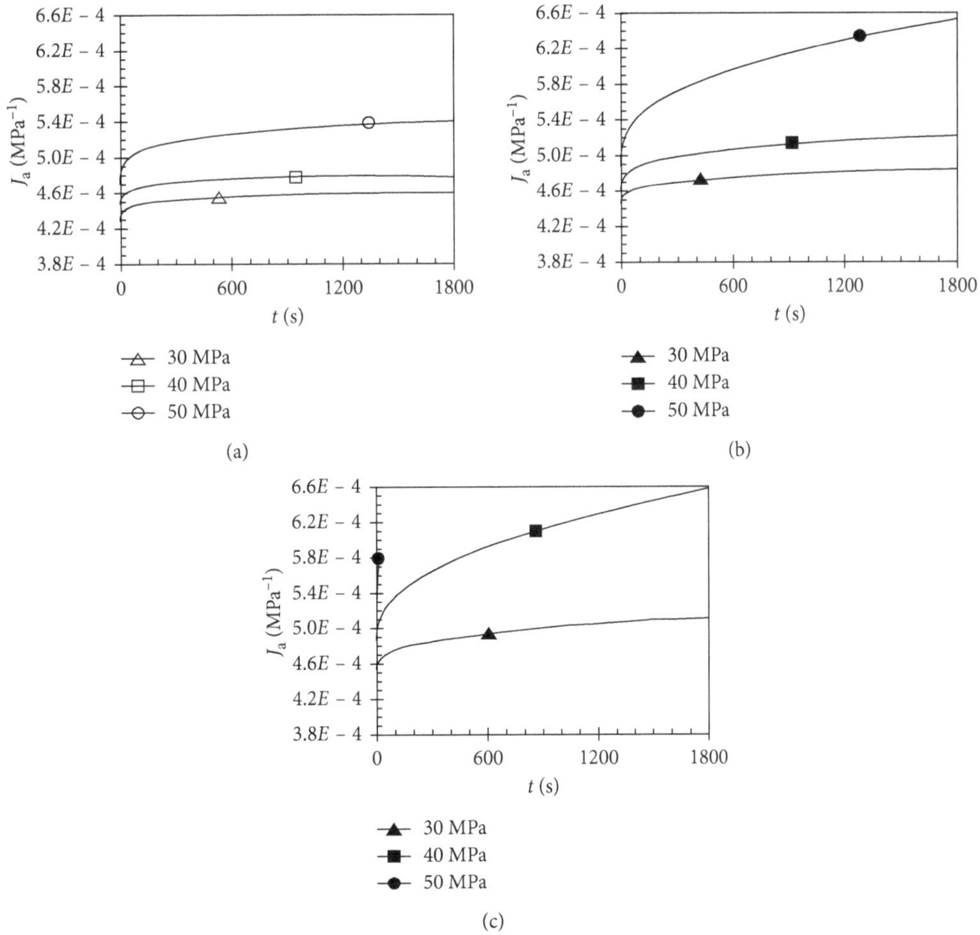

FIGURE 6: Axial creep compliance J_a vs. time curves in tension at 50°C (a), 70°C (b), and 90°C (c).

The master curve in Figure 3(a) that was inferred in monotonic tests at 25°C is easily explained by the hypoelastic-like behavior (nonlinear quasielastic behavior). As shown in Figure 10(b), the stress-strain curves obtained in Figure 3(a) in tension are not significantly different from that using the 3%/s strain rate applied in the first loading stage of the cyclic test. In Figure 11, the inferred stress vs. axial strain master curve obtained during a monotonic test using a 3%/s strain rate is shown; the results obtained in compression are included in this figure. At lower stress rates than the slowest one considered, higher viscous strains would appear. At higher temperatures, viscous phenomena are more significant, and a rate-dependent behavior is shown in Figures 3(b), 3(c), and 3(d). Nevertheless, a master

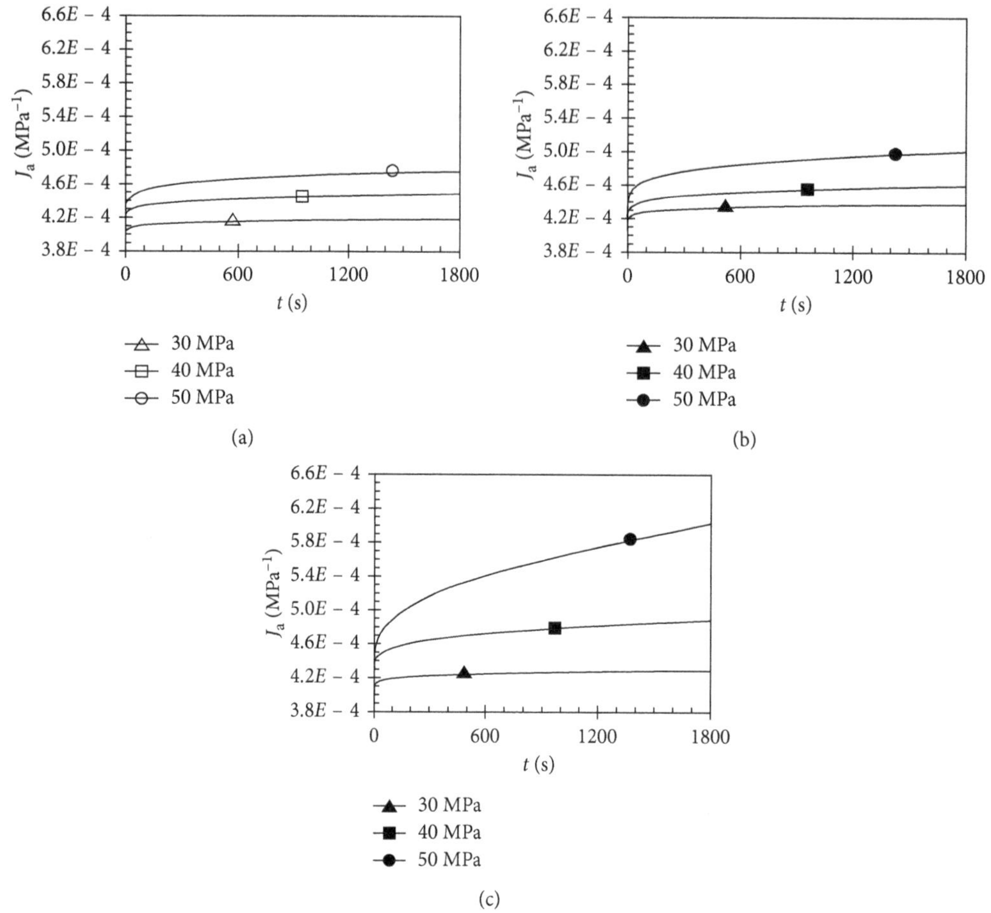

FIGURE 7: Axial creep compliance J_a vs. time curves in compression at 50°C (a), 70°C (b), and 90°C (c).

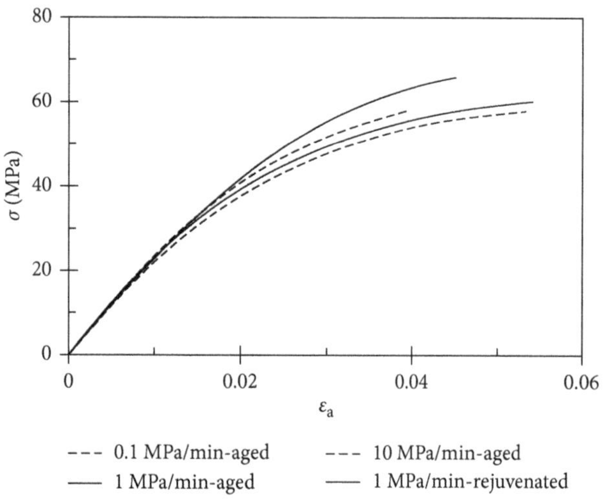

FIGURE 8: σ vs ε_a curves in tensile tests at 25°C with aged and rejuvenated specimens.

curve exists and is obtained by carrying out the tests at higher load rates as the 3%/s applied in the cyclic test.

It is also interesting to notice the effects of the hypoelastic-like behavior of PC at high strain rates in the axial creep compliance J_a curves obtained in Figures 5, 6, and 7. At the same temperature, in tensile and compressive tests, the sudden load provokes an initial axial strain that is greater than the predicted linear elastic strain due to the hypoelastic-like behavior. It is for this reason that the initial creep compliance J_a increases as the applied stress increases.

4.3. Transverse versus Axial Strains Master Curve. Figures 2(a), 2(b), and 2(c) proved that in monotonic tests, the transverse versus axial strains curves obtained at different temperatures merged into a master curve for a given load rate. Let us now plot the curves for all the load rates and temperatures in Figure 12. In this figure, it can clearly be seen that theses curves do not depend on the load rate and temperature and that they all describe a same master curve. The curve is linear in tension but nonlinear in compression. For this reason, in compression, time-dependent Poisson's ratios would be obtained.

It is natural to ask oneself if this master curve keeps its uniqueness for the creep tests performed at different stress levels and temperatures. In Figure 13, the transverse versus axial strains curves obtained at 25°C are plotted with the master curve obtained in the monotonic tests. In this figure, only the evolution of strains in the nonzero stress stage is considered so as to solely visualize the evolution due to viscous strains. The master curve matches perfectly with the

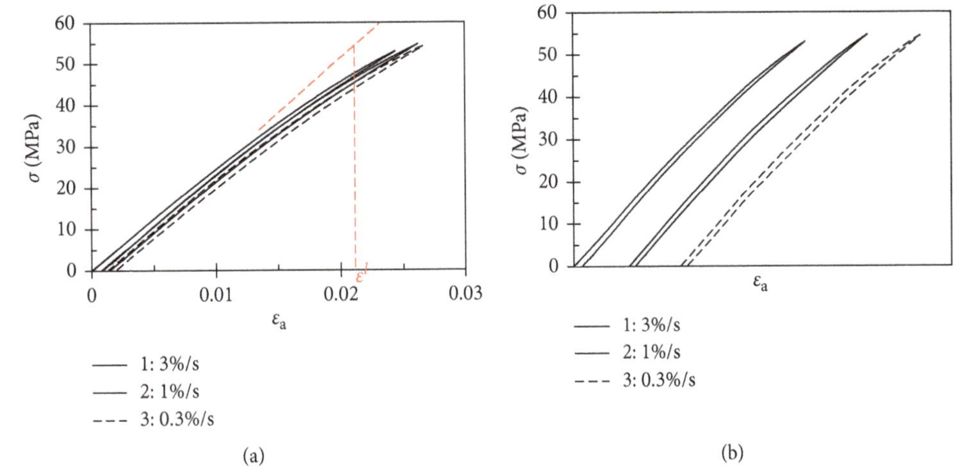

FIGURE 9: σ vs ε_a curves obtained during a 3-cycle tensile test at 25°C (a) and with a horizontal shifting to distinguish the curve in each cycle (b).

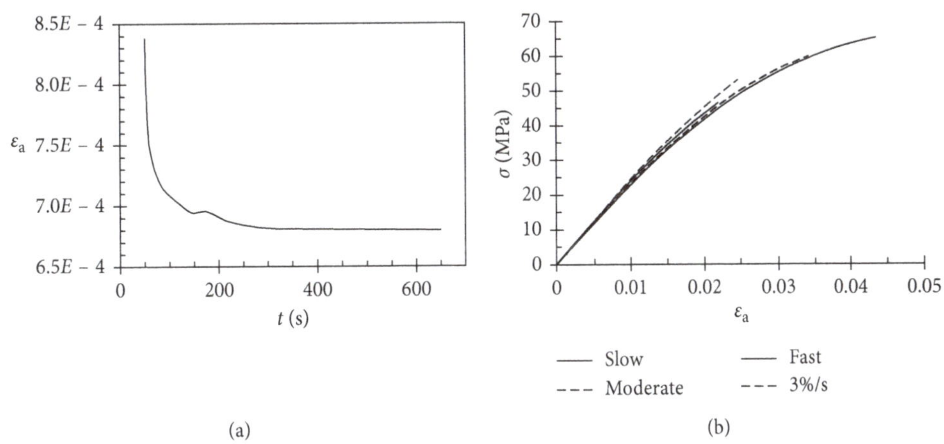

FIGURE 10: ε_a vs t curve in the first recovery stage (a); σ vs ε_a curves obtained in slow, moderate, fast, and ultrafast monotonic tensile tests at 25°C (b).

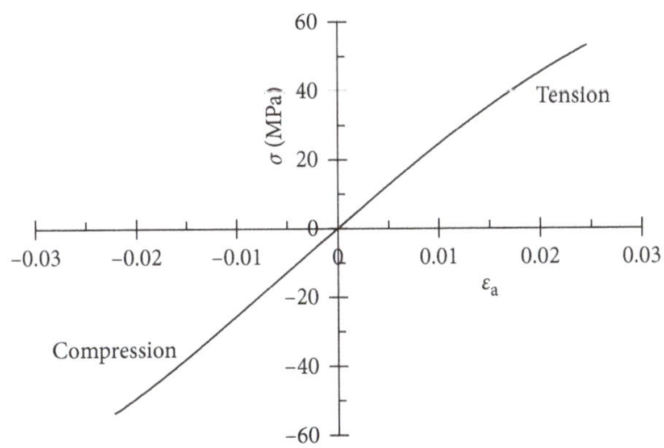

FIGURE 11: σ vs ε_a master curve at high strain rates.

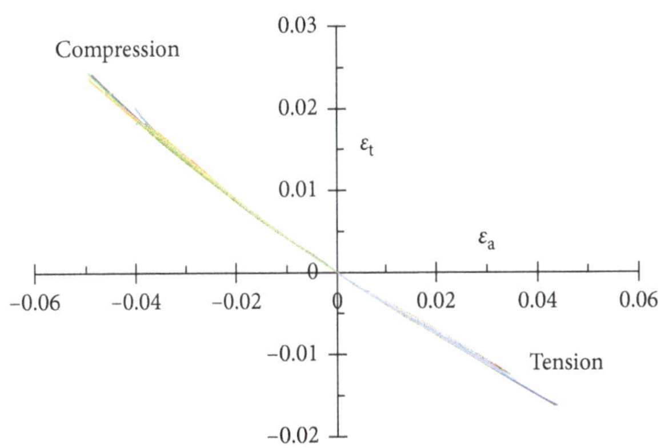

FIGURE 12: ε_t vs ε_a curves obtained in all monotonic tests and merge into a single master curve.

experimental creep test curves. In Figures 14(a), 14(b), and 14(c), the transverse versus axial strains curves obtained at 50, 70, and 90°C are plotted with the master curve, respectively. Once again, the master curve reproduces perfectly the experimental curves. Additionally, one can argue that a small increase in the viscous strains can hardly provoke that the experimental creep curves separate from

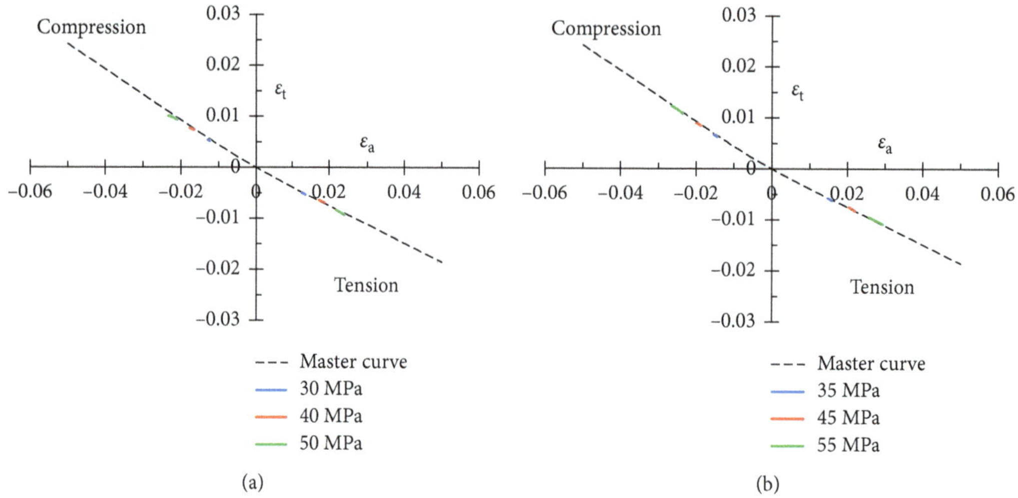

FIGURE 13: ε_t vs ε_a master and experimental curves obtained in 30, 40, and 50 MPa (a) creep tests and 35, 45, and 55 MPa (b) creep tests at 25°C.

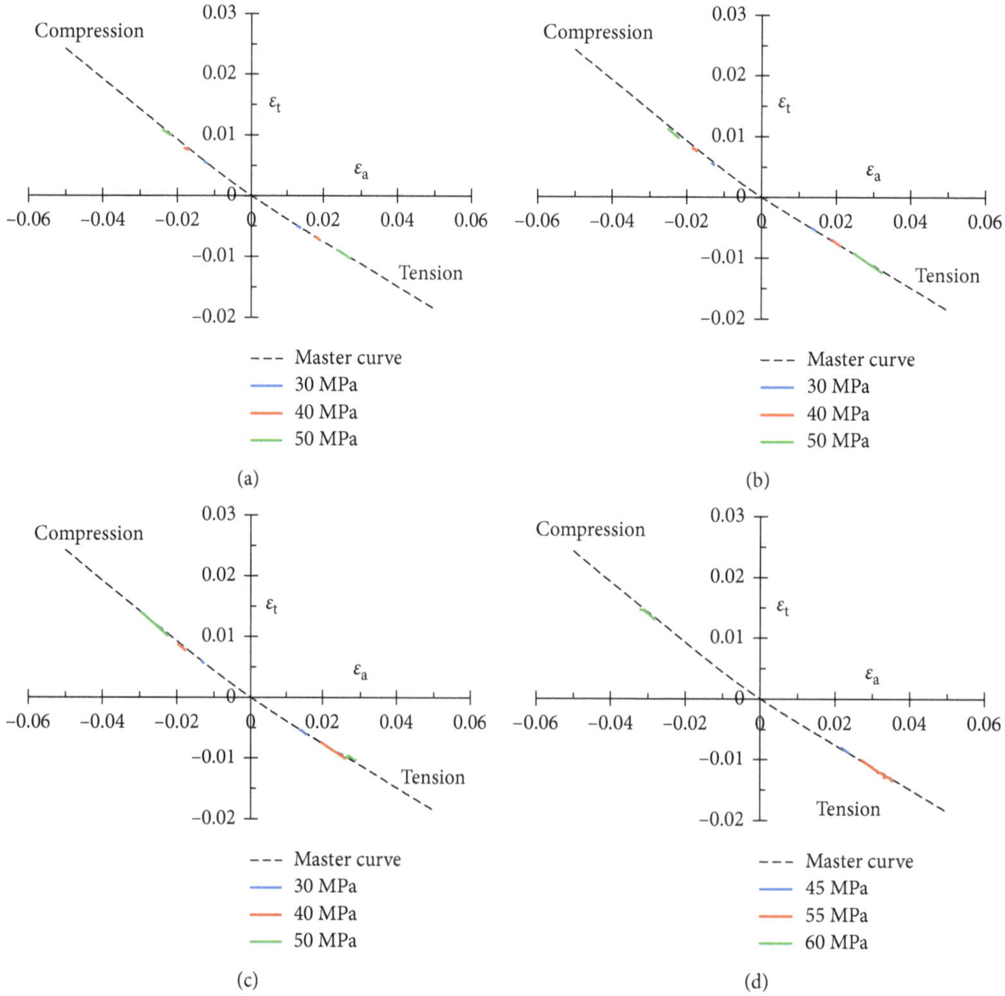

FIGURE 14: ε_t vs ε_a master and experimental curves obtained in creep tests at 50°C (a), 70°C (b), and 90°C (c); long-term creep tests at 25°C (d).

the master one. For this reason, additional 24 h creep tests followed by a 48 h recovery stage were carried out at 25°C using three stress levels: 45, 55, and 60 MPa. The first two tests were tensile tests whilst the latter was a compressive one. In Figure 14(d), the transverse versus axial strains curves obtained in the long-term creep tests during the first step (nonzero stress stage) are plotted with the master curve. Again, the master curve matches perfectly with the long-term creep curves. Finally, let us verify that the master curve remains valid during the recovery stage of the creep-recovery tests. In Figure 15, the transverse versus axial strains curves obtained during the recovery stage in all the short-term tests are plotted with the master curve. In Figure 16, the transverse versus axial strains curves obtained in the long-term tests are shown. The master curve matches perfectly with the recovery curves. Consequently, the ε_t versus ε_a master curve seems to be a property of PC in uniaxial tests that does not depend on temperature, nor on the loading rate. An additional analysis, out of the scope of this paper, is required to express this original result in terms of strain invariants so as to include it in a model aiming to predict the mechanical behavior of PC in a 3D stress state.

It is worth mentioning that since the volumetric strain ε_v is equal to $\varepsilon_a + 2\varepsilon_t$, a master ε_v versus ε_a curve exists for the uniaxial tests considered herein; this curve is shown in Figure 17. It can be observed that in tension, as ε_a increases, the volume increases proportionally. In compression, for $|\varepsilon_a| < \sim 1\%$, as ε_a decreases, the volume decreases, but for $|\varepsilon_a| > \sim 2\%$, the volumetric strain remains virtually constant. Let us compare these results to those obtained by Colucci et al. in [35] for two types of PC (PC1 and PC2) tested in stress relaxation tests. Colucci et al. showed an isochronal ε_v versus ε_a curve at 0.5 s analogue to our curve in Figure 17. In tension, our material has a similar behavior to that of PC1; PC2 exhibits a densification beyond a certain axial strain level which does not appear herein. In compression, this time, our material behaves like PC2; PC1 exhibits a monotonic volume reduction as ε_a decreases. As suggested by Colucci et al. it is possible that the differences in the molecular weights and their distribution may be the cause of the dissimilar behavior of these materials. The determination of the molecular weight of the PC tested herein and its influence on the material behavior are out of the scope of this paper.

4.4. Implications for Modeling PC Behavior. The experimental results obtained in this paper imply that any modeling aiming to predict accurately the mechanical behavior of PC in a 3D stress state at temperatures below Tg and above room temperature must take into account the following:

(i) A nonlinear instantaneous hypoelastic-like response, the effects of hydrostatic stresses (asymmetry), temperature, and load rates on viscoelastic behavior.

(ii) Poisson's ratio may be considered constant in tension but not in compression.

(iii) The existence of a master transverse strain/axial strain curve (a material property) in uniaxial tests. This curve must be interpreted in terms of strain invariants.

(iv) The persistence of small strains that might be due to plasticity. Depending on the accuracy level of the model, these strains may be neglected.

Flory and McKenna [13] tested a PC and proved that the strain energy density virtually does not depend on the second invariant of the deformation tensor and the first invariant does have a significant effect on the strain energy density for the tests considered by these authors. A neo-Hookean behavior may thus be inferred for this material, and if one adopts a neo-Hookean model, the expression of the strain energy density as a function of the first invariant of the deformation tensor has to be determined in such a manner that the model could fit the experimental curves. Another interesting approach which can be adopted to model PC behavior is the one proposed by Caruthers et al. in [36]. The approach is developed on a rational mechanics framework to ensure thermodynamic consistency. This approach requires a proposal for the expression of the Helmholtz free energy density to obtain the constitutive equations. For the PC tested herein, the application of such models to the uniaxial tests has to reproduce the master transverse strain/axial strain curve and the instantaneous response shown in Figure 11. This is not a simple task and represents an interesting challenge that will be tackled in a future paper.

5. Conclusions

In this study, the mechanical behavior of polycarbonate (PC) was investigated by means of uniaxial tensile and compressive tests at various load rates and temperatures below Tg. From the experimental work, the following conclusions can be drawn:

(1) As observed by other authors, a marked asymmetry effect on PC's behavior is confirmed.

(2) An asymmetry effect exists on the viscous Poisson's ratio of PC: in tension it remains constant and equal to 0.38, but in compression it increases up to 0.5 approximately.

(3) A master transverse strain/axial strain curve is revealed. Whatever the test, load rate, and temperature, the experimental curve is perfectly described by the master curve. The inclusion of this original result in a 3D stress state model requires an additional analysis to express the meaning of the master curve in terms of strain invariants.

(4) Within the stress and strain ranges considered in this paper, PC exhibits a viscoelastic behavior for low load rates and an instantaneous, nonlinear, hypoelastic-like behavior for high load rates.

(5) Tests' results are reliable, because the material was subjected to thermal rejuvenation before the tests, so

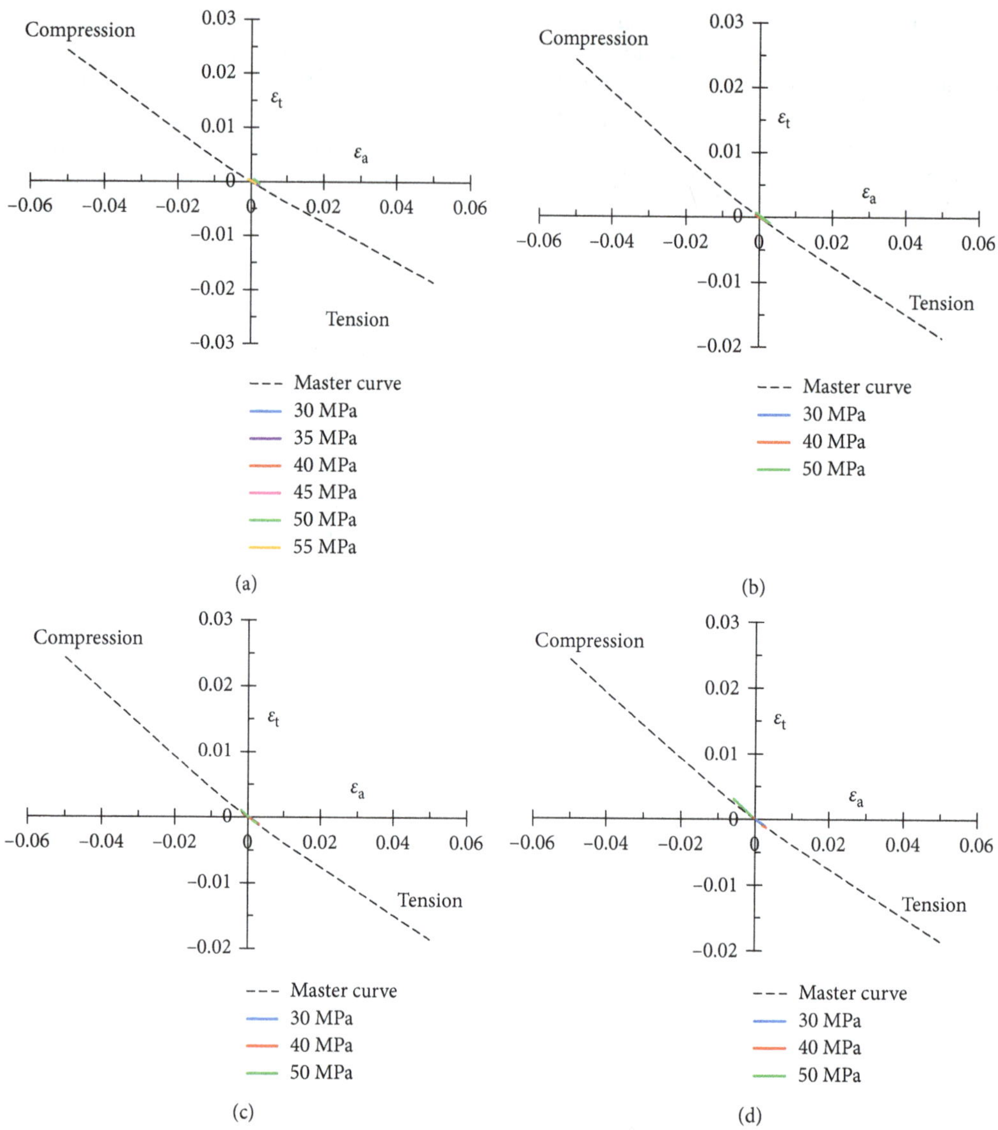

FIGURE 15: ε_t vs ε_a master and experimental curves obtained in creep-recovery tests at (a) 25°C, (b) 50°C, (c) 70°C, and (d) 90°C.

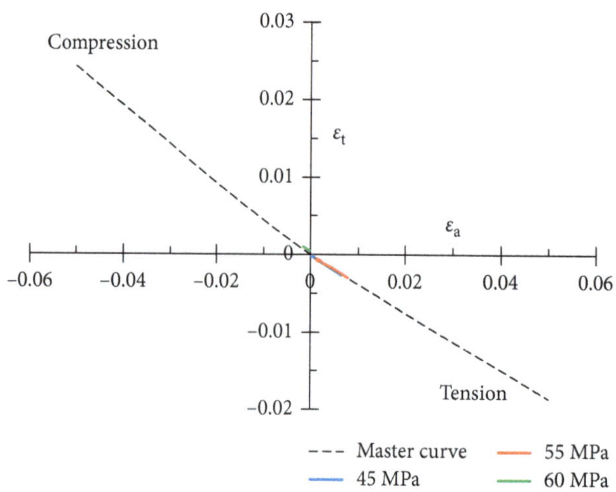

FIGURE 16: ε_t vs ε_a master and experimental curves obtained in long-term creep-recovery tests at 25°C.

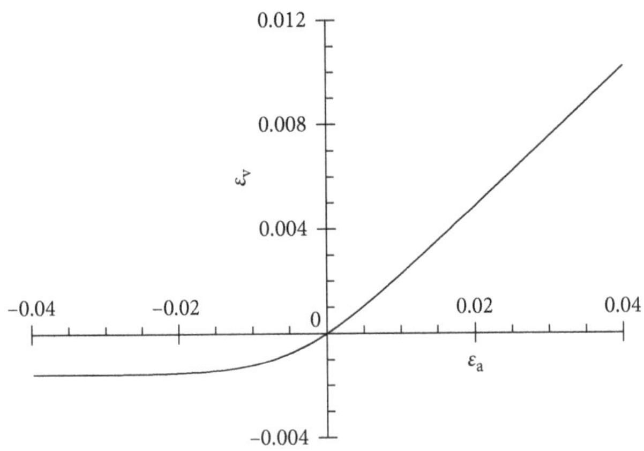

FIGURE 17: ε_v versus ε_a master curve.

the previous history of material was erased, reestablishing its original properties and eliminating changes due to physical aging.

These results should be considered in any model aiming to predict accurately PC behavior. The results of other tests such as combined torsion/tension tests and cyclic uniaxial tests with reverse loadings (tension/compression) should also be considered in a modeling stage so as to obtain reliable predictions for PC subjected to multiaxial stress states with cyclic loading. In a future paper, a model including these recommendations will be developed.

It is not sure that the observed phenomena for PC are valid for other glassy state polymers as well. PC is a very particular polymer exhibiting global thermorheological simplicity [37] and for which β-relaxations are negligible [13]. To confirm this, the methods and tests proposed in this paper can be replicated with other glassy polymers.

Conflicts of Interest

The authors declare that there are no conflicts of interest regarding the publication of this paper.

Acknowledgments

The authors gratefully acknowledge Rubén Castañeda Balderas for carrying out the tests exhibited in this paper. The authors would also like to acknowledge CONACYT for its financial support.

References

[1] L. Cangemi, S. Elkoun, C. G'Sell, and Y. Meimon, "Volume strain changes of plasticized poly(vinylidene fluoride) during tensile and creep tests," *Journal of Applied Polymer Science*, vol. 91, no. 3, pp. 1784–1791, 2004.

[2] A. D. Drozdov and Q. Yuan, "The viscoelastic and viscoplastic behavior of low-density polyethylene," *International Journal of Solids and Structures*, vol. 40, no. 1, pp. 2321–2342, 2003.

[3] O. U. Colak, "Modeling deformation behavior of polymers with viscoplasticity theory based on overstress," *International Journal of Plasticity*, vol. 21, no. 1, pp. 145–160, 2005.

[4] K. Cao, X. Ma, B. Zhang, Y. Wang, and Y. Wang, "Tensile behavior of polycarbonate over a wide range of strain rates," *Materials Science and Engineering A*, vol. 527, no. 16-17, pp. 4056–4061, 2010.

[5] A. Schiavi, R. Cuccaro, and A. Troia, "Strain-rate and temperature dependent material properties of Agar and Gellan Gum used in biomedical applications," *Journal of the Mechanical Behavior of Biomedical Materials*, vol. 53, pp. 119–130, 2016.

[6] J. Richeton, S. Ahzi, K. Vecchio, F. Jiang, and R. Adharapurapu, "Influence of temperature and strain rate on the mechanical behavior of tree amorphous polymers: characterization and modeling of the compressive yield stress," *International Journal of Solids and Structures*, vol. 43, no. 7–8, pp. 2318–2335, 2006.

[7] J. Mohanraj, D. Barton, I. Ward, A. Dahoun, J. Hiver, and C. G'Sell, "Plastic deformation and damage of polyoxymethylene in the large strain range at elevated temperatures," *Polymer*, vol. 47, no. 16, pp. 5852–5861, 2006.

[8] K. Cao, Y. Wang, and Y. Wang, "Effects of strain rate and temperature on the tension behayior of polycarbonate," *Materials and Design*, vol. 38, pp. 53–58, 2012.

[9] L. Zhang, X. Yao, S. Zang, and Q. Han, "Temperature and strain rate dependent tensile behavior of a transparent polyurethane interlayer," *Materials and Design*, vol. 65, pp. 1181–1188, 2015.

[10] I. Ward and J. Sweeney, *Mechanical Properties of Solid Polymers*, Wiley, Hoboken, NJ, USA, 3rd edition, 2013.

[11] A. Shaban, R. Mahnken, L. Wilke, H. Potente, and H. Ridder, "Simulation of rate dependent plasticity for polymers wit asymmetric effects," *International Journal of Solids and Structures*, vol. 44, no. 18-19, pp. 6148–6162, 2007.

[12] G. B. McKenna and L. J. Zapas, "Nonlinear viscoelastic behavior of poly (methyl methacrylate) in torsion," *Journal of Rheology*, vol. 23, no. 2, pp. 151–166, 1979.

[13] A. L. Flory and G. B. McKenna, "Chemical structure - normal force relationships in polymer glasses," *Polymer*, vol. 46, no. 14, pp. 5211–5217, 2005.

[14] W. Knauss and W. Zhu, "Nonlinearly viscoelastic behavior of polycarbonate. I. Response under pure shear," *Mechanics of Time-Dependent Materials*, vol. 6, no. 3, pp. 231–269, 2002.

[15] W. Knauss and W. Zhu, "Nonlinearly viscoelastic behavior of polycarbonate. II. The role of volumetric strain," *Mechanics of Time-Dependent Materials*, vol. 6, no. 4, pp. 301–322, 2002.

[16] K. Ravi-Chandar and Z. Ma, "Confined compression: a stable homogeneous deformation for constitutive characterization," *Experimental Mechanics*, vol. 40, no. 1, pp. 38–45, 2000.

[17] K. Ravi-Chandar and Z. Ma, "Inelastic deformation in polymers under multiaxial compression," *Mechanics of Time-Dependent Materials*, vol. 4, no. 4, pp. 333–357, 2000.

[18] L. Zapas, "Nonlinear behavior of polyisobutylene solutions," in *Deformation and Fracture of High Polymers*, H. Henning Kausch, J. A. Hassell, and R. I. Jaffee, Eds., pp. 381–395, 1973.

[19] D. C. Venerus and H. Kahvand, "Doi—Edwards theory evaluation in double—step strain flows," *Journal of Polymer Science. Part B. Polymer Physics*, vol. 32, no. 8, pp. 1531–1542, 1994.

[20] D. C. Venerus, E. F. Brown, and W. R. Burghardt, "The nonlinear response of a polydisperse polymer solution to step strain deformations," *Macromolecules*, vol. 31, no. 26, pp. 9206–9212, 1998.

[21] B. Bernstein, E. A. Kearsley, and L. J. Zapas, "A study of stress relaxation with finite strain," *Transactions of the Society of Rheology*, vol. 7, no. 1, pp. 391–410, 1963.

[22] D. Tscharnuter, M. Jerabek, Z. Major, and R. W. Lang, "Time-dependent Poisson´s ratio of polypropylene compounds for various strain histories," *Mechanics of Time-Dependent Materials*, vol. 15, no. 1, pp. 15–28, 2011.

[23] S. Pandini and A. Pegoretti, "Time and temperature effects on Poisson's ratio of poly(butylene terephthalate)," *Express Polymer Letters*, vol. 5, no. 8, pp. 685–697, 2011.

[24] S. Pandini and A. Pegoretti, "Time, temperature, and strain effects on viscoelastic Poisson´s ratio of epoxy resins," *Polymer Engineering and Science*, vol. 48, no. 7, pp. 1434–1441, 2008.

[25] N. W. Tschoegl, W. G. Knauss, and I. Emri, "Poisson's ratio in linear viscoelasticity—a critical review," *Mechanics of Time-Dependent Materials*, vol. 6, no. 1, pp. 3–51, 2002.

[26] P. Mallon, D. McCammond, and P. Benham, "Strain ratio and volume change during tension and compression creep of thermpoplastics," *Polymer Engineering and Science*, vol. 12, no. 6, pp. 420–424, 1972.

[27] ASTM D638-14, "ASTM compass," August 2016, https://compass.astm.org/download/D638.39258.pdf.

[28] ASTM D695-15, "ASTM compass," August 2016, https://compass.astm.org/download/D695.3083.pdf.

[29] G. Odergard and A. Bandyopadhyay, "Physical aging of epoxy polymers and their composites," *Journal of Polymer Science Part B: Polymer Physics*, vol. 49, no. 24, pp. 1695–1716, 2011.

[30] D. A. Skoog, F. J. Holler, and S. R. Crouch, *Principios de Análisis Instrumental*, Cengage Learning Editores, Mexico D. F., Mexico, 6th edition, 2008.

[31] G. B. McKenna, "Mechanical rejuvenation in polymer glasses: fact or fallacy?," *Journal of Physics: Condensed Matter*, vol. 15, no. 11, pp. 737–763, 2003.

[32] L. C. E. Struik, *Physical Aging in Amorphous Polymers and Other Materials*, 1977.

[33] M. J. Mindel and N. Brown, "Creep and recovery of polycarbonate," *Journal of Materials Science*, vol. 8, no. 6, pp. 863–870, 1973.

[34] S. Jazouli, W. Luo, F. Brémand, and T. Vu-Khanh, "Nonlinear creep behavior of viscoelastic polycarbonate," *Journal of Materials Science*, vol. 41, no. 2, pp. 531–536, 2006.

[35] D. Colucci, P. O'Connell, and G. McKenna, "Stress relaxation experiments in polycarbonate: a comparison of volume changes for two commercial grades," *Polymer Engineering and Science*, vol. 37, no. 9, pp. 1469–1474, 1997.

[36] J. M. Caruthers, D. B. Adolf, R. S. Chambers, and P. Shrikhande, "A thermodynamically consistent, nonlinear viscoelastic approach for modeling glassy polymers," *Polymer*, vol. 45, no. 13, pp. 4577–4597, 2004.

[37] M. L. Cerrada and G. B. McKenna, "Physical aging of amorphous PEN: isothermal, isochronal and isostructural results," *Macromolecules*, vol. 33, no. 8, pp. 3065–3076, 2000.

12

Kinetics of Free Radical Polymerization of N-Substituted Amides and Their Structural Implications

Anca Aldea,[1] Ana-Maria Albu,[2] Alina Nicolescu,[3,4] and Victorita Tecuceanu[3]

[1]*Department of General Chemistry, University Politehnica Bucharest, 1-7 Gheorghe Polizu Sector 1, 011061 Bucharest, Romania*
[2]*Department of Bioresources and Polymer Science, University Politehnica Bucharest, 1-7 Gheorghe Polizu Sector 1, 011061 Bucharest, Romania*
[3]*Centre for Organic Chemistry Costin D. Nenitescu, Romanian Academy, 202B Splaiul Independentei, CP 35-108, 060023 Bucharest, Romania*
[4]*Institute of Macromolecular Chemistry "Petru Poni" Iasi, 41A Aleea Grigore Ghica Voda, 700487 Iasi, Romania*

Correspondence should be addressed to Ana-Maria Albu; anamarialbu@yahoo.com

Academic Editor: Peter Chang

Two N-substituted amides (N-acryloyl morpholine and N-methyl-N-vinylacetamide) were polymerized in different solvents using radical initiator. The tacticity of obtained polymers was determined by 400 MHz ^{1}H-NMR and ^{13}C-NMR. At a given temperature, the syndiotacticity increased with increasing the solvent polarity. This solvent effect may be related to the hydrogen bonding interaction among solvent, monomer, and/or growing species. A peculiar aspect regards the steric hindrance at the nitrogen atom.

1. Introduction

In the area of smart materials, a peculiar attention is paid to the stimuli-responsive material architectures based on water-soluble polymers. They have been widely investigated for their applications in medicine, biotechnology, food, or cosmetics. An actual orientation of materials science field is designing and developing the stimuli-responsive nanostructured materials [1].

The approaches of modern polymer chemistry accept the challenges of increasing the precision and the complexity of these material systems to drive their properties beyond those of established responsive polymers [2–8]. The possibility to alter polymer properties through external stimuli could be successfully exploited to obtain "smart" block copolymers, polymer aggregates, polymer gels, or polymer surface brushes. Indeed, these methods can provide strong advantages in obtaining active polymer materials.

A special attention was paid to the N-substituted amides (co)polymers such N-isopropyl acrylamide (NIPAM), N-methyl methacrylamide (NMMAM), N-acryloyl morpholine (NAM), and so forth [9, 10].

Recognized for its reactivity, NAM copolymerizes with many different monomers and the synthesized materials, like all the materials derived from disubstituted acrylamides, are characterized by solubility in a wide range of solvents, from aqueous to organic ones. Due to its ability to give high molecular weight polymers with a virtual lack of toxicity, NAM is used in the last years to synthesize crosslinked networks for gel phase, synthesis of peptides, or drug delivery applications, protecting liposomes from blood clearance or increasing the lifetime of enzymes both *in vitro* and *in vivo* [11–13]. Alongside NAM, materials based on N-methyl-N-vinylacetamide (NMNVA) can be an alternative for water-soluble multifunctional materials for immobilization of single stranded DNA probes. Nowadays, the NMNVA copolymers are used for charge density modelling in relation to the molecular weight; the most important means for applications are polymer liquid crystals, DNA architecture, multilayer/expanded films, and so forth [9, 14, 15]. A specific property of the materials derived from structural isomers of N-alkyl acrylamide class is their lower critical solution temperature (LCST) [16]. The most poignant example for how structural configuration of these materials affected the LCST is poly(N-isopropyl acrylamide) (PNIPAM): a gradual decrease is recorded with increasing the isotacticity, and, for an isotactic (mm) triad, the insolubility in water appears to be over 72%, although the atactic (mr) PNIPAM is one of the representative water-soluble polymers

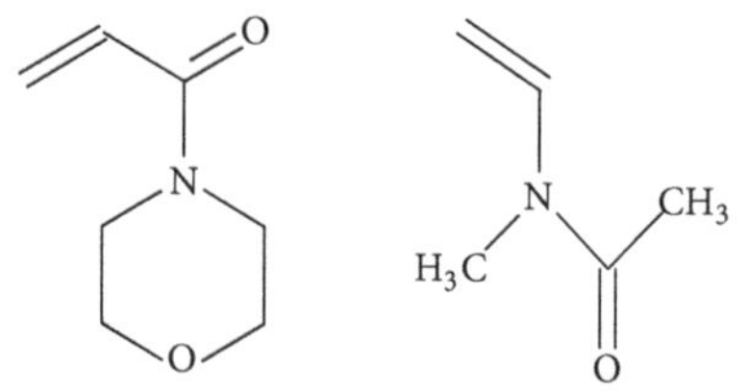

NAM NMNVA

Figure 1: Monomers structures: N-acryloyl-morpholine (NAM) and N-methyl-N-vinylacetamide (NMNVA).

[17]. Therefore, the stereoregulation of the structure has a strong influence on the solubility of the product.

Among the many techniques for structural control of polymers, a particular attention is focused on the tacticity control in the radical synthesis of polymers. This method has many advantages, but an important one is the possibility of leading this mechanism at industrial level. Under chiral circumstances, this approach has been applied even to radical polymerization known to be difficult to setup for stereocontrol, due to the flat orbital at the end of the growing polymer chain [18]. The best method for controlling the micromolecular organization, in this case, attained for several types of monomers [19–21], is using the solvent and temperature effects. Another influence in this technical direction is given by the monomer concentration, whose effect was revealed in the radical stereosynthesis of methacrylate and acrylic polymers [22, 23].

Recently, the study of parameters, which affects the microtacticity, discloses that the hydrogen bonding interactions are an important way to control the stereospecificity in the radical polymerization of N-alkyl acrylamides and substituted vinyl ethers [16].

According to these research directions, the goal of our paper is to discuss the influence of the solvent on the structure of NAM and, respectively, NMNVA (Figure 1) polymers, without neglecting the kinetics of the process on which there is not much information in the literature.

2. Experimental

2.1. Materials. NAM and NMNVA were purchased from Merck and purified by distillation under reduced pressure. The initiator used was 2, 2′-azobisisobutyronitrile (AIBN) and benzoyl peroxide (BPO) purchased from Aldrich and purified by recrystallization procedure. Solvents used were as follows: tetrahydrofuran (THF) and 1,4-dioxane from Aldrich, toluene and diethyl ether (as precipitation agent) from Chimopar SA, anisole from Merck-Schuchardt, diphenyl ether (DPE) from Fluka-AG, and methanol from SC PAM Corporation SRL. All the solvents were purified by usual methods.

2.2. Synthesis. For this study, two synthesis methods were used to investigate the materials from different points of view.

The first one was the dilatometric technique to determine the kinetics of the polymerization. For this case, the monomer concentration in the feed was 2 mole/L; Initiator used was benzoyl peroxide and the initial concentration $[BPO]_0$ was 5×10^{-3} mole/L. The solvents used in this method were tetrahydrofuran (THF), 1,4-dioxane (D), and toluene (T). The working temperatures were 65 and 75°C, respectively. The polymers have been separated by diethyl ether precipitation, washed, and dried at 40°C under vacuum for minimum of 12 hours.

The second procedure was classic radical polymerization: the needed quantities of monomer, solvent, and initiator were placed in standard Pyrex glass tubes. The reaction mixture was flushed with nitrogen for 2 minutes and then soldered and placed in a thermostated silicon oil bath at the working temperature (75°C). The monomer concentration in the feed was 2 mole/L; AIBN concentration was 5×10^{-3} mole/L. The used solvents were as follows: tetrahydrofuran (THF), 1,4-dioxane (D), toluene (T), anisole (An), and diphenyl ether (DPE). The polymers (poly(NAM) and poly(NMNVA)) have been isolated by precipitation in diethyl ether, washed several times with nonsolvent, and dried under vacuum at 40°C for minimum of 12 hours.

2.3. Characterization. To accomplish the goal of this work and to underline the structural peculiarities for obtained polymers, structure and composition of the samples have been determined by FT-IR, ^{1}H (^{13}C) NMR spectroscopy, and elemental analysis, respectively.

^{1}H-NMR and ^{13}C-NMR spectra have been obtained in dimethylsulphoxide with a Bruker Avance DRX 400 Instrument.

For FT-IR spectroscopy, a Brucker VERTEX 70 instrument, equipped with a Harrick MVP2 diamond ATR device apparatus, has been used.

3. Results and Discussions

3.1. Kinetics Investigations. The first aspect of our study proposes a preliminary investigation of chemical reactivity of the two monomers, by experimental determination of the initial rate of polymerization in different solvents. We used the dilatometric technique to measure conversions and reaction rates.

An idealized kinetic model for free radical polymerizations expresses the rate of polymerization as

$$R_p = \frac{k_p}{\sqrt{k_t}} [M]_0 \sqrt{2fk_d [I]_0}, \quad (1)$$

where k_d, k_p, and k_t are the initiator decomposition, propagation, and termination rate constants, respectively; f is the initiator efficiency factor (a constant for an initiator-monomer-solvent system); $[M]_0$ and $[I]_0$ are the initial monomer and initiator concentrations, respectively; a factor of 2 arises from generation of 2 moles of primary radicals from the decomposition of a mole of initiator. $k_p/(k_t)^{1/2}$ is a characteristic ratio which is a measure of a given monomer's propensity for kinetic growth. In real systems, often free radical polymerizations are far removed from this ideality.

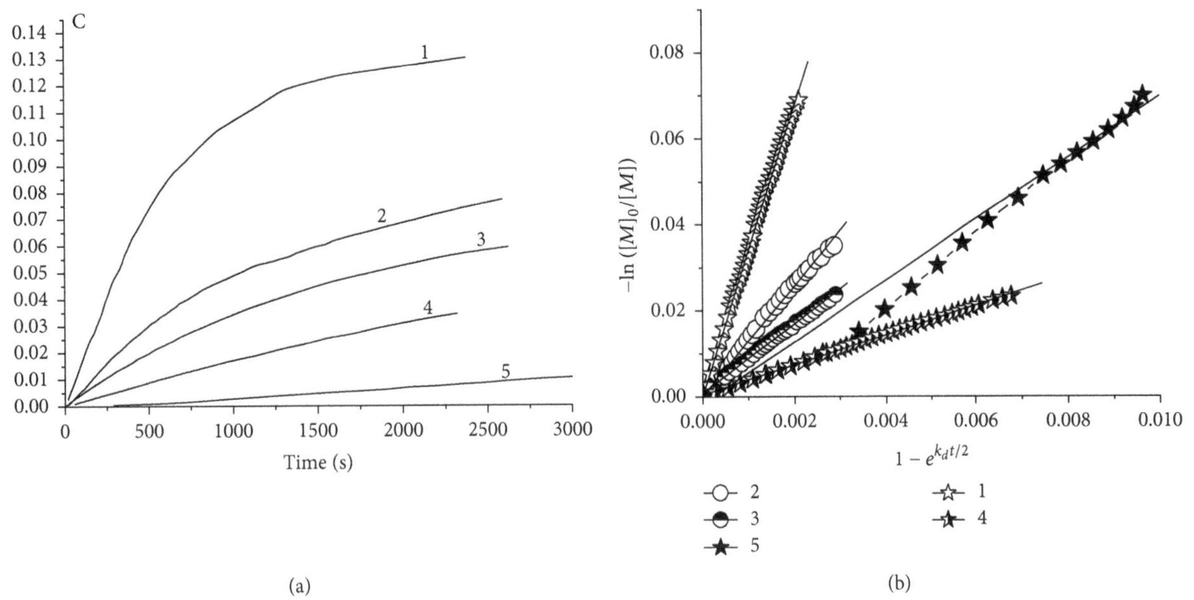

FIGURE 2: Conversion-time curves based on dilatometric results (a) and data of $\ln [M]_0/[M]$ as a function of $1-(1-e^{k_d t/2})$ according to (5) (b), where 2-poly(NMNVA) in 1,4-dioxane; 3-poly(NMNVA) in toluene; 5-poly(NAM) in 1,4-dioxane; 1-poly(NAM) in THF; 4-poly(NAM) in toluene.

Chain transfer reactions with H-abstraction occur at the active site of a primary radical or a macroradical, potentially involving any species present in the system: initiator, solvent, monomer, polymer, and so forth. These may prematurely terminate a growing macroradical, limiting the kinetic chain growth.

Taking into account the initial value of the height of the liquid in the capillary and because the polymerization process is accompanied by volume contraction, the polymerization rate can be written as

$$R_p = \alpha \frac{\Delta h \, [M]_0}{\Delta t}, \quad (2)$$

where Δt is time lapse from $t = 0$; Δh is change in the capillary height at Δt; and $[M]_0$ is initial concentration of the monomer (mole/L). Accurate determination of contraction factor $d[M]_0/\Delta h$ in mole/mm for a given capillary and the change in capillary height corresponding to full conversion (Δh^{∞}) enable the expression of monomer conversions in moles $L^{-1}\,s^{-1}$ and % conversions, respectively. Therefore, the curves height-time can be converted into conversion-time variation using the following relation:

$$c_i = K_D \Delta H_i, \quad (3)$$

where c_i, ΔH_i is the conversion of respective dilatometric height at current time and K_D (Table 1) is contraction factor, a specific constant strictly lied to the pair solvent//monomer//polymer.

The process of evolution in different solvents for both monomers, shown in Figure 2, reveals the absence of the induction period. The polymerization rate is relatively high and influenced by the polarity of reaction medium (Table 1): the global polymerization rate increases by increasing the solvent polarity index. The recorded decreasing of R_p for NAM polymerization in dioxane (row 3) can be explained only by particular interaction of (like hydrogen bonding) vinyl group—solvent.

The collected data from these experiments were used to assess $k_p/(k_t)^{0.5}$ ratio, based on the following equations:

$$R_p = -\frac{d[M]}{dt} = k_p \left(\frac{f k_d [I]}{k_t} \right)^{0.5},$$
$$[I] = [I]_0 \, e^{-k_d t}, \quad (4)$$

by integration of the following:

$$-\ln \frac{[M]_0}{[M]} = 2 \left(\frac{k_p}{\sqrt{k_t}} \right) \left(\frac{2 f [I]_0}{k_d} \right)^{0.5} \left(1 - e^{-k_d t/2} \right), \quad (5)$$

where $[M]$ is the monomer concentration at experimental time t, k_p, k_t, k_d denoting the absolute propagation, termination, and initiator decomposition rate, and f is the initiator efficiency.

Applying (3) the momentary monomer, concentration was obtained. Based on these data, dependence $\ln [M]_0/[M]$ versus $(1-e^{k_d t/2})$ was drawn and it is illustrated in Figure 2(b). From the slope of linear plot ($k_d = 1.5 \times 10^{-6}\,s^{-1}$ and $f = 0.75$ for BPO at 60°C [24]), $k_p/(k_t)^{0.5}$ are given for each homopolymerization experiment (Table 1).

The obtained values are in the same range with those of acrylamide or methacrylamide derivatives (Table 2). $k_p/(k_t)^{0.5}$ is relatively high for acrylamide but is a noticeable diminution with the nature of reaction medium and the

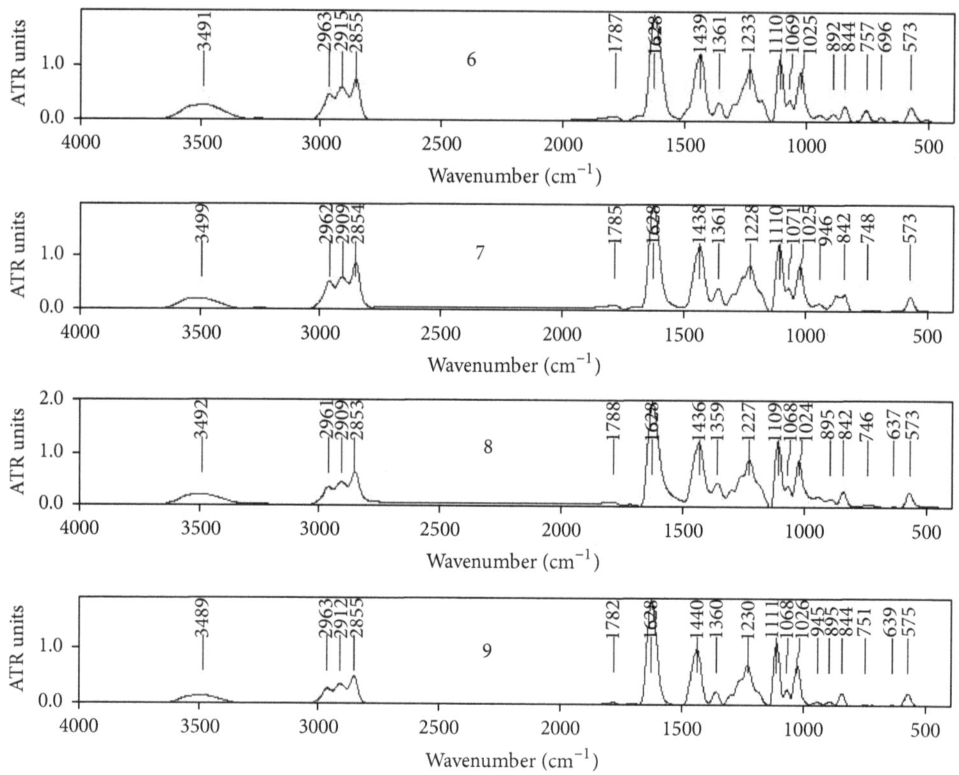

FIGURE 3: FT-IR spectra of poly(NAM) synthesized in different solvents by radical polymerization: 6 in anisole, 7 in 1,4-dioxane, 8 in toluene, and 9 in THF.

TABLE 1: Features of dilatometric measurements for poly(NAM) and for poly(NMNVA).

Code	$10^2 K_D$	C_f (%)	R_p (mol/L s)	$k_p/\sqrt{k_t}$
Poly(NMNVA) in 1,4-dioxane	2.60	7.7	1.033×10^{-4}	0.233
Poly(NMNVA) in toluene	2.20	5.9	7.100×10^{-5}	0.150
Poly(NAM) in 1,4-dioxane	10.00	1.1	7.000×10^{-6}	0.130
Poly(NAM) in THF	5.22	13.0	3.440×10^{-4}	0.611
Poly(NAM) in toluene	2.00	3.4	3.500×10^{-5}	0.064

In all experiments, $[M]_0$ is 2 mole/L; $[I]_0$ is 5×10^{-3} mole/L; T is 60 °C. K_D is contraction factor; C_f is final conversion; R_p is polymerization rate; $k_p/\sqrt{k_t}$ is the ratio between the propagation and termination constants.

degree of nitrogen substitution. On the contrary, methacrylamide is about the increase of value $k_p/(k_t)^{0.5}$ for the N-substituted derivatives. The nature of solvent affects these values [24].

All these behaviours can be attributed to the influence of the side chain as it was reported in the case of alkyl (meth)acrylates: $k_p/k_t^{0.5}$ value decrease was more likely due to a decreasing of termination constant ($k_p/k_t^{0.5} = 0.59$ for lauryl methyl methacrylate instead of 0.04 for methyl methacrylate), as a consequence of increasing the steric hindrance. Such an effect may also appear in the case of NMNVA and NAM homopolymerization related to the medium polarity: this can be provided by the two complexes, monomer-solvent and reactive polymer-solvent, respectively. A maximum value is obtained for NAM in THF. The first consequence of these observations can be retrieving the molar mass value of product (Table 3). Indeed, the increasing of the polarity index of medium causes an increasing of the molar weight. Generally, at moderate value of polarity index, NAM seems to be more reactive, perhaps due to the inexistence of a conjugation of the vinyl group with the substituent.

3.2. Physicochemical Characterizations. The polymers obtained through classical radical polymerization have been characterized in terms of chemical structure, molecular weight, and thermal behaviour.

3.2.1. Structural Investigations. The FT-IR analyses (see Figure 3 as an example) mark out the specific absorption of the structural atomic groups [29].

Thereby, analyzing the spectra from Figure 3, the broad absorption at 3491–3500 cm^{-1} was assigned to complexed ternary amide for both monomers. At 2850–2960 cm^{-1} (s-m) are recorded the antisymmetrical and symmetrical stretching of CH and CH_2 from the backbone and CH_2 and CH_3 side-chain groups. The weak band around 1700 cm^{-1} is assigned to C=O stretching (amide I) and the intense absorption from 1628 cm^{-1} to C=O stretching (amide II). CH_2 scissor vibrations are situated at 1437 cm^{-1} and the weak wagging and

TABLE 2: $k_p/k_t^{0.5}$ values for different acrylamide and methacrylamide.

Monomer	$k_p/k_t^{0.5}$ $(L/moles)^{0.5}$	Process parameters T (°C)	Solvent	Reference
Acrylamide	3.50	19	H_2O	[24]
	1.98	25	$H_2O/D = 9:1$	
	1.82	30	$H_2O/DMF = 9:1$	
	0.22	26	DMSO	
Methacrylamide	0.20	25	H_2O	[25]
N, N-Dimethyl acrylamide	0.46	30	H_2O	[26]
	1.78	50	H_2O	
N-(2,2-Dimethoxy ethyl)-N-methyl acrylamide	1.00	50	DMF	[27]
6-Deoxy-6-methacryloylamido-D-glucopyranose	1.00	50	H_2O	[28]

TABLE 3: Dimensional characteristics of poly(NAM) and poly(NMNVA).

(a) Poly(NAM)

Code	IP	C (%)	M_n (DP)	M_w	I_D
6 (anisole)	3.8	99.0	28 165(200)	42 327(45 611)[1]	1.50
7 (1,4 dioxane)	4.8	90.6	12 279(87)	58 065(48 572)[1]	4.73
8 (toluene)	2.4	71.0	31 709(225)	126 879(148 572)[1]	4.00
9 (THF)	4.0	79.0	3 633(26)	11 349(19 350)[1]	3.12

(b) Poly(NMNVA)

Code	IP	C (%)	M_n (DP)	M_w	I_D
10 (methanol)	5.2	64.2	—	(21 947)[1]	—
11 (1,4 dioxane)	4.8	75.9	5 587(56)	23 947(24 719)[1]	4.28
12 (THF)	4.0	73.4	4 387(44)	20 894(21 452)[1]	4.76
13 (toluene)	2.4	53.3	17 729(179)	63 748(61 650)[1]	3.40
14 (diphenyl ether)	2.2	21.1[2]	—	(19 769)[1]	—
		52.5[3]	—	(24 719)[1]	—

[1]In the brackets the estimated value from the elemental analysis.
[2]Liquid phase. [3]Solid phase.
IP is polarity index; C is conversion; M_n (DP) is number average molecular weight degree of polymerization; M_w is molecular weight; I_D is dispersity index.

twist CH_2, CH, at 1360 cm^{-1}. For NAM are recorded weak signals for transamide III at 1299 cm^{-1} and intense ones for ester bond at 1026 cm^{-1}.

For both polymers, the absorption from 1110 cm^{-1} is assigned to the skeleton modes, coupled for NMNVA with CH_3 rocking. The median band at 841–843 cm^{-1} is assigned to CH situated out of the plane and CH_2 wagging and twisting on morpholine ring. At 573 cm^{-1} can be noticed C=O bending out of the plane.

It is know that the solvent polarity induces the tacticity in polymerization products. This is an interesting and important aspect especially because the solvent effect in radical polymerization is neglected except in some systems where the influence of the solvent relates to the transfer process.

The literature underlines that the polarity of solvent may induce the selectivity at conformational level: syndiotactic (rr) configuration increases with increasing the solvent polarity. This is strictly related to the interaction among solvent, monomer, and growing polymer chain [19–23].

In both proposed cases for study, the nitrogen atom from amide group together with the carbonyl group might contribute to the appearance of hydrogen bonding interactions. Although the solvent effect cannot be completely rationalized, it can be assumed that the presence of the hydrogen bonding between monomer and growing chain decreases syndiotactic specificity.

Reproducing same experiment in a large range of solvent, we obtained information regarding the molecular dimension and stereochemical structure (Tables 3 and 4). All experiments developed as a homogeneous process, excepting the DPE for the NMNVA case. Usually, the heterogeneous evolution of a homogeneous system can be due to the existence of a critical molecular dimension, decreasing the compatibility polymer//mixture solvent-unreacted monomer, or to a peculiarly molecular configuration of the monomer units in the chain. In fact, in both cases, the polymerization changes the chain polarity. Particularly, in this case, considering the small size differences between the polymers, the two-phase evolution seems to be a consequence of the configuration of the monomer units in the polymer chains.

The elemental analysis allows a theoretical estimation of the molecular dimensions. The calculated fractions (%) of AIBN sequence in the polymer are in the accepted limit, under 1%, and we could estimate the dimension of materials. The obtained results (Table 3) are in good accordance with monomer composition and molar mass values obtained by GPC (polystyrene standard):

(i) For NAM, the lowest molecular weight value was obtained for THF that can be construed as activation process by monomer complexation. Consequently, the number of the initial propagation radicals is higher than in other cases. A particular evolution is recorded for anisole solvent: the polydispersity index is low.

(ii) For NMNVA is respected the same tendency reported in the case of NAM: the value of molar weight increases with decreasing the polarity of the solvent

Table 4: Evolution of poly(NMNVA) tacticity.

Solvent	IP		α			β		Pr/m		Pm/r	
		rr	mr	mm	rr	mr	mm	α	β	α	β
1,4-Dioxane	4.8	39	37	24	36	41	23	0.322	0.363	0.435	0.471
THF	4.0	31	40	29	36	37	27	0.392	0.339	0.408	0.407
Toluene	2.4	48	30	22	44	37	19	0.238	0.296	0.405	0.493

IP is polarity index; α and β are the two regions from NMR spectra used to estimate the tacticity; rr is isotactic fraction; mr is atactic fraction; mm is syndiotactic fraction; Pr/m is the conditional probability; and Pm/r is the reverse probability from Bernoullian statistics.

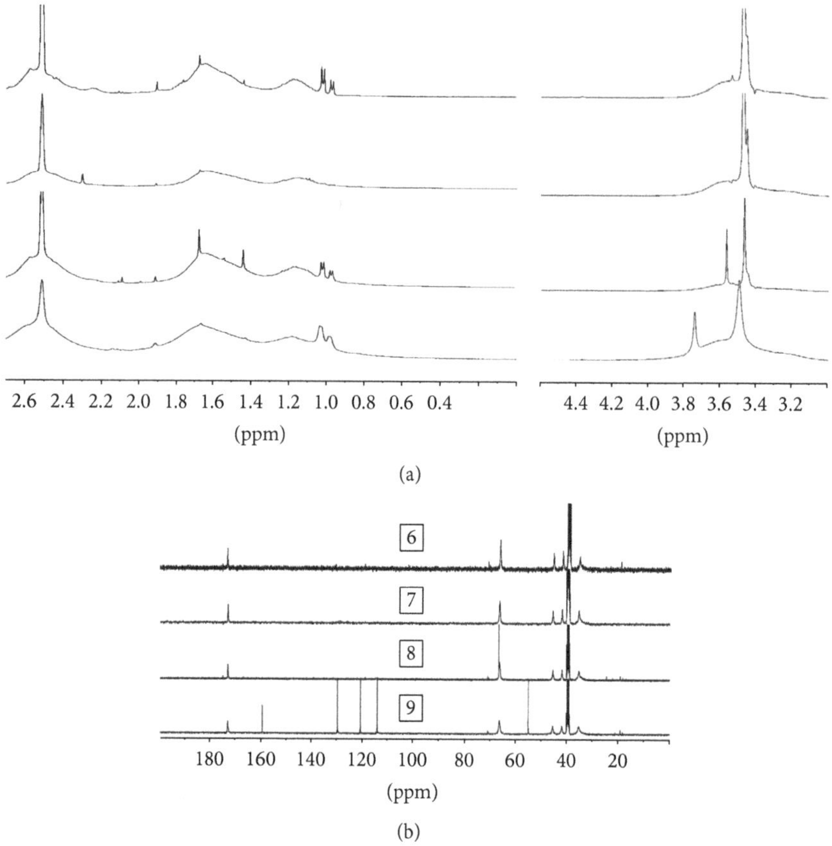

Figure 4: NMR spectra (solvent DMSO) for poly(NAM) synthesized in different solvents: 6 in anisole, 7 in 1,4-dioxane, 8 in toluene, and 9 in THF: (a) ^{1}H-NMR; (b) ^{13}C-NMR.

used in synthesis. Specifically, for this case is the same range for the magnitude order of the molar weight. The dispersion index values are in accordance with the free radical polymerization specific, a high number of the dimensional species. Therefore, the small differences between the two polymer species separated in the case of DPE solvent sustain a growing process at the surface of the separate polymer; process that is kinetic limited by diffusion parameters.

By NMR analyses, the effect of hydrogen bonding interactions on the control of the stereospecificity in radical polymerization can be investigated. It should be noticed that the peak shape and position varied slightly with the tacticity in the NMR spectra. This may be due to the same difference in conformation or to the intra/intermolecular hydrogen bonds in/between polymers with different configuration.

Figures 4(a) and 4(b) show ^{1}H NMR and ^{13}C NMR spectra for poly(NAM). Much simpler, ^{13}C NMR spectra (Figure 4(b)) highlight the specific signals for methylene and methine from the backbone at 35.2 ppm, and the methylene bonded on the nitrogen (H_2C-N-CH_2) at 45.4–41.8 ppm and those bonded on the oxygen (H_2C-O-CH_2) at 66.2 ppm. All the considered quaternary carbonyl carbons exhibit the same relaxation found at 172.6 ppm. When poly(NAM) was synthesized in D, there is a split of the ether carbon (66.4 and 66.2 ppm) that can be because of the changes in the morpholine ring conformation.

The proton spectra (Figure 4(a)) reveal the hydrogen signals of methylene groups in the backbone at 1–2.2 ppm and for those in morpholine ring at 3.4–3.5 ppm. Because the morpholine ring is relatively bulky and flexible, we were expecting a preferential arrangement of the main chain.

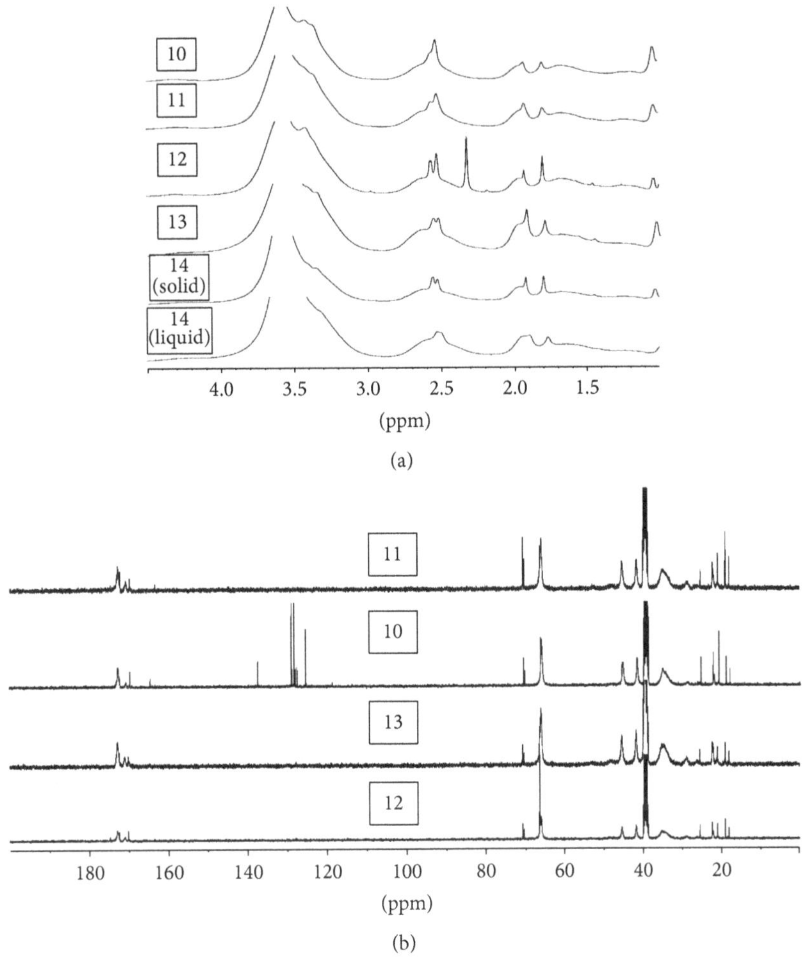

FIGURE 5: NMR spectra (solvent DMSO) for poly(NMNVA) synthesized in different solvents: 10 in methanol, 11 in 1,4-dioxane, 12 in THF, 13 in toluene, and 14 in diphenyl ether (solid and liquid phase): (a) ^{1}H-NMR; (b) ^{13}C-NMR.

Analyzing the region of the main chain of methylene groups, no remark of splitting or variation of the signals was observed. This can be interpreted as a lack of influence of the solvent on monomer unit conformation. However, increasing the polarity of the solvent the signal for methylene in morpholine ring is splitting, showing the distinct displacements for H_2C-N-CH_2 (3.45 ppm) and H_2C-O-CH_2 (3.47 ppm).

For NMNVA, we can talk about the potential stereospecific regions corresponding to the two-side chain methyl: α-region for the one bonded to the carbonyl group and β-region for the one bonded to the nitrogen atom (Figure 5(a)).

^{13}C NMR spectra (Figure 5(b)) confirm the elemental analyses results and show the specific signals for α-CH_3 at 18–21 ppm; β-CH_3 at 22–26 ppm, CH_2 from the backbone at 35 ppm; CH from the backbone at 42 and 45 ppm and at 172 C=O. Unfortunately, this is not enough to determine the tacticity dyad.

The complexity of ^{1}H NMR spectra discloses a broadened signal located around 3.5 ppm. This peak includes the protons from the backbone (CH_2 and CH) and two H_3C distinct regions for α- and β-CH_3: 1.1–2 ppm for β-CH_3 and 2.5–3 ppm for α-CH_3. The specific signal for AIBN methyl appears around 0.9–1 ppm. It should be noted that the shape and position of the peaks for α and β regions varied slightly depending on the tacticity.

Thereby, for β-CH_3, we can delimit the specific signal for rr triade: 1.6 ppm at broadening, mr at 1.7–1.8 ppm, and mm around 1.9 ppm. For both methylene regions, a relative fraction of the isotactic orientation can be observed. At the end of the polarity interval, distinct triads cannot be separated because of the overlapped mr signal (Figure 5(a)). Table 4 summarizes the evolution of stereospecificity for both methylene regions. For both α and β regions, an evolution with maximum for THF is observed.

In order to see if they obey the Bernoullian statistics, we tested three samples by using the triad tacticity information. Conditional probability Pr/m that a monomer is added to the growing end having a racemo configuration in mesofashion and reverse probability Pm/r are given by the following relations:

$$P_{rm} = \frac{mr}{(2rr + mr)}, \qquad P_{mr} = \frac{mr}{(2mm + mr)}. \tag{6}$$

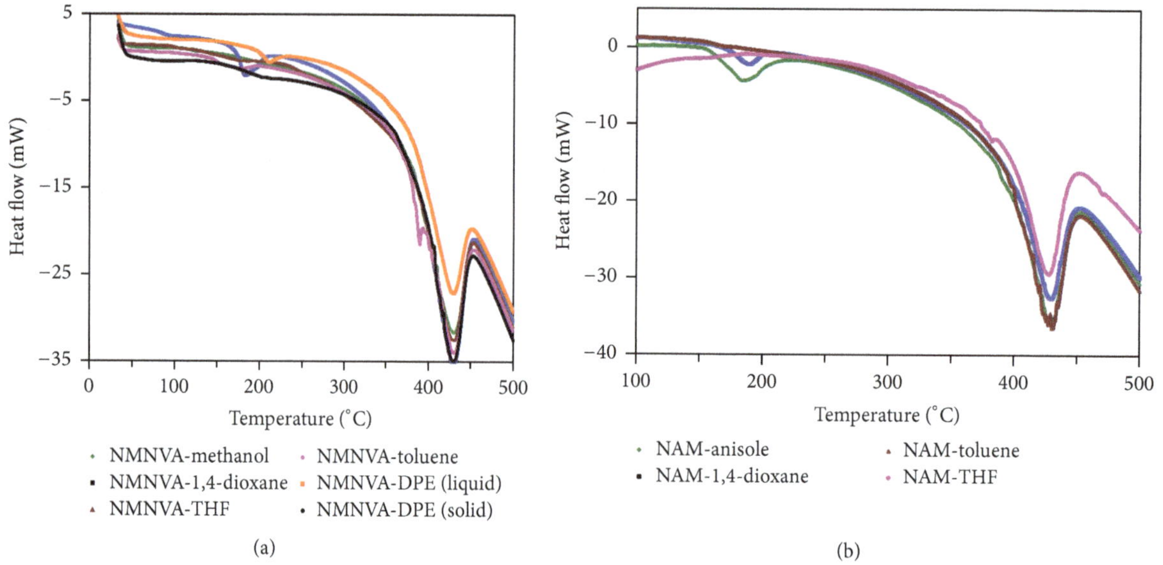

FIGURE 6: DSC curves of poly(NMNVA) (a) and poly(NAM) (b) [2nd heating run, heating rate 5°C/min; N_2 flow; 30 mL/min].

If $P_{rm} + P_{mr} = 1$, then chain propagation obeys Bernoullian statistics. The sum for both regions is less than the unity, pointing a non-Bernoullian model. Only for β_{CH3} in 1,4-dioxane and α_{CH3} in THF, the values are closed to unity (≥0.8). This points out that, with the increase of the polarity of the reaction medium, the evolution of the process approaches the Bernoullian model. In fact, the different evolutions can be explained by simultaneous different spatial orientation due to the steric hindrance on the account of the molar volume of each sequence.

3.2.2. Thermal Investigations. Thermal analysis of all samples, shown in Figure 6, shows one small peak between 50–210°C. Taking into account the associate weight loss (1 ÷ 20%), this is more probable due to the water evaporation. Although we might have expected a different thermal comportment related to rr content, all samples exhibit one broad endothermic peak at 427°C, similar to poly(N-methyl methacrylamide) with a maximum 51% rr [20]. It can be concluded that, from low to 50% tacticity, no changes are recorded from amorphous to crystalline state.

4. Conclusions

Radical polymerization of NAM and NMNVA was carried out using different solvents as reaction medium. The nature of solvent, understanding as polarity, affects the polymerization of the proposed monomers kinetically, dimensionally, and structurally. Although the stereochemistry is not so spectacular, it is clear that, using free radical polymerization and the adequate solvent, a structural modification of poly(NMNVA) can be induced. The information obtained in this work is useful in devising stereoregulation methods for radical polymerization of NMNVA.

Competing Interests

The authors declare that they have no competing interests.

Acknowledgments

The work of Anca Aldea has been funded by the Sectorial Operational Program Human Resources Development 2007–2013 of the Ministry of European Funds through the Financial Agreement POSDRU/159/1.5/S/132397.

References

[1] H. G. Börner, H. Kühnle, and J. Hentschel, "Making 'smart polymers' smarter: modern concepts to regulate functions in polymer science," *Journal of Polymer Science, Part A: Polymer Chemistry*, vol. 48, no. 1, pp. 1–14, 2010.

[2] H. G. Börner and H. Schlaad, "Bioinspired functional block copolymers," *Soft Matter*, vol. 3, no. 4, pp. 394–408, 2007.

[3] H.-A. Klok and H. Schlaad, Eds., *Peptide Hybrid Polymers*, Advances in Polymer Science, Springer, Berlin, Germany, 2006.

[4] J. C. M. Van Hest and D. A. Tirrell, "Protein-based materials, toward a new level of structural control," *Chemical Communications*, no. 19, pp. 1897–1904, 2001.

[5] H.-A. Klok, "Biological-synthetic hybrid block copolymers: combining the best from two worlds," *Journal of Polymer Science, Part A: Polymer Chemistry*, vol. 43, no. 1, pp. 1–17, 2005.

[6] Z. Guan, "Supramolecular design in biopolymers and biomimetic polymers for advanced mechanical properties," *Polymer International*, vol. 56, no. 4, pp. 467–473, 2007.

[7] E. S. Gil and S. M. Hudson, "Stimuli-reponsive polymers and their bioconjugates," *Progress in Polymer Science*, vol. 29, no. 12, pp. 1173–1222, 2004.

[8] I. Y. Galaev and B. Mattiasson, "'Smart' polymers and what they could do in biotechnology and medicine," *Trends in Biotechnology*, vol. 17, no. 8, pp. 335–340, 1999.

[9] W. Jaeger, J. Bohrisch, and A. Laschewsky, "Synthetic polymers with quaternary nitrogen atoms–synthesis and structure of the most used type of cationic polyelectrolytes," *Progress in Polymer Science*, vol. 35, no. 5, pp. 511–577, 2010.

[10] Z. M. O. Rzaev, S. Dinçer, and E. Pişkin, "Functional copolymers of *N*-isopropylacrylamide for bioengineering applications," *Progress in Polymer Science*, vol. 32, no. 5, pp. 534–595, 2007.

[11] F. D'Agosto, M.-T. Charreyre, L. Veron, M.-F. Llauro, and C. Pichot, "Kinetic study of free-radical solution copolymerization of N-acryloylmorpholine with an activated ester-type monomer, N-acryloxysuccinimide," *Macromolecular Chemistry and Physics*, vol. 202, no. 9, pp. 1689–1699, 2001.

[12] F. D'Agosto, M.-T. Charreyre, F. Mélis, B. Mandrand, and C. Pichot, "High molecular weight hydrophilic functional copolymers by free-radical copolymerization of acrylamide and of *N*-acryloylmorpholine with *N*-acryloxysuccinimide: application to the synthesis of a graft copolymer," *Journal of Applied Polymer Science*, vol. 88, no. 7, pp. 1808–1816, 2003.

[13] F. D'Agosto, R. Hughes, M.-T. Charreyre, C. Pichot, and R. G. Gilbert, "Molecular weight and functional end group control by RAFT polymerization of a bisubstituted acrylamide derivative," *Macromolecules*, vol. 36, no. 3, pp. 621–629, 2003.

[14] U. Voigt, V. Khrenov, K. Tauer, M. Hahn, W. Jaeger, and R. von Klitzing, "The effect of polymer charge density and charge distribution on the formation of multilayers," *Journal of Physics: Condensed Matter*, vol. 15, no. 1, pp. S213–S218, 2003.

[15] D. Fischer, H. Dautzenberg, K. Kunath, and T. Kissel, "Poly(diallyldimethylammonium chlorides) and their N-methyl-N-vinylacetamide copolymer-based DNA-polyplexes: role of molecular weight and charge density in complex formation, stability, and in vitro activity," *International Journal of Pharmaceutics*, vol. 280, no. 1-2, pp. 253–269, 2004.

[16] T. Hirano, Y. Okumura, M. Seno, and T. Sato, "Remarkable effect of hydrogen-bonding interaction on stereospecificity in the radical polymerization of N-vinylacetamide," *European Polymer Journal*, vol. 42, no. 9, pp. 2114–2124, 2006.

[17] B. Ray, Y. Okamoto, M. Kamigaito et al., "Effect of tacticity of poly(N-isopropylacrylamide) on the phase separation temperature of its aqueous solutions," *Polymer Journal*, vol. 37, no. 3, pp. 234–237, 2005.

[18] H. Ajiro and M. Akashi, "Radical polymerization of novel *N*-substituted-*N*-vinylformamide derivatives with bulky chiral substitutents," *Journal of Polymer Science, Part A: Polymer Chemistry*, vol. 50, no. 1, pp. 134–141, 2012.

[19] T. Nakano, K.-I. Makita, and Y. Okamoto, "Stereochemistry of free-radical polymerization of bulky vinyl esters," *Polymer Journal*, vol. 30, no. 8, pp. 681–683, 1998.

[20] W. Liu, T. Nakano, and Y. Okamoto, "Stereocontrol in radical polymerization of N,N-dimethylacrylamide and N,N-diphenylacrylamide and thermal properties of syndiotactic poly(methyl acrylate)s derived from the obtained polymers," *Polymer Journal*, vol. 32, no. 9, pp. 771–777, 2000.

[21] J. Zhang, W. Liu, T. Nakano, and Y. Okamoto, "Stereospecific radical polymerization of *N*-methyl methacrylamide," *Polymer Journal*, vol. 32, no. 8, pp. 694–699, 2000.

[22] Y. Okamoto and E. Yashima, "Asymmetric polymerization of methacrylates," *Progress in Polymer Science*, vol. 15, no. 2, pp. 263–298, 1990.

[23] Y. Okamoto, "Chiral polymers for resolution of enantiomers," *Journal of Polymer Science, Part A: Polymer Chemistry*, vol. 47, no. 7, pp. 1731–1739, 2009.

[24] A. Abe and D. R. Bloch, *Polymer Handbook*, Edited by J. Brandrup, E. H. Immergut, and E. A. Grulke, Wiley-Interscience, 4th edition, 1999.

[25] F. S. Dainton and W. D. Sisley, "Polymerization of methacrylamide in aqueous solution. Part 1.—Hydrogen-peroxide photosensitized reaction," *Transactions of the Faraday Society*, vol. 59, pp. 1369–1376, 1963.

[26] D. N. Schulz, "Water-soluble polymer synthesis: theory and practice," in *Water-Soluble Polymers: Synthesis, Solution Properties, and Applications*, vol. 467 of *ACS Symposium Series*, chapter 3, pp. 57–73, American Chemical Society, 1991.

[27] A. Gallardo and J. San Román, "Kinetic and microstructural parameters of the free-radical copolymerization of 2-hydroxyethyl methacrylate with methacrylic monomers bearing bulky polar side-groups," *Polymer*, vol. 34, no. 3, pp. 567–573, 1993.

[28] B. Badey, P. Boullanger, A. Domard, P. Cros, T. Delair, and C. Pichot, "Radically initiated polymerization of a methacryloylamido-terminated saccharide, 1. Monomer synthesis, homopolymerization and characterizations," *Macromolecular Chemistry and Physics*, vol. 197, no. 11, pp. 3711–3728, 1996.

[29] E. Pretsch, Ph. Buhlmann, and M. Badertscher, *Structure Determination of Organic Compounds: Tables of Spectral Data*, Springer, Berlin, Germany, 2009, Revised and Enlarged Edition: chapter 4: ^{13}C NMR Spectroscopy, pp. 69–156, chapter 5: ^{1}H NMR Spectroscopy, pp. 157–242, chapter 7: IR Spectroscopy, pp. 26–337.

13

Experimental Analysis of Tensile Mechanical Properties of Sprayed FRP

Zhao Yang and Kun Wu

Institute of Urban Construction, Wuhan University of Science and Technology, Wuhan 430065, China

Correspondence should be addressed to Zhao Yang; yzwh77@163.com

Academic Editor: Kaveh Edalati

To study the tensile mechanical properties of sprayed FRP, 13 groups of specimens were tested through uniaxial tensile experiments, being analyzed about stress-strain curve, tensile strength, elastic modulus, breaking elongation, and other mechanical properties. Influencing factors on tensile mechanical properties of sprayed FRP such as fiber type, resin type, fiber volume ratio, fiber length, and composite thickness were studied in the paper too. The results show that both fiber type and resin type have an obvious influence on tensile mechanical properties of sprayed FRP. There will be a specific fiber volume ratio for sprayed FRP to obtain the best tensile mechanical property. The increase of fiber length can lead to better tensile performance, while that of composite thickness results in property degradation. The study can provide reference to popularization and application of sprayed FRP material used in structure reinforcement.

1. Introduction

Fiber reinforced polymers (FRP) composite materials have been largely used as strengthening and rehabilitation materials for decades in civil engineering. The main forms of these materials are sheet and plate, which are usually externally bonded on the surfaces of structure members. These materials have been proven to be very efficient and convenient in structure strengthening due to their high strength-to-weight ratio with excellent corrosion resistance [1–5]. But there are still some disadvantages in the use of external bonded FRP. The first is that the interfacial debonding between FRP sheet or plate and structure member will occur for the degradation of the bonding quality, which will lead to a premature failure of FRP [6–13]. Although some anchorage devices or mechanical strategies have been studied to delay such kind of debonding failure, this method will make the construction process more difficult. The second is that the ultimate strain of FRP sheets or plates used in structure strengthening cannot reach the ultimate strain in the material tensile test, which means the high strength of FRP sheets or plates cannot be fully used [14–16]. The third is that the external bonding method of FRP sheets or plates will be hardly conducted in some structure members such as beam-column joints, which cannot provide enough flat surfaces for FRP to be pasted continuously and the core areas of which are difficult for such forms of FRP to wrap efficiently [17–22].

To solve the abovementioned problems, the advanced strengthening method using sprayed FRP consisting of randomly oriented reinforcing fibers in a matrix has been suggested. Spray operation was conducted through special equipment (Figure 1). Resins were transferred to a spray gun and sprayed onto the structure surface. Simultaneously, fiber filaments were chopped to short ones in the spray gun and sprayed onto the structure surface, mixed with the resins. Then, the sprayed FRP material can be formed continuously on the structure surface. This kind of FRP has some advantages compared with the external bonded ones. Firstly, when being used on an uneven concrete surface, external bonded FRP needs some material to level the surface, and then the resin is used to bond the FRP and the leveled substrate. Thus, there are resins, leveling materials, and concrete substrate bonded together. This will increase the debonding possibility and make the application process complicated. While using sprayed FRP, the sprayed FRP material can level the surface as well, and thus there are only sprayed materials and concrete substrates that are bonded together and decrease the debonding possibility.

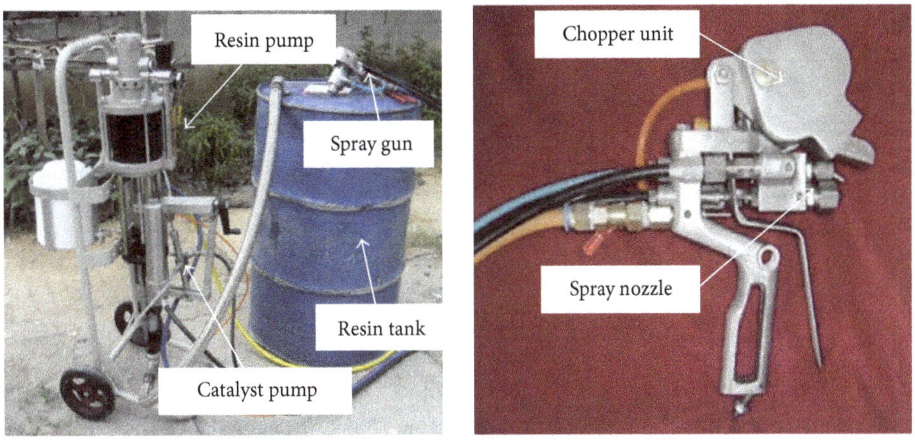

(a) Complete spray equipment

(b) Spray gun

FIGURE 1: Special equipment of sprayed FRP.

The application process is simpler as well without a special leveling process [23]. Secondly, when being used to reinforce a beam-column joint, multilayers of external bonded FRP are required to be bonded on the joint along a different loading direction, which makes the application process much complicated and difficult to assure the construction quality. While using the sprayed FRP, the construction process can be much easier than external bonded method due to the spraying method.

In 2000, sprayed FRP was used for the first time to strengthen the reinforced concrete beams by Boyd. The study showed that the flexural reinforcing effect of sprayed glass FRP was similar to that of FRP fabric wrap or plate bonding onto the beams, while the shear strengthening effect of sprayed FRP was better than other techniques [24]. In 2004, Lee and Hausmann investigated the load capacity, ductility, and energy absorption aspect of RC beams retrofitted with sprayed FRP. The study showed that sprayed FRP was capable of substantially increasing the performance of the beams and was effective in repairing damaged RC beams [25]. Then, Soleimani SM studied sprayed FRP in the shear strengthening and enhancement of impact resistance of reinforced concrete beams [26]. Boyd et al. studied sprayed FRP used in prestressed concrete girders [27]. Ha et al. studied the bond characteristics of sprayed FRP composites bonded to concrete substrate considering various concrete surface conditions [28]. Though the studies above got satisfying results, the use of sprayed FRP for structure strengthening has never been fully investigated.

This paper is written to study the tensile properties and influencing factors of sprayed FRP. A series of specimens of sprayed FRP were tested through tensile experiments. Tensile strength, elastic modulus, breaking elongation, and Poisson ratio of the specimens were studied through the tests. And fiber type, resin type, fiber length, fiber volume ratio, composite thickness, and specimen cutting orientation were compared as influencing factors in the paper too. The results can provide theoretical basis for tensile property calculation and actual application of sprayed FRP.

2. Tensile Tests

2.1. Specimens and Experiment Method. In this paper, 3 types of fiber, which are glass fiber, carbon fiber, and basalt fiber, and 2 kinds of resin, which are vinyl ester resin (901) and unsaturated polyester resin (196), were used to make sprayed FRP specimens. The fiber yarns and resins used in the study are shown in Figures 2 and 3. The properties of the fibers and resins are shown in Tables 1 and 2, where the values were obtained from material suppliers. The special equipment used to conduct the spraying process is shown in Figure 1.

The specimens of sprayed FRP were designed in accordance with the shape recommended by Boyd [24] because no specific test method has been published by ASTM so far. However, the dimensions have been changed a little considering the length of the chopped fibers and the fit for the experiment equipment. The detailed dimensions of the actual specimens are shown in Figure 4, in which all dimensions are in millimeters (mm).

The fabrication process of the tensile specimens was divided into several steps. Firstly, a 1000 mm × 800 mm rectangle glass plate was put on the flat ground with a thin sheet of polyethylene terephthalate paved on it to prevent the adhesion of resin and glass. Then, the chopped fibers and resins were sprayed onto the glass plate to form a sprayed FRP plate. After spraying, a ribbed aluminum compaction roller was used to eliminate the entrapped air voids and make the composite thickness consistent. After hardening, the sprayed FRP plate can be removed from the glass plate and the polyethylene terephthalate sheet can be peeled off. Then, the sprayed FRP plate was cured for seven days in open air at 25°C. After that, the plate was polished and cut to the designed dimensions. In order to study the isotropic behavior of this sprayed composite, the specimens were cut from the plate in three different orientations, shown in Figure 5.

There are 13 groups of specimens fabricated in the tensile test to study the tensile properties and influencing factors of sprayed FRP material. Each group has 6 specimens to compare the test results and the influencing factors are

Table 1: Properties of fibers.

Type	Tensile strength, σ (MPa)	Elastic modulus, E (GPa)	Breaking elongation, ε (%)	Poisson ratio, μ	Density, ρ (kg/m^3)
Glass fiber[a]	1970	79	4.8	0.20	2550
Carbon fiber[b]	3500	205	1.5	0.31	1750
Basalt fiber[c]	2200	93	2.6	0.25	2630

[a]Data from Jilin Carbon Co., Ltd., in China.
[b]Data from Zhejiang GBF Basalt Fiber Co., Ltd., in China.
[c]Data from Jushi Group Co., Ltd., in China.

Table 2: Properties of resins.

Type	Tensile strength, σ (MPa)	Elastic modulus, E (GPa)	Breaking elongation, ε (%)	Poisson ratio, μ	Density, ρ (kg/m^3)
901[a]	85	3.4	5.5	0.3	1050
196[b]	60	3.0	3.0	0.35	1230

[a]Data from Swancor Renewable Energy Co., Ltd., in China.
[b]Data from Shenzhen Tonglai Material Technology Co., Ltd., in China.

(a) Glass fiber (b) Carbon fiber (c) Basalt fiber

Figure 2: Fiber yarns used in the experiments.

(a) Vinyl ester resin 901 (b) Unsaturated polyester resin 196

Figure 3: Resins used in the experiments.

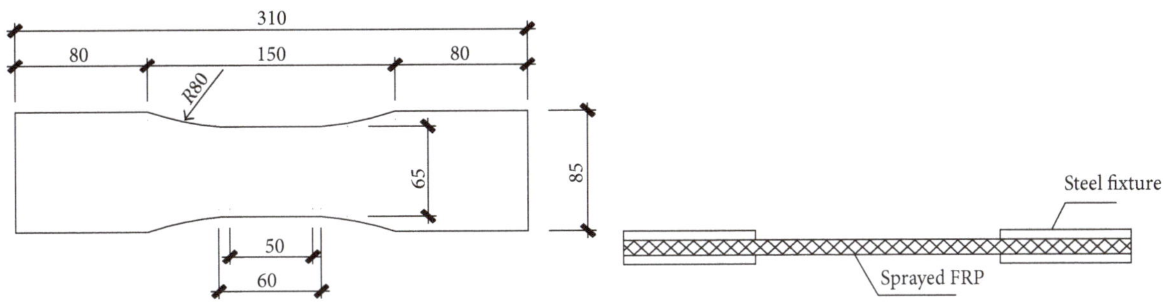

(a) Top view (b) Front view

FIGURE 4: Dimensions of the actual specimens.

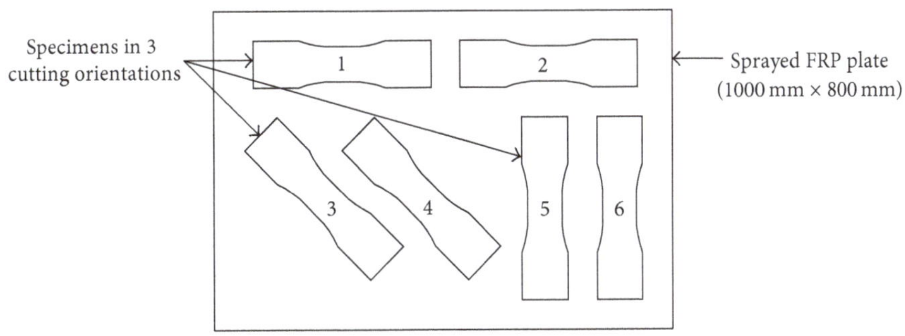

FIGURE 5: Specimen cutting orientation.

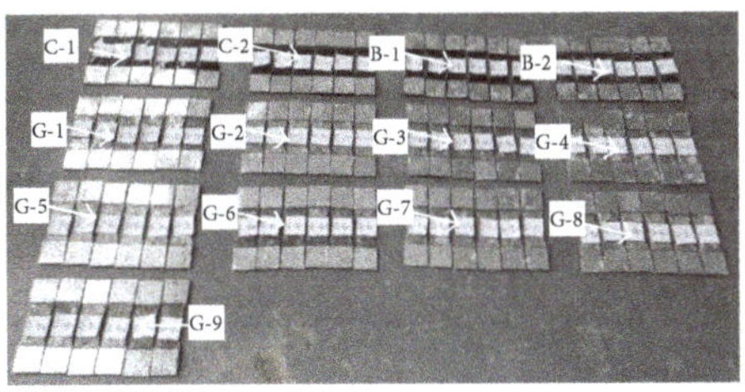

FIGURE 6: 13 groups of specimens.

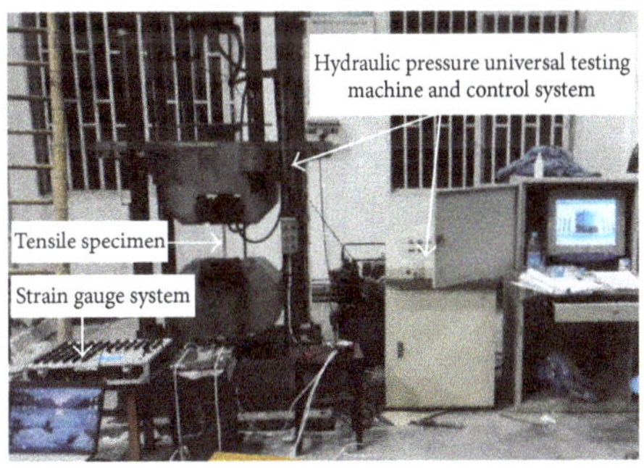

FIGURE 7: Loading and measurement apparatus.

considered as fiber type, resin type, fiber volume ratio, fiber length, and composite thickness in the paper. The specimen plan is shown in Table 3 and the pictures of the 13 specimen groups are shown in Figure 6.

The composite thickness was controlled by using some plastic plates surrounding the rectangle glass plate on the ground, and the required thickness was marked on the plastic plates, and then the mark could be used to control the composite thickness when spraying. Some concrete blocks were used to fix the plastic plates and prevent deformation. In addition, after cutting the specimens, the thickness of each specimen was measured at both sides of the middle section. The measurement errors were observed to be below 3% for all specimens, which indicated that the actual thickness met well the designed value.

The fiber length can be easily controlled by adjusting the chopped unit. The fiber volume ratio can be well controlled by adjusting the spraying speed of chopped fibers and resins, respectively, as well. It was observed that error percentages were both lower than 3%, which could meet the design requirements well.

The tensile tests were conducted using a hydraulic pressure universal testing machine, which is shown in Figure 7. In the middle of every tensile specimen, 2 strain gauges in horizontal and vertical direction, respectively, were bonded on the specimen surface to measure the strain of the composite during loading process through a strain gauge system. The loading speed was set to be 2 mm/min in accordance with Chinese code GB/T1447 (2005) [29].

3. Test Results

3.1. Failure Mode. Through the observation of failure processes of the specimens, all the 13 groups of specimens are similar in failure mode. The failure process can be divided

Table 3: Specimen plan.

Group number	Group name	Fiber type	Resin type	Fiber volume ratio	Fiber length (mm)	Composite thickness (mm)
1	C-1	Carbon	901	20%	30	4
2	C-2	Carbon	196	20%	30	4
3	B-1	Basalt	901	20%	30	4
4	B-2	Basalt	196	20%	30	4
5	G-1	Glass	901	20%	30	4
6	G-2	Glass	196	20%	30	4
7	G-3	Glass	901	15%	30	4
8	G-4	Glass	901	25%	30	4
9	G-5	Glass	901	30%	30	4
10	G-6	Glass	901	20%	15	4
11	G-7	Glass	901	20%	20	4
12	G-8	Glass	901	20%	40	4
13	G-9	Glass	901	20%	30	7

(a) Glass fiber (b) Basalt fiber (c) Carbon fiber

Figure 8: Through cracks in sprayed FRP specimens of different fiber types.

into 3 stages. In the first stage, the stress in the composite increased quickly with the increase of tensile load, while strain increased slowly. It was observed that there are no signs of cracking or debonding inside the specimens, which always appeared like some small white lines. In the second stage, initial crack could be observed. The strain in composite increased fast, while stress increased slowly. Cracks usually started at the edge of the composite and progressed rapidly to the center. Fracture always began in resin and then debonding occurred between fiber and resin, which made the fiber pull out of the resin and led to crack development. In the third stage, the specimen reached the ultimate load. The stress began to decrease rapidly and strain increased a little. Through cracks were formed (Figure 8) and resulted in a total failure of the specimens. Some fibers along the through cracks were pulled out and some still remained in the resin (Figure 9). There were 3 main modes for the through cracks, which are shown in Figure 10.

3.2. Stress-Strain Relationship. The tensile stress of sprayed FRP can be calculated by dividing the load value by the cross-sectional area of the specimen. The strain of sprayed FRP can be collected automatically by the strain gauge system. Thus, the experimental stress-strain relationship can be achieved. The relationship curves of the 13 groups are very similar and 3 groups of them are shown as examples in Figure 11, from which we can see the 3 stages of the failure process clearly. From Figure 11, we can see that, in the first stage, fibers and resins were working together to bear the tensile load, no cracking or debonding appeared, the strain increased slower than stress, and the first stage of the curve was formed. While the initial cracks appeared, some resins ruptured and debonding between fibers and resins appeared. The deformation of specimens increased more quickly than the loads, and thus the second stage of the curve was formed. When the cracks developed, more and more resins ruptured, the loads were mainly borne by fibers until reaching the bearing capacity, and the ultimate stress of specimens was reached. Then, the stress started to decrease and strain continued to increase. The third stage of the curve was formed. So, the sprayed FRP composites showed a trilinear constitutive behavior.

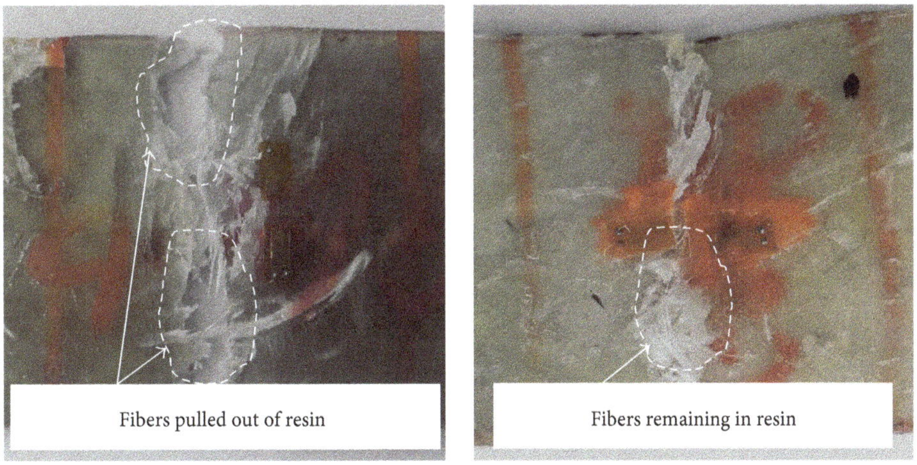

(a) (b)

FIGURE 9: Failure modes of fibers along the through crack.

(a) (b) (c)

FIGURE 10: Failure modes of through cracks.

3.3. Tensile Mechanical Properties of Sprayed FRP. Through the data gained from the tensile experiment, several tensile mechanical properties can be achieved using the calculation method recommended in Chinese code GB/T1447 (2005) [29], as shown in Table 4. Due to the difficulty in obtaining the actual initial modulus, the values of elastic modulus reported in Table 4 referred to a secant modulus. The secant endpoints were the endpoints of 50 $\mu\varepsilon$ and 40% of ultimate stress [24]. Note that the values reported in the table represent the averages of all the six specimens in each group and all the error percentages for each group were lower than 3%.

4. Influencing Factors on Tensile Mechanical Properties of Sprayed FRP

4.1. Fiber Type. Carbon fiber, glass fiber, and basalt fiber were used, respectively, to make sprayed FRP in the study, and their tensile performances were different. As is shown in Table 3, C-1, B-1, and G-1 are made of different fibers but the same resin 901 and other influencing factors are all the same. So, these 3 groups can be used to compare the effect of different fiber types using resin 901. In the same way, C-2, B-2, and G-2 are made of different fibers but the same resin 169 and other influencing factors are all the same too. So, these 3 groups can be used to compare the effects of different fiber types using resin 169.

As is shown in Table 4, C-1 using carbon fiber has the highest strength value of 105.4 MPa, and the strength of G-1 using glass fiber is 91.3 MPa, while B-1 using basalt fiber has the lowest strength value of 87.2 MPa. Among the 3 groups, C-1 also has the highest tensile elastic modulus value of 49.1 GPa, which is 54.4% higher than G-1 and 74.1% higher than B-1. As to breaking elongation, the case is just the opposite. B-1 has the highest value of 2.6%, which is 8.3% higher than G-1 and 23.8% higher than C-1.

For C-2, B-2, and G-2, the comparative results are similar to the above 3 groups. The strength value of C-2 is the highest

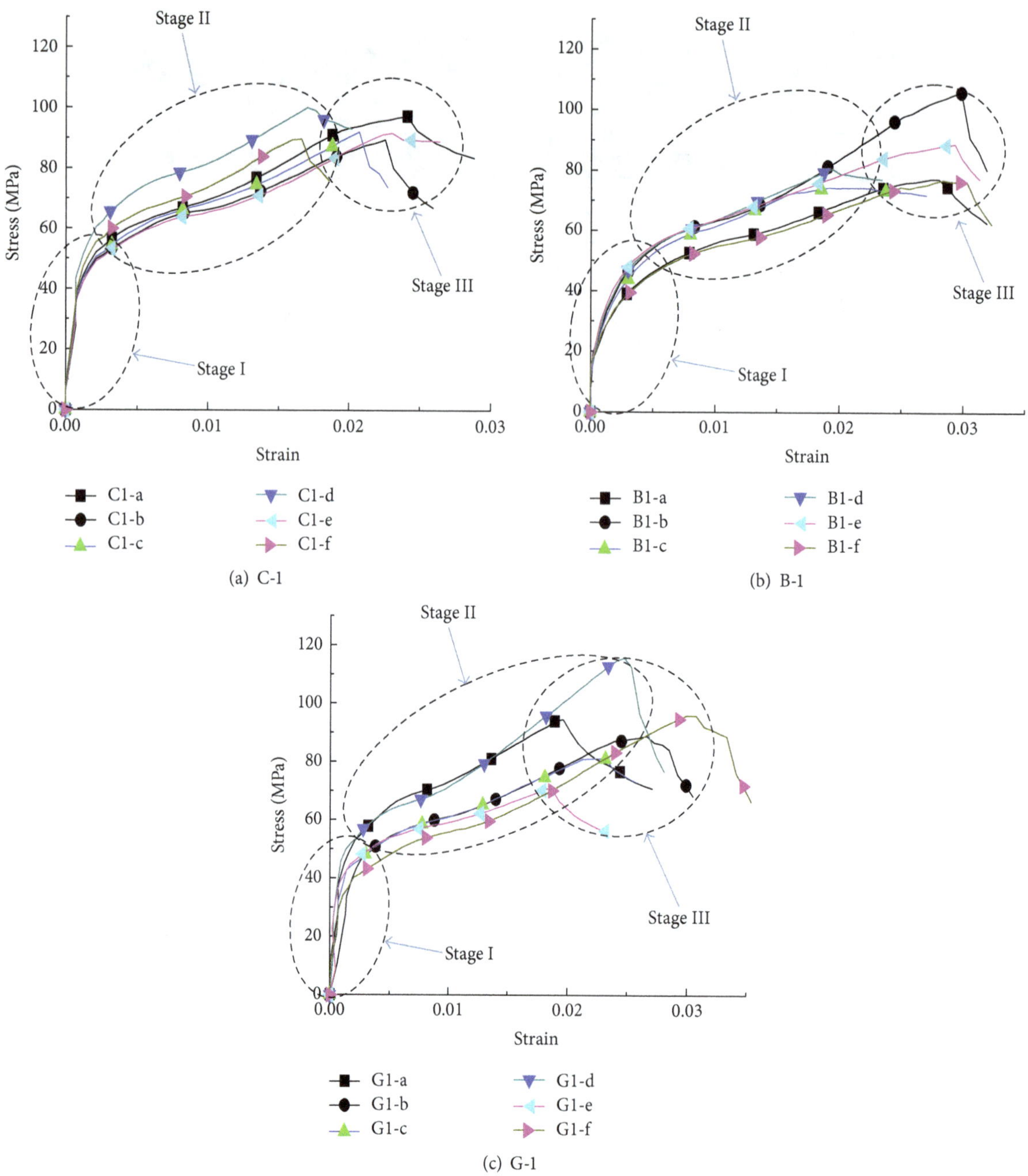

FIGURE 11: Stress-strain relationship curves.

at 89.3 MPa and G-2 is the lowest at 75.2 MPa. The differences in tensile elastic modulus are more significant than the above 3 groups. The value of C-2 is up to 50.7 GPa, G-2 follows by 28.8 GPa, and B-2 is the minimum at 26.1 GPa. While the case of breaking elongation is also opposite, B-2 has the highest value of 3.1%, G-2 follows by 2.4%, and C-2 has the lowest value at 1.9%.

According to the data in Table 4, the effect of different fiber types on tensile properties of sprayed FRP is compared in Figure 12. It can be seen that tensile strength and tensile elastic modulus of sprayed FRP using carbon fiber are higher than those sprayed FRP using glass fiber and basalt fiber, no matter what kind of resin is used. But the breaking elongation of sprayed FRP using carbon fiber is the lowest among the three. These indicate that the tensile performance of sprayed FRP is affected greatly by fiber type and the performance feature of sprayed FRP is always similar to the fiber used, which is in accordance with the conclusion of Lee [30]. It should be noted that the differences in the strength of sprayed FRP using glass and basalt fiber are very small with the maximum value of only 4.9%.

4.2. Resin Type. The effect of resin type on tensile properties of sprayed FRP was studied by comparing C-1 and C-2, B-1

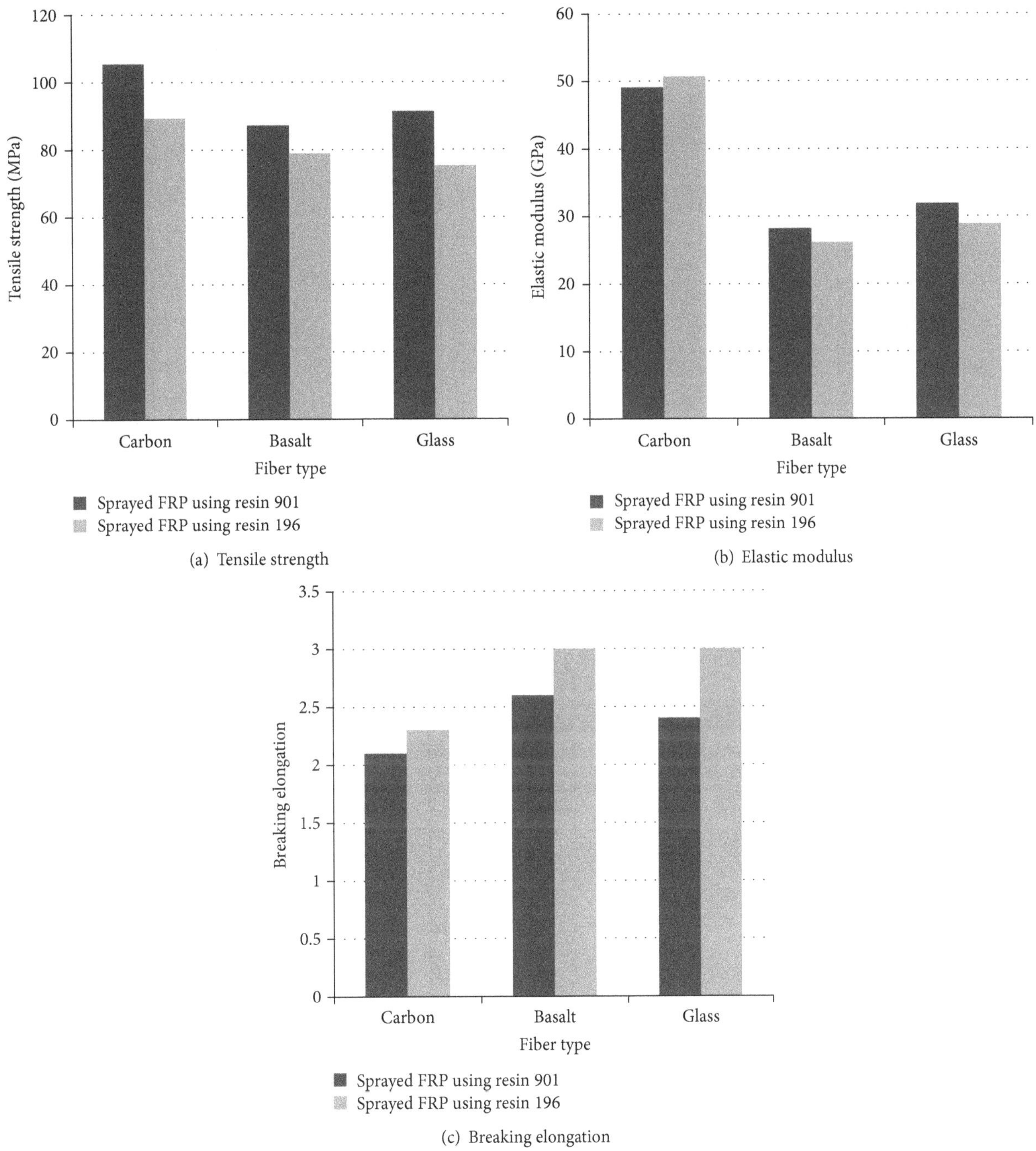

FIGURE 12: Comparison of tensile properties of sprayed FRP using different fiber types.

and B-2, and G-1 and G-2, separately, which are different in resin type but the same in all other factors. C-1, B-1, and G-1 used vinyl ester resins. C-2, B-2, and G-2 used unsaturated polyester resins. As is shown in Table 4, tensile strength of C-1 is 18.2% higher than C-2, B-1 is 6.3% higher than B-2, and G-1 is 21.4% higher than G-2. Regarding tensile elastic modulus, the values of C-1, G-1, and B-1 are also higher than those of C-2, B-2, and G-2. But, in the comparison of breaking elongation, the cases are different. The value of C-1 is 10.5% higher than C-2, the value of B-1 is 8.3% higher than B-2, and the value of G-1 is 20% lower than G-2.

The above results show that using vinyl ester resin as matrix and adhesive for sprayed FRP can get better tensile mechanical properties generally, which is similar to the result of Lee [30]. But it is worth noting that the breaking elongation of the sprayed FRP using glass fiber will be reduced when using this kind of resin.

4.3. Fiber Volume Ratio. From Table 3, there are 4 groups of specimens used to study the effect of fiber volume ratio on tensile properties of sprayed FRP, which are G-3, G-1, G-4,

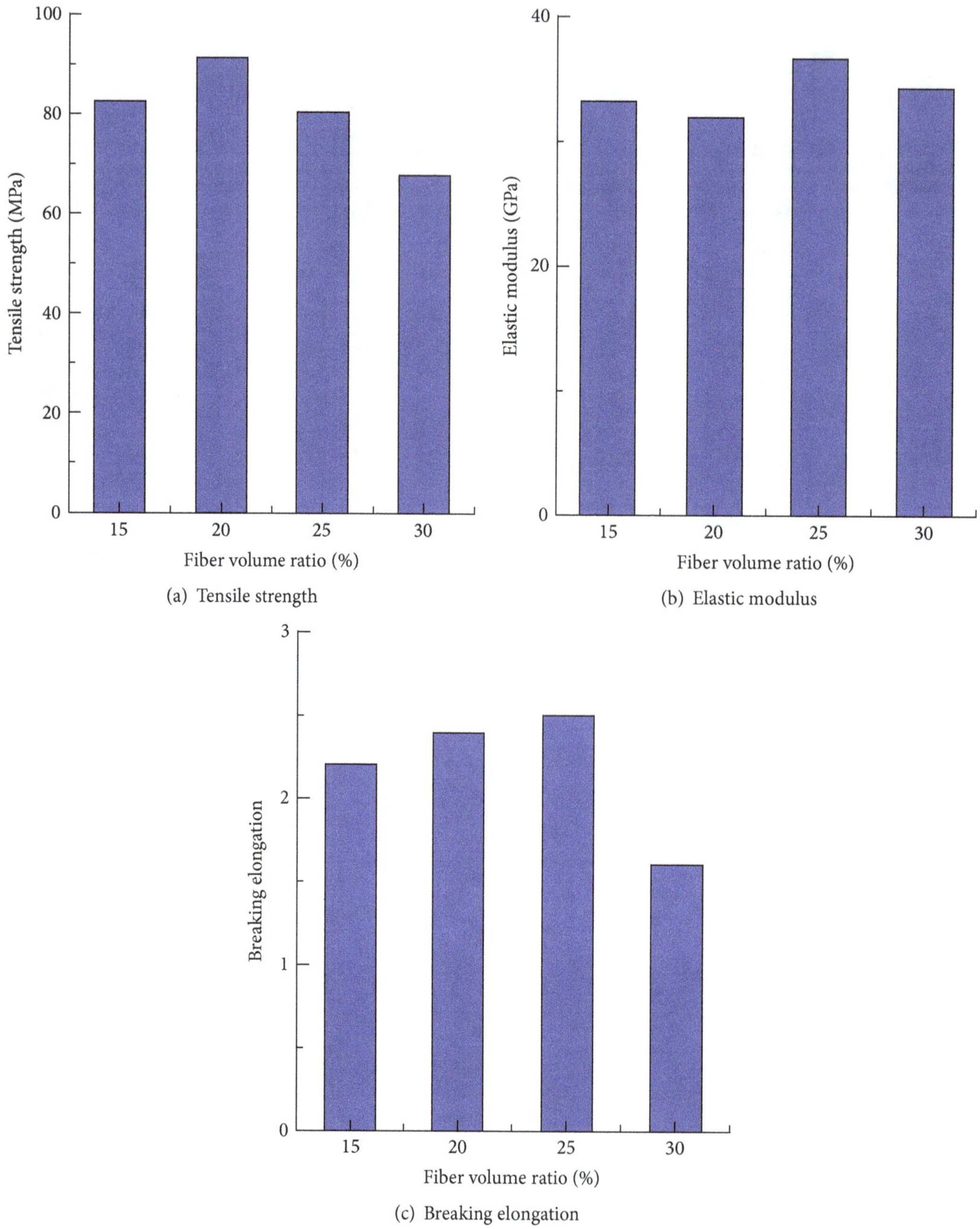

FIGURE 13: Comparison of tensile properties of sprayed FRP in different fiber volume ratios.

and G-5, with a fiber volume ratio of 15%, 20%, 25%, and 30%, respectively.

It can be seen from Table 4 that the tensile strengths of the 4 groups from high to low are G-1, G-3, G-4, and G-5, with a value of 91.3 MPa, 82.6 MPa, 80.3 MPa, and 67.7 MPa, respectively. The highest value of G-1 is 34.9% higher than the lowest one G-5. As to tensile elastic modulus, the differences of the 4 groups became very small. The differences between G-4 with the highest value and the other 3 groups ranged from 6.7% to 15.4%. In the case of breaking elongation, the highest value comes from G-4 too, which is 56.3% higher than that of G-5 with the lowest value.

From the data in Table 4, Figure 13 was drawn to compare the effects of different fiber volume ratios on tensile properties of sprayed FRP. It shows that there will be a specific value of fiber volume ratio that can lead to a maximum value of a tensile property. And when the ratio is lower than the specific value, the tensile property value will increase with the increase of volume ratio. While the fiber volume ratio is higher than the specific value, the case is just the opposite. This is because when the fiber volume ratio is low, there will be more fibers participating in bearing tensile stress with the increase of ratio, which results in an increase of tensile capacity of sprayed FRP. While the ratio is higher than the

Table 4: Summary of tensile properties of sprayed FRP.

Group number	Group name	Tensile strength, σ (MPa)	Elastic modulus, E (GPa)	Breaking elongation (%)	Poisson ratio, μ
1	C-1	105.4	49.1	2.1	0.30
2	C-2	89.3	50.7	2.3	0.31
3	B-1	87.2	28.2	2.6	0.22
4	B-2	78.9	26.1	3.0	0.25
5	G-1	91.3	31.8	2.4	0.24
6	G-2	75.2	28.8	3.0	0.26
7	G-3	82.6	33.3	2.2	0.22
8	G-4	80.3	36.7	2.5	0.25
9	G-5	67.7	34.4	1.6	0.24
10	G-6	78.1	27.1	2.1	0.21
11	G-7	87.1	30.8	2.2	0.26
12	G-8	101.2	38.3	2.4	0.24
13	G-9	66.1	23.7	1.9	0.24

specific value, fibers will be too much, and thus the resin around the fibers will not be enough to transfer and balance the stress of fibers effectively, which will lead to partial failure in some fibers or resins earlier and make the tensile property reduced.

4.4. Fiber Length. From Table 3, there are 4 groups of specimens used to study the effect of fiber length on tensile properties of sprayed FRP, which are G-6, G-7, G-1, and G-8, with a fiber length of 15 mm, 20 mm, 30 mm, and 40 mm, respectively.

From Table 4, tensile strengths of the 4 groups from high to low are G-8, G-1, G-7, and G-6, with a value of 101.2 MPa, 91.3 MPa, 87.1 MPa, and 78.1 MPa, respectively. For tensile elastic modulus, the highest value also belongs to G-8, which is 41.3% higher than the lowest one G-6. As to breaking elongation, the highest value is 2.4% which comes from G-8 and G-1, followed by 2.2% from G-7, and the lowest one is 2.1% from G-6.

The relationship curves of tensile stress and strain of the 4 groups are compared in Figure 14. It shows that the longer fiber can bear more stress than the shorter ones when tensile strain is the same, which is in accordance with Boyd's result [24]. This can be explained as the longer length fibers can obtain a larger bonding area with resins, and thus the bonding effect will be better than the shorter ones, which can lead to better tensile properties.

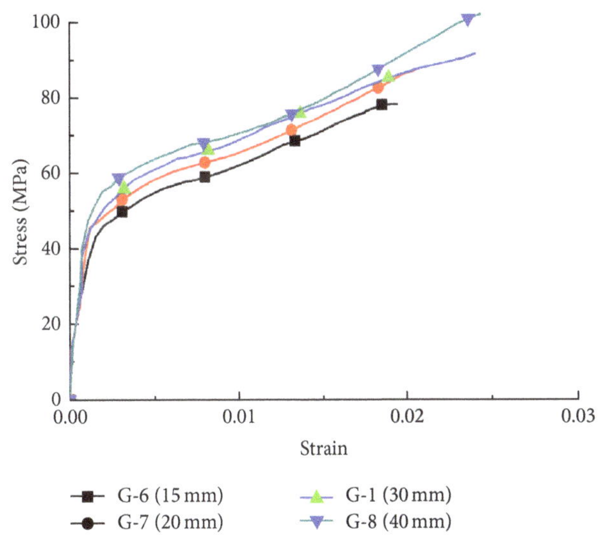

Figure 14: Comparison of stress-strain curves of sprayed FRP with different fiber lengths.

4.5. Composite Thickness. From Table 3, there are 2 groups of specimens used to study the effect of composite thickness on tensile properties of sprayed FRP, which are G-1 with a 4 mm thickness and G-9 with a 7 mm thickness.

From Table 4, the tensile strength of G-1 is 91.3 MPa and that of G-9 is 66.1 MPa. The tensile elastic modulus of G-1 is 31.8 GPa and that of G-9 is 23.7 GPa. As to breaking elongation, G-1 is 2.4% and G-9 is 1.9%. It is shown that tensile strength, elastic modulus, and breaking elongation will all be reduced with the increase of composite thickness. This is because the initial defects in sprayed FRP materials will become more with the increase of composite thickness, which will lead to degradation of tensile properties.

4.6. Specimen Cutting Orientation. All the specimens were cut from the sprayed FRP plate in three different orientations, which is shown in Figure 5. As mentioned before, all the error percentages for each group in Table 4 were lower than 3%, so there was not any obvious difference in tensile properties of the specimens in different cutting orientations in each group. Comparing the Poisson ratios of the 13 groups of specimens in Table 4, most values are close to the average value of 0.25. The above results indicate that the sprayed FRP composite can be seen as an isotropic material.

5. Summary and Conclusions

This paper has studied the tensile properties of sprayed FRP through tensile experiments of 13 groups of specimens, which include tensile strength, elastic modulus, breaking elongation, and Poisson ratio. The effects of several influencing factors are compared too. The following conclusions are drawn:

(1) Fiber type has an important influence on tensile properties of sprayed FRP material. The composites made of carbon fiber have higher values of tensile strength and elastic modulus than those made of glass fiber or basalt fiber, while the breaking elongation of carbon fiber composites is lower. The results equate the tensile properties of the three types of fibers.

(2) The test results of the 2 resins used in this paper show that using vinyl ester resins to make sprayed FRP can get better tensile properties than using unsaturated polyester resins generally. But the breaking elongation of glass fiber specimens using vinyl ester resins is lower compared with using unsaturated polyester resin.

(3) There will be a specific value of fiber volume ratio for every kind of sprayed FRP material, which can lead to the highest value of a tensile property. Less or more than this volume ratio can make the tensile property of this sprayed FRP degrade.

(4) The increase of fiber length can make the tensile properties of sprayed FRP better, while the increase of composite thickness will lead to property degradation.

(5) Cutting orientation of sprayed FRP specimens has no obvious influence on tensile properties and the Poisson ratios are almost the same for all the specimens, which shows that sprayed FRP can be looked at as a kind of isotropic material.

Competing Interests

The authors declare that there are no competing interests regarding the publication of this paper.

Acknowledgments

This research received specific grant from the Research Foundation of Wuhan Construction Committee and was supported by China Scholarship Council as well.

References

[1] W. Li, "The study and application of FRP material in concrete structures," *Research of Materials Science*, vol. 3, no. 1, pp. 17–21, 2014.

[2] H. Karpate, H. Wheat, J. Jirsa, D. Fowler, and D. Whitney, "Repair and rehabilitation of corrosiondamaged concrete elements using FRP composite wraps," *World Journal of Engineering*, vol. 8, no. 2, pp. 147–150, 2011.

[3] G. Portnov, C. E. Bakis, E. Lackey, and V. Kulakov, "FRP Reinforcing bars—designs and methods of manufacture (Review of Patents)," *Mechanics of Composite Materials*, vol. 49, no. 4, pp. 381–400, 2013.

[4] A. Prota, G. Manfredi, and F. P. Nardone, "Assessment of design formulas for in-plane FRP strengthening of masonry walls," *Journal of Composites for Construction*, vol. 12, no. 6, pp. 643–649, 2008.

[5] S. S. Pendhari, T. Kant, and Y. M. Desai, "Application of polymer composites in civil construction: a general review," *Composite Structures*, vol. 84, no. 2, pp. 114–124, 2008.

[6] J. G. Teng, H. Yuan, and J. F. Chen, "FRP-to-concrete interfaces between two adjacent cracks: theoretical model for debonding failure," *International Journal of Solids and Structures*, vol. 43, no. 18-19, pp. 5750–5778, 2006.

[7] B. Ferracuti, M. Savoia, and C. Mazzotti, "Interface law for FRP-concrete delamination," *Composite Structures*, vol. 80, no. 4, pp. 523–531, 2007.

[8] C. Mazzotti and M. Savoia, "FRP-Concrete bond behaviour under cyclic debonding force," *Advances in Structural Engineering*, vol. 12, no. 6, pp. 771–780, 2009.

[9] H. C. Biscaia, I. S. Borba, C. Silva, and C. Chastre, "A nonlinear analytical model to predict the full-range debonding process of FRP-to-parent material interfaces free of any mechanical anchorage devices," *Composite Structures*, vol. 138, pp. 52–63, 2016.

[10] H. W. Zhang, S. T. Smith, and R. J. Gravina, "Analysis of FRP-to-concrete interfaces using a displacement driven partial interaction model," *International Journal of Mechanical Sciences*, vol. 117, pp. 210–217, 2016.

[11] T. Xu, Z. J. He, C. A. Tang, W. C. Zhu, and P. G. Ranjith, "Finite element analysis of width effect in interface debonding of FRP plate bonded to concrete," *Finite Elements in Analysis and Design*, vol. 93, pp. 30–41, 2015.

[12] M. Coelho, A. Caggiano, J. Sena-Cruz, and L. Neves, "Fracture-based interface model for NSM FRP systems in concrete," *Composite Structures*, vol. 152, pp. 816–828, 2016.

[13] H. C. Biscaia, D. Cruz, and C. Chastre, "Analysis of the debonding process of CFRP-to-timber interfaces," *Construction and Building Materials*, vol. 113, pp. 96–112, 2016.

[14] S. F. Breña and G. N. McGuirk, "Advances on the behavior characterization of FRP-Anchored Carbon Fiber-Reinforced Polymer (CFRP) sheets used to strengthen concrete elements," *International Journal of Concrete Structures and Materials*, vol. 7, no. 1, pp. 3–16, 2013.

[15] S. C. Chin, N. Shafiq, and M. F. Nuruddin, "FRP as strengthening material for Reinforced Concrete beams with openings—a review," *KSCE Journal of Civil Engineering*, vol. 19, no. 1, pp. 213–219, 2014.

[16] W.-W. Wang, J.-G. Dai, and K. A. Harries, "Performance evaluation of RC beams strengthened with an externally bonded FRP system under simulated vehicle loads," *Journal of Bridge Engineering*, vol. 18, no. 1, pp. 76–82, 2013.

[17] S. S. Mahini and H. R. Ronagh, "Strength and ductility of FRP web-bonded RC beams for the assessment of retrofitted beam-column joints," *Composite Structures*, vol. 92, no. 6, pp. 1325–1332, 2010.

[18] K. Amin, "Investigation on the effects of L-shaped FRPs for strengthening of exterior RC joints to relocate the plastic hinge place away from the column face," *International Journal of Advances in Engineering Sciences*, vol. 3, no. 2, pp. 13–18, 2013.

[19] Z. Yang and C. Wu, "Analysis on the seismic performance of RC frame beam-column joints strengthened by sprayed FRP," *Applied Mechanics and Materials*, vol. 578-579, pp. 835–838, 2014.

[20] P. Agarwal, A. Gupta, and R. G. Angadi, "Effect of FRP wrapping on axial behavior of concrete and cyclic behavior of external RC beam column joints," *KSCE Journal of Civil Engineering*, vol. 18, no. 2, pp. 566–573, 2014.

[21] A. Bossio, F. Fabbrocino, G. P. Lignola, A. Prota, and G. Manfredi, "Simplified model for strengthening design of beam-column internal joints in reinforced concrete frames," *Polymers*, vol. 7, no. 9, pp. 1732–1754, 2015.

[22] E. Z. Beydokhty and H. Shariatmadar, "Behavior of damaged exterior RC beam-column joints strengthened by CFRP composites," *Latin American Journal of Solids and Structures*, vol. 13, no. 5, pp. 880–896, 2016.

[23] H. K. Lee, "Effectiveness of anchorage in concrete beams retrofitted with sprayed fiber-reinforced polymers," *Journal of Reinforced Plastics and Composites*, vol. 23, no. 12, pp. 1285–1300, 2004.

[24] A. J. Boyd, *Rehabilitation of reinforced concrete beams with sprayed glass giber reinforced polymers [Ph.D. thesis]*, University of British Columbia, British Columbia, Canada, 2000.

[25] H. K. Lee and L. R. Hausmann, "Structural repair and strengthening of damaged RC beams with sprayed FRP," *Composite Structures*, vol. 63, no. 2, pp. 201–209, 2004.

[26] S. M. Soleimani, *Sprayed glass fibre reinforced polymers in shear strengthening and enhancement of impact resistance of reinforced concrete beams [Ph.D. thesis]*, University of British Columbia, Vancouver, Canada, 2006.

[27] A. J. Boyd, N. F. Liang, P. S. Green, and K. Lammert, "Sprayed FRP repair of simulated impact in prestressed concrete girders," *Construction and Building Materials*, vol. 22, no. 3, pp. 411–416, 2008.

[28] S. K. Ha, S. Na, and H. K. Lee, "Bond characteristics of sprayed FRP composites bonded to concrete substrate considering various concrete surface conditions," *Composite Structures*, vol. 100, pp. 270–279, 2013.

[29] GB/T 1447, *Fiber-Reinforced Plastics Composites- Determination of Tensile Properties*, 2005.

[30] K. S. Lee, "A seismic strengthening technique for reinforced concrete columns using Sprayed FRP," *Journal of Advanced Concrete Technology*, vol. 10, pp. 219–230, 2012.

14

Optimization of Process Parameters for Vertical-Faced Polypropylene Bottle Injection Molding

Youmin Wang, Zhichao Yan, and Xuejun Shan

School of Mechanical and Automotive Engineering, Anhui Polytechnic University, Wuhu, China

Correspondence should be addressed to Zhichao Yan; 1148078812@qq.com

Academic Editor: Marco Cannas

In order to obtain the optimal combination of process parameters for vertical-faced polypropylene bottle injection molding, with UG, the model of the bottle was drawn, and then, one module and sixteen-cavity injection molding system was established and analyzed using Moldflow. For filling and maintaining pressure during the process of infusion bottle injection molding, the orthogonal test table L25 (5^6) using CAE was designed for injection molding of the bottle, with six parameters such as melt temperature, mold temperature, injection pressure, injection time, dwell pressure, and dwell time as orthogonal test factors. By finding the best combination of process parameters, the orthogonal experiment was completed, the results were analyzed by range analysis, and the order of influence of each process parameter on each direction of optimization was obtained. The prediction dates of the infusion bottle were gained under various parameters, a comprehensive quality evaluation index of the bottle was formulated, and the multiobjective optimization problem of injection molding process was transformed into a single-objective optimization problem by the integrated weighted score method. The bottle parameters were optimized by analyzing the range date of the weighted scoring method, and the best parameter combination such as melt temperature 200°C, mold temperature 80°C, injection pressure 40 MPa, injection time 2.1 S, dwell pressure 40 MPa, and dwell time 40 S was gained.

1. Introduction

Polypropylene infusion bottles are currently widely used to hold medical liquids since they have advantages such as better sealing properties, lighter weight, better heat resistance, easy to transport and preserve, and so on. Vertical-faced polypropylene bottles are the preforms used for blow molding, which are mainly produced and processed by injection molding process.

The best combination of process parameter for infusion preform injection of the infusion bottle can be obtained through the orthogonal experiment method, and the injection molding process of the bottle was simulated by using Moldflow to obtain the best combination of process parameters. Jiang-dong and Li-xin [1] completed the selection of injection molding machines, checked for parameters, and so on by Moldflow but did not analyze the impact of each factor to the mold. Ke et al. [2] used Moldflow to establish and analyze the injection mold casting system and obtained the best system but did not point out the factors that affect the casting. Zili et al. [3] applied the orthogonal algorithm to the injection molding process parameter design, selected the parameters for the green performance index, and also proved the feasibility of this method as an example but did not indicate how to optimize the parameters. Wei-ging et al. [4] used Moldflow to determine the best gating position of the mold and set the process parameters but did not indicate the impact of various parameters on the gate. Jin et al. [5] proposed a multiobjective robust design method for injection molding, which is based on the design of internal and external parameters, SNR analysis, and approximate ideal solution but did not draw the specific parameters of the injection molding process. Bo et al. [6] used warpage as quality index, obtained the optimum process parameters by the orthogonal experiment and range analysis methods, but only limited to thin-walled plastic parts. Fengli et al. [7] proposed a model of injection molding process multiobjective robust design and optimization algorithm but did not verify it. The quality of preforms directly affects the quality of blow-molded products; hence, studying the quality of preforms is crucial. The optimization of injection molding process parameters of vertical polypropylene preforms was studied, the optimal combination of process parameters was determined after the optimization of each evaluation index, and the injection molding process parameters of synthetic weighted method were simulated and verified.

2. Infusion Bottle Injection Mold Design

2.1. Infusion Bottle Injection Mold Structure Design

2.1.1. Parting Surface Selection. The parting surface of the mold is the surface formed by the contact of the closed-type core and the cavity. The selection of the parting surface has a direct influence on the determination of the mold structure. It is inseparable from the mold structure and the manufacturing process of the mold, and it directly affects the flow filling process of the molten plastic and the demolding process of the plastic parts. The choice of parting surface is crucial in the design of injection molds. When determining the parting surface of a plastic injection mold, various factors in the molding process need to be taken into consideration. For the parting surface selection, the following factors should be considered: the forming condition, pouring in mold system form, the structure of manufacturability and precision, insert position, shape, exit routes, and mold manufacturing, and the factors such as exhaust and operating process generally need to follow the following basic design principles when choosing a parting surface [9]: (1) the selection of parting surface is based on the contour shape; (2) the projected area of the plastic parts on the parting surface should be minimized; (3) the parting surface structure should be as simple as possible so as to design and manufacture the mold; (4) the selection of parting surface shall ensure the shape, size accuracy, and appearance requirements of the plastic parts; and (5) the parting surface shall be conducive to the stripping, side pumping, and exhaust of plastic parts. In the actual design process, the above principles should be satisfied as far as possible, and the parting surface should be reasonably selected.

After considering the gating system state, the structure of plastic parts technology and accuracy, insert position, shape and withdrawal style, mold manufacturing, gas exhaust, operating process of the plastic parts in the mold, the graph of the bottle parting surface could be gained, as shown in Figure 1 [10].

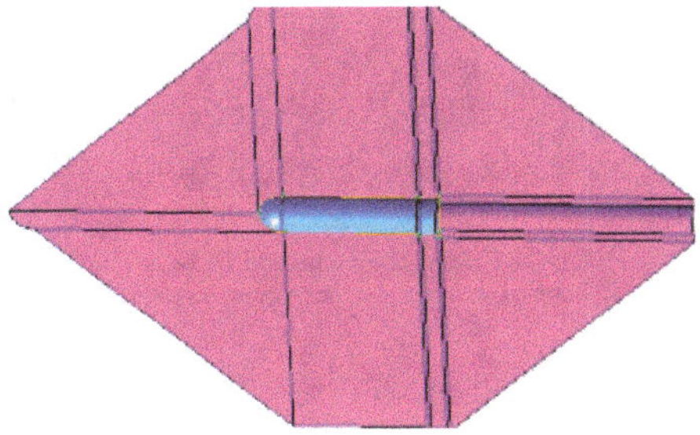

FIGURE 1: Parting surface of the infusion bottle.

After determining the parting surface, the number and arrangement of cavities need to be taken into account. Figure 2 is the infusion bottle distributed in two-row and eight-column layout [11]. In order to ensure the quality of plastic products and the complexity of the molding process, after comprehensive analysis, the gating location was chosen as the bottom, and the best gating system was determined. Figure 3 shows the infusion bottle gating system [12].

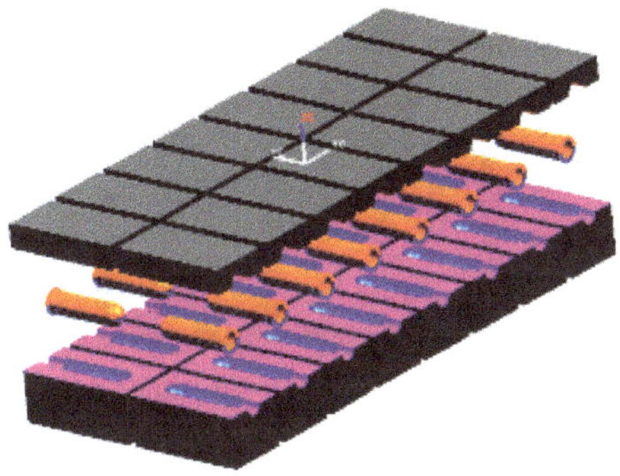

FIGURE 2: Infusion bottle embryo in two-row and eight-column layout.

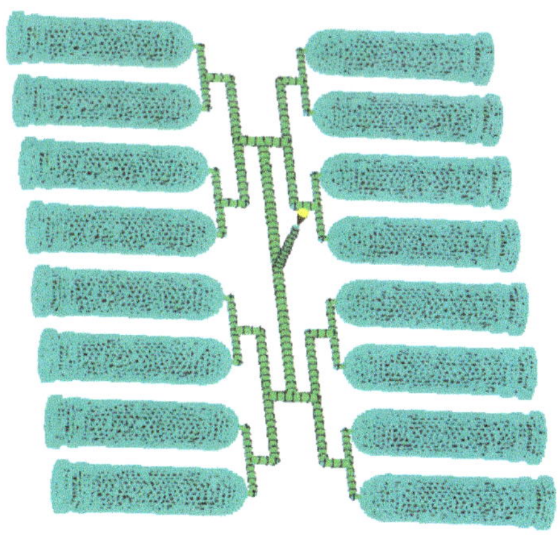

FIGURE 3: Infusion bottle gating system.

2.2. Design of Molding Parts for Injection Mold of Infusion Bottle. The molding parts of injection molds refer to the parts that form the mold cavity, usually including the core, the cavity, and various forming rods and forming inserts. The forming parts can be divided into the installation part and the working part. The installation part mainly supports and fixed the forming part, its surface roughness is generally Ra 1.6. The working part is in direct contact with the plastic parts, which is used to form the plastic parts [13]. The surface roughness is determined according to the requirements of the plastic parts, generally not higher than Ra 0.4 mm. When making the structure design of molding parts, on the one hand, one should check whether the qualified parts were obtained and, on the other hand, whether processing and manufacturing were facilitated and also note that as far as possible to save valuable mold material, in order to reduce the mold cost [14].

Based on the personal experience of the engineer in mapping the structure of the shape and then in designing the injection mold of molding parts, traditional injection mold design is time-consuming and not economic. In the modern injection molding design, the structure of the plastic parts is

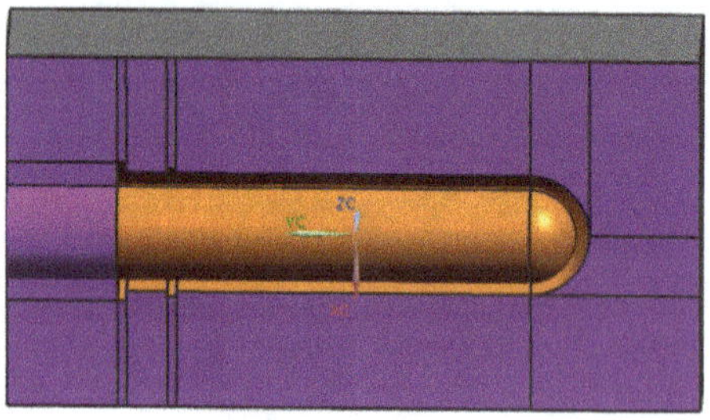

Figure 4: Infusion bottle embryo core cavity structure.

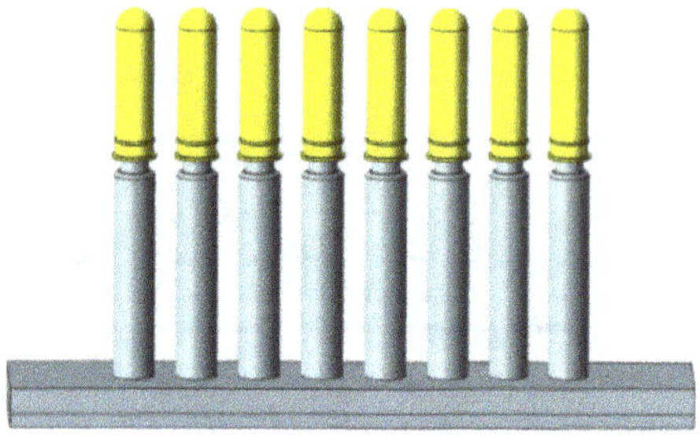

Figure 5: Infusion bottle injection mold side core-pulling mechanism.

usually designed and drawn by using 3D software. Then, make a design drawing of its molded parts. The molding parts of injection molds designed by the modern methods are more accurate, time-saving, and cost-effective [15]. There are two methods for the design of forming parts by using 3D software. One is to automatically model and adopt the modules in the 3D software, such as the Wizard in UG. The second is manual modeling, according to the structure of the plastic parts.

In this topic, the structure of infusion bottleneck parts is relatively simple, and the core, cavity, ejection mechanism, and core-pulling mechanism of the infusion bottle injection mold are mainly designed, and the molded parts are designed using UG software illustration. First of all, using UG to infusion bottle embryo structure modeling 3D model, its shrinkage rate is set at 3%; Through manual modeling method, mold design is performed, various modeling commands in UG are invoked, and the plastic parts are divided and molded by the summation and difference. In the end, the core and cavity are obtained, because the upper and lower parts of the infusion bottle are symmetrical, and the core cavity is the same [16]. The structure of the core cavity of the infusion bottle embryo core obtained by using UG is shown in Figure 4.

2.3. Infusion Bottle Injection Mold Side Core-Pulling Mechanism and Auxiliary Hydraulic System. Side core-pulling mechanism was designed to ensure smooth demolding, and this is shown in Figure 5. This paper used hydraulic loose-core system, as shown in Figure 6 [17].

2.4. The Design of Infusion Bottle Mold Assembly Drawing. It not only needs core, cavity, and core-pulling mechanism for a complete injection molding pattern, but also needs mold frame, gating system, cooling system, ejection mechanism, exhaust system to support it, and the final two-dimensional engineering drawing is shown in Figure 7 [18].

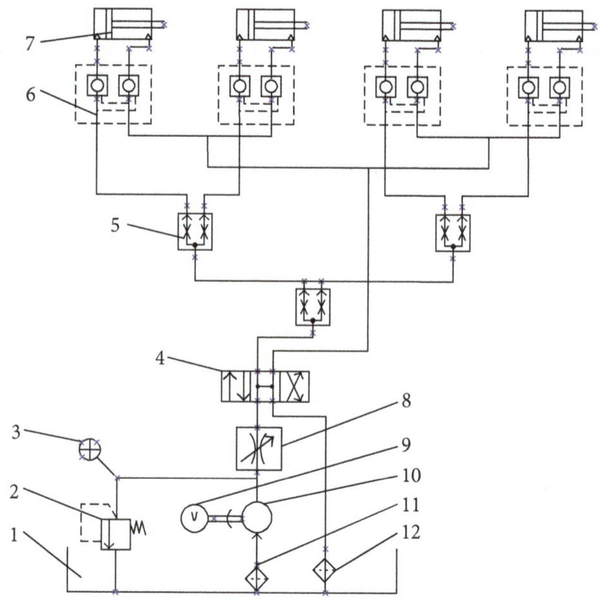

Figure 6: The hydraulic core-pulling system of infusion bottle injection molding: 1, tank; 2, relief valve; 3, pressure gauge; 4, reversing valve; 5, dividing-combining valve; 6, hydraulic lock; 7, hydraulic cylinder; 8, speed control valve; 9, explosion-proof motor; 10, gear pump; 11, oil filter; and 12, refined oil filter.

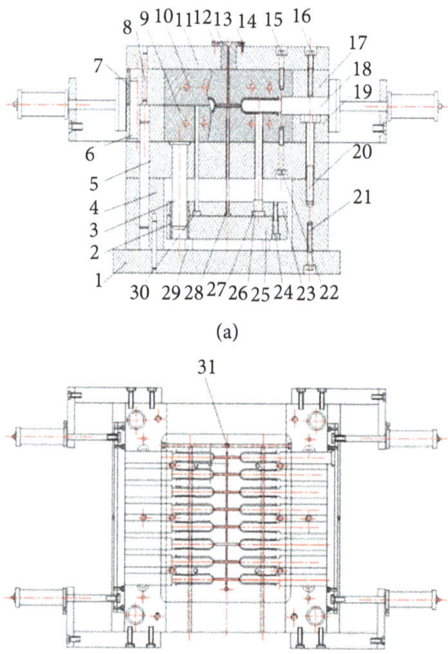

Figure 7: Infusion bottle embryo assembly diagram: 1, bottom clamp plate; 2, ejector guide pillar; 3, ejector guide pin; 4, spacer parallel; 5, bearing plate; 6, guide pillar; 7, guide bush; 8, dowel; 9, cores ejector side; 10, fixed die insert; 11, fixed clamping plate; 12, locating ring; 13, gate; 17, fixed plate fixed die; 18, side core-drawing; 19, dynamic model of fixed plate; 23, push plate fixed plate; 25, stripper; 26, pushrod; 27, pushrod positioning pin; 28, sprue puller; 29, banking pin; 30, release link; 31, panel board dowel; 14, 15, 16, 20, 21, 22, 24, screws.

TABLE 1: Orthogonal test of six factors and five levels.

Levels	Factors					
	Melt temperature, A (°C)	Mold temperature, B (°C)	Injection pressure, C (MPa)	Injection time, D (s)	Dwell pressure, E (MPa)	Dwell time, F (s)
1	200	20	30	1.3	28	20
2	220	35	35	1.5	32	25
3	240	50	40	1.7	36	30
4	260	65	45	1.9	40	35
5	280	80	50	2.1	44	40

TABLE 2: The results of the orthogonal test and the comprehensive weighted score of embryo injection molding of the infusion bottle.

Number	Shrinkage index (%)		Volumetric shrinkage (%)		Residual stress in cavity in the first direction (MPa)		Residual stress in cavity in the second direction (MPa)		Comprehensive weighted score Y^*
	Value	Score	Value	Score	Value	Score	Value	Score	
1	−1.435	48.46	14.25	55.16	46.81	86.65	61.07	78.28	65.94
2	−1.841	62.17	14.44	56.16	48.18	89.19	58.53	75.03	71.76
3	−1.960	66.19	14.13	54.53	49.30	91.26	62.52	80.14	73.43
4	−1.039	35.09	14.36	55.74	51.84	95.96	68.62	87.96	62.72
5	0	0.00	10.22	0	54.02	100.00	74.18	95.09	41.07
6	−2.102	70.99	15.42	71.33	48.42	89.63	53.61	48.72	76.76
7	−2.145	72.44	15.86	73.65	52.92	97.96	78.01	100.00	82.23
8	−1.712	57.82	15.04	79.32	44.78	82.90	60.99	78.18	70.97
9	−2.094	70.72	15.26	80.49	46.62	86.30	67.40	86.40	77.75
10	−1.893	63.93	15.50	81.75	48.89	90.50	69.72	89.37	76.26
11	−1.892	63.90	12.66	36.77	43.99	71.43	53.56	48.66	67.28
12	−2.233	75.41	16.17	85.28	45.87	84.91	59.56	76.35	80.41
13	−2.281	77.03	16.07	84.76	50.07	92.69	73.90	94.73	83.46
14	−2.085	70.42	16.48	86.92	53.76	99.52	76.92	98.60	82.75
15	−2.226	75.18	16.68	87.97	45.64	84.49	70.93	90.92	82.80
16	−2.380	80.38	17.46	92.09	53.80	99.59	55.95	51.72	86.12
17	−2.604	87.94	17.34	91.46	41.50	46.82	55.38	50.99	86.54
18	−2.286	77.20	17.95	94.67	42.65	68.95	60.18	77.14	84.36
19	−1.697	57.31	18.01	94.99	43.60	70.71	67.56	86.60	77.65
20	−0.9201	31.07	17.16	90.51	48.83	90.39	71.95	92.23	66.89
21	−2.961	100.00	18.96	100.00	41.74	47.27	51.38	0	94.31
22	−2.425	81.90	18.88	99.58	43.29	70.14	55.83	51.57	87.76
23	−2.412	81.46	17.95	94.67	45.32	83.89	62.43	80.03	86.85
24	−2.457	82.98	18.38	96.94	40.96	0	61.48	78.81	87.43
25	−2.471	83.45	18.86	99.47	41.14	46.16	64.35	82.49	89.03

3. Orthogonal Experiment

In science experiment investigation, sometimes there are only one or two factors. Because there are few factors, the design of experiment, implementation, and analysis are relatively simple. However, in practical situation, there will be more influencing factors so that the scale of the experiment can be quite extensive [19].

Orthogonal test is a method for seeking the best level combination with multiple factors and the minimum test times [16]. After considering the requirements of quality of the infusion bottle comprehensively, six process parameters were selected as the factors of orthogonal test: melt temperature, mold temperature, injection pressure, injection time, dwell pressure, and dwell time, which was marked as A, B, C, D, E, and F, respectively, with five levels (Table 1).

After selecting the factors and the level of orthogonal test, the final orthogonal test table was selected according to the orthogonal experiment factors and the interaction between the factors. This paper selects the orthogonal test table of six factors and five levels. Table 2 shows the factor levels of the experiment.

3.1. The Analysis of Injection Molding Parameter Effect Tendency towards Infusion Bottle Sink Index. According to the injection molding results shown in Table 2, the influence level order of each process parameter towards the sink index was obtained as follows: melt temperature > mold temperature > injection time > dwell pressure > dwell time > injection pressure. Table 3 shows the range analysis of the sink index for each parameter.

K_1, K_2, K_3, K_4, and K_5 denote, respectively, the sum of the sink index of each factor at five levels, k_1, k_2, k_3, k_4, and k_5 denote, respectively, the average of the sum of the sink index of each factor at five levels, and the range is R. The trend of

Table 3: The range analysis of the sink index for each process parameter.

	A	B	C	D	E	F
$k_1 = K_1/5$	−1.255	−2.154	−2.025	−1.819	−2.087	−2.006
$k_2 = K_2/5$	−1.989	−2.250	−2.056	−2.170	−2.117	−2.150
$k_3 = K_3/5$	−2.143	−2.130	−1.875	−2.244	−2.149	−2.001
$k_4 = K_4/5$	−1.977	−1.874	−1.967	−2.131	−1.753	−1.968
$k_5 = K_5/5$	−2.545	−1.502	−1.988	−1.545	−1.804	−1.816
R	1.290	0.748	0.181	0.699	0.396	0.334
Order	1	2	6	3	4	5

Table 4: The range analysis of volumetric shrinkage for process parameter.

	A	B	C	D	E	F
$k_1 = K_1/5$	13.480	15.750	16.612	16.158	16.338	16.610
$k_2 = K_2/5$	15.416	16.538	16.500	16.370	15.834	16.230
$k_3 = K_3/5$	15.612	16.228	15.638	16.446	16.554	16.595
$k_4 = K_4/5$	17.584	16.498	16.378	16.762	16.378	15.562
$k_5 = K_5/5$	18.606	15.684	15.570	14.962	15.594	15.628
R	5.126	0.854	1.042	1.800	0.960	1.048
Order	1	6	4	2	5	3

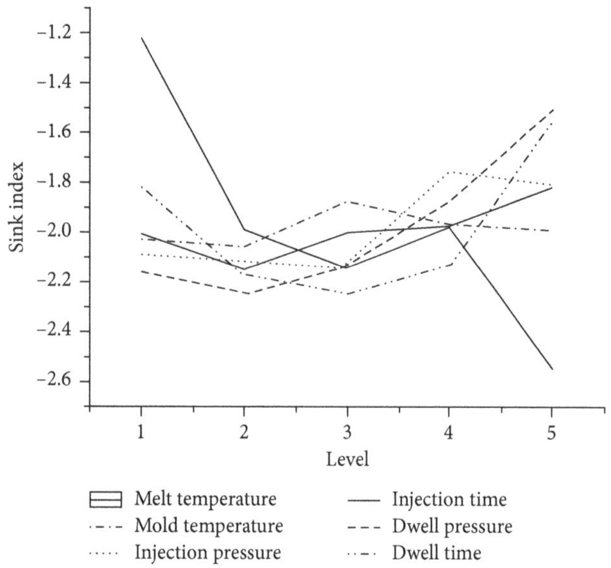

Figure 8: The influence of various factors of the orthogonal test on the sink index of infusion bottle embryo parts.

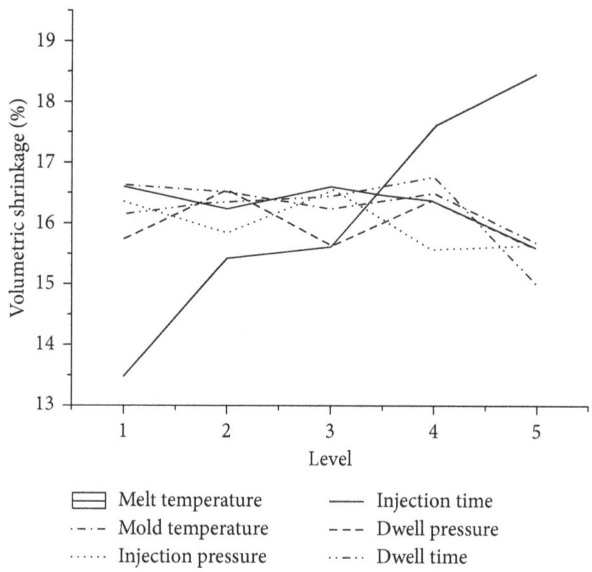

Figure 9: The influence of each factor of the orthogonal test on volumetric shrinkage of infusion bottle.

each factor on the shrinkage index was drawn, as shown in Figure 8. The optimal combination was obtained as follows: *A*1*B*5*C*3*D*5*E*4*F*5, when the melt temperature reached 200°C, mold temperature reached 80°C, the injection pressure was 40 MPa, the injection time was 2.1 S, the dwell pressure was 40 MPa, and the dwell time was 40 S. This combination of process parameters that approved in Moldflow is not included in the existing tests. The shrinkage index is −0.0709%, which shows the results have been optimized.

3.2. The Analysis of Injection Molding Parameter Effect Tendency towards Infusion Bottle Volumetric Shrinkage. The range analysis of volumetric shrinkage for process parameter is shown in Table 4, and the influence level order of each process parameter towards volumetric shrinkage was obtained as follows: melt temperature > injection time > dwell time > injection pressure > dwell pressure > mold temperature.

The trend of each factor on volumetric shrinkage was drawn as shown in Figure 9. The optimal combination was obtained as follows: *A*1*B*5*C*5*D*5*E*5*F*4, when the melt temperature reached 200°C, the mold temperature reached 80°C, the injection pressure was 50 MPa, the injection time was 2.1 S, the dwell pressure was 44 MPa, and the dwell time was 35 S. Moldflow is used to verify whether the volumetric shrinkage is 10.27%, which is close to the minimum value in the experimental group.

Table 5: The range analysis of the residual stress in cavity in the first direction by various process parameters.

	A	B	C	D	E	F
$k_1 = K_1/5$	50.030	46.952	47.080	46.690	43.938	46.908
$k_2 = K_2/5$	48.326	46.352	46.232	48.380	44.516	47.115
$k_3 = K_3/5$	47.866	46.424	47.200	46.824	45.880	47.913
$k_4 = K_4/5$	46.076	47.356	47.486	46.958	48.490	46.308
$k_5 = K_5/5$	42.490	47.704	46.790	45.936	51.964	46.384
R	7.540	1.352	1.254	2.444	8.026	1.605
Order	2	5	6	3	1	4

3.3. The Analysis of the Injection Molding Parameter Effect Tendency towards Infusion Bottle Residual Stress in Cavity in the First Direction. The range analysis of the residual stress in cavity in the first direction by various process parameters is shown in Table 5, and the influence level order of each process parameter towards the residual stress in cavity in the first direction was obtained as follows: dwell pressure > melt temperature > injection time > dwell time > mold temperature > injection pressure.

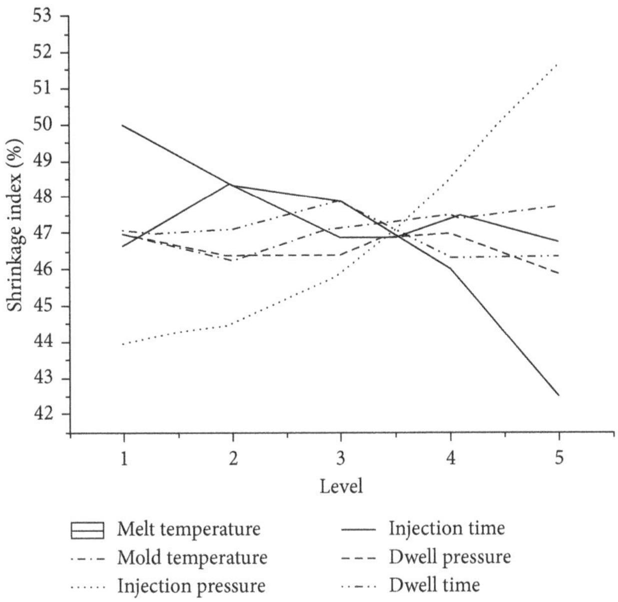

FIGURE 10: The influence of various factors on the residual stress in cavity in the first direction of infusion bottle embryo parts by the orthogonal test.

The effect tendency graph of each factor towards residual stress in cavity in the first direction is shown in Figure 10. The optimal combination was obtained as follows: $A5B2C2D5E1F4$, when the melt temperature reached 280°C, the mold temperature reached 35°C, the injection pressure was 28 MPa, the injection time was 2.1 S, the dwell pressure was 44 MPa, and the dwell time was 35 S. Moldflow is used to verify whether the result is 39.98 MPa, and the stress is smaller than any of the group of the orthogonal test.

3.4. The Analysis of Injection Molding Parameter Effect Tendency towards Infusion Bottle Residual Stress in Cavity in the Second Direction. The range analysis of the residual stress in cavity in the second direction by various process parameters is shown in Table 6, and the influence level order of each process parameter towards the residual stress in cavity in the second direction was obtained as follows: mold temperature > dwell pressure > melt temperature > dwell time > injection pressure > injection time.

The effect tendency graph of each factor towards residual stress in cavity in the second direction is shown in Figure 11. The optimal combination was obtained as follows: $A5B1C4D5E2F5$, when the melt temperature reached 280°C, the mold temperature reached 20°C, the injection pressure was 45 MPa, the injection time was 2.1 S, the dwell pressure was 32 MPa, and dwell time 40 S. Moldflow is used to verify whether the result is 50.50 MPa, and the stress is smaller than any of the group by the orthogonal test.

4. Optimization of Injection Molding Parameters of Infusion Bottle

4.1. Design Variable. The sink index, volumetric shrinkage, residual stress in cavity in the first direction, residual stress in cavity in the second direction as evaluation index were selected, which is a single-objective function of parameter optimization. The index of each parameter was evaluated $M_j = (0.4, 0.4, 0.1, 0.1)$ [20].

TABLE 6: The range analysis of the residual stress in cavity in the second direction by various process parameters.

	A	*B*	*C*	*D*	*E*	*F*
$k_1 = K_1/5$	64.984	55.114	64.744	64.482	61.970	68.978
$k_2 = K_2/5$	65.946	61.462	62.612	63.916	60.804	61.955
$k_3 = K_3/5$	66.974	64.004	65.504	62.556	62.148	64.097
$k_4 = K_4/5$	62.204	68.396	61.894	65.824	64.782	61.942
$k_5 = K_5/5$	59.094	70.226	64.448	62.424	69.498	61.802
R	7.880	15.112	3.610	3.400	8.694	7.176
Order	3	1	5	6	2	4

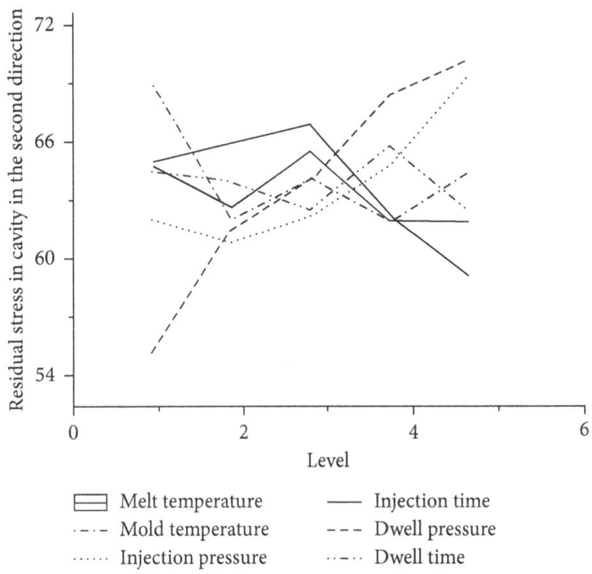

FIGURE 11: The influence of various factors on the residual stress in cavity in the second direction of infusion bottle embryo parts by the orthogonal test.

4.2. Comprehensive Objective Function. A comprehensive weighted objective function was established as follows:

$$Y_i^* = \sum_{i=1}^{25} \left[Y(i, j) \times M_j \right], \tag{1}$$

where $Y(i, j)$ is the evaluation index score value, Y_i^* is the comprehensive weighted score values, and M_j is the experimental evaluation index weight.

From the results of the comprehensive weighted score as shown in Table 2, it was found that, in the trial, the fifth group of trials had the lowest overall score, and Group 21 had the highest score. It can be seen that, in the bottle orthogonal experiment, the best parameters' combination was when the melt temperature reached 200°C, the mold temperature reached 80°C, the injection pressure was 50 MPa, the injection time was 2.1 S, the dwell pressure was 44 MPa, and the dwell time was 40 S.

Table 7: The analysis table of comprehensive weighted score by orthogonal test.

	A	B	C	D	E	F
$K_2 = K_2/5$	76.794	81.74	79.164	81.006	78.036	77.336
$K_3 = K_3/5$	79.34	79.814	75.452	81.702	80.412	81.572
$K_4 = K_4/5$	80.312	77.66	77.85	81.284	75.518	75.93
$K_5 = K_5/5$	89.076	71.21	76.626	68.946	75.804	74.006
R	26.092	10.53	3.962	12.756	4.894	7,566
Order	1	3	6	2	5	4

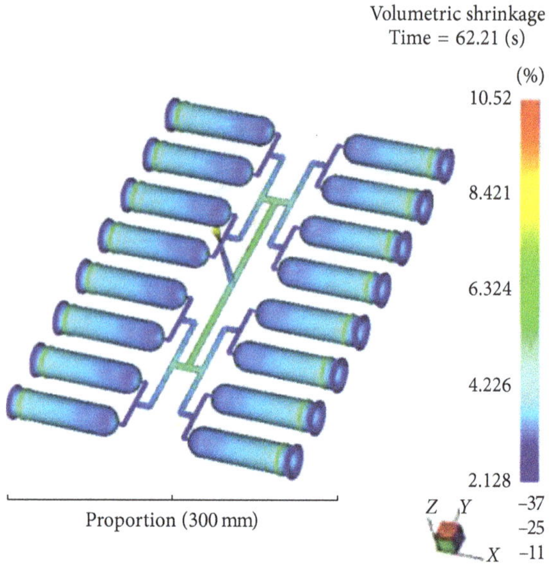

Figure 12: Volume shrinkage after optimization by weighted scoring.

4.3. Analysis of the Range of Comprehensive Weighted Grading Method for Infusion Bottle Injection Molding. The comprehensive weighted score values for each group by orthogonal experiments were calculated as shown in Table 2, and thus, it could be obtained that Group 5 had the best combination property. As this was the only best combination of these 25 trials without taking into account these 25 groups outside the case, the range analysis of comprehensive weighted score of each level would be necessary, so as to obtain various levels of factors contributing to the overall performance. Table 7 shows the influence degree of each factor towards injection molding combination property.

According to the results of the above analysis, the influence level order of each process parameter on shrinkage index was follows as follows: melt temperature $A >$ injection time $D >$ mold temperature $B >$ dwell time $F >$ dwell pressure $E >$ injection pressure C. From the analysis of range in Table 7, the optimum combination of process parameters for injection molding of infusion bottle is when the melt temperature reached 200°C, the mold temperature reached 80°C, the injection pressure was 40 MPa, the inject time was 2.1 S, the dwell pressure was 40 MPa, and the dwell time was 40 S.

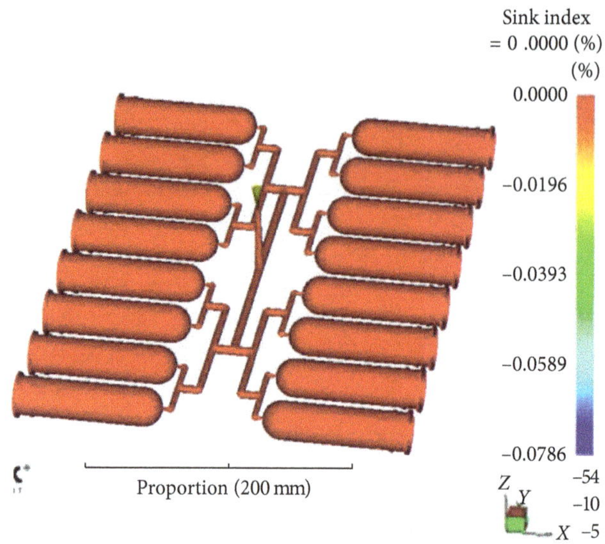

Figure 13: Sink index after optimization by weighted scoring.

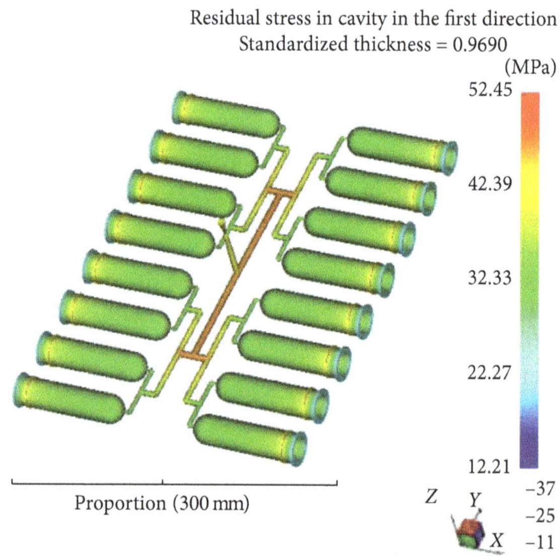

Figure 14: Residual stress in first cavity in the first direction after optimization by weighted scoring.

5. Simulation Verification

Setting the relevant process parameters of infusion bottle injection molding in Moldflow, the results of the injection molding simulation test can be obtained as the sink index was 0%, the volumetric shrinkage was 10.52%, the residual stress in cavity in the first direction was 52.45 MPa, and the residual stress in cavity in the second direction was 73.25 MPa, at which point the comprehensive performance is the best, as shown in Figures 12–15.

6. Conclusions

From this study, the following conclusions can be drawn:

(1) According to the structure of infusion bottle embryo and process requirements, an infusion bottle

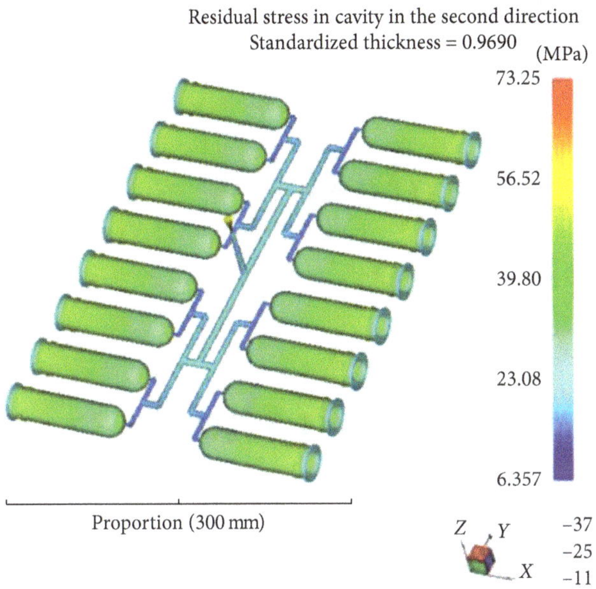

FIGURE 15: Residual stress in first cavity in the second direction after optimization by weighted scoring.

preform mold was designed, the plastic parts were distributed by two rows and eight columns, and the hydraulics-assisted method was used to process the side core-pulling.

(2) The orthogonal experiment table of L25 (5~6) was designed, and the sink index, the volumetric shrinkage, the residual stress in cavity in the first direction, and the residual stress in cavity in the second direction as its performance evaluation index were selected. The influence level order of each process parameter on volume shrinkage was obtained as follows: melt temperature > injection time > dwell time > injection pressure > dwell pressure > mold temperature.

(3) The final optimal process parameters were obtained by the comprehensive weighted method: melt temperature 200°C, injection time 2.1 S, dwell time 40 S, injection pressure 40 MPa, dwell pressure 40 MPa, and mold temperature 80°C.

Conflicts of Interest

The authors declare that they have no conflicts of interest.

Acknowledgments

This article belongs to the major projects of the "Natural Science Foundation of Anhui Province Education Department (KJ2016SD05 and KJ2017ZD12)." This article belongs to the project of the "Wuhu Scientific Research Project (2016cxy01)."

References

[1] C. Jiang-dong and L. Li-xin, "Optimum design of injection mold for hair dryer shell based on moldflow," *Plastic*, vol. 39, pp. 122–125, 2010.

[2] S. Ke, H. Qing-chun, and J. Xiao-ping, "Optimization design of injection mold cooling channels arrangement," *Plastics Industry*, vol. 40, pp. 25-26, 2014.

[3] W. Zili, Z. Shuyou, and Q. Lemiao, "Optimization method for process parameters coupling design oriented to green injection molding," *Computer Integrated Computing System*, vol. 12, pp. 48–50, 2015.

[4] C. Wei-qing, W. Sheng-kai, and R. Wei, "Optimized design of printer on the cover of injection mold based on moldflow," *Plastic*, vol. 16, pp. 101-102, 2012.

[5] C. Jin, T. Jianrong, and Y. Jiahong, "Multi-objective robust optimization of injection molding process parameters based on TOPSIS," *Journal of Mechanical Engineering*, vol. 27, pp. 39-40, 2011.

[6] W. Bo, S. Wenxue, L. Qingtao, and D. Haiping, "Optimization of injection process parameters for thin walled plastic parts of battery cover based on orthogonal experiment," *Light Industry Machinery*, vol. 4, pp. 50-51, 2014.

[7] H. Fengli, L. Ianping, Z. Meipeng, and X. Jinhong, *Multi-Objective Robust Design and Optimum Algorithm in Injection Molding Process*, Journal of Tongji University, Shanghai, China, 2011.

[8] H. Zhang, "Structural design of mouse injection mold," *Mechanical Research and Application*, vol. 2, pp. 76-77, 2010.

[9] X. Zheng, B. Meng, and H. huang, "Design of injection mould for rear door handle based on CAE analysis," *Die industry*, vol. 10, pp. 43-46, 2010.

[10] L. Qiong, *A Practical Guide to Plastic Injection Moldflow*, Mechanical Industry Press, Beijing, China, 2008.

[11] H. Ming-Shyan and C. Chin-Feng, "Injection molding and injection compression molding of thin-walled light-guided plates with V-grooved microfeatures," *Applied Polymer Science*, vol. 12, pp. 1151–1159, 2011.

[12] Q. Huachang, *Plastic Forming Process and Mold Design*, Higher Education Press, Beijing, China, 2006.

[13] H. Jianming, "Improvement of air trap defects of power buttons based on flow simulation of moldflow," *China Plastics*, vol. 7, pp. 60-61, 2013.

[14] W. Guilong, Z. Guoqun, L. Huiping, G. Yanjin, and C. Liang, "Heat response simulation of variotherm injection molding and optimization of mold structure," *Journal of Mechanical Engineering*, vol. 2, pp. 18–20, 2009.

[15] C. Sheng-rong, X. Yong, Y. Guo-tai, and H. Cheng-hong, Research methods of parametric optimization for injection molding, China Plactics, vol. 4, 2003.

[16] F. Daver and B. Demirel, "A simulation study of the effect of preform cooling time in injection stretch blow molding," *Journal of Materials Processing Technology*, vol. 11, 2012.

[17] B. Demirel and F. Daver, "Experimental study of preform reheat temperature in two-stage injection stretch blow molding," *Polymer Engineering & Science*, vol. 4, 2013.

[18] B. Demirel and F. Daver, "Effects of preform deformation behavior on the properties of the poly(ethylene terephthalate) bottles," *Journal of Applied Polymer Science*, vol. 4, 2012.

15

Synthesis and Thermomechanical Characterization of Nylon 6/Cu Nanocomposites Produced by an Ultrasound-Assisted Extrusion Method

Rubén Sierra-Ávila,[1] Marissa Pérez-Alvarez,[2] Janett Valdez-Garza,[1] Carlos Alberto Avila-Orta,[1] Enrique Javier Jiménez-Regalado,[1] José M. Mata-Padilla,[3] Enrique Soto-Castruita,[2] and Gregorio Cadenas-Pliego[1]

[1]*Centro de Investigación en Química Aplicada, Blvd. Enrique Reyna 140, 25294 Saltillo, Coahuila, Mexico*
[2]*CONACyT Research Fellow, Instituto Mexicano del Petróleo, Eje Central Lázaro Cárdenas Norte 152, 07730 Ciudad de México, Mexico*
[3]*CONACyT Research Fellow, Centro de Investigación en Química Aplicada, 25294 Saltillo, Mexico*

Correspondence should be addressed to Carlos Alberto Avila-Orta; carlos.avila@ciqa.edu.mx and Gregorio Cadenas-Pliego; gregorio.cadenas@ciqa.edu.mx

Academic Editor: Raul Arenal

A nylon 6 nanocomposite with copper nanoparticles processed by ultrasound-assisted extrusion was prepared at concentrations between 0.01 and 0.50 wt.%, and its thermal and mechanical properties were determined. The presence of the crystalline phase α (α_1 and α_2) in the polymer matrix was confirmed by X-ray diffraction, and the presence of the α_2 phase showed a greater increase than the α_1 phase as a function of the copper nanoparticle concentration. This process was attributed to secondary crystallization. Furthermore, it was determined that the chemical composition of the nanoparticles is a blend of metallic copper and cupric oxide. The formation of copper nanowires was observed by scanning electron microscopy, and the concentration of 0.10% exhibited the best dispersion in comparison with the other concentrations. The melting temperature of the nanocomposites underwent a slight decrease in comparison with the nylon 6, while thermal stability, crystallization temperature, and crystallinity were increased in relation to the pure polymer. This behavior is attributed to an efficient dispersion of the nanoparticles and to their functionality as crystal nucleation sites. For the 0.10% concentration nanocomposite, higher mechanical properties were obtained; tensile strength increased by 8.9%, and the tensile modulus increased by 25.4%; as a consequence, elongation at break was 62% less than that of the polymer matrix.

1. Introduction

Currently, one of the ways to improve the properties of polymers is to disperse inorganic nanoparticles to form polymer nanocomposites; to mention some of the properties that can be improved and that are of interest for technological development, we have mechanical properties, thermal and dimensional stability, antifriction characteristics, catalytic capability, electrical and thermal conductivity, protection against UV radiation, flame retardancy, and gas barrier properties, among others [1, 2]. The improvement depends not only on the chemical nature of the nanoparticles, but on concentration, particle size, morphology, dispersion, and spatial distribution, as well as the crystallinity of the matrix, usually influenced by the presence of the nanoparticles [3, 4]. The improvement of the properties has been so successful that nanocomposites have been used in different industrial applications in many fields such as automotive, electrical, electronic, food packaging, and textiles [5, 6].

In particular, nanocomposites based on nylon 6 formulated with copper nanoparticles have aroused the interest of the academic and industrial scientific community since they exhibit excellent properties such as high tensile strength and tensile strain modulus [7–9], high electrical conductivity [10] and thermal stability [11], attractive optical properties [12], improved tribological characteristics [13, 14], and many others. For this reason, these nanocomposites are attractive in applications where high-performance materials are required, such as electrical and electronic markets, because they perform better than the alternative materials [11, 15].

However, only a few papers regarding this composite have been published. In 2015, Xu et al. obtained nylon 6 nanocomposites with nanoparticles of metallic copper [14]. The polymer was synthesized by ring-opening polymerization of ε-caprolactam and copper oxide as a metallic source at the concentrations of 0.1, 0.3, 0.5, 0.7, and 0.9 wt.%. After polymerization, the nanoparticles with an initial size of around 4-5 nm formed clusters with a size of 30 to 50 nm. The authors claimed the presence of copper metallic nanoparticles by X-ray diffraction (XRD), but small peaks at around $2\theta = 40^\circ$ may be associated with copper oxide and a change of the crystalline phase α to the γ phase was observed as the concentration of copper oxide increased. The tensile strength of the 0.3 wt.% sample increased by 47.5% compared to that of the nylon 6.

In 2013, Komeily-Nia et al. reported the preparation of copper nanowires in polyamide 6 composites at 0.5 wt.%, by the chemical reduction of copper sulfate; in this case, the authors claimed the presence of metallic copper and the presence of the crystalline phase α of the nylon 6, as confirmed by the X-ray diffractograms and the energy disperse spectrometers. Scanning electron microscopy showed that the nanoparticles had an average size of 85 nm, and the tensile strength of the nylon 6 increased by 9% with the addition of the nanoparticles [7].

Copper nanowires in nylon 6 were also observed by Xu et al. in 2013; the polymerization in situ was kept in a reducing atmosphere in order to prevent the oxidation of the nanowires found at the concentration of 0.5 wt.%. In this case, the presence of the nanowires did not affect thermal stability or melting temperature, but crystallinity increased by 1.38%, crystallization temperature decreased by 1.79°C, and tensile strength increased by 77.4% compared to the pure polymer [8].

Martelli et al. synthesized the nylon 6 nanocomposite with metallic copper nanoparticles, with a diameter of around 8 nm at 5 wt.%, through a ring-opening polymerization of the ε-caprolactam monomer. The reaction was carried out under high vacuum conditions in order to avoid the oxidation of the copper and the polymer. In spite of the presence of the nanoparticles, no significant changes in thermal conductivity or polymer crystallinity were observed [12].

In 2009, Rusu and Rusu synthesized nylon 6-copper microcomposites at concentrations from 0 to 25 wt.% through anionic ring-opening polymerization of ε-caprolactam. The X-ray diffractograms confirmed the formation of metallic copper, and the transition from phase α to phase γ of the nylon 6 was associated with the increase in copper concentration; as long as melting temperature decreased, crystallinity temperature increased and the degree of crystallinity decreased as well. The microcomposites at concentrations of 6 and 15% were analyzed by X-ray diffraction and only the metallic copper peak at $2\theta = 43.4^\circ$ was found. For the specific concentration of 25%, the decomposition temperature of the composite was 450.5°C, higher than the 428.0°C reported for the nylon 6 [11].

As previously mentioned, most authors have used in situ polymerization to obtain the nylon 6 composite. Despite the improvement in the mechanical properties reported in these works, the synthesis involves expensive reactants, specialized equipment, very specific reaction conditions, long reaction time, and low production capacity.

As an alternative, this work presents synthesis by ultrasound-assisted reactive extrusion. The procedure has the following advantages: (i) the reaction is held in the melting state without the use of solvents, (ii) it is possible to handle many polymers or copolymers branched or grafted with distribution of molecular weight, even to recycle polymers, (iii) the reaction time is in the order of minutes and more product can be processed simultaneously and continuously, and (iv) the required equipment has standard characteristics and flexibility in the control of the process variables [4, 16]. It is important to note that the production of the nanocomposite without the formation of copper oxide nanoparticles is quite complicated due to the strong tendency of the metallic copper to form the oxide metal under environmental conditions [17].

The aggregation of the nanoparticles inside the polymer matrix is related to the high superficial energy of the particles [14]. Usually the presence of the aggregates implies a decrease in the mechanical and thermal properties of the nanocomposite. Furthermore, it is well known that aggregation can be avoided by ultrasonic waves at a specific power and frequency without degradation of the polymer matrix [18, 19]; for this reason, ultrasound is implemented during the reactive extrusion in this work.

This document reports the synthesis of nylon 6 nanocomposites formulated with different contents of metallic copper nanoparticles (nylon 6/nano-Cu), using the melt mixing route in a single-screw extruder. The preparation was carried out using an ultrasound system to disperse the nanoparticles inside the nylon 6 matrix. The physicochemical characterization of the raw and obtained products was done by different experimental techniques such as X-ray diffraction (XRD), scanning electron microscopy (SEM), differential scanning calorimetry (DSC), and the tensile strength test. It should be mentioned that currently, in the scientific literature, there is no report of a similar work and that only a few studies of nylon 6 nanocomposites with metallic copper particles (described above) are available.

2. Experimental Methods and Materials

2.1. Chemical Products. Granulated nylon 6 commercialized as Zytel®7301 NC010 by Dupont™ with a density of 1.13 g/cm^3 and a melting point of 230°C [20] was used. Likewise, metallic copper nanoparticles with an average diameter of 21 nm obtained by a previously reported procedure were used [21, 22].

2.2. Preparation of the Nylon 6 and Synthesis of the Nanocomposites. First, the nylon 6 was dried in a vacuum oven at 80°C for four hours according to the data sheet provided by the supplier [20] Then, the pure polymer and the samples with a concentration of metallic copper nanoparticles of 0.01, 0.05, 0.10, and 0.50 wt.% were processed under the same conditions in the single spindle Moreli extruder with a length/diameter ratio of 13.0, set to 230°C at 23 rpm, a feed of 400 g/h, and a residence time of 15 minutes. An ultrasonic device with a power of 450 W and a frequency of 35 kHz was coupled with the extruder. In every sample, 100 g was taken and used for the physicochemical characterization.

2.3. Characterization Techniques

2.3.1. X-Ray Diffraction (XRD). To determine the crystalline structure of the nylon 6 and the nanocomposites, X-ray diffraction was used. The samples were evaluated in a diffractometer model D-5000 manufactured by Siemens, operated at a current intensity of 25 mA and a voltage of 35 kV to obtain the radiation corresponding to the Cu $K\alpha$ at a longitude wave of 1.54056 Å. The scanning in the 2θ scale was from 15° to 80° with a sample velocity of 0.02°/s.

2.3.2. Scanning Electron Microscopy (SEM). The morphology of the polymer and its nanocomposites were determined by means of a field-emission scanning electron microscope model JSM-7401F manufactured by Jeol. A secondary electron detector, LEI, was fixed with a voltage of 3.0 kV and a working distance of 3 mm. The samples were frozen in liquid nitrogen and fractured. In order to make their surface conductive, the samples were coated with a gold-palladium alloy using a sputtering technique. All micrographs were taken at a magnification of 30,000x.

2.3.3. Differential Scanning Calorimetry (DSC). The polymorphic structure, melting temperature, crystallization temperature, and degree of crystallinity of all the samples were determined by means of a calorimeter model DSC-2920 Modulated DSC provided by TA Instruments. The heating and cooling rate were set at 10°C/min, and a high purity nitrogen (UAP, 99.999%) with a flow rate of 50 mL/min was used to avoid oxidation of the samples, and between 10 and 15 mg of each sample was tested from 0 to 260°C. The degree of crystallinity was calculated using the following equation [23]:

$$X_C(\%) = \frac{\Delta H_f}{(1-\varnothing)\Delta H^*} \times 100, \qquad (1)$$

where ΔH_f is the heat of fusion of every sample and ΔH^* is the heat of fusion of the pure nylon 6 with 100% crystallinity, equal to 188 J/g for the specific case of our polymer, and $\varnothing$ is the weight fraction of the copper nanoparticles.

2.3.4. Mechanical Properties. Mechanical properties were determined according to the *Standard Test Method for Tensile Properties of Plastics* ASTM D638-10 [24] in a universal materials testing machine model 4301 supplied by INSTRON. The operating conditions were as follows: the length between jaws (L) was 25.4 mm, a load cell of 100 kg, and an elongation velocity of 5 mm/min. All tests were carried out at room temperature, and 5 specimens were tested for every polymer or nanocomposite. To obtain the testing probes, the processed polymer and the nanocomposites were injected in a proper mold with an injection machine manufactured by Dynisco. Before the probes were tested, they were vacuum dried at 80°C for four hours in order to eliminate the residual humidity in the samples. The statistical analysis indicated that the standard deviation was below 2.6% for tensile strength stress, 2.2% for tensile strength modulus, and 3.0% for tensile elongation at break.

3. Results and Discussion

3.1. X-Ray Diffraction of the Polymer and Composites. The diffractograms of the pure nylon 6 and the nanocomposites for the concentrations of 0.01, 0.05, 0.10, and 0.50 wt.% are shown in Figure 1. It can be seen that the pure nylon 6 shows peaks at 20 and 23.7° with crystalline planes (200) and (002, 202), attributed to the crystalline phases α_1 and α_2, respectively [25]; both signals have nearly the same intensity. In the case of the nanocomposites in Figures 1(b)–1(e), the same peaks remain after the incorporation of the copper nanoparticles whose presence is confirmed for the new peaks observed. The intensity of the crystalline plane (002, 202) is higher than the intensity of the reflection (200) which indicates that the nanoparticles favor the formation of the α_2 phase in the polymer matrix. This behavior has been previously observed by other authors in spherical nanoparticles dispersed in nylon 6; these authors concluded that this phenomenon is closely related to a secondary crystallization process, where crystal growth is held onto the crystalline plane with the higher energy level [26–29].

The new peaks located at 37.1° and 43.3° were associated with the nanoparticles. The first corresponds to the crystalline plane (111) assigned to the copper II oxide (Cu_2O) with a body-centered cubic structure (BCC) [29], while the second is related to a crystalline plane (111) assigned to the metallic copper (Cu^0) with a face-centered cubic (FCC) [7, 8]. As the concentration of nanoparticles increases those peaks increase as well, the peak corresponding to the oxide increases in a higher proportion which means that the increase in concentration increases the oxidation of the metallic copper nanoparticles. The transition from metallic copper to the oxidized state is attributed to the conditions during the preparation of the nanoparticles, for example, the presence of air, the residence time, and the process temperature.

3.2. Morphology and Nanoparticle Distribution. The surface of the pure nylon 6 after the fracture at low temperatures is presented in Figure 2. It can be seen that the fracture was

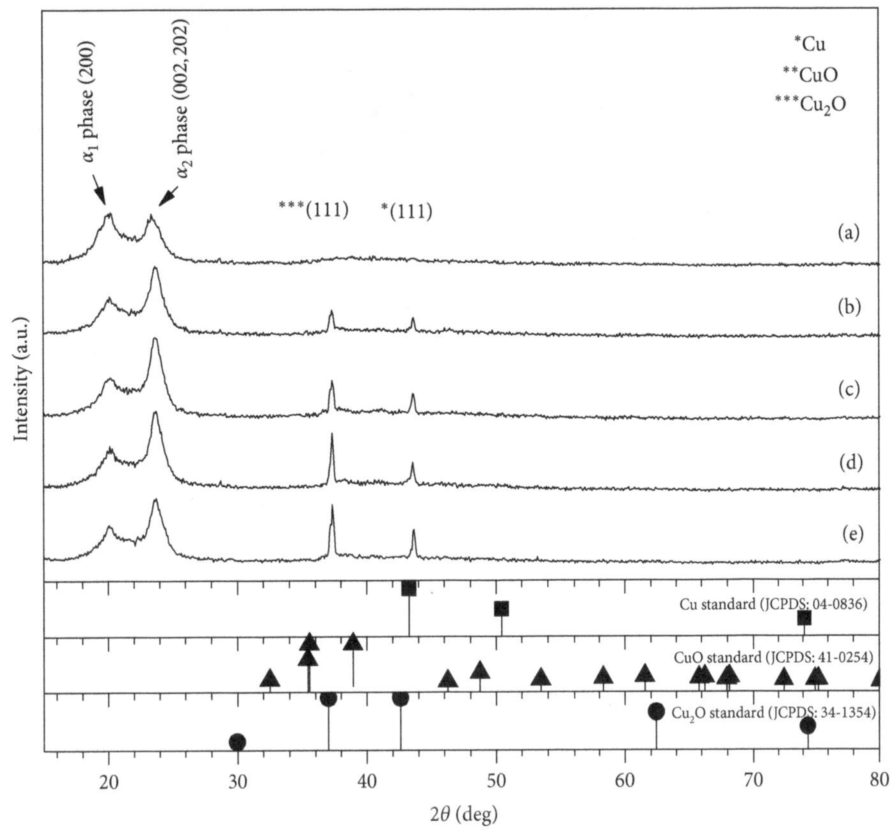

FIGURE 1: XRD diffractograms for all the samples: (a) pure nylon 6 and nanocomposites at the evaluated concentrations: (b) 0.01, (c) 0.05, (d) 0.10, and (e) 0.50 wt.% of copper nanoparticles.

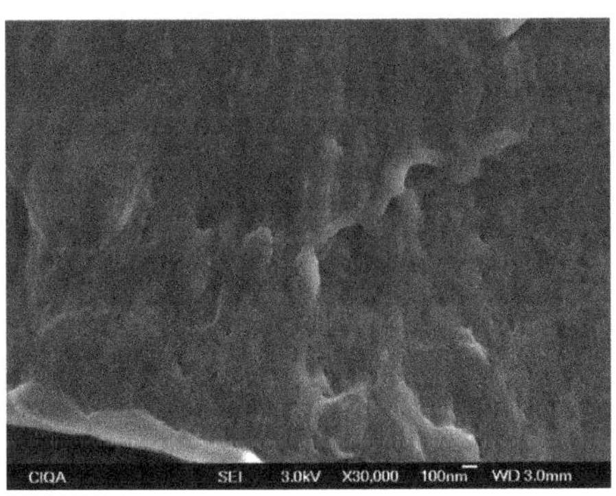

FIGURE 2: Fractured surface of the nylon 6.

ductile, due to the fact that it undergoes a plastic deformation and there is no evidence of tear and the surface remains smooth and uniform.

Figure 3 shows the micrographs corresponding to the obtained nanocomposites. At the low concentration of nanoparticles (0.01 and 0.05%), the matrix of the polymer has a smooth appearance without the presence of tears or hollows, and some spherical nanoparticles smaller than 100 nm can be observed. In the case of 0.05%, some aggregates have the same morphology as that exhibited by the monodispersed nanoparticles.

In particular, Figure 3(c) shows the surface of nanocomposite with a weight percent equal to 0.10 of copper nanoparticles. The polymer matrix of this nanocomposite exhibits zones with smooth features and others with the presence of cracks. As can be seen, scattered spherical particles with sizes less than 100 nm and aggregates in form of wires are observed. The particles and nanowires of the nanocomposite exhibited are brighter than those of the matrix, perhaps due to the good electrical conductivity of the copper nanoparticles. Chen et al. [30] prepared nylon 6 nanocomposites with different amounts of multiwalled nanocarbon tubes (MWNTs) using the fusion blend method. By SEM analysis of a fractured sample with liquid nitrogen, they saw high brightness in the MWNTs, attributed to their high electrical conductivity.

Figure 3(d) illustrates the micrograph of the nanocomposite with 0.50 wt.% of copper nanoparticles. It shows that the polymer matrix has smooth characteristics and also some cracks. Additionally, monodisperse spherical particles and aggregates of subparticles of smaller size were observed. The distribution of both isolated particles and agglomerates is relatively uniform. In addition, a higher content of nanoparticles is visualized compared to the other nanocomposites, and this agrees with the XRD results shown in Section 3.1.

3.3. Thermal Characterization. The formation of crystals of different sizes of the nylon 6 polymer matrix on the bulk of the nanocomposites during the heating-cooling process can

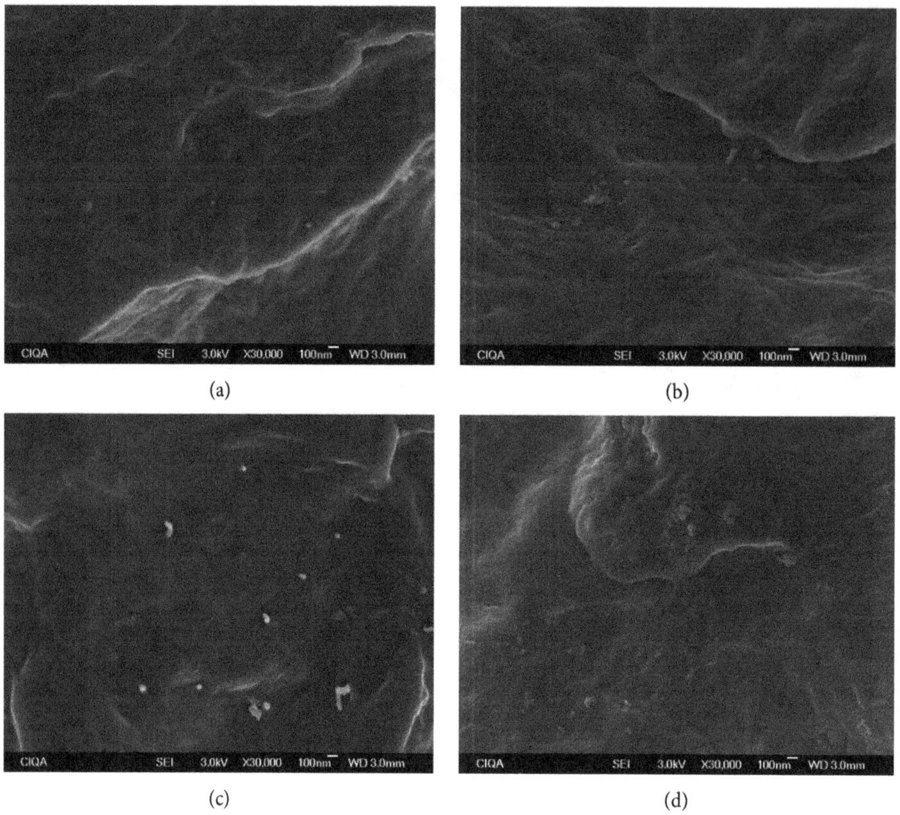

Figure 3: Fractured surface of the nanocomposites: (a) 0.01, (b) 0.05, (c) 0.10, and (d) 0.50 wt.% of copper nanoparticles.

show significant dependence on the content of copper nanoparticles. In this case, it is essential to study the processes of melting and crystallization of the polymer matrix in nanocomposites using DSC thermal analysis.

3.3.1. Effect of the Nanoparticles in the Fusion Process. The endothermic thermograms of the nylon 6 and the nanocomposites at the concentrations of 0.01, 0.05, 0.10, and 0.50 wt.% of nanoparticles are presented in Figure 4.

In Figure 4(a), for the nylon 6, the fusion endotherm exhibits a single peak with a melting temperature (T_m) of 222.3°C attributed to the crystalline phase α, in agreement with the scientific literature [28]. On the contrary, for the nanocomposites, two well-defined peaks were found. For the nanocomposites at 0.01, 0.05, and 0.1 wt. %, in Figures 4(b)–4(d), the melting temperature of peak 1 decreases compared to the pure polymer with values of 222.1, 221.5, and 220.8°C, respectively. The decrease in the melting point may be related to the presence of the copper nanoparticles which acts as a nucleation promoter [31]. Figure 4(e) shows the endotherm of nanocomposite at 0.50 wt.% with a melting temperature of 221.8°C, quite similar to that of the 0.05 wt.% concentration. This behavior is evidence of the possible rearrangement of the polymer crystals which undergo a slight growth in size [32]. Thus, such variations in the melting temperature support the idea that the nanoparticles participate in the nucleation process of the polymer matrix [11].

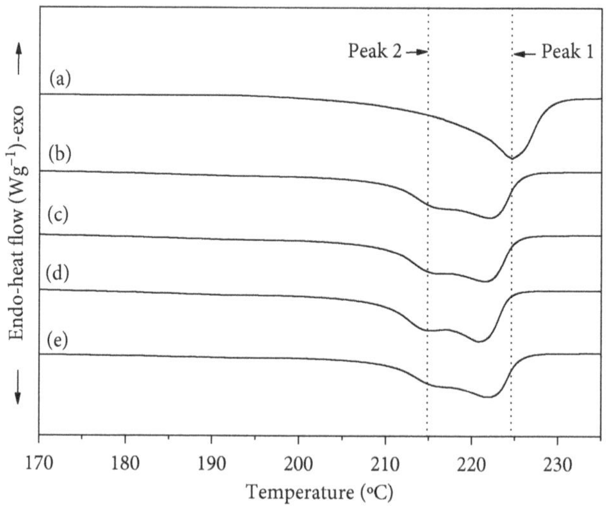

Figure 4: Endothermic thermograms of the nylon 6 (a) and the nanocomposites at the concentrations of 0.01 (b), 0.05 (c), 0.10 (d), and 0.50 (e) wt.%.

The second peak observed for the nanocomposites shown in Figure 4 and found at approximately 215°C has a lower intensity than that of peak 1. In general, the presence of a second peak can reveal that the crystals have imperfections or defects in the crystalline structure and in their spatial distribution [33]. This peak has also been observed in other nanocomposites prepared with different nanoparticles in the development of engineering plastics such as polyethylene terephthalate and polycarbonate. The authors of those works attributed the presence of the second peak to a secondary crystallization process in the polymer matrix that occurs after the formation of the primary crystals which are more thermodynamically stable [34, 35]. Other authors have associated this behavior with the formation of the crystalline phase γ [11, 36].

3.3.2. Effect of the Nanoparticles on the Crystallization Process. The exothermic thermograms of the nylon 6 and the nanocomposites at the concentrations of 0.01, 0.05, 0.10, and 0.50 wt.% of nanoparticles are shown in Figure 5.

The crystallization exotherm of the nylon 6 is shown in Figure 5(a); the wide crystallization peaks have a crystallization temperature (T_c) of 188.7°C and a degree of crystallinity of 42.7%. On the contrary, for the nanocomposites with concentrations of 0.01, 0.05, and 0.1 wt.% of copper nanoparticles, a single peak slightly more narrower than that of the nylon 6 with an increase in the crystallization temperature at values of 189.9, 190.5, and 190.8°C for the respective concentrations was observed. Additionally, the degree of crystallinity slightly increases, producing the corresponding values of 43.1, 45.9, and 46.9, respectively. In this case, it is difficult to draw conclusions about the origin of this behavior because opposite behaviors have been reported in the literature: an increase in the degree of crystallinity [8], a decrease in this variable [11], and an insignificant effect of the nanoparticles on this variable [12]. However, our findings do support the idea that the copper nanoparticles change the nucleation process in the polymer matrix, as has been reported by the other authors [27, 32, 37].

The exotherm of the sample with a concentration of 0.5 wt.%, presented in Figure 5(e), exhibits a completely different behavior with a crystallization temperature of 189.3°C and a degree of crystallinity of 40.1%. In these cases, the crystallization temperature decreases to a value closer to that corresponding to the concentration of 0.01%, the lowest crystallinity degree of all of the nanocomposites, even considering the pure polymer. It can also be seen that the peak is wider and less intense than the others. This behavior can be associated with the formation of aggregates of particles that are not homogeneously dispersed in the continuous phase as can be seen in Figure 3(d); the nucleation process is less favorable, and crystalline growth is limited. Furthermore, the aggregates restrict the movement of the polymer chain which hinders the orientation and packing of the chains; this mechanism is supported by the reduced degree of crystallinity of the polymer matrix [23]. Table 1 gives a summary of the results observed in Figures 4 and 5.

The thermal stability of the nanocomposites determined by thermogravimetric analysis indicates that the presence of the nanoparticles does not produce a considerable effect; for example, the maximum increase in this variable ranges from 452.4°C for nylon 6 to 453.4°C for the nanocomposite at 0.5 wt.% of copper nanoparticles. This result is similar to the cases reported previously [8, 11].

As previously described, there are only a few reports regarding the synthesis of nylon 6/copper nanoparticles

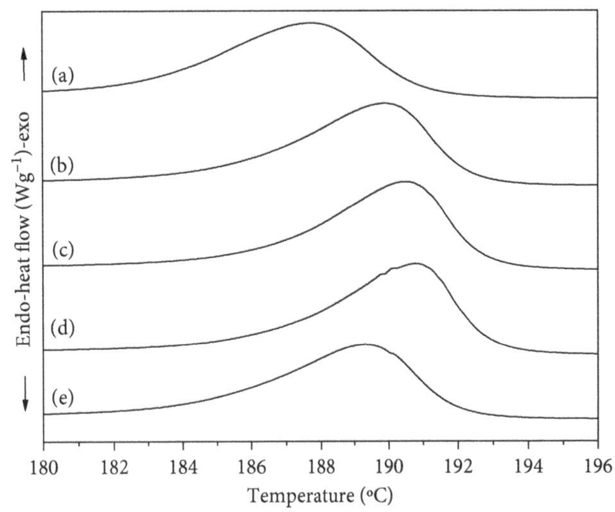

FIGURE 5: Exothermic thermograms of the nylon 6 (a) and the nanocomposites at concentrations of 0.01 (b), 0.05 (c), 0.10 (d), and 0.50 (e) wt.%.

[8, 11, 12] in addition to the information presented in this work. It is difficult to find good agreement among all the works with regard to the mechanical or thermal behavior of the nylon 6/Cu nanocomposites; this may be attributed to the fact that the characteristics of the nanoparticles depend on the conditions of synthesis and processing. Unfortunately, in the works mentioned, not all the parameters are equal, and a detailed comparison cannot be made. In the present work, it can be said that the physicochemical and mechanical properties of nylon 6 are improved by the presence of the copper nanoparticles for concentrations below 0.5 wt.%.

3.4. Mechanical Behavior of Nylon 6 and Composites. Nylon 6 is an engineered polymer that exhibits excellent mechanical properties [15]. In nanocomposites, these characteristics can be improved or diminished depending on the chemical composition, content, shape, and dispersion of the particles embedded in the polymer matrix [38]. In this section, the tensile strain properties determined experimentally are presented.

3.4.1. Tensile Strength. Figure 6 shows the results of the tensile stress test performed with pure nylon 6 and nylon 6 nanocomposites with a copper nanoparticle weight content of 0.01, 0.05, 0.10, and 0.50 wt.%.

Tensile strength increased with the concentration of the copper nanoparticles by 8.9% compared to the pure nylon 6, which corresponds to a magnitude of 71 MPa instead of 65.13 corresponding to the polymer. Again, this behavior correlates with the good dispersion of the nanoparticles or their aggregates in the polymer matrix, which increases the interfacial adhesion and the mechanical performance of the material [39, 40]. It is important to note that the tensile strength values determined for the nanocomposites via extrusion are superior to those of similar compounds prepared by the in situ polymerization method [8, 14].

Table 1: Thermal properties of nylon 6 and its nanocomposites with copper nanoparticles.

Nylon 6/nano-Cu	Melting temperature, T_m (°C)	Crystallization temperature, T_c (°C)	Degree of crystallinity, X_c (%)
Pure nylon 6	222.3	187.7	42.7
Nylon 6/0.01	222.1	189.9	43.1
Nylon 6/0.05	221.5	190.5	45.9
Nylon 6/0.10	220.8	190.8	46.9
Nylon 6/0.50	221.8	189.3	40.1

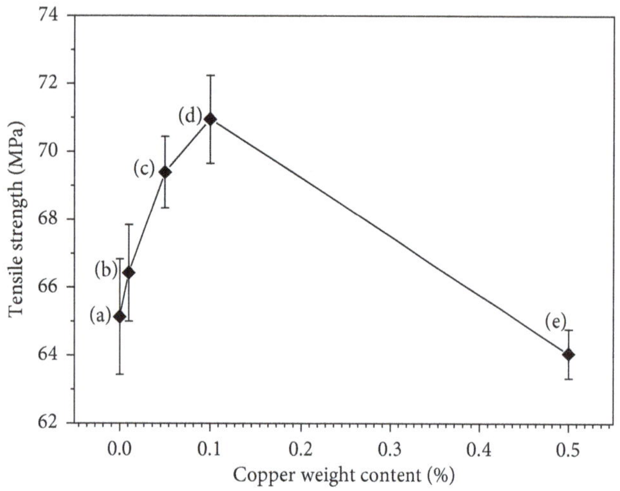

Figure 6: Tensile strength: (a) nylon 6 and nanocomposites with (b) 0.01, (c) 0.05, (d) 0.10, and (e) 0.50 wt.% of copper nanoparticles.

At the 0.50 wt.% concentration of the nanoparticles, a decrease of 1.7% in tensile strength was observed (Figure 6). This behavior has been attributed to the formation of particle aggregates [41]. In Figure 3(d), such aggregates can be observed, and in Figure 7, the composition of the particles was confirmed by elemental mapping determination using energy dispersive spectroscopy (EDS); groups of particles concentrate in small spaces, and the formation of some elongated aggregates quite similar to nanowires were observed as well.

These aggregates can act as stress concentration centers, causing a drastic reduction in the mechanical properties due to the fact that there is no homogenous distribution of the nanoparticles within the nylon 6 matrix [42]. Another possible cause of this behavior is the low crystallinity of the polymer matrix in the compound (Table 1), which favors a decrease in the rigidity of the polymer, and as a consequence, a reduction in its resistance to the tension [43].

In polymers such as polypropylene, it is known that ultrasound-assisted extrusion may have a slightly degradative effect on the polymer matrix [44], and this effect can be greater at higher nanoparticle concentrations [45]. The use of ultrasound and nanoparticles with polyamides is quite controversial [46–48], and further studies are necessary in order to clarify these points.

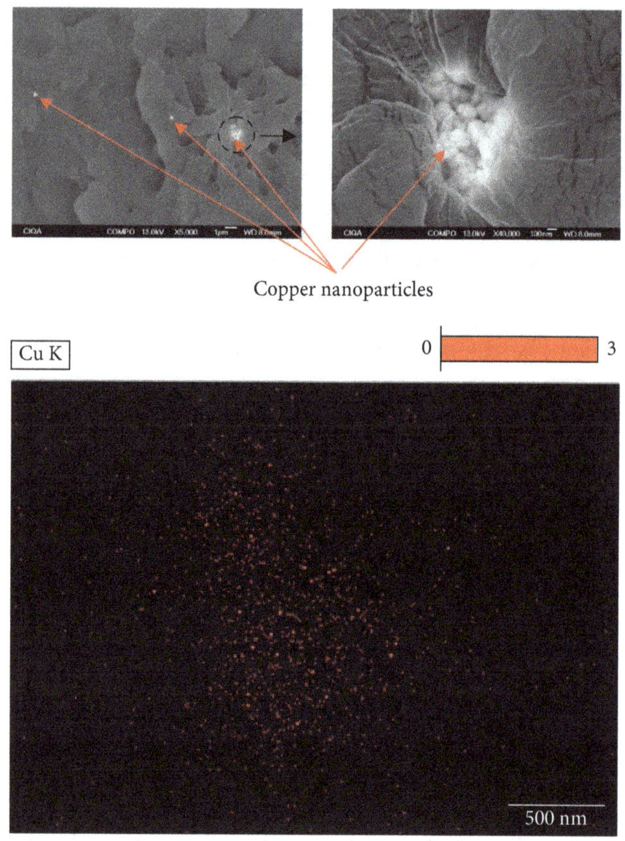

Figure 7: Copper elemental mapping by energy dispersive spectroscopy (EDS).

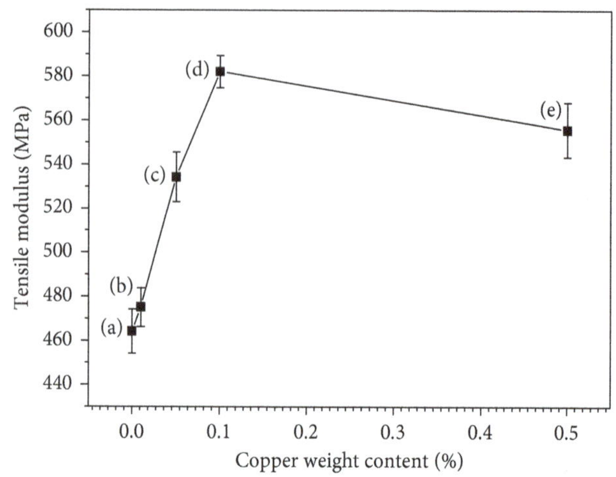

Figure 8: Tensile modulus: (a) nylon 6 and nanocomposites with (b) 0.01, (c) 0.05, (d) 0.10, and (e) 0.50 wt.% of copper nanoparticles.

3.4.2. Tensile Modulus. Figure 8 shows the tensile modulus for the nylon 6 and the nanocomposites prepared with the copper nanoparticles.

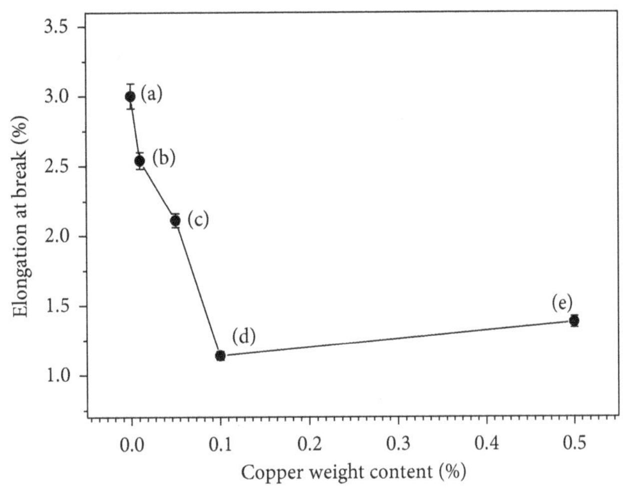

FIGURE 9: Tensile elongation at break: (a) nylon 6 and nanocomposites with (b) 0.01, (c) 0.05, (d) 0.10, and (e) 0.50 wt.% of copper nanoparticles.

TABLE 2: Mechanical properties and standard deviation of Nylon 6 and the nanocomposites.

Nylon 6/nano-Cu*	Tensile strength (MPa)	Tensile modulus (MPa)	Tensile elongation at break (%)
Nylon 6	65.13 ± 1.70	464.22 ± 10.12	3.00 ± 0.09
99.09/0.01	66.42 ± 1.42	475.08 ± 8.85	2.54 ± 0.06
99.05/0.05	69.39 ± 1.05	534.27 ± 11.26	2.11 ± 0.05
99.90/0.10	70.95 ± 1.30	582.13 ± 7.34	1.14 ± 0.03
99.95/0.50	64.05 ± 0.73	555.82 ± 12.41	1.38 ± 0.04

*The total weight of each sample is of 100 g. Values shown with a ± symbol belong to the standard deviation.

The tensile modulus of the nylon 6 had a magnitude of 464.22 MPa. For the nanocomposites (Figures 8(b)–8(d)), the modulus of the concentrations of 0.01, 0.05, and 0.10 wt.% increased by 2.3, 15.1, and 25.5%, respectively. As noted above, the incorporation of nanoparticles in the compound favors increased rigidity and hardness of the polymer matrix [49, 50].

Regarding the compound with 0.50 wt.% of nanoparticles whose tensile modulus is 19.7% higher than of the nylon 6 but lower that of the nanocomposite with 0.10 wt. % of nanoparticles, this mechanical behavior can be explained by the fact that the nanoparticles in the compound are not homogeneously distributed and dispersed properly throughout the continuous phase, which causes zones where the polymer matrix exclusively remains, in which stresses causing discontinuities are concentrated which weakens the structure [42].

3.4.3. Tensile Elongation at Break. Figure 9 shows the elongation at break of the nylon 6 and the nanocomposites with concentrations of 0.01, 0.05, 0.10, and 0.50 wt.% content of nanoparticles.

In contrast to the other mechanical properties, the tensile elongation at break decreases with the presence of nanoparticles. For the concentrations of 0.01, 0.05, and 0.10 wt.% of nanoparticles, decrease in elongation at break by 15.3, 29.7, and 62%, respectively, was observed. This is in agreement with the information shown in Figure 6 on tensile strength for these nanocomposites; since the content of nanoparticles within the nanocomposites increases, their continuous phase or polymer matrix acquires more rigid characteristics, due to a synergistic effect between the particles and the polymer. The theoretical explanation of this observation is based on the fact that frequently polymer compounds that exhibit high tensile strength usually acquire fewer elongations at rupture because their hardness tends to increase [51].

The sample at 0.50 wt.% of nanoparticles showed reduced elongation at break by 54% compared to the nylon 6, as shown in Figure 9(e), but it is slightly higher than the elongation at break of the 0.10% sample. This can be explained by the fact that the formation of aggregates leaves some regions free of nanoparticles in the polymer matrix which are susceptible to more deformation before breakage [41].

These aggregates of particles can act as centers of stress concentration that can be considered possible points of failure, which contributes to the deformation of the material. Additionally, they may change the spatial configuration of the hydrogen bonds between the polymer chains, preventing the reinforcing action that these bonds provide to the nylon 6 matrix [52]. Table 2 shows the results of the mechanical tests shown in the previous graphs accompanied by their standard deviation (error bars shown in the graphs).

4. Conclusions

Through the ultrasound-assisted reactive extrusion method, it was possible to synthesize nylon 6 nanocomposites with copper nanoparticles. The XRD studies showed the presence of Cu and Cu_2O nanoparticles, and this result confirms the oxidation of nanoparticles at the reaction conditions used. It was also demonstrated that the presence of nanoparticles favors the formation of the α_2 crystalline phase in the nylon 6 matrix.

The results of the scanning electron micrographs showed that the compound with 0.10% had the highest degree of dispersed particles with spherical and elongated shaped nanowires. This nanocomposite showed the greatest tensile strength and tensile modulus, as well as a lower elongation at break. In general, the incorporation of copper nanoparticles into nylon 6 improves its physical and mechanical properties as the concentration increases; 0.10 wt. % produces the best results while higher percentages cause agglomeration and decreased mechanical properties.

The nylon 6 nanocomposites with copper nanoparticles are prominent candidates for applications in which high-performance materials are required, mainly those in which the materials must withstand high mechanical stresses and maintain their dimensional stability such as automotive, aviation, and aerospace parts.

Conflicts of Interest

The authors declare that they have no conflicts of interest.

Acknowledgments

The authors gratefully acknowledge the financial support of CONACYT/Innovate UK through ACTin Project 268003 (Mexico)/102729 (UK) and CONACYT/EU 7 CuVito Project 127151 (Mexico)/263963 (EU). R. Sierra-Avila wishes to thank CONACYT for grant no. 23415. The authors would like to thank V. E. Comparán-Padilla, G. Mendez-Padilla, and Rodrigo Cedillo García for their valuable technical support.

References

[1] A. K. Kaw, *Mechanics of Composite Materials*, CRC Taylor & Francis Group, Boca Raton, FL, USA, 2nd edition, 2006.

[2] C. A. Harper, *Handbook of Plastics, Elastomers, and Composites*, McGraw-Hill Companies, Inc., New York, NY, USA, 4th edition, 2002.

[3] K. I. Winey and R. A. Vaia, “Polymer nanocomposites,” *MRS Bulletin*, vol. 32, no. 4, pp. 314–322, 2007.

[4] S. Thomas and R. Stephen, *Rubber Nanocomposites: Preparation, Properties and Applications*, John Wiley & Sons, Hoboken, NJ, USA, 1st edition, 2010.

[5] S. K. Mazumdar, *Composites Manufacturing: Materials, Product, and Process Engineering*, CRC Press LLC, Boca Raton, FL, USA, 1st edition, 2002.

[6] R. S. Davé and A. C. Loos, *Processing of Composites*, Hanser/Gardner Publications, Inc., Cincinnati, OH, USA, 1st edition, 2000.

[7] Z. Komeily-Nia, M. Montazer, and M. Latifi, “Synthesis of nano copper/nylon composite using ascorbic acid and CTAB,” *Colloids and Surfaces A: Physicochemical and Engineering Aspects*, vol. 439, pp. 167–175, 2013.

[8] Q. Xu, X. Li, S. Zhang, Y. Hao, and Z. Zhang, “Copper nanowire/PA6 composites prepared by in situ polymerization and its properties,” *Journal of Polymer Research*, vol. 20, no. 10, pp. 1–6, 2013.

[9] Q. J. Xu, X. H. Li, S. M. Zhang, and Z. J. Zhang, “Preparation and characterization of copper nanowire/polyamide 6 nanocomposites and its properties,” *Journal of Macromolecular Science, Part A*, vol. 51, no. 7, pp. 598–603, 2014.

[10] A. Larena and G. Pinto, “Electroconductive nylon 6 and copper composites,” *Polymer Composites*, vol. 16, no. 6, pp. 536–541, 1995.

[11] G. Rusu and E. Rusu, “Nylon 6/copper composites by in situ polymerization,” *Journal of Optoelectronics and Advanced Materials*, vol. 11, pp. 673–680, 2009.

[12] V. Martelli, N. Toccafondi, and G. Ventura, “Low-temperature thermal conductivity of nylon-6/Cu nanoparticles,” *Physica B: Condensed Matter*, vol. 405, no. 20, pp. 4247–4249, 2010.

[13] Q. Xu, X. Li, and Z. Zhang, “Preparation of Cu nanoparticles-modified PA6 composites using CuO as filler,” *Journal of the Iranian Chemical Society*, vol. 11, no. 6, pp. 1717–1721, 2014.

[14] Q. Xu, X. Li, and Z. Zhang, “Preparation of copper nanoparticle-improved polyamide 6 composites by an in situ solution route with cupric oxide as the metallic copper source and investigation of their properties,” *New Journal of Chemistry*, vol. 39, no. 4, pp. 3015–3020, 2015.

[15] M. Kohan, *Nylon Plastics Handbook*, Hanser, New York, NY, USA, 1st edition, 1995.

[16] E. Dimitriou and M. Petralia, *Ceramic and Polymer Matrix Composites: Properties, Performance and Applications*, Nova Science Publishers, Inc., New York, NY, USA, 2010.

[17] N. Jardón, M. Pérez, R. Sierra et al., “Oxidation of copper nanoparticles protected with different coatings and stored under ambient conditions,” *Journal of Nanomaterials*, vol. 2018, Article ID 9512768, 8 pages, 2018.

[18] A. I. Isayev, R. Kumar, and T. M. Lewis, “Ultrasound assisted twin screw extrusion of polymer–nanocomposites containing carbon nanotubes,” *Polymer*, vol. 50, no. 1, pp. 250–260, 2009.

[19] S. K. Swain and A. I. Isayev, “PA6/clay nanocomposites by continuous sonication process,” *Journal of Applied Polymer Science*, vol. 114, no. 4, pp. 2378–2387, 2009.

[20] DuPont TM Zytel® 7301 NC010 Nylon Resin, 2017.

[21] R. Sierra, M. Pérez, G. Cadenas et al., “Synthesis of copper nanoparticles coated with nitrogen ligands,” *Journal of Nanomaterials*, vol. 2014, Article ID 361791, 8 pages, 2014.

[22] R. Sierra, M. Pérez, G. Cadenas et al., “Synthesis of copper nanoparticles using mixture of allylamine and polyallylamine,” *Journal of Nanomaterials*, vol. 2015, Article ID 367341, 9 pages, 2015.

[23] F. Zohra and N. Vasanthan, “Crystallization, crystal structure, and isothermal melt crystallization kinetics of novel polyamide 6/SiO_2 nanocomposites prepared using the sol–gel technique,” *Journal of Physical Chemistry B*, vol. 118, no. 31, pp. 9486–9495, 2014.

[24] ASTM, *Standard Test Method for Tensile Properties of Plastics*, ASTM, West Conshohocken, PA, USA, 2010.

[25] F. J. Medellín, L. Larios, A. Zapata, O. Dávalos, P. J. Phillips, and J. S. Lin, “Melting behavior of polymorphics: molecular weight dependence and steplike mechanisms in nylon-6,” *Macromolecules*, vol. 37, no. 5, pp. 1799–1809, 2004.

[26] M. García, J. Barsema, R. E. Galindo et al., “Hybrid organic inorganic nylon-6/SiO_2 nanocomposites: transport proper-

ties," *Polymer Engineering & Science*, vol. 44, no. 7, pp. 1240–1246, 2004.

[27] D. W. Chae, S. G. Oh, and B. C. Kim, "Effect of silver nanoparticles on the dynamic crystallization behavior of nylon-6," *Journal of Polymer Science:Part B: Polymer Physics*, vol. 42, no. 5, pp. 790–799, 2004.

[28] J. Zheng, R. W. Siegel, and C. G. Toney, "Polymer crystalline structure and morphology changes in nylon-6/ZnO nanocomposites," *Journal of Polymer Science:Part B: Polymer Physics*, vol. 41, no. 10, pp. 1033–1050, 2003.

[29] K. Akamatsu and S. Deki, "Formation process and characterization of copper oxide nanoparticles dispersed in polymer thin films," *Scripta Materialia*, vol. 44, no. 8-9, pp. 2149–2152, 2001.

[30] G. X. Chen, H. S. Kim, B. H. Park, and J. S. Yoon, "Multi-walled carbon nanotubes reinforced nylon 6 composites," *Polymer*, vol. 47, no. 13, pp. 4760–4767, 2006.

[31] D. M. Lincoln, R. A. Vaia, Z. G. Wang, and B. S. Hsiao, "Secondary structure and elevated temperature crystallite morphology of nylon-6/layered silicate nanocomposites," *Polymer*, vol. 42, no. 2, pp. 1621–1631, 2001.

[32] V. J. Cruz, B. L. España, C. A. Avila, and F. J. Medellín, "Nanocomposites based on plasma-polymerized carbon nanotubes and nylon-6," *Polymer Journal*, vol. 44, no. 9, pp. 952–958, 2012.

[33] A. R. Allafi and M. A. Pascall, "The effect of different percent loadings of nanoparticles on the barrier and thermal properties of nylon 6 films," *Innovative Food Science & Emerging Technologies*, vol. 20, pp. 276–280, 2013.

[34] R. Roberts, "The melting behavior of bulk crystallized polymers," *Journal of Polymer Science:Part B: Polymer Letters*, vol. 8, no. 5, pp. 381–384, 1970.

[35] F. N. Liberti and B. Wunderlich, "Melting of polycaprolactam," *Journal of Polymer Science: Part A-2: Polymer Physics*, vol. 6, no. 5, pp. 833–848, 1968.

[36] M. García, G. Van Vliet, M. G. J. ten Cate et al., "Large-scale extrusion processing and characterization of hybrid nylon-6/SiO_2 nanocomposites," *Polymers for Advanced Technologies*, vol. 15, no. 4, pp. 164–172, 2004.

[37] Q. Xu, X. Feng, J. Liu, and P. Gong, "Preparation of Ag nanoparticles-reinforced polyamide 6 nanocomposites by in situ polymerization and investigation of its properties," *Polish Journal of Chemical Technology*, vol. 19, no. 4, pp. 36–40, 2017.

[38] Y. W. Mai and Z. Z. Yu, *Polymer Nanocomposites*, Woodhead Publishing, Cambridge, UK, 1st edition, 2006.

[39] M. M. Hasan, Y. Zhou, H. Mahfuz, and S. Jeelani, "Effect of SiO_2 nanoparticle on thermal and tensile behavior of nylon-6," *Materials Science and Engineering: A*, vol. 429, no. 1-2, pp. 181–188, 2006.

[40] B. Finnigan, K. Jack, K. Campbell et al., "Segmented polyurethane nanocomposites: impact of controlled particle size nanofillers on the morphological response to uniaxial deformation," *Macromolecules*, vol. 38, no. 17, pp. 7386–7396, 2005.

[41] A. D. Erem, G. Ozcan, M. Skrifvars, and M. Cakmak, "In vitro assesment of antimicrobial activity and characteristics of polyamide 6/silver nanocomposite fibers," *Fibers and Polymers*, vol. 14, no. 9, pp. 1415–1421, 2013.

[42] C. Radheshkumar and H. Münstedt, "Morphology and mechanical properties of antimicrobial polyamide/silver composites," *Materials Letters*, vol. 59, no. 14-15, pp. 1949–1953, 2005.

[43] T. D. Fornes, P. J. Yoon, H. Keskkula, and D. R. Paul, "Nylon 6 nanocomposites: the effect of matrix molecular weight," *Polymer*, vol. 42, no. 25, pp. 9929–9940, 2001.

[44] J. Zhong and A. I. Isayev, "Ultrasonically assisted compounding of CNT with polypropylenes of different molecular weights," *Polymer*, vol. 107, pp. 130–146, 2016.

[45] S. He, J. Zhang, X. Xiao, and X. Hong, "Effects of ultrasound vibration on the structure and properties of polypropylene/graphene nanoplatelets composites," *Polymer Engineering & Science*, vol. 58, no. 3, pp. 377–386, 2017.

[46] P. Cerruti and C. Carfagna, "Thermal-oxidative degradation of polyamide 6,6 containing metal salts," *Polymer Degradation and Stability*, vol. 95, no. 12, pp. 2405–2412, 2010.

[47] W. Bao, H. Wu, T. Xie, and S. Guo, "Improved interfacial properties of PA6/DE blends by DE-g-MAH prepared through ultrasound assisted extrusion," *International Polymer Processing*, vol. 29, no. 3, pp. 371–378, 2014.

[48] M. A. Mohamed, A. El-Maghraby, M. El-Latif, H. Farag, and K. Kalaitzidou, "Nanocrystalline Fe-Ni alloy/polyamide 6 composites of high mechanical performance made by ultrasound-assisted master batch technique," *Polymer Composites*, vol. 35, no. 12, pp. 2343–2352, 2014.

[49] W. E. Van Zyl, M. García, B. A. G. Schrauwen, B. J. Kooi, and J. T. M. De Hosson, "Hybrid polyamide/silica nanocomposites: synthesis and mechanical testing," *Macromolecular Materials and Engineering*, vol. 287, no. 2, pp. 106–110, 2002.

[50] E. M. Araujo, A. M. D. Leite, R. A. Da Paz, V. Da Medeiros, T. J. A. De Melo, and H. De Lira, "Polyamide 6 nanocomposites with inorganic particles modified with three quaternary ammonium salts," *Materials*, vol. 4, no. 11, pp. 1956–1966, 2011.

[51] M. W. Wilbrink, A. S. Argon, R. E. Cohen, and M. Weinberg, "Toughenability of nylon-6 with $CaCO_3$ filler particles : new findings and general principles," *Polymer*, vol. 42, no. 26, pp. 10155–10180, 2001.

[52] R. Damerchely, M. E. Yazdanshenas, A. S. Rashidi, and R. Khajavi, "Morphology and mechanical properties of antibacterial nylon 6/nano-silver nano-composite multifilament yarns," *Textile Research Journal*, vol. 81, no. 16, pp. 1694–1701, 2011.

16

Property Optimisation of EPDM Rubber Composites Using Mathematical and Statistical Strategies

Sivaldo Leite Correia,[1] Denilso Palaoro,[1] and Ana Maria Segadães[2]

[1]*State University of Santa Catarina (UDESC), Center of Technology Sciences (CCT), 89223-100 Joinville, SC, Brazil*
[2]*Department of Materials and Ceramics Engineering (CICECO), University of Aveiro, 3810-193 Aveiro, Portugal*

Correspondence should be addressed to Ana Maria Segadães; segadaes@ua.pt

Academic Editor: Francesco Ruffino

This paper describes a study in which EPDM-based rubber composites were investigated aiming at developing formulations subjected to restrictions on cost and the properties of the material. The contents of components other than calcium carbonate, paraffinic oil, and CBS vulcanising accelerator, as well as additives and processing conditions, were kept constant. Fractional factorial design coupled with computational numerical optimisation was used to minimise the number of mixtures. The results demonstrate that statistical design of experiments and particle swarm optimisation (PSO) algorithms are promising methods to design composition variables. Mixture costs as low as 1.92 US$/kg can be achieved in compositions containing, for example, 107 phr of calcium carbonate, 95 phr of paraffinic oil, and 1.13 phr of CBS accelerator. The corresponding composite property-predicted values were 66.8 Shore A for hardness, tensile strength of 7.8 MPa, 570.8% elongation at break, and 23.0% rebound resilience. This demonstrates that, in this way, the desired product with specified characteristics can be comfortably manufactured at minimum cost.

1. Introduction

The modern rubber industry offers a very wide variety of technological products derived from synthetic elastomers like ethylene–propylene–diene M-class rubber (EPDM rubber). These products find applications in different fields, namely, in automotive, naval, and mechanical industries. Such rubber compounds (composites) are manufactured from complex mixtures of different raw materials (different kinds of EPDM elastomers, fillers, process oil, vulcanising and protecting agents, etc.), and the production steps involve a variety of processes (e.g., mixing, extruding, cutting, moulding, and vulcanising) [1–6].

The performance and manufacture of rubber products has been receiving more and more attention, and the industry has successfully introduced and applied the usage of a series of quality certification standards. Also, stringent market and price competition demand shorter product development cycles and reduced costs, which include raw materials and processing, as well as research and development costs. All of that makes it difficult to define an adequate new formulation by simple adjustment of older ones, based on rule of thumb or virtue of experience.

The application of statistical design of experiments (DoE) to the industrial formulation of rubber composites can be a convenient and accurate means of obtaining reliable quantitative estimates of properties as the result of any change in contents of raw materials [7, 8]. The modelling of a given property using the design of mixture experiments is becoming common practice [9–17] and was proven, in all cases reported, to lead to greater efficiency and confidence in the results obtained, and to be less demanding in time and both material and human resources.

Many studies on the effects of raw material changes on the physical properties of rubber composites coupled with DoE can be found in the literature, but few data are available on research carried out using the cost characteristics of rubber compounds. However, although standard requirements for physical and mechanical properties of a rubber compound are mandatory, high costs might preclude the product's competitiveness in the market.

Thus, the rubber compound engineer most often needs to produce an optimised formulation, which fits the requirements of physical and mechanical properties, while subjected to processing and cost constraints [18]. In industrially oriented applications of materials like rubbers or ceramics, the technique generally used to optimise equality and inequality property constraints is the graphical overlay of contour plots generated by the regression models of the properties [10, 15, 19–21]. However, as the number of functions and constraints increases, so do the difficulties to handle and design the different contour plots, and these strategies begin to have a rather limited performance.

Computational optimisation of nonlinear programming problems, which include numerical analysis of continuous and discrete variables, has been an active and important engineering research issue. The optimisation problem consists in finding out a solution for the objective function and related constraints. The use of particle swarm optimisation (PSO) algorithms for solving nonlinear, multimodal, and nondifferentiable optimisation problems, which are not well fitted for conventional optimisation algorithms, has gained increasing attention in recent years [22, 23]. PSO algorithms fundamentals result from the observation, interpretation, and modelling of the movements of individuals in bird flocks or fish schools, as well as their group behaviour as a swarm. It is a simple algorithm, so only a few lines of the computer program based on simple mathematical operations are needed to deploy the basic tool of PSO. The computer-aided optimisation method provides an efficient way to predict the optimum formulation without using those awkward contour plot graphical overlays.

In this work, a fractional factorial design of experiments was used to study the effect of filler, process oil, and vulcanising accelerator contents on the mechanical properties (hardness, tensile strength, elongation at break, and rebound resilience) of EPDM rubber composites. Regression models were calculated from the results of the measured properties, under constant processing conditions and contents of other raw materials and additives. The regression models were then used in a PSO algorithm to obtain optimised EPDM rubber formulations subjected to property constraints and cost requirements.

2. Experimental Procedure

2.1. Compound Ingredients and Base Composition. The elastomer used in this work was a commercial M-class ethylene–propylene–diene (EPDM) monomer (Keltan, supplied by Branco Indústria e Comércio Ltda). Other ingredients included carbon black (Spheron 5000, Cabot Brasil Indústria e Comércio), calcium carbonate (Mineração São Judas Ltda), and paraffinic oil (Ipiranga Indústria Quimica Ltda). Besides these ingredients, the mixtures include special additives, namely, vulcanising agent (sulphur, Basile Química), vulcanising accelerator (CBS, N-cyclohexylbenzothiazol-2-sulphenamide, Basile Química), vulcanising activator (ZnO, Brazinco Indústria e Comércio), aging inhibitor (Naugard 495, 4-5-methyl-2-mercaptobenzimidazole, Chemtura Indústria Química do Brasil Ltda), and stearic acid (Basile Química).

Table 1: Base composition of industrial EPDM rubber composites.

Ingredients	Content [phr][a]
EPDM	100.00
Carbon black	115.00
Calcium carbonate (CC)	Variable
Paraffinic oil (PO)	Variable
Sulphur, vulcanising agent	0.40
CBS, vulcanising accelerator (VA)	Variable
ZnO, vulcanising activator	5.00
Naugard 495, aging inhibitor	1.00
Stearic acid	1.00

[a]Per hundred of rubber, by weight.

The rubber formulations were based on that of heat- and air-resistant products, such as the automotive hoses manufactured by NSO Borrachas, Joinville, SC, Brazil. As shown in Table 1, the contents (phr, per hundred of rubber, by weight) of calcium carbonate filler (CC), process (paraffinic) oil (PO), and vulcanising accelerator (VA) were varied in the compositions, but the contents of the other raw materials and additives, as well as processing conditions, were kept constant. The chosen processing conditions closely followed the conventional laboratory rubber compound procedure used in industrial practice [5, 6].

2.2. Experiment Design. A 3^{3-1} fractional factorial design was chosen to model the effect of varying contents of the three factors CC, PO, and VA on the composite properties because it required the minimum number of experiments (nine mixture compositions) for which nonlinear effects and interactions of all the factors could be investigated [7, 8]. Given that the contents of all raw materials and additives other than the three factors were kept constant, a new calculation basis was defined to translate contents of the factors from their usual phr base values, m_i, into fractions, X_i, as needed in the factorial design, and vice versa. Among the factors limiting phr values, the CC content is the highest of them all (125 phr in the reference product). Hence, this was chosen as the base reference value, M, and contents of all the factors were then expressed as weight fraction relative to that of the CC content, that is, $X_i = m_i/M$. To design the matrix of mixture experiments, three weight fraction content levels were chosen within the usual ranges for the manufacture of general-purpose products, as shown in Table 2.

STATISTICA statistical software (StatSoft Inc., 2010) was used to determine the geometric and coded notations as well as randomise the treatment combinations, resulting in a standard experiment order. Table 3 shows the mixing ratios of chosen factors for the nine compounds, obtained from the 3^{3-1} fractional factorial design.

2.3. Mixture Preparation, Moulding, and Property Evaluation. For each of the nine different formulations, in two replications, the selected amounts of raw materials were mixed in a two-roll laboratory mill (Equipabor, Brazil) at 70°C and 1 : 1.20 speed ratio, as recommended by the ASTM

TABLE 2: Factors and levels (weight fraction) adopted for the 3^{3-1} fractional factorial design used to define EPDM rubber composites.

Factor	Level (weight fraction)		
	Low	Medium	High
Calcium carbonate (CC)	0.000	0.500	1.000
Paraffinic oil (PO)	0.400	0.600	0.800
Vulcanising accelerator (VA)	0.008	0.014	0.020

TABLE 3: Mixture compositions (weight fractions of chosen factors only) in the 3^{3-1} fractional factorial design.

Mixture	Weight fraction		
	CC (X_1)	PO (X_2)	VA (X_3)
1	0.000	0.400	0.008
2	0.000	0.600	0.020
3	0.000	0.800	0.014
4	0.500	0.400	0.020
5	0.500	0.600	0.014
6	0.500	0.800	0.008
7	1.000	0.400	0.014
8	1.000	0.600	0.008
9	1.000	0.800	0.020

D 15 Standard [24]. The sheets obtained were conditioned at 25 ± 2°C for 24 h in a sealed container before the estimation of optimum curing time. Batches were then compression-moulded to a 90% cure using an electrical resistance heated hydraulic press (model EMIC) at ~10 MPa and 160°C during 10 minutes.

The hardness (HD) test was carried out according to ASTM D 676 Standard [25] using a Zwick durometer. The tensile strength (TS) and elongation at break (EB) tests were carried out according to ASTM D 412 Standard [26] using an EMIC DL 2000 testing machine. The rebound resiliency (RS) was determined in accordance with ASTM D2632 using a rebound tester [27]. For each mixture, in each replication, the property final value was taken as the average of the test results obtained for five different test pieces.

2.4. Optimisation Strategy. The experimental results obtained for each property were used to iteratively calculate, with STATISTICA, the coefficients of a regression equation, until a statistically relevant model and response surface was obtained, relating that property value with the weight fractions of calcium carbonate (CC), paraffinic oil (PO), and vulcanising accelerator (VA) present in the corresponding mixture of raw materials.

A PSO algorithm was developed using the property model equations obtained with STATISTICA and common limiting property values (those of the reference automotive hoses manufactured by NSO Borrachas, Joinville, SC, Brazil), aiming at finding the best composition range (weight fractions) that meets the property limits while minimising costs of the composites.

3. Results and Discussion

3.1. Measured Properties and Statistical Analysis. Table 4 presents the values of hardness (HD), tensile strength (TS), elongation at break (EB), and rebound resiliency (RS) obtained for the nine mixtures in two replications. Material costs for the nine mixtures in replication 1 are also shown in Table 4.

Table 5 shows the results of the variance analysis of the regression equations obtained for HD, TS, EB, and RS, using the nomenclature commonly found in the literature (major statistical properties: p values and coefficient of multiple determination R^2) [7, 8]. It can be seen that, in all cases, the nonlinear models are statistically significant at the required level (p value ≤ significance level) and present small variability (high coefficients of multiple determination). Although only effects with p value lower than 0.10 were considered significant, p values higher than 0.10 were kept in Table 5 because those effects should appear in the models. In all cases, the errors could be considered randomly distributed around a zero mean value (i.e., they are uncorrelated), which suggests a common constant variance. On the basis of this analysis, the regression models obtained were accepted to describe the effect of contents of raw materials (CC, PO, and VA) on HD, TS, EB, and RS, and the final results are (1). These equations are all referred to the weight fractions of the components calcium carbonate (X_1), paraffinic oil (X_2), and vulcanising accelerator (X_3), so that mixing of raw materials can easily be carried out.

$$\begin{aligned} HD &= 91.45 - 42.13X_2 + 252.07X_3 + 10.64X_1X_2^2, \\ TS &= 12.39 - 0.06X_1 + 1.72X_1^2 - 1.76X_2^2 - 890.02X_3^2 \\ &\quad - 19.51X_1X_2 + 16.22X_1X_2^2, \\ EB &= 475.55 - 56.58X_1 - 103.51X_2 + 471.24X_2^2 \\ &\quad - 1040.76X_3 - 77.94X_1X_2^2, \\ RS &= 29.49 - 4.94X_1 - 132.57X_3 - 2.44X_1X_2^2. \end{aligned} \tag{1}$$

3.2. Experimental Validation of the Models. Three extra mixtures, F_1, F_2, and F_3 (check-point mixtures), were used to validate the calculated statistical models (the mixtures and their test pieces were prepared following the same procedure as before). Table 6 presents the compositions of those three mixtures and the corresponding measured and predicted values for HD, TS, EB, and RS. It can be seen that the estimates calculated using (1) can be higher or lower than, but are always very close to, the corresponding experimental value (low error), which validates the calculated models.

3.3. Cost and Property Optimisation Using PSO Algorithm. Following the same procedure and reasoning described above for the compounds mechanical properties, a valid and significant regression model was also obtained for the cost of mixtures (CT), which can be described by the following equation:

$$CT = 2.68 - 0.66X_1 - 0.26X_2 + 0.28X_3. \tag{2}$$

TABLE 4: Measured values of hardness (HD), tensile strength (TS), elongation at break (EB), rebound resilience (RS), and material costs (CT) for the nine designed mixtures in two replications.

Mixture		HD (Shore A)	TS (MPa)	EB (%)	RS (%)	Cost (US$/kg)
Replication 1	1	77.0 ± 0.7	12.5 ± 0.1	496.7 ± 08.2	27.8 ± 1.1	2.63
	2	70.8 ± 0.5	10.7 ± 0.2	588.0 ± 10.9	26.6 ± 0.6	2.55
	3	59.4 ± 0.6	11.2 ± 0.6	670.0 ± 11.6	26.8 ± 0.5	2.47
	4	79.0 ± 0.0	9.5 ± 0.2	453.3 ± 10.3	25.0 ± 0.7	2.22
	5	70.0 ± 0.7	9.0 ± 0.2	550.0 ± 10.0	24.4 ± 0.6	2.17
	6	61.4 ± 0.6	8.7 ± 0.2	636.0 ± 08.9	26.4 ± 0.6	2.11
	7	80.8 ± 0.5	8.1 ± 0.2	403.3 ± 08.2	22.6 ± 0.9	1.91
	8	72.4 ± 0.9	7.2 ± 0.2	468.0 ± 10.9	23.8 ± 0.5	1.89
	9	65.2 ± 1.3	7.4 ± 0.2	565.0 ± 10.5	24.6 ± 0.6	1.87
Replication 2	1	79.8 ± 0.5	11.6 ± 1.8	496.6 ± 08.2	29.6 ± 0.9	—
	2	70.8 ± 0.8	11.8 ± 0.4	552.0 ± 10.6	26.0 ± 0.0	—
	3	63.4 ± 0.9	11.1 ± 0.4	670.0 ± 11.6	29.0 ± 1.2	—
	4	82.4 ± 1.1	9.3 ± 0.2	436.0 ± 08.9	25.2 ± 1.1	—
	5	76.4 ± 0.9	9.6 ± 0.2	512.5 ± 09.6	26.2 ± 0.5	—
	6	61.2 ± 0.8	9.0 ± 0.2	647.5 ± 09.6	26.6 ± 0.6	—
	7	73.0 ± 0.7	8.8 ± 0.2	470.0 ± 11.6	*	—
	8	73.0 ± 1.2	7.5 ± 0.2	486.6 ± 10.3	25.2 ± 0.5	—
	9	75.6 ± 0.6	7.4 ± 0.2	566.7 ± 10.3	22.6 ± 0.6	—

*Value could not be measured.

TABLE 5: Major statistical properties [7, 8] relevant for variance analysis (R^2, p value for the significance and lack-of-fit tests) of the calculated regression models (X_1 calcium carbonate, X_2 paraffinic oil, and X_3 vulcanising accelerator).

Effect	p value*			
	HD	TS	EB	RS
X_1	—	0.00	0.00	0.00
X_1^2	—	0.07	—	0.07
X_2	0.00	—	0.00	—
X_2^2	—	0.00	0.13	0.00
X_3	0.06	0.10	0.12	0.10
X_3^2	—	0.20	—	0.20
X_1X_2	—	0.27	—	0.27
$X_1X_2^2$	0.03	0.26	0.09	0.25
R^2	0.85	0.82	0.96	0.96
Lack-of-fit	0.99	0.89	0.77	0.55

*p values of the effects used in the regression models. Other insignificant p values (>0.10) are not presented.

In mathematical language, the optimisation problem consists in minimising this objective function (CT) with respect to the design variables X_1, X_2, and X_3, subjected to the nonlinear inequality constraints posed on HD, TS, EB, and RS, as presented in Table 7. These property ranges (optimisation goals) were chosen having in mind the standard property specification range commonly required for the heat- and air-resistant products used as reference. From an industrial competitiveness point of view, property values outside (above) that range are not so interesting, as they certainly imply extra cost.

The result of the PSO algorithm procedure (minimum costs) is returned as a composition range (weight fraction) for each of the raw materials CC, PO, and VA. These ranges are also presented in Table 7.

An endless number of mixture composition points can be found which meet the requirements of the properties at low cost values. Alternatively a 2D graphical visualisation can be used, by keeping constant one of the factors (feasibility curve). In the present case, the CC content was chosen as the base reference value, and contents of all the factors were expressed as weight fraction relative to that CC content. Thus, feasibility curves can be obtained as VA versus PO for various constant CC contents, as shown in Figure 1.

Figure 1 clearly shows that the function describing the infinite number of mixtures with properties within the specified ranges is complex and nonlinear. For a constant CC weight fraction equal to 1.00, optimised costs vary from 1.82 to 1.87 US$/kg, and the PO weight fraction can vary between 0.60 and 0.78, whereas the VA weight fraction has a much narrower range, varying from 0.008 to 0.009. Similarly, when the CC level is kept constant at 0.50, the composite costs vary between 2.17 and 2.23 US$/kg. The optimum VA weight fraction varies nearly in the same range, but the optimum PO weight fractions have a different range, in this case from 0.47 to 0.68. Calcium carbonate at lower levels has a more complex effect on the optimisation and results in higher costs, ranging from 2.53 to 2.58 US$/kg. Although the PO weight fraction range is narrower (0.40 to 0.60), the VA weight fraction has a broader range, from 0.008 to 0.017.

Table 8 presents the raw materials' weight fractions and property-predicted values for three illustrative mixtures within the optimum range and shows the corresponding costs, ranging from 1.92 to 2.37 US$/kg. From those weight fractions' compositions, the formulations of the corresponding optimised EPDM rubber composites can now be calculated back from the reference CC content (125 phr). The full formulations corresponding to the mixtures in Table 8, meeting the

Table 6: Composition (weight fraction) of the check-point mixtures F_1, F_2, and F_3 and corresponding measured (M) and predicted (P) values of hardness (HD), tensile strength (TS), elongation at break (EB), and rebound resilience (RS).

Wt. fraction	F_1		F_2		F_3	
X_1 (CC)	0.303		0.677		0.860	
X_2 (PO)	0.438		0.616		0.755	
X_3 (DV)	0.008		0.015		0.009	
Property	P	M	P	M	P	M
HD (Shore A)	68.5	69.8 ± 0.7	76.2	78.4 ± 0.7	69.5	70.6 ± 0.7
RS (%)	26.5	27.0 ± 0.0	25.0	25.0 ± 0.0	24.3	24.0 ± 0.0
EB (%)	585.3	668.3 ± 9.8	470.5	522.0 ± 19.2	551.0	526.7 ± 12.1
TR (MPa)	10.47	10.10 ± 0.3	8.92	9.32 ± 0.3	7.65	7.43 ± 0.2

Table 7: Constraint requirements posed on mechanical properties (HD, TS, EB, and RS) and raw materials' (CC, PO, and VA) weight fraction ranges that minimise costs of the composites, as returned by the PSO algorithm.

Name	Goal	Lower limit	Upper limit
HD (Shore A)	In the range	65	75
TS (MPa)	In the range	7.5	12.5
EB (%)	In the range	490	560
RS (%)	In the range	23	28
CT (US$/kg)	Minimise	—	—
X_1 (CC)	In the range	0.00	1.00
X_2 (PO)	In the range	0.40	0.80
X_3 (VA)	In the range	0.008	0.020

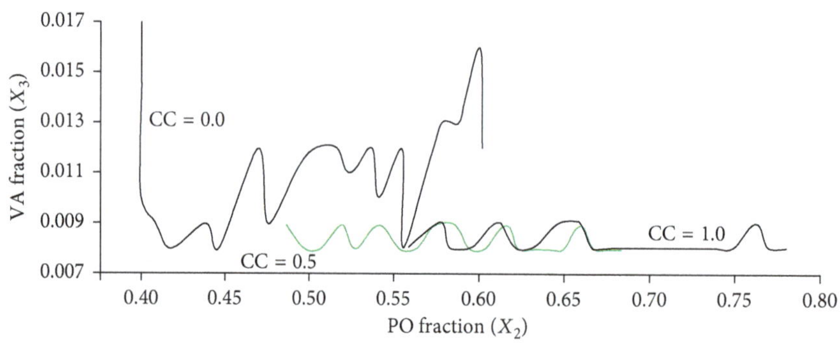

Figure 1: Feasibility curves relating VA and PO weight fractions for optimised cost of composites with properties within the specified ranges, when the CC weight fraction is kept constant at 1.0, 0.5, and 0.0.

Table 8: Examples of optimised mixture compositions and predicted values of the corresponding properties.

Mixture point	Weight fractions X_1 (CC)	X_2 (PO)	X_3 (VA)	HD (Shore A)	RS (%)	EB (%)	TS (MPa)	Cost (US$/kg)
1	0.860	0.755	0.009	66.8	23.0	570.8	7.8	1.92
2	0.677	0.616	0.015	72.0	23.5	516.7	8.3	2.08
3	0.303	0.438	0.008	75.6	26.8	490.6	10.5	2.37

requirements of all the mechanical properties, are presented in Table 9.

4. Conclusions

This study showed that fractional factorial design of experiments in two replicates based on three mixture ingredients generally used on EPDM rubber composites and a particle swarm optimisation (PSO) algorithm are promising methods to design composition variables.

For the chosen key raw materials (calcium carbonate, paraffinic oil, and CBS vulcanising accelerator) and the processing conditions under consideration, the optimisation results readily show that there is an infinite number

TABLE 9: Examples of optimised EPDM rubber composites with costs between 1.92 and 2.37 US$/kg.

Ingredients (phr)[a]	Formulation 1	Formulation 2	Formulation 3
EPDM	100.00	100.00	100.00
Carbon black	115.00	115.00	115.00
Calcium carbonate (CC)	107.00	85.00	38.00
Paraffinic oil (PO)	95.00	77.00	55.00
Sulphur, vulcanising agent	0.40	0.40	0.40
CBS, vulcanising accelerator (VA)	1.13	1.88	1.00
ZnO, vulcanising activator	5.00	5.00	5.00
Naugard 495, aging inhibitor	1.00	1.00	1.00
Stearic acid	1.00	1.00	1.00

[a]Per hundred of rubber, by weight.

of compositions that meet specified property values with minimum costs. For example, within the base composition range investigated, mixture costs as low as 1.92 US$/kg can be achieved with the use of 107 phr of calcium carbonate, 95 phr of paraffinic oil, and 1.13 phr of CBS vulcanising accelerator. The predicted values for the corresponding compound properties are 66.8 Shore A for hardness, tensile strength of 7.8 MPa, 570.8% elongation at break, and 23.0% rebound resilience.

In this way, this investigation showed that the specified characteristics of the desired product can be subjected to restrictions typical of the manufacture process, and a broad range of compositions can still be selected so that the final product has minimum cost and can be comfortably manufactured.

Conflicts of Interest

The authors declare that they have no conflicts of interest.

Acknowledgments

This research was supported by UDESC-Joinville Project 15085/2011 (Sivaldo Leite Correia). The authors appreciate the financial support received from the Brazilian Research Agencies CNPq (Sivaldo Leite Correia) and CAPES (Denilso Palaoro) and from the Santa Catarina State Research Agency FAPESC (Project 3338/2013) and are thankful to NSO Borrachas Ltda (Joinville-SC, Brazil) for providing the raw materials and access to its laboratories for the development of this work.

References

[1] P. Tangudom, S. Thongsang, and N. Sombatsompop, "Cure and mechanical properties and abrasive behaviour of natural rubber, styrene-butadiene rubber and their blends reinforced with silica hybrid fillers," *Materials & Design*, vol. 53, pp. 856–864, 2014.

[2] E. Demirhan, F. Kandemirli, and M. Kandemirli, "The effects of furnace carbon blacks on the mechanical and the rheological properties of SBR1502 styrene butadiene rubber," *Materials & Design*, vol. 28, no. 4, pp. 1326–1329, 2007.

[3] A. Mostafa, A. Abouelk-Kasem, M. R. Bayoumi, and M. G. El-Sebaie, "Effect of carbon black loading on the swelling and compression set behavior of SBR and NBR rubber compounds," *Materials & Design*, vol. 30, no. 5, pp. 1561–1568, 2009.

[4] A. Mostafa, A. Abouelk-Kasem, M. R. Bayoumi, and M. G. El-Sebaie, "Insight to the effect of CB loading on tension, compression, hardness and abrasion properties of SBR and NBR filled rubber compounds," *Materials & Design*, vol. 30, no. 5, pp. 1785–1791, 2009.

[5] M. Morton, *Rubber Technology*, American Chemical Society, Van Nostrand Reinhold, New York, NY, USA, 3rd edition, 1987.

[6] C. M. Blow and C. Hepburn, *Rubber Technology and Manufacture*, Butterworth Scientific, London, UK, 1982.

[7] R. M. Myers and D. C. Montgomery, *Response Surface Methodology: Process and Product Optimization Using Designed Experiments*, John Wiley & Sons, New York, NY, USA, 2002.

[8] D. C. Montgomery, *Design and Analysis of Experiments*, John Wiley & Sons, New York, NY, USA, 3rd edition, 2001.

[9] H. M. Costa, T. A. S Abrantes, R. C. R. Nunes, L. L. Y. Visconte, and C. R. G. Furtado, "Design and analysis of experiments in silica filled natural rubber compounds—effect of castor oil," *Polymer Testing*, vol. 22, no. 7, pp. 769–777, 2003.

[10] S. Moonchai and D. Moonchai, "Modelling and optimization of rebound resilience and hardness of defatted rice bran and calcium carbonate-filled NR vulcanisates," *Polymer Testing*, vol. 32, no. 8, pp. 1472–1478, 2013.

[11] L. Y. Chan, "Optimal designs for experiments with mixtures: a survey," *Communications in Statistics–Theory and Methods*, vol. 29, no. 9-10, pp. 2281–2312, 2000.

[12] E. K. Nambiar and K. Ramamurthy, "Models relating mixture composition to the density and strength of foam concrete using response surface methodology," *Cement and Concrete Composites*, vol. 28, no. 9, pp. 752–760, 2006.

[13] S. L. Correia, A. P. N. Oliveira, D. Hotza, and A. M. Segadães, "Properties of triaxial porcelain bodies: interpretation of statistical modelling," *Journal of the American Ceramic Society*, vol. 89, no. 11, pp. 3356–3365, 2006.

[14] S. L. Correia, G. Dienstmann, M. V. Folgueras, and A. M. Segadães, "Effect of quartz sand replacement by agate rejects in triaxial porcelain," *Journal of Hazardous Materials*, vol. 163, no. 1, pp. 315–322, 2009.

[15] S. L. Correia, T. Partala, F. C. Loch, and A. M. Segadães, "Factorial design used to model the compressive strength of mortars containing recycled rubber," *Composite Structures*, vol. 92, no. 9, pp. 2047–2051, 2010.

[16] Y. Cheng, J. Wang, P. G. Jönsson, and Z. Zhao, "Optimization of high-quality vertically aligned ZnO rod arrays by the response surface methodology," *Journal of Alloys and Compounds*, vol. 626, pp. 180–188, 2015.

[17] D. Guan, J. H. Wu, and L. Jing, "A statistical method for predicting sound absorbing property of porous metal materials by using quartet structure generation set," *Journal of Alloys and Compounds*, vol. 626, pp. 29–34, 2015.

[18] B. Cheng, N. Zhu, R. Fan et al., "Computer aided optimum design of rubber recipe using uniform design," *Polymer Testing*, vol. 21, no. 1, pp. 83–88, 2002.

[19] M. Balachandran, S. Devanathan, R. Muraleekrishnan, and S. S. Bhagawan, "Optimizing properties of nanoclay-nitrile rubber (NBR) composites using face centred central composite design," *Materials & Design*, vol. 35, pp. 854–862, 2012.

[20] S. L. Correia, D. Hotza, and A. M. Segadães, "Application of mathematical and statistical strategies to optimize ceramic bodies: effects of raw materials on the technological properties," *Ceramic Forum International*, vol. 82, pp. 39–43, 2005.

[21] S. L. Correia, D. Hotza, and A. M. Segadães, "Simultaneous optimisation of linear firing shrinkage and water absorption of triaxial ceramic bodies using experiments design," *Ceramics International*, vol. 30, no. 6, pp. 917–922, 2004.

[22] A. Banks, J. Vincent, and C. Anyakoha, "A review of particle swarm optimization. Part I: background and development," *Natural Computing*, vol. 6, no. 4, pp. 467–484, 2007.

[23] L. Yiqing, Y. Xigang, and L. Yongjian, "An improved PSO algorithm for solving non-convex NLP/MINLP problems with equality constraints," *Computers & Chemical Engineering*, vol. 31, no. 3, pp. 153–162, 2007.

[24] ASTM D15, *Standard Test Method for Rubber Property. Sample Preparation for Physical Testing of Rubber Products*, Annual Book of ASTM Standards, Philadelphia, PA, USA, 1987.

[25] ASTM D676, *Standard Test Method for Indentation of Rubber by Means of a Durometer*, Annual Book of ASTM Standards, Philadelphia, PA, USA, 1987.

[26] ASTM D412, *Standard Test Method for Tension Test of Vulcanized Rubber*, Annual Book of ASTM Standards, Philadelphia, PA, USA, 1987.

[27] ASTM D676, *Standard Test Method for Rubber Property—Resilience by Vertical Rebound*, Annual Book of ASTM Standards, Philadelphia, PA, USA, 1987.

17

Finite Element Modeling of FRP-Strengthened RC Beam under Sustained Load

Shiyong Jiang, Weilai Yao, Jin Chen, and Tao Cai

Department of Military Infrastructure Engineering, Army Logistics University of PLA, Chongqing, China

Correspondence should be addressed to Shiyong Jiang; jiangshiy@163.com

Academic Editor: Antonio Boccaccio

External bonding of FRP laminates to the tension soffit of concrete members has become a popular method for flexural strengthening. However, the long-term field performance of FRP-strengthened RC members under service conditions is still a concern, and more work needs to be done. Based on concrete smeared-crack approach, this paper presents a finite-element (FE) model for predicting long-term behavior of FRP-strengthened RC beam, which considers the time-dependent properties of all components including the aging effect of concrete. According to the comparison between theoretical predictions and test results, the validity of the FE model is verified. The interfacial edge stresses in adhesive layer were determined through appropriate mesh refinement near the plate end, and their time-dependent characteristics were investigated. The results show that creep of concrete and epoxy resin cause significant variations of the edge stresses with time. According to the research in this paper, the FE approach is found to be able to properly simulate the long-term behavior of the FRP-strengthened beam and help us better understand the complex changes in the stress state occurring over time.

1. Introduction

Using fiber-reinforced polymer (FRP) composites as externally bonded reinforcement is in the forefront of retrofitting technology. Most of previous studies have been rightfully focused on the ultimate limit state behavior [1–3]. Relatively few studies have been performed on the behavior of FRP-strengthened RC structures under long-term services conditions. In fact, under sustained load, concrete exhibits clear creep response, and the viscous property of adhesive is remarkable [4–7]. Related to the kind of fiber used and fiber content, some types of FRP products may also present noticeable viscoelasticity [8–10]. As time goes on, structure deformation increases and cracks further develop, which negatively affects the serviceability [11]. In addition, it has been shown that the creep deformation of very old concrete can still be substantial [12, 13]. Thus, if a structure is strengthened because the service load will be increased, although the concrete may be aged, creep response caused by the additional load should still receive critical attention. Nevertheless, the long-term response of strengthened members is not fully explored, and further research is essential in order to enhance the safety and design of FRP-strengthened members [14, 15].

So far, previous studies have proposed some analytical methods to investigate time-dependent characteristic of RC structure strengthened by external reinforcement [11, 16–19]. It can be found that most of them are based on cross section analyses, which satisfy strain compatibility and equilibrium of forces. One of the drawbacks of the analytical method is that it may be difficult to consider some significant characteristics. For instance, the basic mechanism for strengthening is the shear deformation and stress transferring of epoxy adhesive between concrete and external FRP. Normally, epoxy presents evident flow property [20, 21], which therefore leads to relaxation of interfacial stress and affects the long-term behavior. Whereas, it is actually not easy to model such shear creep characteristic of adhesive in cross section analyses. Besides, the tension stiffening of concrete needs to be reasonably taken into account when predicting deflections. Nevertheless, for the method of section analysis, it is also a challenging task to model such tension-stiffening effect appropriately and accurately. Thus, it can be noted that the above influential factors are usually roughly approximated and even neglected [11, 16, 19]. On the other hand, in order to improve the efficiency of solving process or reduce the difficulty in mathematics, simplifications have been ordinarily introduced in analytical method. Based on axial strain reduction coefficient and curvature reduction coefficient, Hong and Park [11] calculated the time-dependent change of the concrete stress in a simple and empirical way. Marí et al. [16] assumed that the stresses of tension bars and FRP laminate remain constant over time. Although, reasonable theoretical results are still obtained in their studies [11, 16], there is no evidence to show those simplifications are suitable for other cases, where the creep responses of materials, the geometry of member, and reinforcement ratio (for steel and FRP) are different. It is therefore implied that the universality of the analytical method applying specific simplification is not favorable enough. Additionally, it should be noted that despite simplifications have been made, large amount of iterative calculations still may be needed to obtain the final results [11]. Moreover, the stresses at the adhesive interfaces near the bonding edges are responsible for the debonding failures, and their redistribution with time is an issue worth studying. Nevertheless, when determining and investigating the time-dependent interfacial stresses, cross section analyzing is no longer competent, and a quite different analysis procedure (particularly to interfacial forces) needs to be carried out [22, 23]. In the related studies, it can be found that although precise and complicated derivations had been conducted, assumption not fully corresponding to the fact was still made in order to make the mathematical process more manageable [22, 23]. The stresses within the adhesive layer were assumed to be constant through the thickness of the adhesive layer in both studies of Benyoucef et al. [22] and Zhang and Wang [23], which in fact vary strongly across the layer thickness [24].

To eliminate the aforementioned deficiencies, finite element (FE) modeling may be a feasible scheme. Comparing to the analytical method, to study the behavior of FRP-strengthened RC member under sustained load, the FE method may provide a more powerful tool, which calculates numerically and does not operate on the basis of cross section analysis. Those specific simplifications mentioned above may also be avoided, which makes the FE method more generally applicable. However, to date, studying the time-dependent performance by FE approach has not been extensively conducted. One of the likely reasons should be that reasonably developing time-dependent constitutive models of materials can be a challenging task.

The major thrust of this paper is to propose a finite element model to investigate the long-term response of FRP-strengthened RC beam under sustained load. The time-dependent characteristics of all components including the evolution of concrete property along with time were modeled as user-defined subroutines in ABAQUS 6.12 [25]. Besides, load sustaining experiment of a strengthened beam was also conducted. By comparing the experiment and FE results, the validity of the FE model is demonstrated. Also, interfacial stresses were accurately determined by the same FE model, with appropriate mesh refinement near the end of CFRP plate. Additional numerical investigations were carried out to study the changes of edge stresses over time induced by the creep of materials.

2. Long-Term Deformation Monitoring

Time-dependent deformation of a FRP-strengthened RC beam under sustained load was monitored for 200 d. The sectional dimension was 250 × 400 mm with a 30 mm thickness. The total and effective span was 3300 mm and 2900 mm, respectively. D28 reinforcement (diameter is 28 mm) was used for reinforcement on the compression side, while D14 reinforcement (diameter is 14 mm) was used for reinforcement on the tension side. Under service state, concrete stress is not high, which presents linear creep. In this study, adequate compression bars were used to control instantaneous stress in compressive concrete at lower level simulating the service condition. CFRP plate with length and width of 2600 and 200 mm was bonded on the tension soffit of beam. D8 (diameter is 8 mm) reinforcement at interval of 200 mm was used for the stirrups. To avoid premature debonding near the end of CFRP plate, CFRP sheets were used to wrap around the test beam. Figure 1 provides the detailed information of the test beam.

The test beam was cast by commercial concrete. The 28-day cylinder compressive strength of the concrete was 34.0 MPa. Other parameters of concrete are listed in Table 1. Table 2 shows the properties of steel reinforcement. Table 3 gives the properties of epoxy resin (applied as adhesive), CFRP sheet (for avoiding premature debonding), and CFRP plate (for flexural strengthening). The reported material strengths above are measured mean values.

The test beam was simply supported and loaded in four point bending. The sustained load F was 75% of the theoretical ultimate capacity F_u (on the basis of cross section analysis) of the same dimensional RC beam without strengthening. The beam was cured for 7 d, then placed indoors for another 69 d before loading. In the test, a dedicated experimental system, made up of vertical reaction frame, screw jack, and steel distribution beam, was designed to apply and sustain the external load, as illustrated in Figure 2. Load was applied to the beam through slowly lifting up the screw jack, meanwhile, its value was monitored by the force sensor at the top of the jack. As the value reached to 0.75 F_u (117.71 kN), screw jack was locked to prevent lifting or declining, which sustained the load. When the deformation of test beam increased due to creep behavior, the applied load might decrease, then the Jack was unlocked and further lifted up to replenish the force to the designed level (0.75F_u). During the long-term test, the error of applied load was controlled within ±1% of the planned value.

As shown in Figure 1, in the midspan, resistance strain gauges were bonded to the tension, compression steel, and CFRP plate to measure the instantaneous strains only, since their long-term working performance is unreliable. Vibrating wire gauges were especially used to obtain both

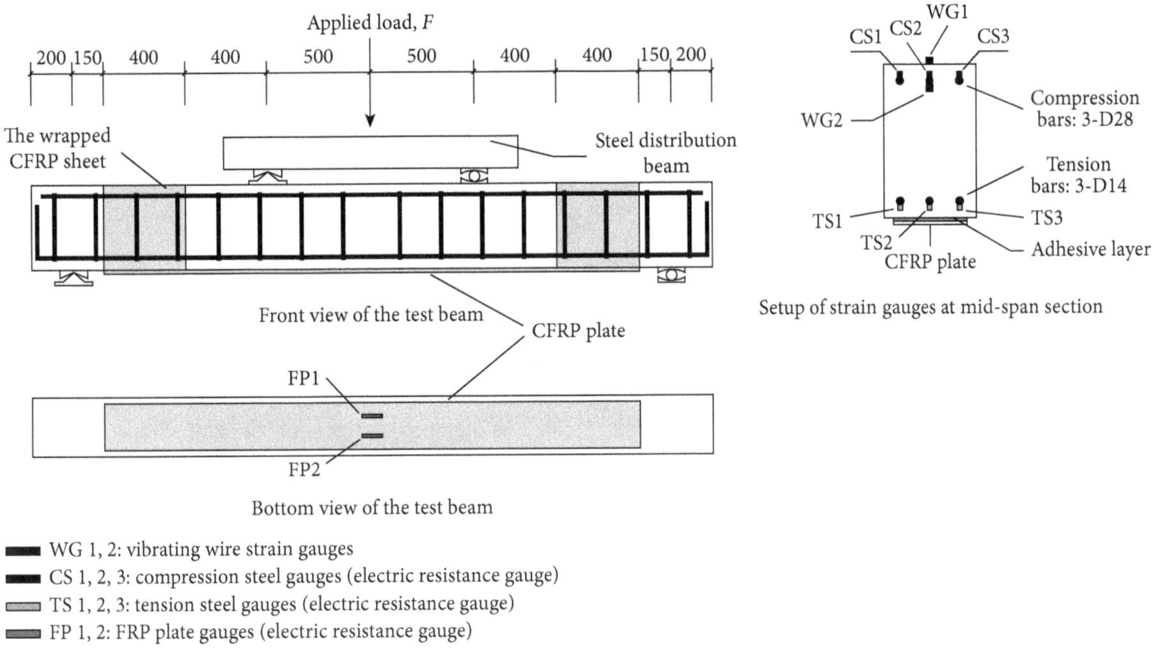

FIGURE 1: Test beam details.

TABLE 1: Concrete properties.

Slump of fresh concrete (mm)	Ratio of fine aggregate to total aggregate by weight (%)	Cement content (kg/m^3)	Fly ash content (kg/m^3)	Water cement ratio
220	38	325	60	0.4

TABLE 2: Materials' properties of steel reinforcement.

Reinforcement type	Yield strength (MPa)	Ultimate strength (MPa)
Tension, D14	461	625
Compression, D28	420	570
Shear, D8	435	610

TABLE 3: Adhesive and CFRP properties.

	Ultimate strength in tension (MPa)	Elastic modulus (GPa)	Thickness (mm)
Epoxy resin layer	39.6	3.468	1.4
CFRP sheet	3510	241	0.15
CFRP plate	2482	174	1.4

instantaneous and long-term strains in some places because of their reliable performance in long period of monitoring. The 1# wire strain gauge (WG1) was bonded on the top surface of the beam. The 2# wire strain gauge (WG2) tied with the middle compression bar was buried in the specimen when casting concrete. One dial gauge was installed at the bottom midspan point and other two dial gauges installed at pedestals were used mainly to check the symmetrical nature of the loaded beam.

The experiment was conducted indoors. Figure 3 shows the recorded temperature and humidity, which presented unavoidable variations with time. It should be noted that the observed creep might be affected by the change of environmental conditions in the laboratory.

3. FE Modeling

Based on ABAQUS 6.12 [25], nonlinear numerical analysis was performed to study the time-dependent characteristics of concrete beam strengthened with external FRP reinforcement. The FE model used herein is on the basis of the one developed by Jiang et al. [26]; however, the current FE model in this paper was further updated and is more advanced: the aging effect of concrete was taken into account and the corresponding calculating method was presented, and also, the method for analyzing the time-dependent interfacial stress was developed.

3.1. Modeling of Instantaneous Material Property. In this study, concrete cracking was simulated by using the smeared-cracked approach. Unlike the discrete-crack approach, there is no need to predefine the crack paths in the smeared-cracked approach. In addition, aiming to overcome the mesh-nonobjectivity problem in the conventional smeared-crack approach, the crack band model was used [27].

The concrete was modeled within the framework of the concrete-damaged plasticity model provided by

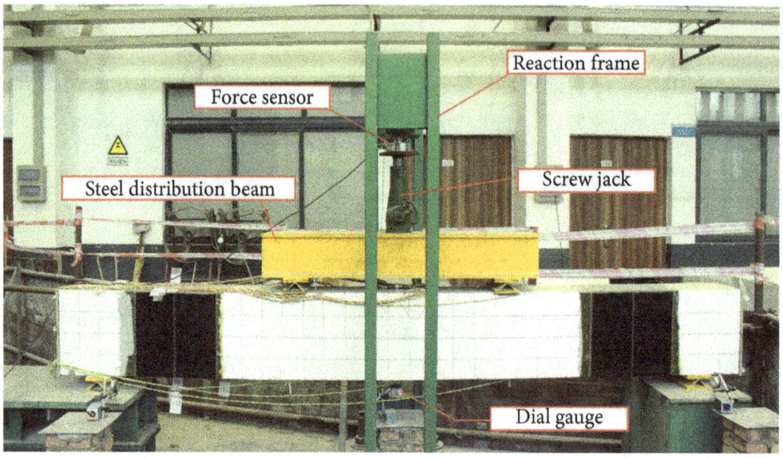

FIGURE 2: Experiment setup in laboratory.

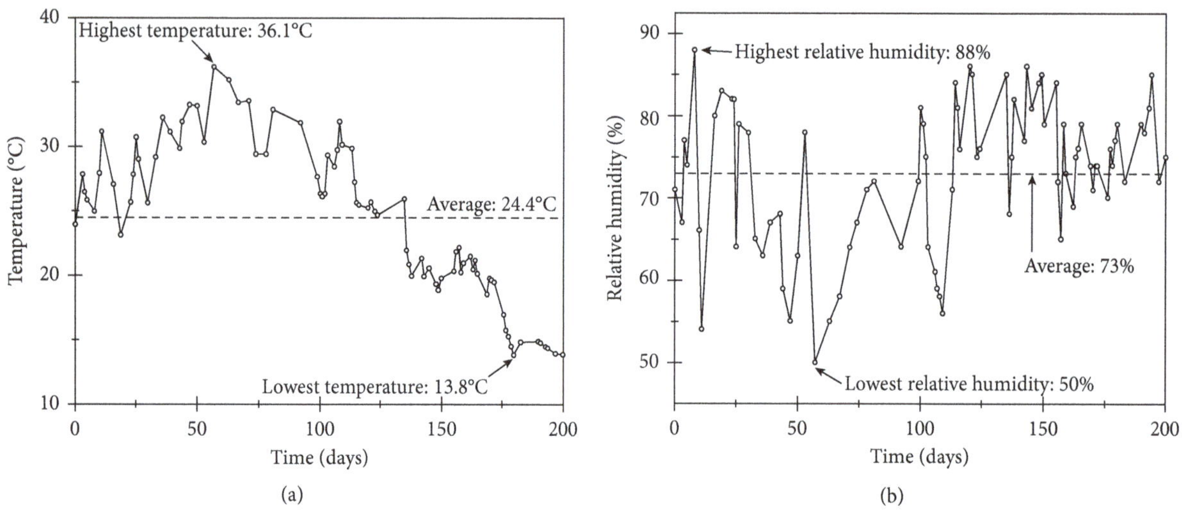

FIGURE 3: Environmental conditions during test. (a) Temperature indoors. (b) Relative humidity indoors.

ABAQUS. The stress-strain relationship under uniaxial compression recommended by Saenz [28] (Equation (1)) was applied. Under uniaxial tension, the problem involves tensile cracking. The stress-crack opening displacement curve recommended by Hordijk [29] (Equations (2) and (3)) was used. The parameters in Equations (2) and (3) can be further determined through Equations (4) and (5) based on CEB-FIP standard [30]. Figure 4 presents the uniaxial mechanical behavior of concrete.

$$\sigma_c = \frac{E_0\varepsilon_c}{1 + \left(\left(\left(E_0\varepsilon_{cp}\right)/f_c\right) - 2\right)\left(\varepsilon_c/\varepsilon_{cp}\right) + \left(\varepsilon_c/\varepsilon_{cp}\right)^2}, \tag{1}$$

$$\frac{\sigma_t}{f_t} = \left[1 + \left(c_1\frac{w_t}{w_{cr}}\right)^3\right]e^{\left(-c_2\left(w_t/w_{cr}\right)\right)} - \frac{w_t}{w_{cr}}\left(1 + c_1^3\right)e^{\left(-c_2\right)}, \tag{2}$$

$$w_{cr} = 5.14\frac{G_F}{f_t}, \tag{3}$$

$$f_t = 1.4\left(\frac{f_c - 8}{10}\right)^{2/3}, \tag{4}$$

$$G_F = \left(0.0469d_a^2 - 0.5d_a + 26\right)\left(\frac{f_c'}{10}\right)^{0.7}, \tag{5}$$

where σ_c and σ_t (in MPa) are the compressive and tensile stress, respectively. The f_c and f_t are the uniaxial compressive and tensile strength, respectively. The ε_c is the compressive strain; ε_{cp} is the peak compressive strain (0.002) corresponding to the strength f_c; E_0 is the elastic modulus, according to ACI 318 [31], $E_0 = 4730\sqrt{f_c}$ (in MPa). The w_t (in mm) is the crack opening displacement; w_{cr} (in mm) is the crack opening displacement at the complete release of stress and fracture energy; $c_1 = 3.0$ and $c_2 = 6.93$ are constants determined by test; G_F is the energy required to create a unit area of stress-free crack; d_a (in mm) is size of the maximum aggregate and is assumed as 20 mm if no experimental data are available.

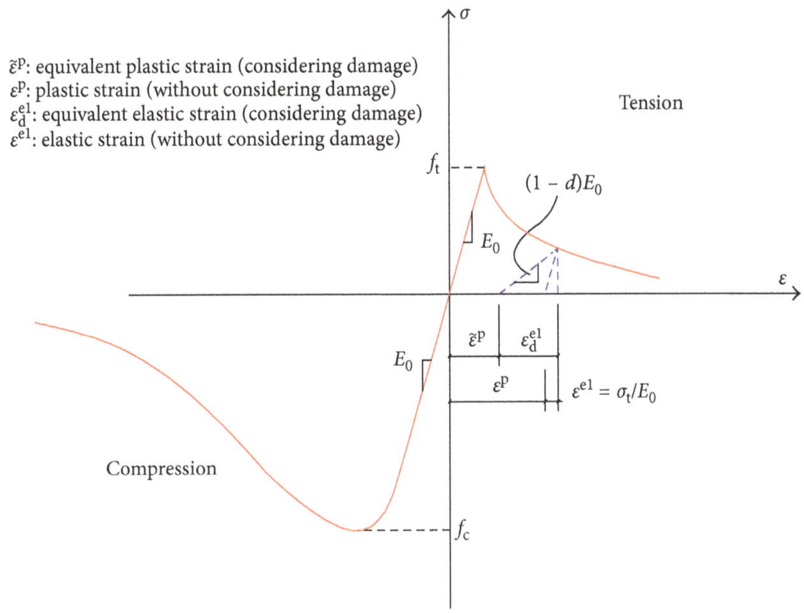

FIGURE 4: Uniaxial mechanical behavior of concrete and the damage pattern.

Based on Bazant and Planas [32], mesh sensitivity problem can be effectively overcome through applying the crack band model [33], in which the displacement of crack opening w equals to the cracking strain ε_{cr} accumulating over the width of the crack band h_c.

$$w = \int_{h_c} \varepsilon_{cr}\, dh. \tag{6}$$

According to Rots [34], when meshing concrete with four-node plane stress element with four integration points, the crack band width is taken to be $\sqrt{2}e$, where e is the side length of the element [27]. Thus, the tensile stress-crack opening displacement relationship given by Equation (2) can be transformed into a stress-strain curve needed by FE program through Equation (6).

The damage evolution was also considered to better describe the nonlinearity of concrete. The relationship between the damage factor d and plastic strain ε^p, which does not take stiffness degradation into account was defined. Damage factor can be determined as Equation (7) according to Tao and Chen [35] for both uniaxial tension and compression. The patterns of damage are illustrated in Figure 4.

$$d = \frac{(1-\sigma/f)\varepsilon^p}{(1-\sigma/f)\varepsilon^p + \sigma/E_0}, \tag{7}$$

where σ is the uniaxial tensile or compressive stress and f is uniaxial tensile or compressive strength.

In this study, the response of FRP-strengthened RC beam under service state is investigated, therefore, the instantaneous material properties of epoxy adhesive, steel reinforcements, and CFRP plate were modeled as linear elastic. Also, perfect bonds were assumed between adhesive-concrete, adhesive-CFRP plate, and concrete-steel. The wrapped CFRP sheets for avoiding premature debonding were reasonably ignored in the FE model based on the relatively low stress level [20].

3.2. Modeling of Time-Dependent Material Property. Under sustained load, concrete presents significant creep behavior, and additional stress-independent shrinkage deformation is also considerable. The different time-dependent characteristics of materials induce complicated stress redistribution in structure along with time.

The loading age is written as t_0, and the load sustaining period is from t_0 to t_n ($t_n > t_0$). The total time period t_0 to t_n is divided by a number of discrete periods: t_0 to t_1, t_1 to $t_2, \ldots, t_{n-2}$ to t_{n-1}, t_{n-1} to t_n. This paper presents a quasielastic constitutive model of concrete considering creep and shrinkage based on age-adjusted effective modulus method (Equations (8)–(10)). The model is in recursion form, which comforts codes programming and improves calculating efficiency. The detailed derivation process of Equations (8)–(10) can be found in Jiang et al. [26], which is not presented herein for brevity.

$$\sigma^*(t_n) = \frac{E(t_{n-1})}{1 + \chi(t_n, t_{n-1})\left[\sum_{j=1}^{m} b_j(t_{n-1})\left[1 - e^{-\lambda_j \cdot \Delta t_n}\right]\right]} \cdot (\varepsilon_n - \Delta\varepsilon), \tag{8}$$

$$\Delta\varepsilon = \varepsilon_{sh}(t_n) - \varepsilon_{sh}(t_{n-1}) + \sum_{j=1}^{m} A_{n,j}^* \cdot \left(1 - e^{-\lambda_j \cdot \Delta t_n}\right), \tag{9}$$

$$A_{n,j}^* = A_{n-1,j}^* \cdot e^{-\lambda_j \cdot \Delta t_{n-1}} + b_j(t_{n-2}) \cdot e^{-\lambda_j \cdot \Delta t_{n-1}} \cdot \frac{\chi(t_n, t_{n-2})}{E(t_{n-2})} \cdot \sigma^*(t_{n-1}) \quad \text{for } n > 1. \tag{10}$$

When $n = 1$:

$$A^*_{n=1,j} = b_j(t_0) \cdot \frac{\sigma_0}{E(t_0)}, \quad (11)$$

where $\sigma^*(t_i)$ is the stress increment of concrete due to the effect of creep and shrinkage from t_{i-1} to t_i, reflecting the redistribution of stress within structure; $E(t_i)$ is the elastic modulus at time t_i; $\varepsilon_{sh}(t_i)$ refers to the shrinkage strain at time t_i; and $\Delta t_i = t_i - t_{i-1}$ $(i = 0, 1, 2, \ldots, n-1, n)$. The b_j, λ_j, and m are material parameters for expanding the creep coefficient $\varphi(t,\tau)$ as Dirichlet series [26] (Equation (12)). The $\chi(t_n, t_{n-1})$ is the aging coefficient [36] (Equation (13)). The σ_0 refers to the instantaneous stress state under loading. The ε_n is the strain increment of concrete during t_{n-1} to t_n, which reasonably considers the effects of stress increment induced by creep and shrinkage.

$$\varphi(t,\tau) = \sum_{j=1}^{m} b_j(\tau)\left[1 - e^{-\lambda_j(t-\tau)}\right], \quad (12)$$

$$\chi(t,\tau) = \frac{1}{1 - e^{-\varphi(t,\tau)}} - \frac{1}{\varphi(t,\tau)}. \quad (13)$$

Due to the progress of hydration, concrete properties including strength and elastic modulus develop with time. The above time-dependent elastic modulus of concrete $E(t_i)(i = 0, 1, 2, \ldots, n-2, n-1, n)$ was determined by the time-dependent concrete strength $f_c(t_i)$ (concrete strength at time t_i) according to ACI 318 [31] ($E_c = 4730\sqrt{f_c}$). The time interval (t_0 to t_1, t_1 to $t_2, \ldots, t_i$ to $t_{i+1}, \ldots, t_{n-2}$ to t_{n-1}) was set small (less than 5 d) to obtain accurate enough result. The elastic modulus of concrete was assumed to be constant in each time interval, but renewed and updated when calculation of the latest time interval was finished. In this way, the aging effect of concrete was considered. This study applied the evolution law of concrete strength proposed by ACI 209 [37].

$$f_c(t_i) = f_{c28} \cdot \left(\frac{t_i}{a + b \cdot t_i}\right), \quad (14)$$

where f_{c28} is the concrete mean compressive strength at 28 d in MPa and t_i is the age of concrete in days. $a = 4.0$ and $b = 0.85$ are material constant for concrete using ordinary Portland cement. In this study, the test beam was loaded at the age of 76 d after casting concrete. Thus, the calculated 76 d strength 37.7 MPa and its corresponding elastic modulus 29.0 GPa were used to determine the instantaneous response of the test beam under loading. The evolution of concrete property along with time is shown in Figure 5.

The above time-dependent property of concrete was programmed as FORTRAN code in user-defined subroutine UEXPAN and USFLD in ABAQUS. Concrete creep coefficient and shrinkage strain was calculated using ACI 209 code [37], and during the calculation, the measured average humidity (73%) was used.

In this study, both shear and axial creep of adhesive resin were modeled in the FE approach. For FRP-strengthened system, the performance of adhesive is of crucial importance in providing effective stress transfer. It has been demonstrated that epoxy adhesive presented evident viscoelastic manner [20, 21, 38]. As a result, the relaxation of interfacial stresses caused by adhesive creep affects the strengthening mechanism, which therefore influences the long-term response of strengthened beam. In this research, to describe the decreased shear relaxation modulus of epoxy $G_a(t)$ with time, Maxwell chain was used [20, 21].

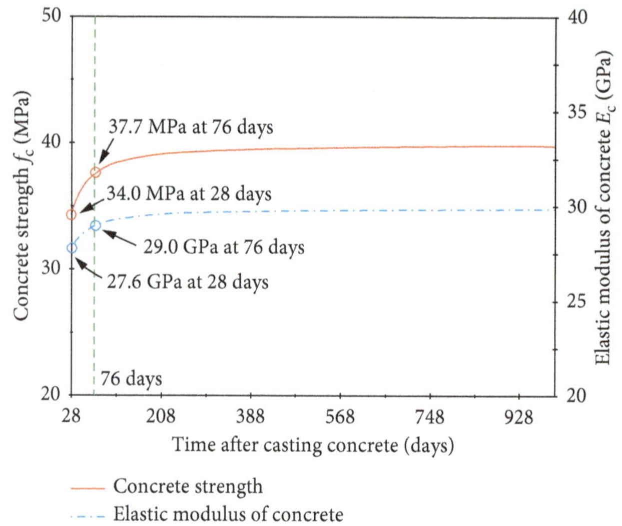

FIGURE 5: The evolution of concrete property with time.

$$G_a(t) = G_u + \sum_{i=1}^{N_G} G_i \cdot e^{-t/\tau_{Gi}}, \quad (15)$$

where G_u is the shear modulus at infinite time, N_G is number of Maxwell elements, G_i refers to the relaxed shear modulus of the ith Maxwell element, and τ_{Gi} is material constant. Following Choi [20], $N_G = 1$, $\tau_{G1} = 2$ d, and $G_u = G_1/5$. In addition, the creep of epoxy in axial direction was also taken into account, since near the midspan, where section moment is large and the instantaneous stress of epoxy resin in axial direction usually reaches a noticeable value, which causes nonignorable creep response. It is assumed that Poisson's ratio of adhesive ν_a remains constant with time [39, 40], and so, the relaxation modulus in axial direction $E_a(t)$ is derived from that in the shear direction or vice versa: $E_a(t) = 2(1 + \nu_a)G_a(t)$ [14, 15].

The viscous property of CFRP was determined as follows [41]:

$$\varepsilon_{CFRP}(t) = \varepsilon_0 \cdot \sinh\left(\frac{\sigma}{\sigma_m}\right) + m \cdot \sinh\left(\frac{\sigma}{\sigma_m}\right) \cdot t^{n_1}, \quad (16)$$

where $\varepsilon_{CFRP}(t)$ = time-dependent strain; t = time after loading, in hour; and σ = applied stress. The $\varepsilon_0 = 0.379$; $\sigma_m = 68950$ MPa; $m = 0.0011$; $n_1 = 0.123$ are material constants [41]. Similar to concrete, the above mechanical properties of adhesive and CFRP related to time were also programmed in user subroutines UEXPAN and USFLD.

3.3. Solution Strategy. In the FE model, only a half of the beam was included by taking advantage of symmetry about midspan plane, which effectively reduces calculating time. The concrete, epoxy adhesive, and CFRP plate were modeled by plane stress element CPS4. The steel reinforcement was modeled using truss elements.

During the FE analyses, step-by-step procedures were applied for calculations. Instantaneous behavior of the beam was calculated first, which was set as the initial condition of the calculation of long-term response afterwards. Then, based on the time-dependent properties of materials, numerical simulation of long-term behavior of the test beam was carried out. In this study, if not otherwise stated, square element with an element size of 10 mm was used to model concrete. Matching element sizes were chosen to represent the steel bars, the epoxy layer, and the FRP plate. The epoxy layer and the FRP plate were each modeled with two layers of elements in their thicknesses. Nevertheless, when discussing the time-dependent interfacial stresses at the bonding edge, mesh refinement is needed, which is particularly mentioned in the later section.

4. Results and Discussion

4.1. Verification of the Recommended FE Model. Under instantaneous loading, comparison of FE-predicted results and experimental results of the specimen is shown in Table 4. As shown in Figure 1, the tension steel gauge, compression steel gauge, and FRP plate gauge are labeled as TS, CS, and FP. Several strain gauges were employed at the place of midspan: TS1, TS2, and TS3 were bonded to each of the tension bar; CS1, CS2, and CS3 were bonded to each of the compression bar; and FP1 and FP2 were adjacently bonded to the CFRP plate. In Table 4, the strain of tension bar, compression bar, and CFRP plate is the average value of the measured data by TS1, TS2, TS3; CS1, CS2, CS3; FP1, and FP2, respectively. The strain of the top concrete of test beam was obtained by the 1# wire strain gauge (WG1). Table 4 shows that the test results are reasonably predicted by the FE model.

From Table 4, the compressive strain of top concrete is 277 $\mu\varepsilon$ under instantaneous loading, indicating the stress level does not exceed 40 percent of strength, thus, applying the linear creep theory for concrete is appropriate in this paper. The time dependent midspan deflection is shown in Figure 6(a). The time-dependent strains of top concrete and the concrete near compression bars are presented in Figure 6(b), which were experimentally obtained by the vibrating wire gauges WG1 and WG2, respectively. Distinct additional long-term deformation was observed during load sustaining period. After time passage of 200 d, the deflection, strain of top concrete, and strain of concrete near compression bars were respectively 56.7%, 98.6%, and 59.2% higher than their initial value under instantaneous loading. The increasing of deformation corresponds to the typical law of RC member under sustained load: quick in early period and slow in later time. From Figure 6(a), it is shown that after sustaining load for 78 d (39% of the total time period), most part of additional long-term deflection (87.2%) is finished. Also, as shown in Figure 6(b), there are no clear increases for the strains of concrete after 78 d of sustaining load. The results (FE1) predicted by the FE model agree well with the experimental results. The predicted time-dependent deflection at the 50, 99, 155, and 200 d after sustaining load is 4.97, 5.24, 5.42, and 5.52 mm, respectively, which is close to the corresponding test values of 5.00, 5.43, 5.52, and 5.59 mm, with differences of only 0.6%, 3.5%, 1.8%, and 1.3%, respectively. The time-dependent strains of top concrete, concrete near compression bars, are also predicted with good accuracy as shown in Figure 6(b). Besides, The predicted results (FE2) without considering the further evolution of concrete property are also displayed in Figure 6. As expected, neglecting the evolution of concrete property leads to overestimated results, however, with slight effect only. The reason is that the maturity of the used commercial concrete has been good enough after 76 d of aging (the time that the long-term test commenced), as shown in Figure 5.

The result of FE1 and FE2, respectively, refers to the case with and without considering the evolution of concrete property. The additional deflection (Δf) is defined as the delayed deflection occurring in load sustaining period. The total additional deflection (Δf_{total}) is the delayed deflection after 200 d of sustaining load. The ration between Δf and Δf_{total} represents the creep progress of the test beam.

As shown in Figure 6, in this study, the test results of both strain and deflection do not present continuing growth along with time, which is likely to be caused by the fluctuations of environmental humidity and temperature [11]. However, effects of climate changing were currently not considered during FE modeling; therefore, the predicted long-term deformation increases continuously and smoothly. Whereas, from the comparison above, the accuracies of the FE results are favorable enough.

The proposed FE model rationally predicts the cracking behavior of concrete. Figure 7(a) shows the experimentally observed cracking pattern of test beam under instantaneous loading. In this study, because adequate stirrups were deployed to prevent the occurrence of shear failure, in the earlier stage of loading, the observed cracks were flexural cracks in vertical direction mainly. With the increasing of load, new cracks gradually emerged in the bending-shear section, propagating vertically first but then obliquely afterwards. As shown in Figure 8, the FE results appropriately simulate this phenomenon. The effect of tension stiffening, which causes the concrete between adjacent cracks to suffer tensile stress, is also simulated by the FE model, as presented in Figure 8(c). During the test, further development of instantaneous cracks during load sustaining period was clearly observed (Figure 7(b)). The combined effect of materials' creep behaviors and decreasing of section stiffness induced by further formation of cracks causes the additional long-term deformation of test beam.

The instantaneous cracks are marked by color grey; the further development of cracks in long-term are marked by color red.

The FE model in this paper has been further validated against the test results in other literatures [42, 43]. Totally, seven specimens are selected herein: four of them are from Sobuz et al. [42] including one control beam (not strengthened) and three CFRP-strengthened beams and the other three are from Tan and Saha [43]. Sobuz et al. [42]

Table 4: Immediate response at midspan of the test beam.

Results	Deflection (mm)	Strain of tension steel bar ($\mu\varepsilon$)	Strain of compression steel bar ($\mu\varepsilon$)	Strain of CFRP plate ($\mu\varepsilon$)	Strain of top concrete ($\mu\varepsilon$)
Test	3.57	878	−129	937	−277
FE model	3.68	785	−137	845	−248
Theor./exp.	1.03	0.89	1.06	0.90	0.89

Note. Theor./exp. in the above table refers to the ratio of result predicted by FE model and test result, which reflects the predicting error.

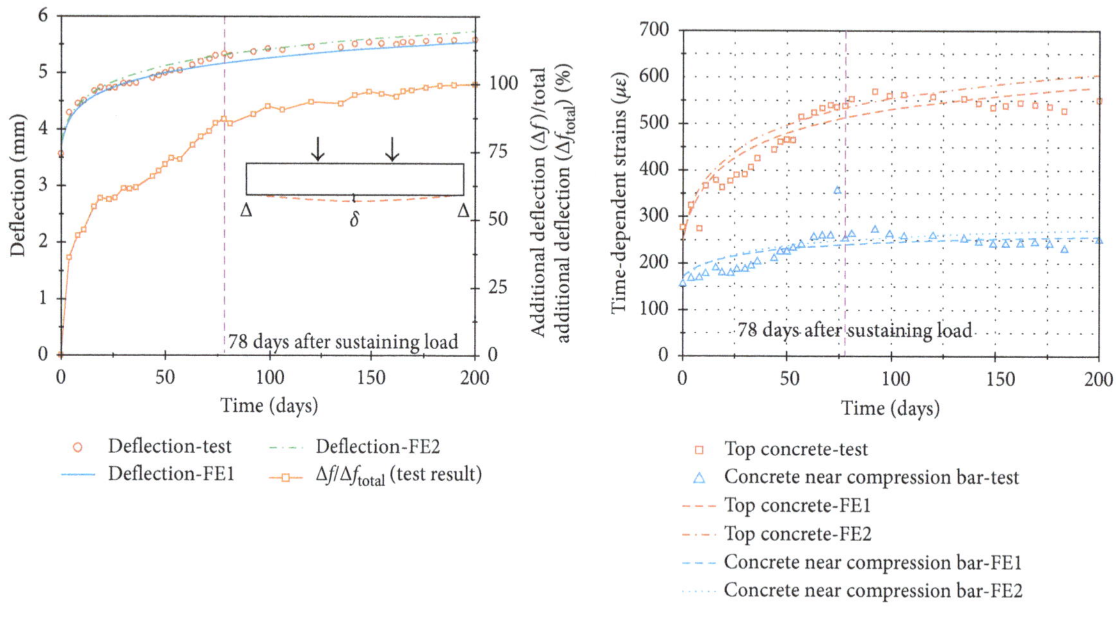

Figure 6: Test and FE results of the specimen. (a) Deflection at the midspan. (b) Concrete strains.

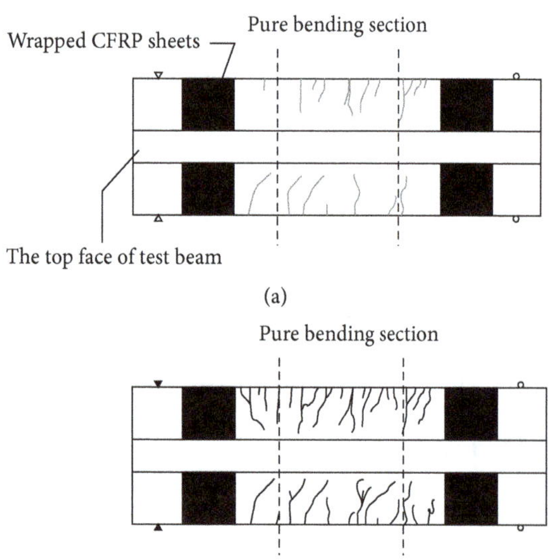

Figure 7: Cracking pattern observed experimentally (unfolded-drawing). (a) At instantaneous loading. (b) After time passage (200 d).

studied the effects of strengthening scheme and level of sustained load on long-term response of RC beam strengthened by CFRP. Four representative specimens are selected herein to demonstrate the accuracy of the proposed FE approach, namely, CBC (not strengthened, the control beam), FBC-1L, FBC-2L, and FBC-3L. Tan and Saha [43] tested the time-dependent responses of nine GFRP-strengthened specimens. Three typical specimens are chosen herein, namely, GB3-40, GB3-49, and GB3-59. Table 5 presents the detailed information of these specimens.

According to Sobuz et al. [42] and Tan and Saha [43], the viscoelasticity of FRP plate was ignored. The aging effect of concrete was considered based on the aforementioned method in Section 3.2. For the specimens from Sobuz's test [42], the ultimate creep coefficient and ultimate shrinkage strain of concrete were 1.87 and 418 $\mu\varepsilon$, which were determined through *3.2.1.1 Compressive creep test* and *3.2.1.2 Drying shrinkage test* in Sobuz's paper [42]. For the specimens from Tan's study [43], the specific laws of concrete creep and shrinkage have been given in their paper [43], which were, therefore, directly used in the FE simulations. From Figure 9, the proposed FE approach well predicts the test results.

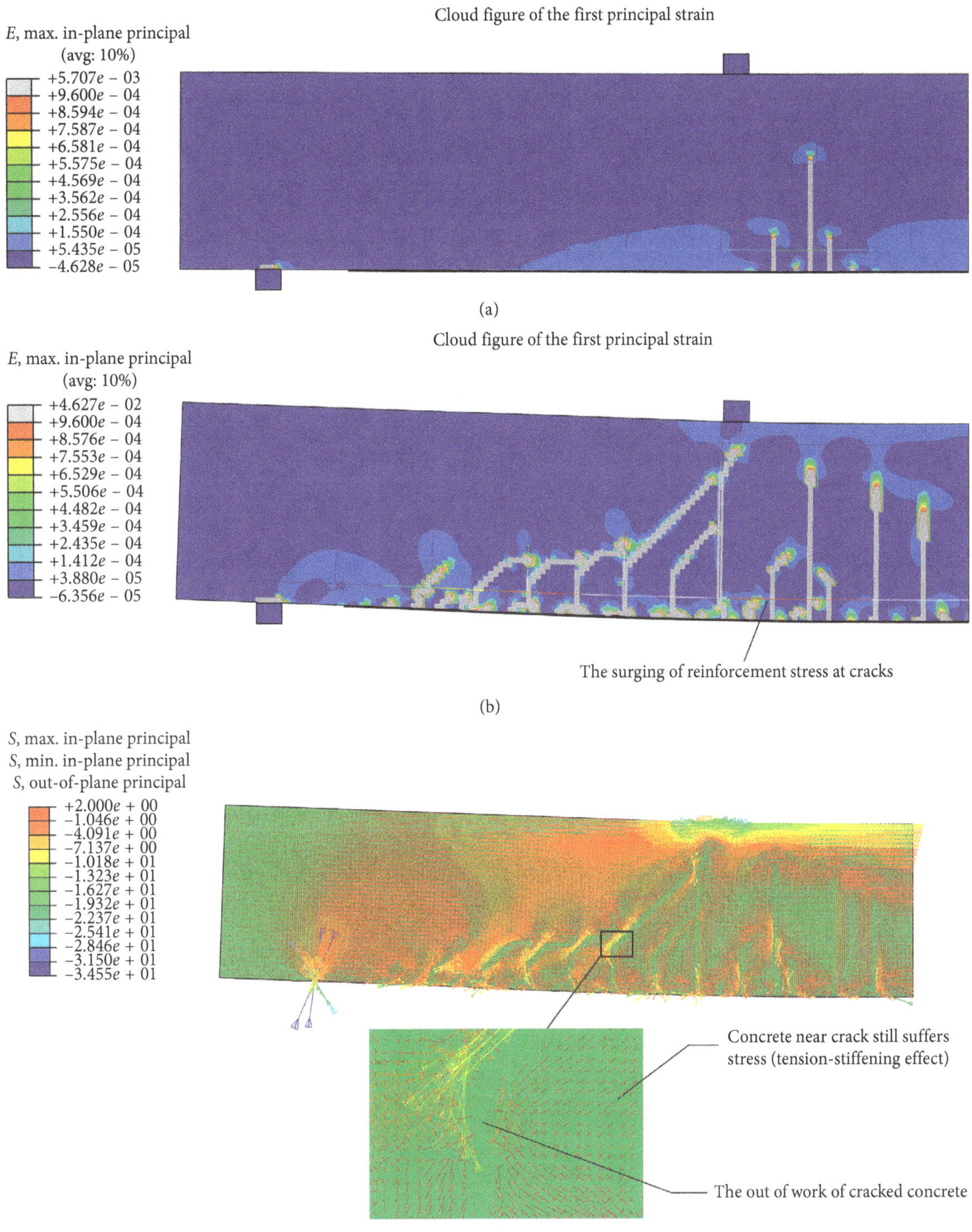

Figure 8: FE cracking pattern. (a) Cracking pattern as 50% of instantaneous load applied. (b) Cracking pattern as 100% of instantaneous load applied. (c) Principle stress presented by vector field (MPa).

4.2. Time-Dependent Interfacial Edge Stresses. The bonding interfaces are critical regions that transfer stresses between concrete and FRP. One of the most concerned issues of RC beams reinforced by FRP is the premature debonding at the plate edges or near cracks, where the stresses (both shear and vertical normal) in the adhesive layer are relatively high. In this paper, numerical investigations were performed to study the time-dependent edge stresses of adhesive interface through the proposed FE approach. The FE model was established on the basis of the test beam in the experimental study of this paper. The responses of specimen under different levels of applied load F (23.54 kN, 47.08 kN, and 70.62 kN) were simulated. Since the wrapped CFRP sheets near the end of beam were not modeled, their

Table 5: Properties of selected specimens for numerical simulation.

Source		Sobuz et al. [42]				Tan and Saha [43]		
	Beam name	CBC	FBC-1L	FBC-2L	FBC-3L	GB3-40	GB3-49	GB3-59
Beam dimensions	Span L (mm)			1900			1800	
	Height h_c/width b_c (mm)			200/150			125/100	
Concrete	28-day cylinder compressive strength f_c (MPa)			36.0			30.4	
	Modulus of elasticity E_c (GPa)			28.6			27.2	
Steel	Tension reinforcement area (mm^2)/yield strength f_{yt} (MPa)/elastic modulus E_{st} (GPa)			157/482/195			157/520/194	
	Compression reinforcement area (mm^2)/yield strength f_{yc} (MPa)/elastic modulus E_{sc} (GPa)			57/470/186			57/525/183	
FRP reinforcement	Nominal thickness ($n \times$ ply thickness t_r)/width b_{frp} (mm)	—	$1.2 \times 1/100$	$1.2 \times 2/100$	$1.2 \times 3/100$		$0.8 \times 3/100$	
	Ultimate strength f_r (MPa)/elastic modulus E_r (GPa)	—		1685/165			1700/39	
Applied moment at midspan (kN·m)				7.144		4.740	5.640	6.840

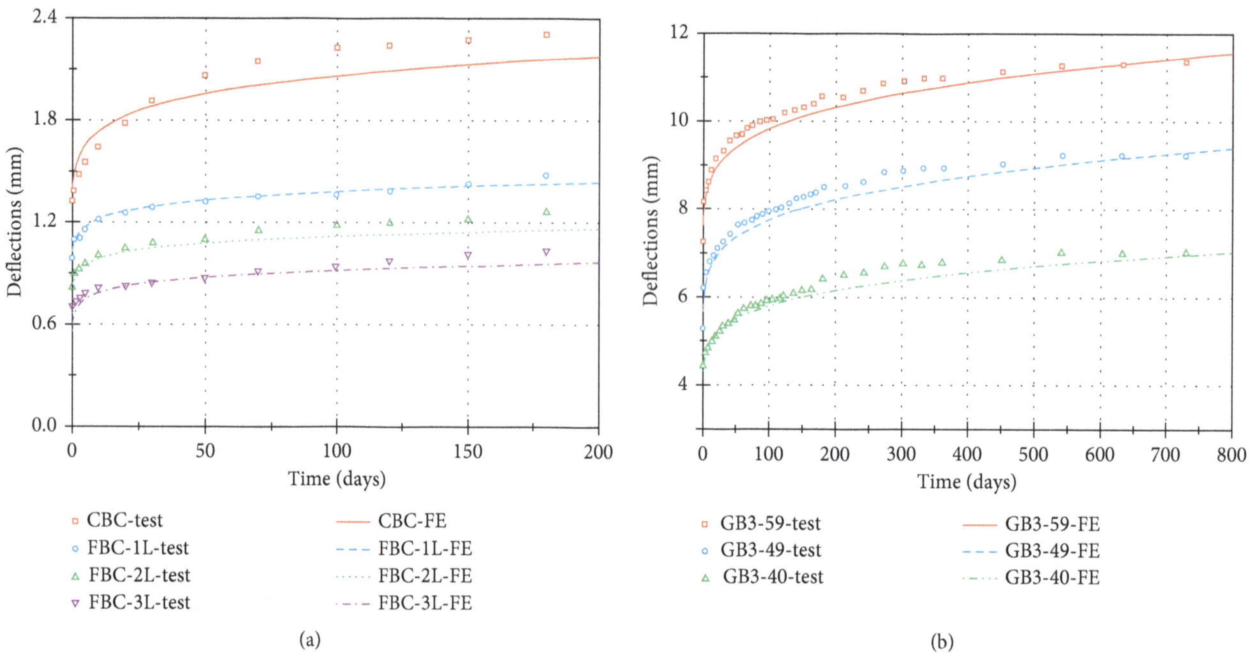

Figure 9: FE versus test deflections at midspan. (a) FE and test results for Sobuz's [42] specimens. (b) FE and test results for Tan's [43] specimens.

influences on the localized interfacial stresses were not discussed currently.

As shown in Figure 10, due to the existence of singular points (Point A and B), in order to obtain accurate stresses, the region near the plate end needs a particularly fine mesh. Following the recommendation of Teng et al. [24], in the vertical direction, for the adhesive layer, the top part (near concrete-adhesive interface) and the bottom part (near adhesive-FRP plate interface) were, respectively, meshed by two layers of elements (the height of each element is 0.1 mm). The rest part of the adhesive layer was evenly meshed by other three layers of elements. In the horizontal direction, a graded mesh was used starting with an aspect ratio of 1 for the minimum height (0.1 mm) elements. The CFRP plate has the same mesh pattern as the adhesive layer. A graded and matching mesh was adopted for the concrete. The stress distribution near the plate end under instantaneous loading is given in Figures 11 and 12 (applied load $F = 70.62$ kN), whose characteristics correspond to the findings of previous researches [24, 44]. It shows that high vertical normal (peeling) and shear stresses exist near the bonding end, which are responsible for the debonding failures widely reported in tests. Also, the normal and shear stresses vary strongly across the thickness of the adhesive layer in the vicinity of the end of the plate. This aspect cannot be captured by existing approximate analytical solutions [22, 23] developed with the assumption that stresses are invariant across the adhesive thickness.

AC interface refers to adhesive-to-concrete interface, PA interface refers to plate-to-adhesive interface, and MA refers

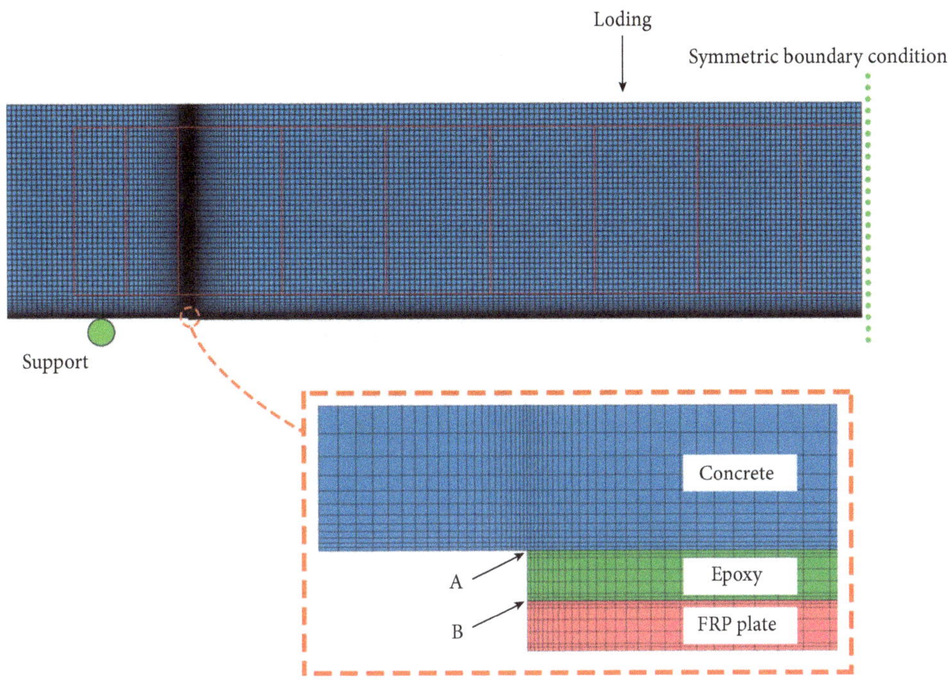

FIGURE 10: The element refinement near the plate end.

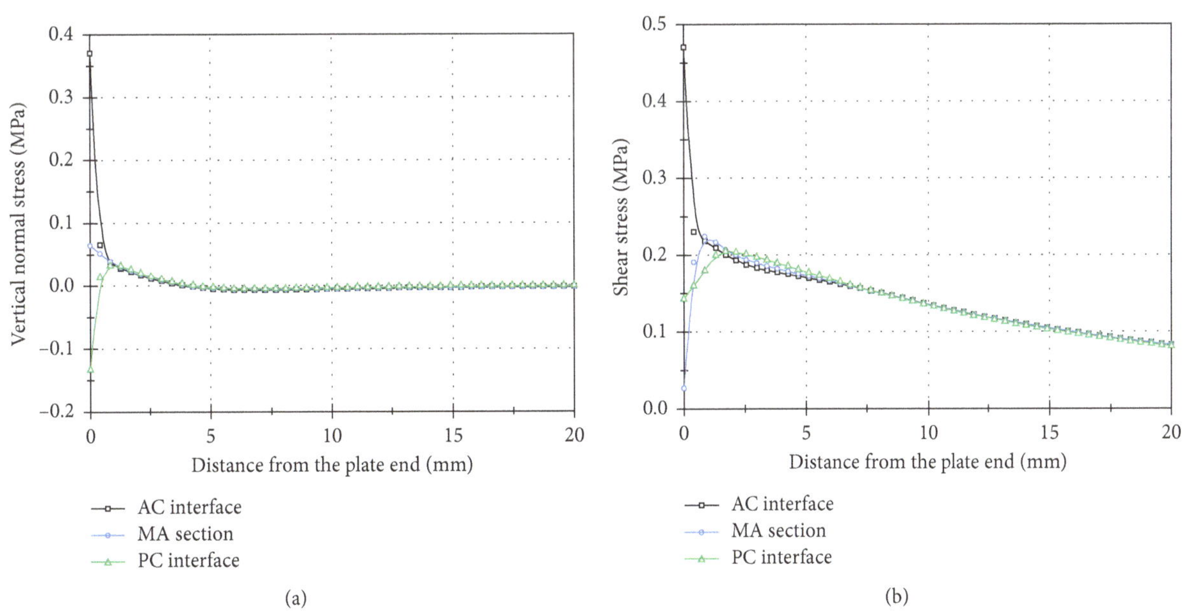

FIGURE 11: Interfacial Stress distributions under instantaneous loading. (a) Vertical normal stress. (b) Shear stress.

to the horizontal section of the adhesive layer obtained by a horizontal cut at midthickness.

In this research, the time-dependent characteristic of interfacial edge stress at point A (in AC interface) is especially discussed, which is one of the most important factors causing debonding failures. The changes of interfacial shear and vertical normal stresses with time are shown in Figure 13. In this case, only the creep of adhesive in shear direction is considered (both FRP and concrete are modeled as time-invariant materials). Figures 13(a) and 13(b) show the viscous flow of epoxy in shear direction leading to significant relaxations for both shear and vertical normal stresses at the plate end. With the increase of sustained load, the level of instantaneous stress in epoxy layer is promoted, which causes more evident creep response and stress relaxation. In Figure 13, the relaxation rate of stress is fast in the earlier stage, but almost slows down to zero afterwards. Such phenomenon is believed to be induced by the shear creep law of adhesive adopted herein, which presents that most of creep deformation develops within relatively short time period compared to the well-known long creep period of concrete [21]. For the results shown in Figure 14, the axial

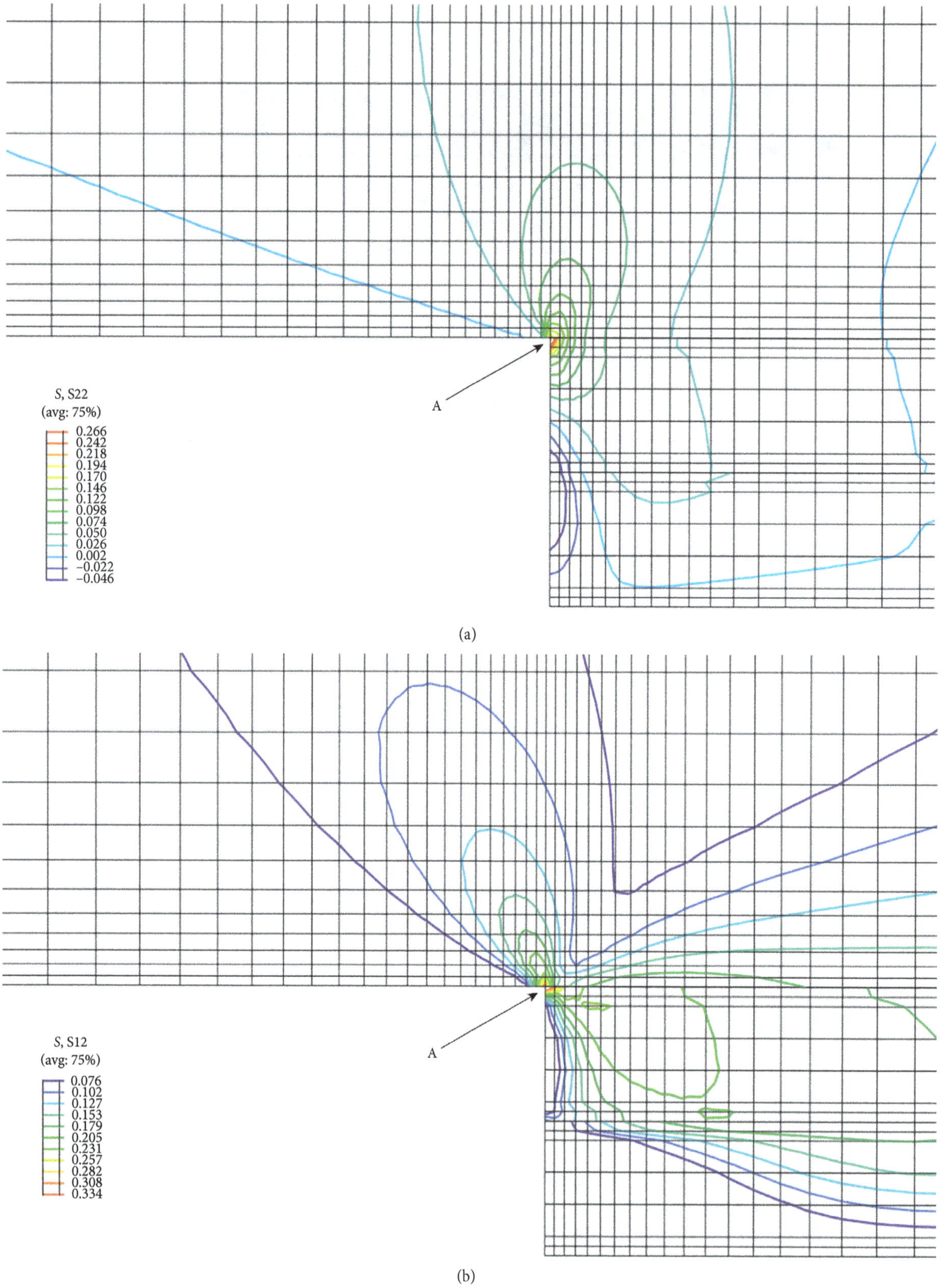

Figure 12: Contours of stress near the plate end (MPa). (a) Vertical normal stress. (b) Shear stress.

creep of epoxy is only considered to separately investigate its effect on the edge stress. It is shown that like the shear creep, the axial creep of adhesive also causes relaxations of the edge stresses, however, with a much more inferior effect. The likely reason is that near the plate end, the section moment is small, and there is low axial stress in the adhesive layer;

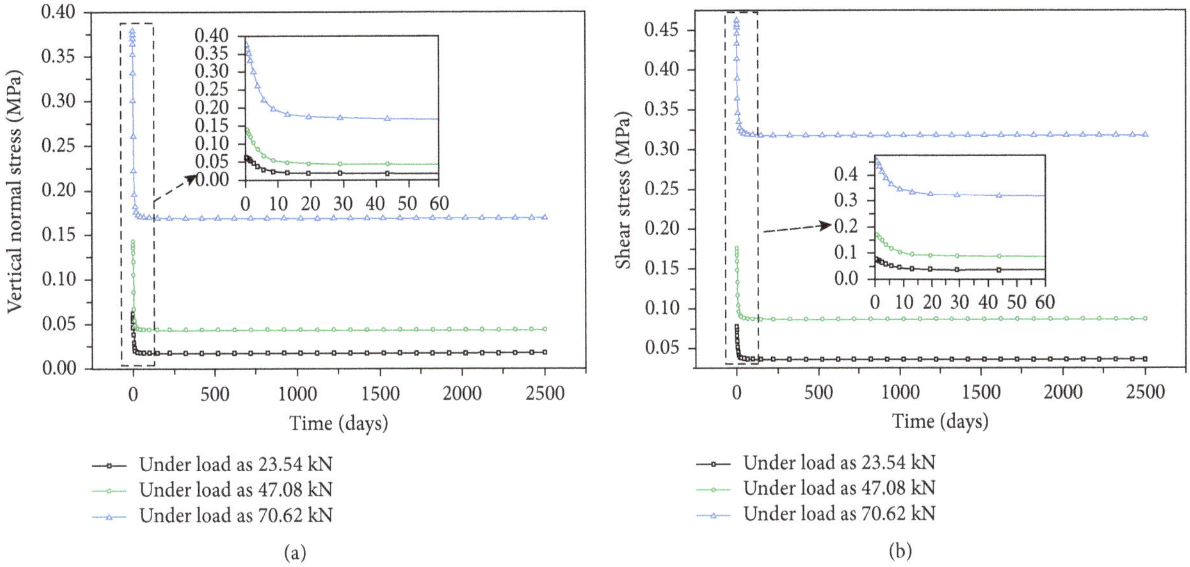

FIGURE 13: The separate effect of shear creep of adhesive on the edge stresses. (a) Variation of vertical normal stress with time. (b) Variation of shear stress with time.

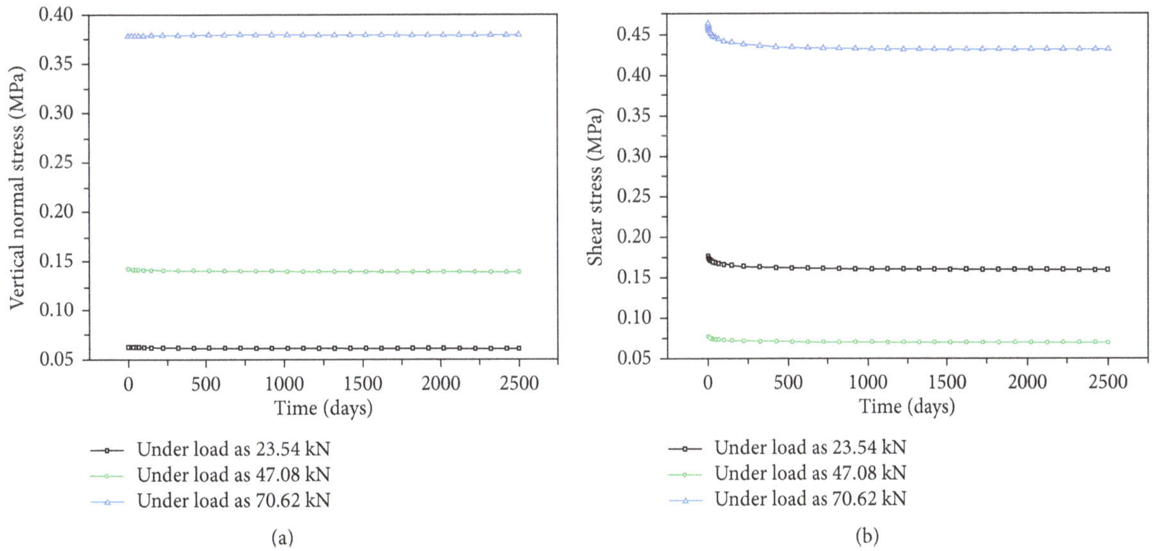

FIGURE 14: The separate effect of axial creep of adhesive on the edge stresses. (a) Variation of vertical normal stress with time. (b) Variation of shear stress with time.

hence, the axial creep is not obvious, which does not clearly influence the edge stresses.

The combined effect of concrete creep and adhesive creep was also investigated, as illustrated in Figure 15. It is found that the creep of concrete individually leads to increases of both interfacial shear and vertical normal stresses along with time, which may eventually lead to premature debonding. This corresponds to the findings in previous studies [14, 15, 45], which demonstrates the rationality of the suggested FE approach. However, as the creep of concrete and adhesive occurs at the same time, the relaxation effect by adhesive creep counteracts the increasing effect by concrete creep on the edge stresses to some extent. Therefore, the edge stresses present lower increases with time, comparing to the case that concrete creep works separately. As the relaxation effect is evident enough, which is larger than the increasing effect of concrete creep, the stresses even show decreases in the earlier stage (curve A, B, and C in Figure 15). Following that, the creep of adhesive is almost completed, while the creep of concrete is still significant; thus, there is an ongoing increase for the edge stresses afterwards. Besides, since the relaxation effect of adhesive creep in axial direction is smaller, its counteraction to the increasing effect of concrete creep on edge stress is less evident comparing to that of adhesive creep in shear direction.

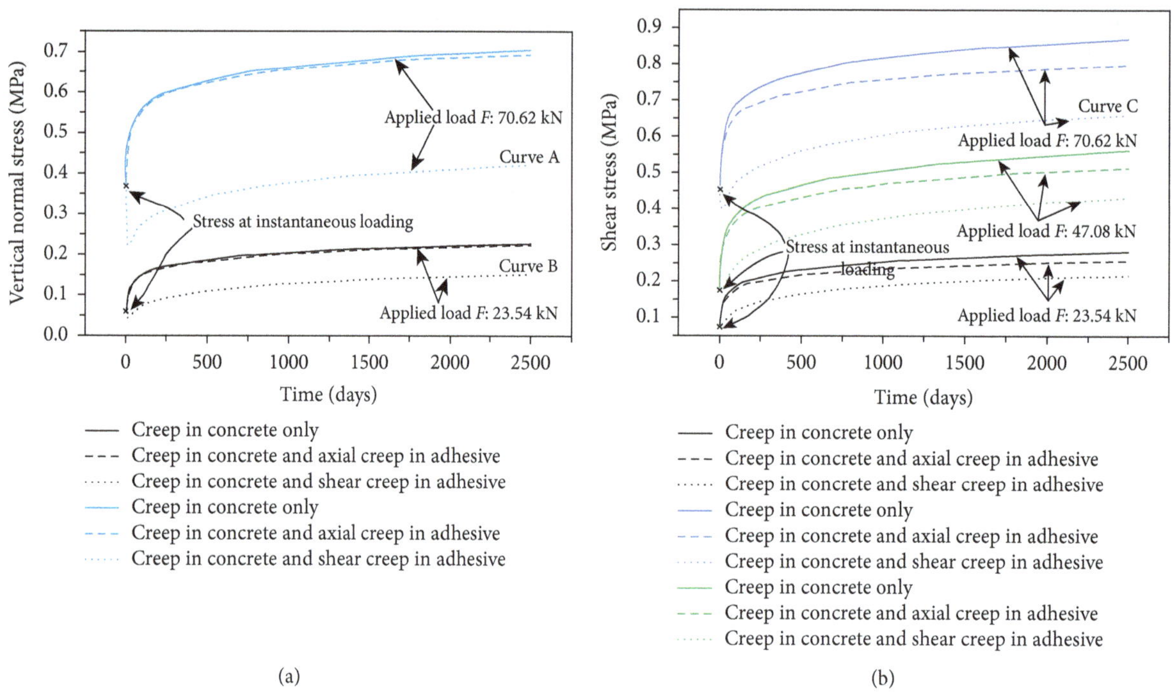

FIGURE 15: The combined effect of adhesive creep and concrete creep on the edge stresses. (a) Variation of vertical normal stress with time. (b) Variation of shear stress with time.

5. Conclusions

In this study, the challenges associated with the time-dependent analysis of RC members strengthened by bonding FRP externally have been discussed, and a finite element model has been developed. The recommended FE model simultaneously considers several aspects of the strengthened structures that the existing analytical methods do not take into account: the time-dependent characteristics of all components including the aging effect of concrete, the cracking behavior, and tension stiffening of concrete. These aspects have been considered through the constitutive laws of materials. User-defined subroutines embedded in the main FE program were used to model the time-dependent material properties. Unlike some existing analytical methods, specific simplifications were not included, which may further improve the general applicability of this FE approach.

The capability and accuracy of the proposed FE model were demonstrated through comparisons of its predictions with the results of test performed by the authors. Although, detailed changes of environmental conditions have not been considered and the average humidity during test was used to generally reflect the environmental effect, the FE-predicted results are in qualitative agreement with the experimental observations. Besides, test data in other research studies were also used to further verify the FE approach.

The analyses of interfacial stresses can be carried out through the same FE model, which only requires mesh refinement near the plate end, rather than to build a quite different analytical model from the beginning. The analyses show that concrete creep causes a significant increase in the shear and vertical normal stresses in the adhesive layer at the edges. This phenomenon may initiate debonding failure, which is of critical importance for the design of FRP-strengthened beams. Whereas, the viscoelastic adhesive may present a favorable effect on alleviating such problem, since it has been presented that the viscous flow of adhesive can relieve the edge stresses.

The FE model developed in this study may set a theoretical platform to better analyze the time-dependent characteristics of FRP-strengthened concrete structures. Further aspects of the structural behavior including effects of environment changes (e.g., variations of humidity and temperature), development of debonded regions, and creep rupture need to be further researched, which can be built on the findings and concepts of this study.

Data Availability

The data in Figures 3 and 6 used to support the findings of this study may be released upon application to the corresponding author (e-mail: jiangshiy@163.com). These prior studies (and datasets) are cited at relevant places within the text as references [42, 43].

Conflicts of Interest

The authors declare that they have no conflicts of interest.

Acknowledgments

The authors are grateful for the financial support from the Chongqing Science and Technology Committee (Project nos: cstc2015shmszx30006 and cstc2011ab0043).

References

[1] S. Hong, "Effects of the amount and shape of carbon fiber-reinforced polymer strengthening elements on the ductile behavior of reinforced concrete beams," *Mechanics of Composite Materials*, vol. 50, no. 4, pp. 427–436, 2014.

[2] S. Hong, "Effect of intermediate crack debonding on the flexural strength of CFRP-strengthened RC beams," *Mechanics of Composite Materials*, vol. 50, no. 4, pp. 523–536, 2014.

[3] M. Rezazadeh, H. Ramezansefat, and J. Barros, "NSM CFRP prestressing techniques with strengthening potential for simultaneously enhancing load capacity and ductility performance," *Journal of Composites for Construction*, vol. 20, no. 5, article 04016029, 2016.

[4] S. H. Aboubakr, U. F. Kandil, and M. R. Taha, "Creep of epoxy-clay nanocomposite adhesive at the FRP interface: a multi-scale investigation," *International Journal of Adhesion and Adhesives*, vol. 54, pp. 1–12, 2014.

[5] Y. Jeong, J. Lee, and W. S. Kim, "Modeling and measurement of sustained loading and temperature-dependent deformation of carbon fiber-reinforced polymer bonded to concrete," *Materials*, vol. 8, no. 2, pp. 435–450, 2015.

[6] Y. Jeong, M. M. Lopez, and C. E. Bakis, "Effects of temperature and sustained loading on the mechanical response of CFRP bonded to concrete," *Construction and Building Materials*, vol. 124, pp. 442–452, 2016.

[7] I. Costa and J. Barros, "Tensile creep of a structural epoxy adhesive: experimental and analytical characterization," *International Journal of Adhesion and Adhesives*, vol. 59, pp. 115–124, 2015.

[8] J. Jeon, J. Kim, and A. Muliana, "Modeling time-dependent and inelastic response of fiber reinforced polymer composites," *Computational Materials Science*, vol. 70, pp. 37–50, 2013.

[9] A. Sorzia, "Modelling of creep and stress relaxation test of a polypropylene microfibre by using fraction-exponential kernel," *Modelling and Simulation in Engineering*, vol. 2016, Article ID 3823047, 7 pages, 2016.

[10] J. Tanks, K. Rader, S. Sharp, and T. Sakai, "Accelerated creep and creep-rupture testing of transverse unidirectional carbon/epoxy lamina based on the stepped isostress method," *Composite Structures*, vol. 159, pp. 455–462, 2017.

[11] S. Hong and S. K. Park, "Long-term behavior of fiber-reinforced-polymer-plated concrete beams under sustained loading: analytical and experimental study," *Composite Structures*, vol. 152, pp. 140–157, 2016.

[12] D. Trost, *Stress Relaxation of Concrete Structures and Relaxation Tests of Very Old Concrete*, Springer, Berlin, Germany, 1982.

[13] K. W. Nasser and A. M. Neville, "Creep of old concrete at normal and elevated temperature," *ACI Journal Proceedings*, vol. 64, no. 2, 1967.

[14] E. Hamed and M. A. Bradford, "Creep in concrete beams strengthened with composite materials," *European Journal of Mechanics-A/Solids*, vol. 29, no. 6, pp. 951–965, 2010.

[15] E. Hamed and M. A. Bradford, "Flexural time-dependent cracking and post-cracking behaviour of FRP strengthened concrete beams," *International Journal of Solids and Structures*, vol. 49, no. 13, pp. 1595–1607, 2012.

[16] A. R. Marí, E. Oller, J. M. Bairán, and N. Duarte, "Simplified method for the calculation of long-term deflections in FRP-strengthened reinforced concrete beams," *Composites Part B: Engineering*, vol. 45, no. 1, pp. 1368–1376, 2013.

[17] E. Oller and A. R. Marí, "Long-term bond stresses and debonding failure of FRP-strengthened RC cracked members," *Composites Part B: Engineering*, vol. 52, pp. 30–39, 2013.

[18] K. Hadjazi, Z. Sereir, and S. Amziane, "Creep response of intermediate flexural cracking behavior of reinforced concrete beam strengthened with an externally bonded FRP plate," *International Journal of Solids and Structures*, vol. 94-95, pp. 196–205, 2016.

[19] A. K. El-Sayed, R. A. Al-Zaid, A. I. Al-Negheimish, A. B. Shuraim, and A. M. Alhozaimy, "Long-term behavior of wide shallow RC beams strengthened with externally bonded CFRP plates," *Construction and Building Materials*, vol. 51, pp. 473–483, 2014.

[20] K. K. Choi, M. M. Reda Taha, M. J. Masia, P. L. Shrive, and N. G. Shrive, "Numerical investigation of creep effects on FRP-strengthened RC beams," *Journal of Composites for Construction*, vol. 14, no. 6, pp. 812–822, 2010.

[21] P. Meshgin, K. K. Choi, and M. M. R. Taha, "Experimental and analytical investigations of creep of epoxy adhesive at the concrete–FRP interfaces," *International Journal of Adhesion and Adhesives*, vol. 29, no. 1, pp. 56–66, 2009.

[22] S. Benyoucef, A. Tounsi, K. H. Benrahou, and E. A. Adda Bedia, "Time-dependent behavior of RC beams strengthened with externally bonded FRP plates: interfacial stresses analysis," *Mechanics of Time-Dependent Materials*, vol. 11, no. 3-4, pp. 231–248, 2007.

[23] C. Zhang and J. Wang, "Viscoelastic analysis of FRP strengthened reinforced concrete beams," *Composite Structures*, vol. 93, no. 12, pp. 3200–3208, 2011.

[24] J. G. Teng, J. W. Zhang, and S. T. Smith, "Interfacial stresses in reinforced concrete beams bonded with a soffit plate: a finite element study," *Construction and Building Materials*, vol. 16, no. 1, pp. 1–14, 2002.

[25] ABAQUS, *ABAQUS 6.12 User's Manual*, 2012.

[26] S. Jiang, W. Yao, J. Chen, and S. Tao, "Time dependent behavior of FRP-strengthened RC beams subjected to preload: experimental study and finite element modeling," *Composite Structures*, vol. 200, pp. 599–613, 2018.

[27] G. M. Chen, J. G. Teng, and J. F. Chen, "Finite element modeling of intermediate crack debonding in FRP-plated RC beams," *Journal of Composites for Construction*, vol. 15, no. 3, pp. 339–353, 2010.

[28] L. P. Saenz, "Discussion of equation for the stress-strain curve of concrete-by Desayi, P. and Krishan, S," *ACI Journal*, vol. 61, no. 9, pp. 1229–1235, 1964.

[29] D. A. Hordijk, "Local approach to fatigue of concrete," *Civil Engineering and Geosciences*, vol. 1, no. 4, pp. 442–448, 1991.

[30] CEB-FIP, *CEB-FIP Model Code 2010*, Ernst & Sohn Publishing House, Berlin, Germany, 2013.

[31] ACI Committee 318, *Building Code Requirements for Structural Concrete and Commentary*, American Concrete Institute, Farmington Hills, MI, USA, 2014.

[32] Z. P. Bazant and J. Planas, *Fracture and Size Effect in Concrete and Other Quasibrittle Materials*, CRC Press, Boca Raton, FL, USA, 1998.

[33] Z. P. Bažant and B. H. Oh, "Crack band theory for fracture of concrete," *Matériaux Et Construction*, vol. 16, no. 3, pp. 155–177, 1983.

[34] J. G. Rots, *Computational modeling of concrete fracture*, Ph.D. thesis, Delft University of Technology, Delft, The Netherlands, 1988.

[35] Y. Tao and J. F. Chen, "Concrete damage plasticity model for modeling FRP-to-concrete bond behavior," *Journal of Composites for Construction*, vol. 19, no. 1, article 04014026, 2015.

[36] Z. P. Bazant, "Prediction of creep effects using age-adjusted effective modulus method," *ACI Journal Proceedings*, vol. 69, no. 4, 1972.

[37] ACI Committee 209, *Prediction of Creep, Shrinkage, and Temperature Effects in Concrete Structures*, American Concrete Institute, Farmington Hills, MI, USA, 2002.

[38] C. W. Feng, C.-W. Keong, Y.-P. Hsueh, Y.-Y. Wang, and H.-J. Sue, "Modeling of long-term creep behavior of structural epoxy adhesives," *International Journal of Adhesion and Adhesives*, vol. 25, no. 5, pp. 427–436, 2005.

[39] W. N. Findley, J. S. Lai, K. Onaran, and R. M. Christensen, "Creep and relaxation of nonlinear viscoelastic materials," *Journal of Applied Mechanics*, vol. 44, no. 2, p. 364, 1977.

[40] Y. F. Wen, R. F. Gibson, and J. L. Sullivan, "Prediction of momentary transverse creep behavior of thermoplastic polymer matrix composites using micromechanical models," *Journal of Composite Materials*, vol. 31, no. 21, pp. 2124–2145, 1997.

[41] W. N. Findley, "Mechanism and mechanics of creep of plastics," *Society of Plastic Engineering Journal*, vol. 16, no. 1, pp. 57–65, 1960.

[42] H. R. Sobuz, E. Ahmed, N. Mohamed Sutan, N. M. Sadiqul Hasan, M. Alhaz Uddin, and M. Jahir Uddin, "Bending and time-dependent responses of RC beams strengthened with bonded carbon fiber composite laminates," *Construction and Building Materials*, vol. 29, pp. 597–611, 2012.

[43] K. H. Tan and M. K. Saha, "Long-term deflections of reinforced concrete beams externally bonded with FRP system," *Journal of Composites for Construction*, vol. 10, no. 6, pp. 474–482, 2006.

[44] O. Rabinovich and Y. Frostig, "Closed-form high-order analysis of RC beams strengthened with FRP strips," *Journal of Composites for Construction*, vol. 4, no. 2, pp. 65–74, 2000.

[45] E. Hamed and Z. T. Chang, "Effect of creep on the edge debonding failure of FRP strengthened RC beams–a theoretical and experimental study," *Composites Science and Technology*, vol. 74, pp. 186–193, 2013.

18

Laboratory Evaluation of Interactions in the Degradation of a Polypropylene Geotextile in Marine Environments

José Ricardo Carneiro ,[1] Paulo Joaquim Almeida,[2] and Maria de Lurdes Lopes[1]

[1]*Construct-Geo, Faculty of Engineering, University of Porto, Rua Dr. Roberto Frias, 4200-465 Porto, Portugal*
[2]*Requimte, Faculty of Sciences, University of Porto, Rua do Campo Alegre, 687, 4169-007 Porto, Portugal*

Correspondence should be addressed to José Ricardo Carneiro; rcarneir@fe.up.pt

Academic Editor: Jun Liu

The long-term behaviour of geosynthetics applied in coastal engineering structures can be adversely affected by many agents. This paper studies the resistance of a nonwoven polypropylene geotextile against some degradation agents present in marine environments and evaluates the existence of interactions between them. For that purpose, the geotextile was exposed to some laboratory degradation tests: immersion tests (in seawater, deionised water, and sodium chloride 35 $g{\cdot}L^{-1}$), thermooxidation, and artificial weathering. The geotextile was (1) exposed separately to each degradation test and (2) exposed successively to combinations of two or three degradation tests. The damage caused by the degradation tests was evaluated by monitoring the tensile properties of the geotextile. Based on the changes occurred in tensile strength, reduction factors were determined. The reduction factors obtained directly in the multiple exposures were compared with those obtained by the traditional methodology for the combined effect of the degradation agents. The results, among other findings, showed the existence of relevant interactions between the degradation agents and showed that the reduction factors obtained by the traditional methodology were unable to represent accurately (by underestimating) the degradation occurred in the geotextile.

1. Introduction

Geosynthetics are polymeric materials that can be used in the construction of many coastal engineering structures, such as breakwaters, dykes, groynes, seawalls, jetties, artificial reefs, or revetments [1, 2]. The advantages of using these materials are the ease of installation (and future removal, if necessary), high versatility and efficiency, low cost, and low environmental impact. The functions of geosynthetics in coastal engineering structures include filtration, drainage, separation, reinforcement, containment, or erosion control [1, 2]. The geotextiles (one of the main groups of geosynthetics) are typically used in coastal engineering structures as filters and for manufacturing sand-filled elements (called geosystems), like geobags, geotubes, or geocontainers. Pilarczyk [1] and Bezuijen and Vastenburg [2] provide an extensive description of the available knowledge about geosynthetics in coastal engineering.

In marine environments, the geosynthetics can be in contact with many degradation agents capable of causing unwanted changes in their properties, affecting their performance. The most common degradation agents in these environments include seawater, oxygen, solar radiation and other weathering agents, biological agents, and the action of waves, currents, and tides [3–5]. The geosynthetics may also suffer mechanical damage during the installation process, where in some cases they are subjected to higher stresses than during service [6].

Polypropylene (PP) (a polymer mostly used for the production of geotextiles) has a good resistance against many liquids (including seawater [7]) and biological degradation. However, its resistance against oxidation is substantially lower [8, 9]. The oxidation process of PP follows a chain reaction mechanism formed by three main steps: initiation, propagation, and termination. The initiation step consists in the formation of free radicals (R•) (Reaction (1)) mainly by the action of temperature (thermooxidation) or by the action of ultraviolet (UV) radiation (photooxidation). The free radicals R• promptly react with oxygen (O_2), forming peroxide radicals (ROO•) (Reaction (2)). The radicals ROO• then extract hydrogen atoms from surrounding polymers, forming hydroperoxides (ROOH) and more free radicals R• (Reaction (3)). The free radicals R• can react again according to Reaction (2), propagating degradation. The hydroperoxides ROOH are very unstable, being easily converted into the free radicals RO•, ROO•, and •OH (Reactions (4) and (5)), which can also extract hydrogen atoms from surrounding polymers and propagate degradation [8–10].

$$\text{RH (polymer)} + \text{energy}\left(\frac{\text{Heat}}{\text{UV radiation}}\right) \longrightarrow \text{R}\bullet \quad (1)$$

$$\text{R}\bullet + \text{O}_2 \longrightarrow \text{ROO}\bullet \quad (2)$$

$$\text{ROO}\bullet + \text{RH} \longrightarrow \text{ROOH} + \text{R}\bullet \quad (3)$$

$$\text{ROOH} \longrightarrow \text{RO}\bullet + \bullet\text{OH} \quad (4)$$

$$2\text{ROOH} \longrightarrow \text{RO}\bullet + \text{ROO}\bullet + \text{H}_2\text{O} \quad (5)$$

The oxidation process (autooxidative mechanism) will continue until all free radicals are transformed into nonreactive compounds (e.g., by reacting between them or by the action of chemical stabilisers) [8–10]. The occurrence of oxidation of PP products is practically unavoidable. However, in the absence of UV radiation, the oxidation process of PP is relatively slow at ambient temperatures (but cannot be neglected when considering products for long-term use) [11].

In many coastal engineering applications, the geosynthetics are only exposed to UV radiation during installation (usually a short period of time), being subsequently covered by sand, by other construction materials or by seawater. However, there are cases (like geobags, geotubes, or geocontainers used in groynes or breakwaters) where the geosynthetics can be exposed for longer periods of time. In addition, unpredictable UV exposure may also occur (e.g., due to the removal of sand by the action of the waves in buried structures) [7]. The damage caused by photooxidation (e.g., reduction in mechanical strength) can be retarded by adding chemical additives (like antioxidants or UV stabilisers) to the geosynthetics [9, 12–14].

The long-term behaviour of the geosynthetics is normally predicted based on data obtained from laboratory tests (degradation under accelerated conditions) [15–18]. In addition, their resistance against degradation can also be evaluated by field tests (degradation under natural conditions) [5, 7, 14]. The damage occurred in geotextiles during the degradation tests is often determined by monitoring changes in their tensile behaviour (in accordance with the recommendations of EN 12226 [19]). The changes occurred in other properties (puncture properties [5, 14], physical properties [7], or hydraulic properties [20]) and other techniques (like scanning electron microscopy [13, 14, 21], infrared spectroscopy [22, 23], or liquid chromatography [23, 24]) can also be used for damage evaluation.

When designing with geosynthetics, reduction factors (RFs) are often introduced to account for the degradation that occurs over time. Each reduction factor normally represents a decrease in resistance (known or estimated) due to the action of one, or more, degradation agent. For example, for reinforcement applications, the tensile strength (T) of the geosynthetics is typically affected by a set of partial reduction factors accounting for the effects of installation damage (RF_{ID}), creep (RF_C), weathering (RF_W), chemical and biological agents (RF_{CB}), and a factor of safety (f_s) (6) [25, 26]. The use of (6) enables the determination of the design tensile strength (T_D):

$$T_D = \frac{T}{RF_{ID} \times RF_C \times RF_W \times RF_{CB} \times f_s}. \quad (6)$$

The actual design methods, the standards available for degradation tests of geosynthetics (e.g., [15–18]), and most studies available in literature consider the isolated action of the degradation agents, not accounting for possible interactions between them. However, the combined effect of the degradation agents can be different (more severe) from the sum of their isolated actions [27–29]. For example, Carneiro et al. [27] showed the existence of interactions between chemical agents in the degradation of a PP geotextile. Interactions have also been found between mechanical degradation agents of geosynthetics [28–30]. The existence of such interactions can lead to inaccurate global reduction factors (used in the design), which are traditionally obtained by multiplying relevant partial reduction factors (each determined in isolation).

This work studies the resistance of a PP geotextile against some degradation agents present in marine environments. For that purpose, laboratory degradation tests were carried out to evaluate the resistance of the geotextile against seawater, thermooxidation, and weathering and to identify interactions between the previous agents. The main goals of the work included (1) determination of the effect of the degradation agents in the tensile properties of the geotextile, (2) identification of interactions between the degradation agents, and (3) comparison of the reduction factors obtained by the traditional methodology (determination of the reduction factors in isolation for each degradation agent and further multiplication) and by an alternative approach (successive exposure) for the combined effect of the degradation agents.

2. Materials and Methods

2.1. Geotextile. This work studied a nonwoven needle-punched geotextile made from PP fibres stabilised with 0.2% (percentage in weight) of the additive Chimassorb 944—a UV stabiliser belonging to the HALS (hindered amine light stabilisers) family. The PP fibres had a linear mass of 8 denier, a length of 75 mm, and a diameter of about 30–40 μm. The main physical, mechanical, and hydraulic properties of the geotextile can be found in Table 1.

The sampling process (for the characterisation and degradation tests) was carried out according to EN ISO 9862 [37]). The specimens (machine direction of production) were collected from positions evenly distributed over the full width and length of the geotextile (supplied in a roll), but not closer than 100 mm to the edges. The specimens for the same

Table 1: Main properties of the geotextile (undamaged sample).

Property	Test standard	Geotextile
Mass per unit area ($g \cdot m^{-2}$)	EN ISO 9864 [31]	502 (±25)
Thickness at 2 kPa (mm)	EN ISO 9863-1 [32]	3.81 (±0.08)
Tensile strength MD[a] ($kN \cdot m^{-1}$)	EN 29073-3 [33]	26.64 (±1.41)
Elongation at the maximum load MD[a] (%)	EN 29073-3 [33]	71.6 (±3.2)
Tensile strength CMD[b] ($kN \cdot m^{-1}$)	EN 29073-3 [33]	13.43 (±1.03)
Elongation at the maximum load CMD[b] (%)	EN 29073-3 [33]	122.9 (±6.2)
CBR puncture resistance (kN)	EN ISO 12236 [34]	2.69 (±0.15)
Permeability (VI_{H50})[c] ($mm \cdot s^{-1}$)	EN ISO 11058 [35]	39.7 (±5.0)
Pore size (O_{90})[d] (μm)	EN ISO 12956 [36]	66.0 (±4.9)

The 95% confidence intervals are given in parentheses; [a]MD, machine direction of production; [b]CMD, cross-machine direction of production; [c]VI_{H50}, velocity index for a head loss of 50 mm; [d]O_{90}, characteristic opening size.

characterisation or degradation tests (5 specimens for each test) were taken from different longitudinal and transverse positions of the roll. The specimens were 50 mm wide and 400 mm long (number and dimensions according to EN 12226 [19]).

2.2. Degradation Tests. First, the geotextile was exposed separately to some laboratory degradation tests (single exposure): immersions in liquids, thermooxidation, and artificial weathering (description of the tests in the following points). Then, the geotextile was exposed consecutively (multiple exposure) to some combinations of two or three of the previous degradation tests. Table 2 summarizes the degradation tests performed.

Table 2: Degradation tests performed.

Single exposure	Multiple exposures
Immersion in seawater	Seawater + TO
Immersion in H_2O	Seawater + QUV
Immersion in NaCl	H_2O + TO
Thermooxidation (TO)	H_2O + QUV
Artificial weathering (QUV)	NaCl + TO
	NaCl + QUV
	QUV + TO
	Seawater + QUV + TO

2.2.1. Immersion Tests. The geotextile was immersed, at room temperature (about 20°C), in seawater (collected directly from the sea and used without any treatment), in deionised water (H_2O), and in an aqueous solution of sodium chloride (NaCl) (concentration of $35\ g \cdot L^{-1}$). The concentration of the solution of sodium chloride intended to simulate the amount of salts (mostly, but not entirely, sodium chloride) existent in seawater is typically $35\ g \cdot L^{-1}$. The immersion tests were carried out in the dark and had a duration of 12 months.

The specimens that were subsequently exposed to thermooxidation or artificial weathering (multiple exposures) were not washed after the immersion tests (in order to be contaminated with remains of the immersion solutions) and were dried at room temperature in the absence of light. The subsequent degradation tests were carried out in a short period of time (a few days or weeks) after the drying process.

2.2.2. Thermooxidation Tests. The thermooxidation tests consisted in exposing the geotextile at 110°C during 56 days. These tests were carried out in an oven (*Heraeus Instruments*, model T6120) with a normal oxygen atmosphere (21% of O_2) and without forced air circulation. With exception for the higher exposure time, the thermooxidation tests were carried out according to method A of EN ISO 13438 [17]. According to this standard, the resistance of PP geotextiles against thermooxidation can be evaluated by exposing the materials at 110°C during 14 days (method A1—for all applications, except reinforcement) or during 28 days (method A2—for reinforcement applications). The increase of the exposure time to 56 days was intended to harshen the degradation conditions and thereby enhance the effects of thermooxidation.

2.2.3. Artificial Weathering Tests. The artificial weathering tests (alternated exposure to UV radiation, water spray, and condensation) were performed in a laboratory weatherometer (*Q-Panel Lab Products*, model QUV/spray). The geotextile was exposed during 500 hours to the following weathering cycle:

Step 1: Exposure to UV radiation (6 hours, 60°C)

Step 2: Water spray (10 minutes, water at room temperature)

Step 3: Condensation (2 hours, 45°C)

(Return to *Step 1*)

The UV radiation was provided by fluorescent UV lamps (type UVA-340). The water spray (flow of $5\ L \cdot min^{-1}$) causes a fast decrease in temperature (from 60°C to about 24°C) and can promote some surface erosion. The water used in the spray step was treated microbiologically and purified by reverse osmosis, followed by deionisation on ionic exchange columns. In the condensation step, the water (from the public supply network) existent in a reservoir located at the bottom of the weatherometer was heated to produce vapour that condensates (distilled water) at the surface of the exposed specimens.

During the 500 test-hours, the geotextile was exposed to about 61 weathering cycles (each cycle had 8 hours and 10 minutes). The irradiance during the UV-exposure step (total duration of about 367 hours) was 0.68 $W{\cdot}m^{-2}$ at 340 nm. The total UV radiant exposure (between 290 and 400 nm) was 51.8 $MJ{\cdot}m^{-2}$.

2.3. Evaluation of the Damage Suffered by the Geotextile. The damage suffered by the geotextile (during the degradation tests) was evaluated by tensile tests according to EN 29073-3 [33] (the method specified in EN 12226 [19] for determining the changes in the tensile properties of aged nonwoven geotextiles). These tests (velocity of 100 $mm{\cdot}min^{-1}$) were carried out in an equipment from *Lloyd Instruments* (model LR 50K) equipped with a load cell of 5 kN (from *Lloyd Instruments*). Each tensile test included the analysis of 5 specimens (in the machine direction of production) with a length of 200 mm (between grips) and a width of 50 mm.

The mechanical parameters obtained in the tensile tests included tensile strength (T in $kN{\cdot}m^{-1}$) and elongation at the maximum load (E_{ML} in %). Elongation was determined by expressing the relative displacement of the grips as a percentage of the original length (200 mm) (as specified in EN 29073-3 [33]). The results obtained for tensile strength and elongation at the maximum load are presented with 95% confidence intervals determined according to the following Montgomery and Runger [38] equation:

$$\mu = x \pm t_{\alpha/2,\, n-1}\frac{s}{\sqrt{n}}, \tag{7}$$

Where μ is the population mean, x is the sample mean, t is Student's t-distribution value for the confidence level α and $n-1$ degrees of freedom, n is the number of specimens tested for each sample, and s is the sample standard deviation. The changes occurred in tensile strength are also presented in terms of retained tensile strength (T_{residual} in %), obtained in accordance with EN 12226 [19] by dividing the tensile strength of the samples exposed to the degradation tests (T_{damaged}) by the tensile strength of unexposed samples ($T_{\text{reference}}$):

$$T_{\text{residual}} = \frac{T_{\text{damaged}}}{T_{\text{reference}}} \times 100. \tag{8}$$

2.4. Determination of Reduction Factors. Reduction factors were determined based on the changes occurred (during the degradation tests) in the tensile strength of the geotextile. Each reduction factor represents a reduction in resistance provoked by the degradation tests (single or multiple exposures). The reduction factors were determined by the following equation:

$$\text{RF} = \frac{T_{\text{reference}}}{T_{\text{damaged}}}, \tag{9}$$

where $T_{\text{reference}}$ and T_{damaged} correspond, respectively, to the tensile strength of the geotextile before and after the degradation tests.

Table 3: Tensile properties of the geotextile before and after the single exposure degradation tests.

Degradation test	T ($kN{\cdot}m^{-1}$)	E_{ML} (%)	T_{residual} (%)	RF
Reference (unexposed)	26.64 (±1.41)	71.6 (±3.2)	—	—
Immersion in seawater	26.20 (±2.50)	72.4 (±7.7)	98.3	1.02
Immersion in H_2O	26.82 (±2.67)	71.5 (±5.6)	100.7	1.00
Immersion in NaCl	25.66 (±0.86)	75.3 (±7.3)	96.3	1.04
Thermooxidation	24.67 (±1.88)	53.3 (±4.5)	92.6	1.08
Artificial weathering	12.42 (±1.85)	34.4 (±2.3)	46.6	2.14

The 95% confidence intervals are given in parentheses.

The reduction factors obtained by the traditional methodology for the combined effect of two (or more) degradation agents ($\text{RF}_{\text{agents A+B trad}}$) (agents denominated generically by "A" and "B") were determined by multiplying the reductions factors obtained individually for each degradation agent ($\text{RF}_{\text{agent A}}$ and $\text{RF}_{\text{agent B}}$, resp.):

$$\text{RF}_{\text{agents A+B trad}} = \text{RF}_{\text{agent A}} \times \text{RF}_{\text{agent B}}. \tag{10}$$

The reduction factors determined in this work correspond to particular degradation conditions and cannot be generalized or applied directly in the design (their determination was intended for allowing a better analysis and understanding of the results). For being used in the design, the reduction factors must be analysed case by case, having into account the specific conditions of the constructions. Moreover, it is necessary to determine if the conditions imposed in the degradation tests effectively represent the expected field degradation conditions.

3. Results and Discussion

3.1. Exposure to One Degradation Test (Single Exposure). The immersion tests did not cause relevant changes in the tensile strength and elongation at the maximum load of the geotextile (Table 3). The minor variations observed in tensile strength (retained tensile strengths between 96.3% and 100.7%) and elongation at the maximum load may be attributed to the heterogeneity of the geotextile. Indeed, the nonwoven geotextiles typically have some heterogeneity (arising from their manufacturing process), which can be responsible for small differences between properties determined in consecutive repetitions of a characterisation test.

The tensile properties of the geotextile were significantly more affected by the thermooxidation and artificial weathering tests than by the immersion tests (Table 3). Despite the inexistence of a considerable reduction in tensile strength (retained tensile strength of 92.6%), the exposure to thermooxidation led to a pronounced decrease in elongation at the maximum load (reduction from 71.6% to 53.3%). The artificial weathering test caused even higher reductions in the tensile strength (retained tensile strength of 46.6%) and elongation at the maximum load (reduction from 71.6% to 34.4%) of the geotextile.

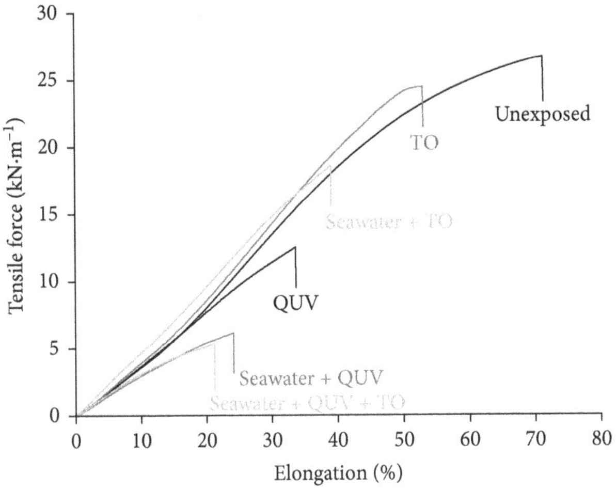

FIGURE 1: Mean curves "tensile force-elongation" obtained before and after the single and multiple exposure degradation tests.

The mean curves "tensile force-elongation" obtained for the geotextile before and after the thermooxidation and artificial weathering tests can be found in Figure 1. The analysis of those curves showed that, besides the changes in tensile strength and elongation at the maximum load, the exposure to thermooxidation also led to an increase in the stiffness of the geotextile. This increase in stiffness (and the reduction in elongation at the maximum load) may be related with the occurrence of some shrinkage (about 2.5%) in the geotextile during the exposure to thermooxidation, which caused a decrease in the deformability of the nonwoven structure. Contrarily to the thermooxidation test, the exposure to artificial weathering did not cause an increase in the stiffness of the geotextile (at lower elongations, the mean curves "tensile force-elongation" were practically coincident before and after artificial weathering). It is worthy to mention that the mean curves "tensile force-elongation" obtained after the immersion tests were omitted from Figure 1 for simplification purposes (those curves were much coincident with the curve presented for the unexposed sample, confirming the inexistence of relevant changes in the tensile behaviour of the geotextile).

The PP geotextile presented a good resistance against the effects of seawater and hydrolysis. The damage occurred during the artificial weathering tests (considerable deterioration of the tensile properties of the geotextile) can be mainly attributed to the effects of UV radiation (it is well known that UV radiation is one of the main degradation agents of PP products). The presence of a small amount of the HALS-type UV stabiliser (only 0.2%, percentage in weight) resulted in a substantial increase of the resistance of the PP geotextile against the effects of thermooxidation and UV radiation. Indeed, in the absence of the additive, the geotextile would be totally destructed (tensile strength reduced to zero) after the thermooxidation and artificial weathering tests [11].

3.2. Exposure to Consecutive Degradation Tests (Multiple Exposures)

3.2.1. Immersion Tests Followed by Thermooxidation. The successive exposure to seawater and thermooxidation (two agents that individually did not cause much relevant changes in the tensile strength of the geotextile) caused a decrease in tensile strength (reduction of 30.3%) (Table 4). In addition, a decrease also occurred in elongation at the maximum load (reduction higher than that observed after the single exposure to thermooxidation). The mean curve "tensile force-elongation" obtained after the successive exposure to seawater and thermooxidation can be seen in Figure 1. Besides the changes in the tensile properties reported above, an increase in stiffness was also observed (like before, the occurrence of some shrinkage in the geotextile during the thermooxidation test may have contributed for this increase).

TABLE 4: Tensile properties of the geotextile after the multiple exposure degradation tests.

Degradation test	T (kN·m^{-1})	E_{ML} (%)	$T_{residual}$ (%)	RF
Seawater + TO	18.56 (±2.35)	39.1 (±5.5)	69.7	1.44
H_2O + TO	25.01 (±2.27)	54.2 (±4.6)	93.9	1.07
NaCl + TO	15.51 (±2.58)	34.2 (±8.1)	58.2	1.72
Seawater + QUV	6.26 (±2.68)	24.0 (±3.7)	23.5	4.26
H_2O + QUV	12.31 (±1.43)	34.5 (±2.1)	46.2	2.16
NaCl + QUV	10.12 (±2.32)	30.4 (±4.1)	38.0	2.63
QUV + TO	12.30 (±1.19)	31.9 (±2.6)	46.2	2.17
Seawater + QUV + TO	5.39 (±1.60)	21.2 (±4.1)	20.2	4.94

The 95% confidence intervals are given in parentheses.

The deterioration of tensile strength in the successive exposure to seawater and thermooxidation shows that two agents that individually did not cause relevant damage to the geotextile (retained tensile strengths of 98.3% and 92.6%, resp., after the single exposure degradation tests), together led to some degradation. Therefore, the interactions that may occur between the different degradation agents must not be neglected when accessing the durability of geotextiles.

The reduction observed in tensile strength may have some possible explanations: (1) the remains of seawater accelerated the oxidation process, (2) occurrence of losses of the HALS-type UV stabiliser during the immersion in seawater, leaving the geotextile less protected against oxidation (as previously mentioned, the oxidative resistance of PP geotextiles can be highly enhanced by the presence of Chimassorb 944 [11]), and (3) the remains of seawater could have prevented the HALS-type UV stabiliser from performing efficiently its protective function against thermooxidation. In order to try to explain the reduction in tensile strength observed after the successive exposure to seawater and thermooxidation, two other combinations of degradation tests were performed: (1) immersion in deionised water followed by thermooxidation and (2) immersion in sodium chloride 35 g·L^{-1} followed by thermooxidation.

The tensile strength (after the thermooxidation tests) of the sample previously immersed in water was similar to the tensile strength of the sample without immersion (retained tensile strengths of 93.9% and 92.6%, resp.) (Table 4). Similarly to tensile strength, elongation at the maximum load also had no relevant differences (54.2% and 53.3%, resp.). Moreover, the mean curve "tensile force-elongation" obtained after the successive exposure to water and thermooxidation was much coincident with the curve presented in Figure 1 for the single exposure to thermooxidation. This way, the thermooxidative resistance of the geotextile was not affected by the previous immersion in water.

Contrarily to what happened with water, the immersion in sodium chloride 35 g·L^{-1} led to a reduction of the resistance of the geotextile against thermooxidation (Table 4). Indeed, the successive immersion in sodium chloride 35 g·L^{-1} and thermooxidation provoked relevant decreases in the tensile strength (retained tensile strength of 58.2%) and elongation at the maximum load (from 71.6% to 34.2%) of the geotextile. The reduction occurred in tensile strength was even more pronounced than that observed in the successive exposure to seawater and thermooxidation (reductions of 41.8% and 30.3%, resp.). The sample previously immersed in sodium chloride 35 g·L^{-1} also suffered an increase in stiffness after the thermooxidation tests. With the obvious differences due to lower tensile strength and elongation at the maximum load, the mean curve "tensile force-elongation" obtained after the successive immersion in sodium chloride 35 g·L^{-1} and thermooxidation was relatively similar to the curve obtained after the multiple exposure to seawater and thermooxidation and therefore is not presented in Figure 1.

The only difference between the solution of sodium chloride 35 g·L^{-1} and water was the presence, or not, of the salt. This way, and having into account that the immersion in water did not cause a decrease in the resistance of the geotextile against thermooxidation, the damage occurred after the successive immersion in sodium chloride 35 g·L^{-1} and thermooxidation can only be ascribed to the presence of sodium chloride. It is worthy to remember that the concentration of sodium chloride was intended to simulate the typical amount of salts (mostly sodium chloride) existent in seawater.

Within the previous hypotheses for explaining the reduction of the thermooxidative resistance of the geotextile caused by seawater, the influence of sodium chloride in the oxidation process (probably by acting as a catalyst) is the most probable one. The inexistence of any interaction between water and thermooxidation helps to support that sodium chloride had a relevant role in the degradation occurred in the multiple exposures to (1) seawater and thermooxidation and (2) sodium chloride 35 g·L^{-1} and thermooxidation.

The occurrence of significant losses of the HALS-type UV stabiliser during the immersion in sodium chloride 35 g·L^{-1} and seawater is unlikely. Indeed, there were no evidences of losses of the additive in the immersion in water (the geotextile had a good resistance against the multiple exposure to water and thermooxidation), and there are no reasons to presume that the additional presence of sodium chloride (and other salts in the case of seawater) could have promoted such losses.

Regarding the third hypothesis, no information was found in literature about the possible interaction between sodium chloride and the HALS-type UV stabiliser. The results obtained in this work indicate that sodium chloride had some influence in the thermooxidative degradation of the geotextile. However, further studies are needed to undoubtedly clarify the reasons for the interactions found in the multiple exposures to (1) seawater and thermooxidation and (2) sodium chloride 35 g·L^{-1} and thermooxidation. These studies are mainly related with polymer chemistry, falling outside the area of civil engineering.

3.2.2. Immersion Tests Followed by Artificial Weathering. The successive immersion in seawater and artificial weathering caused a considerable decrease in the tensile strength of the geotextile (reduction of 76.5%) (Table 4). This decrease was higher than the decrease that occurred after the single exposure to artificial weathering (reduction of 53.4%). The elongation at the maximum load (after the artificial weathering tests) was also lower for the sample previously immersed in seawater than for the sample without immersion (reductions from 71.6% to 24.0% and 34.4%, resp.). When compared with the unexposed sample or with the sample exposed uniquely to artificial weathering, the stiffness of the geotextile suffered a relevant decrease for elongations higher than 5% (Figure 1).

Having into account that the immersion in seawater (single exposure) had no relevant influence in the tensile properties of the geotextile, the degradation occurred in the multiple exposure to seawater and artificial weathering was higher than the sum of the individual effects of both single exposures. This way, and similarly to what happened with thermooxidation, an interaction also occurred between seawater and artificial weathering.

The hypotheses for explaining the interaction found between seawater and artificial weathering are the same as previously mentioned for the interaction observed between seawater and thermooxidation. As before, two more combinations of degradation tests were considered: (1) immersion in water followed by artificial weathering and (2) immersion in sodium chloride 35 g·L^{-1} followed by artificial weathering.

The immersion in water did not provoke a reduction of the weathering resistance of the geotextile (Table 4). Indeed, the sample previously immersed in water and the sample without immersion (exposed only to artificial weathering) had similar tensile properties after artificial weathering (retained tensile strengths of 46.2% and 46.6%, resp.). Moreover, the mean curve "tensile force-elongation" obtained after the successive exposure to water and artificial weathering was practically coincident with the curve obtained after the single exposure to artificial weathering (presented in Figure 1). Therefore, as the resistance against thermooxidation, the weathering resistance of the geotextile was not affected by the previous immersion in water.

Similarly to what happened when combined with thermooxidation, the immersion in sodium chloride 35 g·L^{-1} also led to a reduction of the resistance of the geotextile against artificial weathering (Table 4). However, the reduction occurred in tensile strength after the successive exposure to sodium chloride 35 g·L^{-1}, and artificial weathering (decrease of 62.0%) was significantly lower than the reduction observed after the successive exposure to seawater and artificial weathering (reduction of 76.5%).

Despite being a lower reduction, this indicates the existence of some effect of sodium chloride in the degradation process of the geotextile by artificial weathering.

The main degradation mechanism during artificial weathering was photooxidation and, as for thermooxidation, the influence of sodium chloride in the oxidative process (by acting as a catalyst) is the most likely hypothesis for explaining the higher degradation of the geotextile. The occurrence of significant losses of the HALS-type UV stabiliser during the immersion tests is unlikely (as discussed before in Section 3.2.1), and no data were found in the literature about possible interactions between sodium chloride and Chimassorb 944. Like for the interactions with thermooxidation, additional studies (mainly related with polymer chemistry) are needed to explain the effect of sodium chloride in the weathering process of the geotextile.

3.2.3. Artificial Weathering Followed by Thermooxidation. The multiple exposure to artificial weathering and thermooxidation induced a decrease in tensile strength identical to the single exposure to artificial weathering (retained tensile strengths of 46.2% and 46.6%, resp.) (Table 4). Similarly to what happened for tensile strength, there were also no relevant differences in elongation at the maximum load (31.9% and 34.4%, resp.). In addition, the mean curves "tensile force-elongation" obtained after the multiple exposure to artificial weathering and thermooxidation and after the single exposure to artificial weathering were almost coincident (for simplification purposes, the curve obtained after the multiple exposure is not represented in Figure 1). This showed that the subsequent thermooxidation test (after artificial weathering) did not cause additional damage to the geotextile (no interaction was found between artificial weathering and thermooxidation).

3.2.4. Successive Exposure to Seawater, Artificial Weathering, and Thermooxidation. The successive exposure to seawater, artificial weathering, and thermooxidation provoked the most pronounced changes in the tensile properties of the geotextile (reduction of 79.2% in tensile strength and reduction of elongation at the maximum load from 71.6% to 21.2%) (Table 4), showing the existence of extensive damage in the nonwoven structure. With the difference due to lower tensile strength and elongation at the maximum load, the mean curve "tensile force-elongation" obtained after the multiple exposure to seawater, artificial weathering, and thermooxidation was relatively similar to the curve obtained after the multiple exposure to seawater and artificial weathering (no relevant changes were found in stiffness for elongations up to about 20%) (Figure 1). When compared to the single exposures to seawater, artificial weathering, and thermooxidation, the stiffness of the geotextile suffered a considerable decrease for elongations higher than about 5%.

The presence of the additional thermooxidation test (when compared to the multiple exposure to seawater and artificial weathering) resulted in a slightly higher deterioration of the tensile properties of the geotextile. Since no interactions were found between artificial weathering and thermooxidation (Section 3.2.3), the further deterioration of the geotextile may be due to traces of sodium chloride (from seawater) that could still be present after the multiple exposure to seawater and artificial weathering (Section 3.2.1 showed the existence of an interaction between sodium chloride and thermooxidation; the possible reasons for such interaction have already been discussed before).

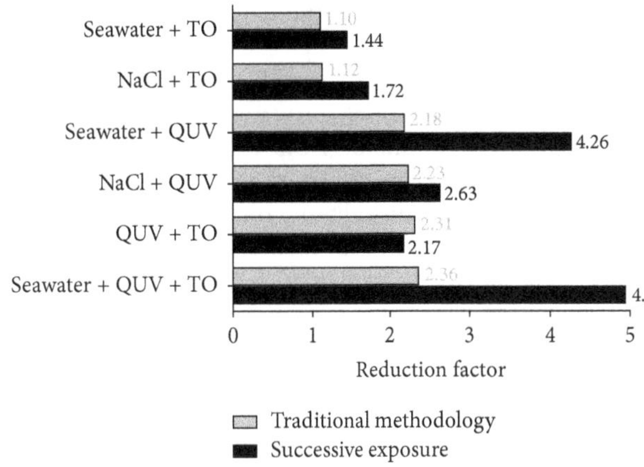

Figure 2: Comparison of the reduction factors obtained by the traditional methodology and by the successive exposure to the degradation agents.

3.3. Reduction Factors. The reduction factors obtained in the successive exposures to two or three degradation agents were compared with the reduction factors determined by the traditional methodology for the combined effect of those agents (determination of reduction factors in separate for each agent and further multiplication) (Figure 2). The reduction factors obtained directly from the single and multiple exposures carried out in this work can be found in Tables 3 and 4, respectively.

The reduction factor determined by the traditional methodology for the combined effect of artificial weathering and thermooxidation was similar (slightly higher) to that obtained in the successive exposure to both degradation agents (reduction factors of 2.31 and 2.17, resp.) (Figure 2). In this case, the traditional methodology provided a good approach to estimate the deterioration occurred in the tensile strength of the geotextile due to the combined effect of artificial weathering and thermooxidation (it is worthy to remember that no interaction was found between these degradation agents).

The reduction factors obtained in the successive exposures to (1) immersion in seawater and thermooxidation, (2) immersion in sodium chloride $35\,g{\cdot}L^{-1}$ and thermooxidation, (3) immersion in seawater and artificial weathering, and (4) immersion in sodium chloride $35\,g{\cdot}L^{-1}$ and artificial weathering were higher than those obtained by the traditional methodology for the combined effect of those degradation agents (Figure 2). For example, the reduction factor obtained in the successive immersion in seawater and artificial weathering (4.26) was about 2-times higher than the reduction factor (2.18) determined by multiplying the reduction factors obtained in the single exposures to seawater and to artificial weathering (1.02 and 2.14, resp.). The previous results indicate that the reduction factors provided by the traditional methodology were unable to represent correctly (by underestimating) the combined effect of the degradation agents. The differences found between the reduction factors (determined by the traditional methodology or obtained directly in the successive exposure to the degradation agents) can be explained by the interactions occurred between the degradation agents (which the traditional methodology is not being able to account for).

Similarly to the previous combinations of two degradation agents, the reduction factor obtained by the traditional methodology for the combined effect of three degradation agents (seawater, artificial weathering, and thermooxidation) was much lower (less than half) than the reduction factor obtained in the successive exposure to the agents (reduction factors of 2.36 and 4.94, resp.) (Figure 2). Once again, the reduction factors obtained by the traditional methodology underestimated the degradation caused by the combined effect of the degradation agents.

The reduction factors determined by the traditional methodology were lower than those obtained directly in the successive exposures to the degradation agents (exception for the combined effect of artificial weathering and thermooxidation, where no interactions were found between the agents). This showed that the multiplication of the reduction factors (each representing the isolated effect of a degradation agent) may not represent with accuracy the combined effect of the agents.

4. Conclusions

The damage (deterioration of tensile properties) suffered by a nonwoven PP geotextile in the successive exposure to degradation agents present in marine environments tended to be much different (higher) from the sum of the individual effects of those agents, showing the existence of interactions between them. Relevant interactions were found between (1) immersion in seawater and thermooxidation, (2) immersion in sodium chloride 35 $g{\cdot}L^{-1}$ and thermooxidation, (3) immersion in seawater and artificial weathering, (4) immersion in sodium chloride 35 $g{\cdot}L^{-1}$ and artificial weathering, and (5) immersion in seawater followed by artificial weathering and then by thermooxidation.

Some hypotheses have been advanced for trying to explain the interactions found between the degradation agents. The results showed that sodium chloride had a key influence in the thermo- or photooxidative process of the geotextile, probably by acting as a catalyst. However, further studies (mainly related with polymer chemistry) are needed to undoubtedly clarify the reasons for the interactions found in this work.

The identification and quantification of interactions between the different degradation agents is important to understand and predict the behaviour of the geotextiles under real conditions (where the agents rarely act in isolation). Recognizing such interactions can help understanding the degradation process of the materials and may enable the development of strategies for enhancing their durability and improving their performance.

The reduction factors obtained directly in the multiple exposures to the degradation agents tended to be different from those calculated by the traditional methodology (determination of reduction factors in separate for each agent and further multiplication) for the combined effect of those agents. The traditional methodology was unable to represent accurately (underestimating) the damage caused by the combined effect of the degradation agents, being incapable of accounting for the interactions occurred between the agents. The most relevant examples of this occurred (1) in the successive immersion in seawater and artificial weathering and (2) in the successive immersion in seawater, followed by artificial weathering and then by thermooxidation. In both cases, the reduction factors obtained by the traditional methodology were much lower (about half) than those obtained directly in the successive exposures to the degradation agents.

The definition of reliable reduction factors (having into consideration the interactions that may occur between the degradation agents) can contribute for a better application of geosynthetics. Indeed, the definition of accurate reduction factors allows a better design and provides a higher guarantee of durability.

Finally, it is worthy to remember that the reduction factors presented in this work correspond to particular degradation conditions and cannot be generalized or applied directly in the design. No studies were carried out in order to find the relation between field conditions and the degradation conditions imposed to the geotextile.

Abbreviations and symbols

α:	Confidence level
CMD:	Cross-machine direction of production
E_{ML}:	Elongation at the maximum load
f_s:	Factor of safety
HALS:	Hindered amine light stabiliser
MD:	Machine direction of production
n:	Number of specimens
O_{90}:	Characteristic opening size
PP:	Polypropylene
RF:	Reduction factor
$RF_{agent\ A}$:	Reduction factor for agent A
$RF_{agent\ B}$:	Reduction factor for agent B
$RF_{agents\ A+B\ trad}$:	Reduction factor for the combined effect of agents A and B (traditional methodology)
RF_C:	Reduction factor for creep
RF_{CB}:	Reduction factor for chemical and biological agents
RF_{ID}:	Reduction factor for installation damage
RF_W:	Reduction factor for weathering
s:	Sample standard deviation
t:	Student's t-distribution value
T:	Tensile strength
T_D:	Design tensile strength
$T_{damaged}$:	Tensile strength of exposed samples
$T_{reference}$:	Tensile strength of reference samples (unexposed)

$T_{residual}$: Retained tensile strength
TO: Thermooxidation
μ: Population mean
UV: Ultraviolet
VI_{H50}: Velocity index for a head loss of 50 mm
x: Sample mean.

Conflicts of Interest

The authors declare that they have no conflicts of interest.

Acknowledgments

The authors would like to thank "Carvalhos Lda." (Lousã, Portugal) for supplying the geotextile studied in this work. This work was financially supported by the Project POCI-01-0145-FEDER-007457, funded by FEDER funds through COMPETE 2020—"Programa Operacional Competitividade e Internacionalização" (POCI) and by national funds through "FCT—Fundação para a Ciência e a Tecnologia". José Ricardo Carneiro would also like to thank FCT for the research grant SFRH/BPD/88730/2012 (grant supported by POPH/POCH/FSE funding).

References

[1] K. W. Pilarczyk, *Geosynthetics and Geosystems in Hydraulic and Coastal Engineering*, A.A. Balkema, Rotterdam, Netherlands, 1st edition, 2000.

[2] A. Bezuijen and E. W. Vastenburg, *Geosystems: Design Rules and Applications*, CRC Press/Balkema, Leiden, Netherlands, 1st edition, 2013.

[3] T. S. Ingold, *The Geotextiles and Geomembranes Manual*, Elsevier Advanced Technology, Oxford, UK, 1st edition, 1994.

[4] R. M. Koerner, *Designing with Geosynthetics*, Prentice-Hall, Upper Saddle River, NJ, USA, 1999.

[5] C. Hsieh, J.-B. Wang, and Y. F. Chiu, "Weathering properties of geotextiles in ocean environments," *Geosynthetics International*, vol. 13, no. 5, pp. 210–217, 2006.

[6] S. K. Shukla and J.-H. Yin, *Fundamentals of Geosynthetics Engineering*, Taylor & Francis/Balkema, Leiden, Netherlands, 2006.

[7] J. R. Carneiro, M. Morais, and M. L. Lopes, "Degradation of polypropylene geotextiles with different chemical stabilisations in marine environments," *Construction and Building Materials*, vol. 165, pp. 877–886, 2018.

[8] C. Maier and T. Calafut, *Polypropylene—the Definitive User's Guide and Databook*, Plastic Design Library, New York, NY, USA, 1998.

[9] D. Feldman, "Polymer weathering: photo-oxidation," *Journal of Polymers and the Environment*, vol. 10, no. 4, pp. 163–173, 2002.

[10] J. H. Greenwood, H. F. Schroeder, and W. Voskamp, *Durability of Geosynthetics*, CRC Press, Boca Raton, FL, USA, 2nd edition, 2016.

[11] J. R. Carneiro, *Durability of Geosynthetics in Environmental Structures—the Importance of Chemical Additives*, Ph.D. thesis in Environmental Engineering, University of Porto, Porto, Portugal, 2009, in Portuguese.

[12] L. D. Suits and Y. G. Hsuan, "Assessing the photo-degradation of geosynthetics by outdoor exposure and laboratory weatherometer," *Geotextiles and Geomembranes*, vol. 21, no. 2, pp. 111–122, 2003.

[13] J. R. Carneiro, P. J. Almeida, and M. L. Lopes, "Accelerated weathering of polypropylene geotextiles," *Science and Engineering of Composite Materials*, vol. 18, no. 4, pp. 241–245, 2011.

[14] J. R. Carneiro and M. L. Lopes, "Natural weathering of polypropylene geotextiles treated with different chemical stabilisers," *Geosynthetics International*, vol. 24, no. 6, pp. 544–553, 2017.

[15] EN 14030, *Geotextiles and Geotextile-Related Products-Screening Test Method for Determining the Resistance to Acid and Alkaline Liquids*, European Committee for Standardization, Brussels, Belgium, 2001.

[16] EN 12447, *Geotextiles and Geotextile-Related Products-Screening Test Method for Determining the Resistance to Hydrolysis in Water*, European Committee for Standardization, Brussels, Belgium, 2001.

[17] EN ISO 13438, *Geotextiles and Geotextile-Related Products-Screening Test Method for Determining the Resistance to Oxidation*, European Committee for Standardization, Brussels, Belgium, 2004.

[18] EN 12224, *Geotextiles and Geotextile-Related Products-Determination of the Resistance to Weathering*, European Committee for Standardization, Brussels, Belgium, 2000.

[19] EN 12226, *Geosynthetics—General Tests for Evaluation following Durability Testing*, European Committee for Standardization, Brussels, Belgium, 2012.

[20] S. P. Moreira, J. R. Carneiro, and M. L. Lopes, "Effect of natural weathering on the water permeability behaviour of nonwoven polypropylene geotextiles," in *Proceedings of the 9th International Conference on Geosynthetics*, vol. 2, pp. 829–832, Guarujá, Brazil, May 2010.

[21] J. R. Carneiro, P. J. Almeida, and M. L. Lopes, "Durability of polypropylene geotextiles exposed to artificial weathering conditions," in *Proceedings of the 5th International Congress on Environmental Geotechnics*, vol. 1, pp. 399–406, Cardiff, UK, 2006.

[22] X. Yang and X. Ding, "Prediction of outdoor weathering performance of polypropylene filaments by accelerated weathering tests," *Geotextiles and Geomembranes*, vol. 24, no. 2, pp. 103–109, 2006.

[23] I. M. Valente, J. R. Carneiro, P. J. Almeida, and M. L. Lopes, "Chemical studies about the durability of polypropylene geotextiles," in *Proceedings of the 9th International Conference on Geosynthetics*, vol. 2, pp. 785–788, Guarujá, Brazil, May 2010.

[24] I. M. Valente, J. R. Carneiro, P. J. Almeida, and M. L. Lopes, "Determination of Chimassorb 944 in polypropylene geotextiles by HPLC-UV," *Analytical Letters*, vol. 44, no. 4, pp. 617–625, 2011.

[25] ISO/TR 20432, *Guidelines for the Determination of the Long-Term Strength of Geosynthetics for Soil Reinforcement*, International Organization for Standardization, Geneva, Switzerland, 2007.

[26] D. M. Carlos, J. R. Carneiro, M. Pinho-Lopes, and M. L. Lopes, "Effect of soil grain size distribution on the mechanical damage of nonwoven geotextiles under repeated loading," *International Journal of Geosynthetics and Ground Engineering*, vol. 1, no. 1, p. 7, 2015.

[27] J. R. Carneiro, P. J. Almeida, and M. L. Lopes, "Some synergisms in the laboratory degradation of a polypropylene geotextile," *Construction and Building Materials*, vol. 73, pp. 586–591, 2014.

[28] J. R. Carneiro and M. L. Lopes, "Resistance of nonwoven geotextiles against mechanical damage under repeated loading and abrasion," in *Proceedings of the 6th European Geosynthetics Congress*, pp. 875–883, Ljubljana, Slovenia, September 2016.

[29] M. Dias, J. R. Carneiro, and M. L. Lopes, "Resistance of a nonwoven geotextile against mechanical damage and abrasion," *Ciência e Tecnologia dos Materiais*, vol. 29, no. 1, pp. 177–181, 2017.

[30] A. Rosete, P. M. Lopes, M. Pinho-Lopes, and M. L. Lopes, "Tensile and hydraulic properties of geosynthetics after mechanical damage and abrasion laboratory tests," *Geosynthetics International*, vol. 20, no. 5, pp. 358–374, 2013.

[31] EN ISO 9864, *Geosynthetics—Test Method for the Determination of Mass per Unit Area of Geotextiles and Geotextile-Related Products*, European Committee for Standardization, Brussels, Belgium, 2005.

[32] EN ISO 9863-1, *Geosynthetics—Determination of Thickness at Specified Pressures, Part 1: Single Layers*, European Committee for Standardization, Brussels, Belgium, 2005.

[33] EN 29073-3, *Textiles—Test Methods for Nonwovens, Part 3: Determination of Tensile Strength and Elongation*, European Committee for Standardization, Brussels, Belgium, 1992.

[34] EN ISO 12236, *Geosynthetics—Static Puncture Test (CBR Test)*, European Committee for Standardization, Brussels, Belgium, 2006.

[35] EN ISO 11058, *Geotextiles and Geotextile Related Products—Determination of Water Permeability Characteristics Normal to the Plane, without Load*, European Committee for Standardization, Brussels, Belgium, 2010.

[36] EN ISO 12956, *Geotextiles and Geotextile Related Products—Determination of the Characteristic Opening Size*, European Committee for Standardization, Brussels, Belgium, 2010.

[37] EN ISO 9862, *Geosynthetics-Sampling and Preparation of Test Specimens*, European Committee for Standardization, Brussels, Belgium, 2005.

[38] D. C. Montgomery and G. C. Runger, *Applied Statistics and Probability for Engineers*, John Wiley & Sons, Hoboken, NJ, USA, 5th edition, 2010.

19

Maillard Reaction in Natural Rubber Latex: Characterization and Physical Properties of Solid Natural Rubber

S. Montha,[1] P. Suwandittakul,[1] A. Poonsrisawat,[2] P. Oungeun,[3] and C. Kongkaew[1]

[1] *MTEC, National Science and Technology Development Agency, 114 Thailand Science Park, Phahonyothin Road, Khlong Nueng, Khlong Luang, Pathum Thani 12120, Thailand*
[2] *BIOTEC, National Science and Technology Development Agency, 113 Thailand Science Park, Phahonyothin Road, Khlong Nueng, Khlong Luang, Pathum Thani 12120, Thailand*
[3] *Department of Chemistry, Faculty of Science, Naresuan University, Phitsanulok 65000, Thailand*

Correspondence should be addressed to C. Kongkaew; chaveer@mtec.or.th

Academic Editor: Luigi Nicolais

Maillard reaction in Natural Rubber (NR) latex was investigated by treating fresh NR latex with glutaraldehyde ($C_5H_8O_2$) in amounts of 0, 50, 100, and 200 mmol/kg of latex. Protein cross-linking in fresh NR latex and solid NR was confirmed by using sodium dodecylsulfate polyacrylamide gel electrophoresis (SDS-PAGE) and attenuated total reflection infrared (ATR-IR) spectroscopy, respectively. It was found that degree of protein cross-linking in NR increased with increasing $C_5H_8O_2$ concentration. Physical properties of untreated and treated NR substances in terms of gel content, initial Wallace plasticity (P_0), plasticity retention index (PRI), Mooney viscosity, and tensile strength were carefully explored. Results clearly showed that the Maillard cross-linking of proteins had remarkable effect on bulk NR properties, that is, solvent resistance, hardness, resistance to oxidation, rheological behavior, and resistance to stretching out.

1. Introduction

Natural Rubber (NR), the naturally occurring elastomer in the form of latex obtained from the *Hevea brasiliensis* tree, is an industrially significant polymer. Its various applications and products cover the fields of medicine, laboratory, chemistry, technology, and engineering, either alone or in combination with other materials. Currently, rubber is harvested mainly in the form of the latex from certain trees by making incisions into the bark and collecting the fluid in vessels in a process called "tapping." Fresh NR latex is composed of about 25–35% rubber fraction and 5% nonrubber components, with water accounting for the remaining [1, 2]. The solid NR is chemically 94% *cis*-1,4-polyisoprene and contained 6% nonrubber components such as lipids, proteins, carbohydrates and inorganic salts as well [3]. The in-depth structural analysis revealed that each NR molecule contains 2-*trans*-isoprene units connected to a long-chain of *cis*-isoprene and two terminal groups. These terminal groups are joined with mono- and/or diphosphate groups of some phospholipids at one position and with a dimethyl allyl group of proteins at the other by hydrogen bonding. It is well established that rubber molecules in NR are composed of long-chain branched structure [4]. Proteins in NR were presumed to form branch-points via hydrogen bonding between the rubber chains. Proteins in NR play an important role in controlling NR properties [5]. The accelerated degradation of saponified skim rubber due to the decrease of nitrogenous substances acting as antioxidants after saponification was also reported [6]. Unpurified NR presents shorter cure time than purified NR because there are some phospholipids and proteins acting as natural accelerators for curing reaction [7]. Maillard reaction is a type of nonenzymatic browning which involves the reaction of carbonyl groups and amines of proteins. The chemistry underlying the Maillard reaction is very complex because a wide range of reaction products are formed [8]. Maillard reaction plays an important role in food processing which results in the formation of aroma, taste, and color and appearance in foods [9–11]. The improved thermal stability of modified soy protein concentrate resin could be attributed to the cross-links formed by glutaraldehyde with soy protein concentrate [12, 13]. Maillard reaction also plays important role in reducing allergenic protein and increasing mechanical properties of NR products by introducing banana skin powder in NR latex due to its good sources of nutrients where Maillard reaction occurred in cross-link with the proteins in NR latex [14]. The formation of brownish compounds by Maillard reaction occurred in NR containing glucose after vulcanization process due to the presence of *N*-cyclohexylbenzothiazole-2-sulfenamide (CBS), a nitrogenous accelerator [15]. In this work, proteins in fresh NR latex were cross-linked by direct application of the Maillard reaction in fresh NR latex before the solid NR was prepared by evaporation technique. Protein cross-linking in fresh NR latex and solid NR was characterized. Changes of physical properties of solid NR in terms of gel content, initial Wallace plasticity, plasticity retention index, Mooney viscosity, and tensile strength were disclosed.

2. Materials and Methods

2.1. Materials. Glutaraldehyde (25% solution), ammonium hydroxide (25% solution), and toluene were purchased from Sigma-Aldrich. The original pH of glutaraldehyde ($C_5H_8O_2$) was adjusted to be a neutral pH by using potassium carbonate (K_2CO_3) before being used. Fresh NR latex was obtained from rubber plantation in Rayong Province, Thailand. Total solid content (TSC), dry rubber content (DRC), and nonrubber content (NRC) of fresh NR latex, determined according to ISO 2004, were 38.16%, 35.79%, and 2.37%, respectively.

2.2. Sample Preparation. Glutaraldehyde ($C_5H_8O_2$) was added to fresh NR latex in amounts of 0, 50, 100, and 200 mmol/kg of latex. The latex mixture was continuously stirred for 1 h at room temperature and then added with ammonium hydroxide (NH_4OH) solution to keep the alkalinity of the latex at 0.7%. Solid NR films were prepared by casting the latex mixture on the glass plate and then putting it in an oven for 48 h at 60°C. Solid NR sheets were also prepared by passing solid NR films through the gap between the rolls of laboratory mill. The rolls were cooled with the running water at room temperature and the final sheet thickness was 1.6 to 1.8 mm.

2.3. Characterization. Protein profiles in the latex mixture were analyzed by using sodium dodecylsulfate polyacrylamide gel electrophoresis (SDS-PAGE) according to Laemmli [16]. Proteins were separated from fresh NR latex by using 15% polyacrylamide gel electrophoresis with a Mini-PROTEAN II cell (Bio-Rad, Hercules, CA) at a constant current of 15 mA per gel for 80 min. After separation, the protein bands were stained with Coomassie Brilliant Blue R-250 and then destained with a solution containing of glacial acetic acid, methanol, and water. A prestained protein marker (14.4–116.0 kDa) was used as molecular weight standard. Protein cross-linking in NR films was also examined by attenuated total reflection infrared (ATR-IR) spectroscopy according to ASTM: E573-01. NR film was put on the sample holder and then was examined using ATR-IR technique. The measurement conditions were as follows: crystal: ZnSe, angle: 45 degrees, light source: middle range infrared (4000–650 cm^{-1}), resolution: 4 cm^{-1}, and detector: TGS. Changes in peak area at wavenumbers 1,480–1,575 cm^{-1}, referring to amide band (C-N stretching and N-H bending), were recorded.

2.4. Physical Properties. Gel content in NR was evaluated according to ISO/TC 45/SC 3 N 1166. About 0.1 g of NR sheets with approximately 1 mm^3 size was immersed into 30 cm^3 of toluene in the centrifuge tube. The tube was capped and shaken by hand for a few seconds and then allowed to stand for 16 to 20 h without stirring at 25°C. After completion of soaking the sample in the solvent, the tube was shaken by hand vertically for 60 sec to disperse the jelly-like precipitate at the bottom and then placed in the centrifuge machine at 14,000 rpm for 2 h. The tube was removed from the machine and the liquid was pipetted from the tube. About 1 mL of acetone was added to the tube to precipitate the gel fraction. The gel fraction was then peeled off and dried for 1 h at 110°C. The procedure was repeated 2 times for each sample. Gel content was given by the expression

$$\text{Gel content (\%)} = \left(\frac{m_1}{m_0}\right) \times 100, \quad (1)$$

where m_0 is the mass of the original test piece and m_1 is the mass of the dried precipitant.

Plasticity of NR was determined by using Wallace rapid plastimeter. The method used to measure the plasticity of NR was modified from the method of ASTM D3194-02. The test pieces were cut from the solid NR sheets with the Wallace punch. The test pieces were inserted into Wallace ageing chamber for 15 min at 140°C, removed, and allowed to cool at room temperature. Plasticity of the unaged sample (P_0) and plasticity of aged sample (P_{15}) were obtained by measuring the thickness change of the unaged and aged test pieces, respectively, before and after they were compressed under constant compressive force of 100 N at 100°C for 15 sec. The median of three pieces was taken as the plasticity value. The plasticity relation index (PRI) was calculated using the expression

$$\text{PRI} = \left(\frac{P_{15}}{P_0}\right) \times 100, \quad (2)$$

where P_0 is the initial Wallace plasticity of the unaged test piece and P_{15} is the plasticity of the aged test piece after ageing for 15 min at 140°C.

Mooney viscosity (MS1 + 4) of NR was examined by using TECHPRO Mooney viscometer. The temperature of testing was 100 ± 1°C and the small size rotor was used. The NR sheets were preheated at 100°C for 1 min with continuous shear for 4 min. The measurement was repeated 3 times for each sample.

Tensile strength of NR was determined by using an Instron universal testing machine model 5566 at room

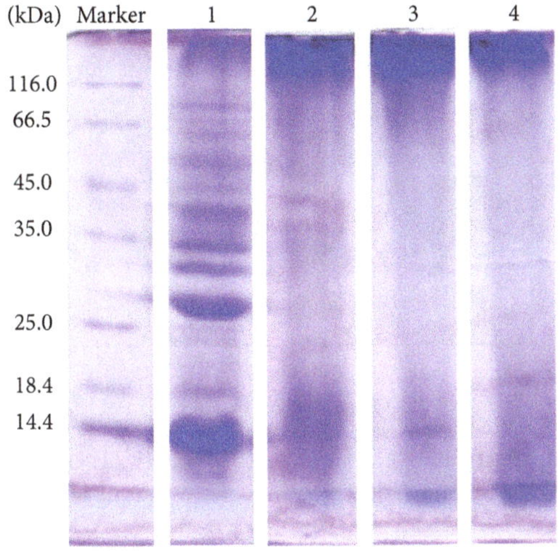

FIGURE 1: SDS-PAGE analysis of protein profiles: (1) untreated fresh NR latex, (2–4) fresh NR latex treated with $C_5H_8O_2$ 50, 100, and 200 mmol/kg of latex.

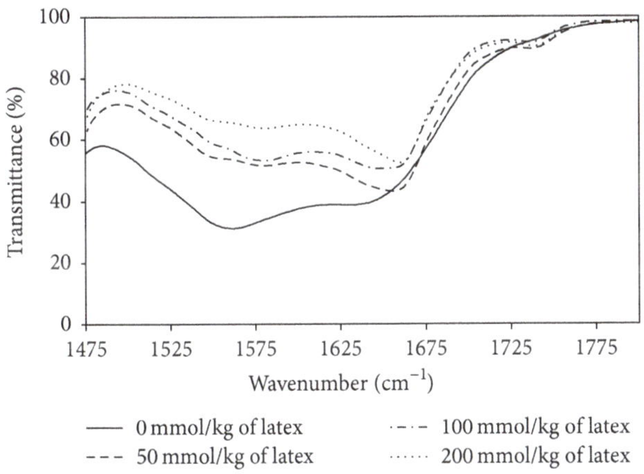

FIGURE 2: ATR-IR spectra of NR films obtained from untreated and treated fresh NR latex.

temperature. The testing crosshead speed of 500 mm/min and the load cell of 1 kg were used. The NR films were stamped out using a type C dumbbell die according to ASTM D412-97. The thickness of test pieces was 0.5–1.5 mm. Five pieces of each sample were used for the test.

3. Results and Discussion

3.1. Characterization of Protein Cross-Linking. Protein profiles in fresh NR latex samples are shown in Figure 1. Lane 1 showed proteins in untreated fresh NR latex containing molecular weight from 14.4 to 116 kDa. It revealed that major protein bands were disappeared after adding $C_5H_8O_2$ to fresh NR latex, suggesting cross-linking of proteins to products with higher molecular weight [9–11, 17–21].

Protein cross-linking was also confirmed in NR films by ATR-IR spectroscopy. The ATR-IR spectra of NR films obtained from untreated and treated fresh NR latex are shown in Figure 2. The most distinctive spectral features for proteins are the strong amide I and amide II bands. Amide I band (1,600–1,690 cm^{-1}) is referring to C=O stretching and amide II band (1,480–1,575 cm^{-1}) is corresponding to C-N stretching and N-H bending [20]. As shown in Table 1, changes in peak area of amide II band at wavenumber 1,561 cm^{-1} in the ATR-IR spectra of the NR films were observed after treating fresh NR latex with $C_5H_8O_2$. Peak area of amide II band was continuously decreased with increasing $C_5H_8O_2$ concentration. It was explained that the aldehyde groups of $C_5H_8O_2$ might combine with nitrogen atoms of proteins, forming Michael addition to double bonds with numerous terminal formyl groups and quaternary pyridinium cross-link [17–19]. Mechanism of protein cross-linking by $C_5H_8O_2$ is shown in Figure 3. The data from SDS-PAGE technique and ATR-IR spectroscopy could be used to confirm the Maillard cross-linking of proteins in the fresh NR latex.

TABLE 1: Peak area of amide II band at wavenumber 1,561 cm^{-1} in ATR-IR spectra of NR films obtained from untreated and treated fresh NR latex.

$C_5H_8O_2$ [mmol/kg of latex]	Peak area [A·cm^{-1}]
0	10.24
50	2.88
100	1.81
200	1.11

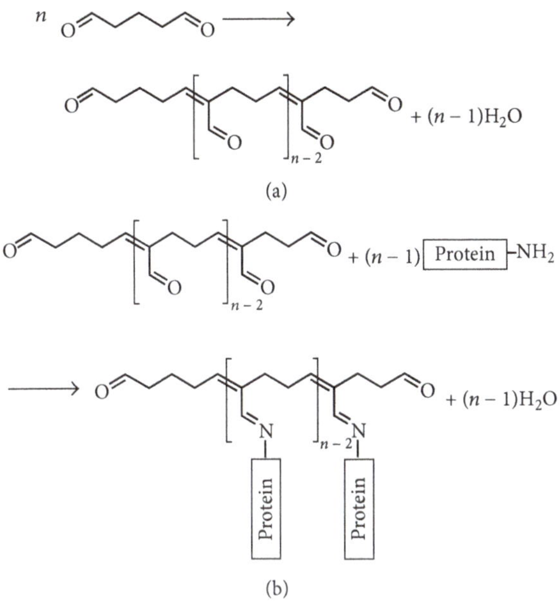

FIGURE 3: Proposed mechanisms for cross-linking of proteins by $C_5H_8O_2$ via Maillard reaction [9], glutaraldehyde.

3.2. Physical Properties of NR. The role of Maillard protein cross-linking in the physical properties of NR was investigated through gel content, initial Wallace plasticity, plasticity retention index, Mooney viscosity, and tensile strength of NR.

Figure 4 shows gel content in NR substances obtained from untreated fresh NR latex and the fresh NR latex treated with different concentrations of $C_5H_8O_2$. In contrast to

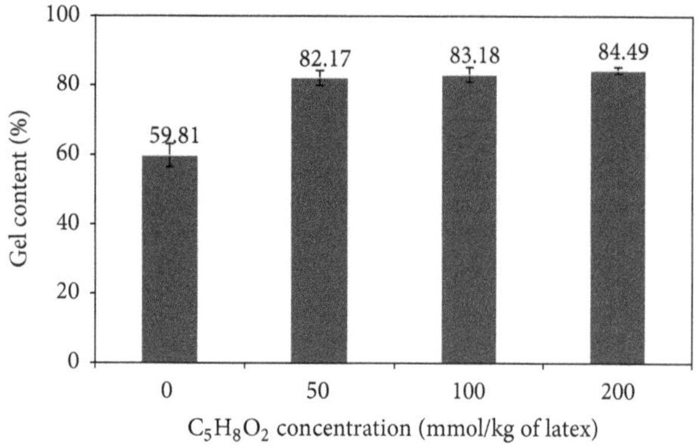

FIGURE 4: Gel content in NR obtained from the fresh NR latex treated with different concentrations of $C_5H_8O_2$.

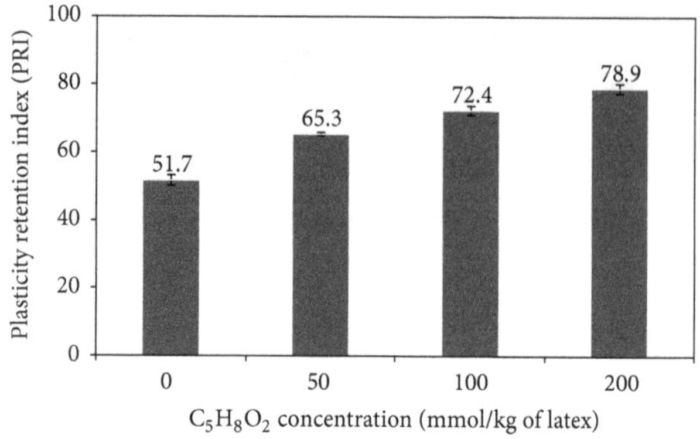

FIGURE 6: Plasticity retention index (PRI) of NR obtained from the fresh NR latex treated with different concentrations of $C_5H_8O_2$.

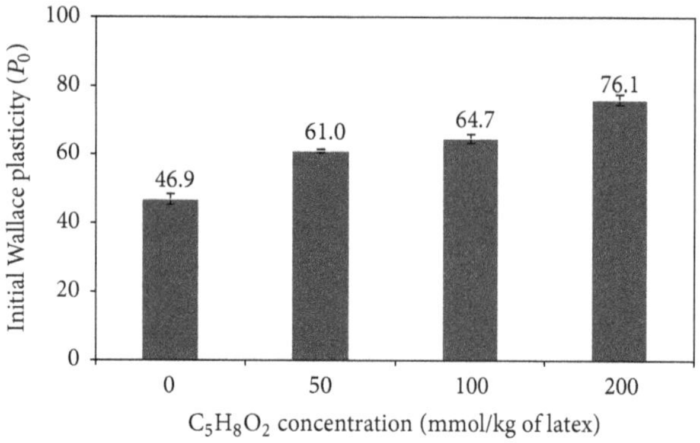

FIGURE 5: Initial Wallace plasticity (P_0) of NR obtained from the fresh NR latex treated with different concentrations of $C_5H_8O_2$.

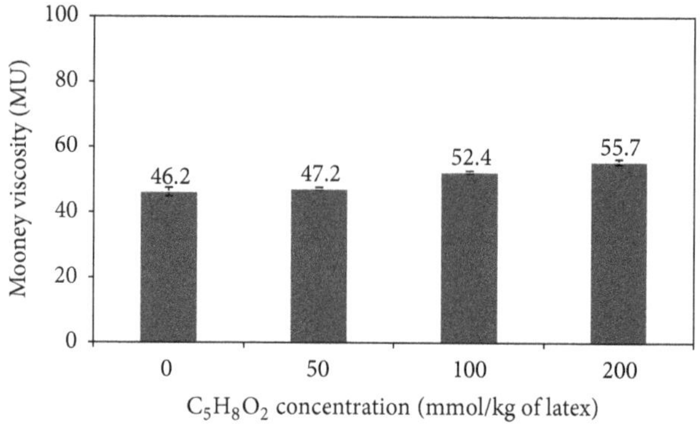

FIGURE 7: Mooney viscosity of NR obtained from the fresh NR latex treated with different concentrations of $C_5H_8O_2$.

untreated NR, treated NR substances contained more gel content values. This implied that treated NR substances were hardly dissolved in a solvent (toluene). It has long been recognized that there are two components called sol and gel in NR [4, 21, 22]. The sol fraction dissolves easily in a good solvent such as toluene, cyclohexane, and tetrahydrofuran. The gel fraction swells without dissolving. From the result in Figure 4, it might be due to the fact that two or more NR molecules might be linked together through cross-linking of proteins at chain-ends of NR molecules (as evidenced by SDS-PAGE technique and ATR-IR spectroscopy), resulting in an increase of molecular weight of NR. It is widely accepted that solubility of NR in a solvent is mainly dependent on its molecular weight [23]. The higher molecular weight values, the lower solubility of NR. However, gel content in treated NR substances was slightly increased with increasing $C_5H_8O_2$ concentrations. It was remarkable that applying Maillard reaction in fresh NR latex could provide NR with more resistantce to solvent solubility.

Figures 5 and 6 present the change of plasticity of NR obtained from untreated fresh NR latex and fresh latex treated with different concentrations of $C_5H_8O_2$. It is found from Figure 5 that initial Wallace plasticity (P_0) of treated NR substances was higher than that of untreated NR. P_0 values tended to increase continuously with rising $C_5H_8O_2$ concentration. This indicated that protein cross-linking in the fresh NR latex led to obtaining NR with more rigid property. In fact, cross-linking of proteins might occur in proteins adsorbed on NR particles as well as proteins dissolved in water phase through the Maillard reaction. Hence, it might lead to the formation of network structure of proteins in NR matrix [24–26], as evidenced by SDS-PAGE technique and ATR-IR spectroscopy. These protein networks might act as rigid-like materials, providing a hardener NR [24]. Interestingly, it is found from Figure 6 that plasticity retention index (PRI) values of treated NR substances were higher than that of untreated NR. The PRI values tended to increase continuously with rising $C_5H_8O_2$ concentration. Normally, PRI is an indicator to evaluate the resistance of raw NR to oxidation [27]. The PRI value has been used to control the quality of NR. The higher PRI values, the better grade of NR. Hence, cross-linking of proteins in fresh NR latex by the Maillard reaction could provide higher resistance of NR to oxidation or lower oxidizability of NR [28, 29].

Change in Mooney viscosity of NR substances received from untreated fresh NR latex and NR latex treated with different concentrations of $C_5H_8O_2$ is shown in Figure 7. Normally, Mooney viscosity is extensively used to measure the viscosity of raw NR. It is found from Figure 7 that Mooney viscosity of NR increased continuously with increasing $C_5H_8O_2$ concentration. This indicated that networks of cross-linked proteins in NR matrix via Maillard reaction,

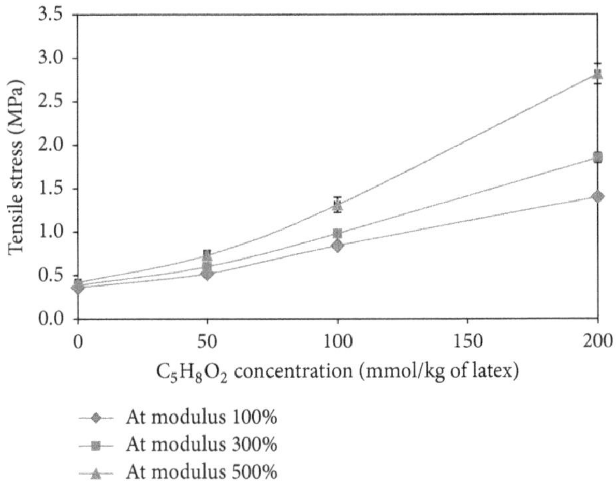

FIGURE 8: Tensile stress at 100%, 300%, and 500% modulus of NR obtained from the fresh NR latex treated with different concentrations of $C_5H_8O_2$.

as confirmed by SDS-PAGE and ATR-IR techniques, might obstruct the flow of NR molecules. The higher degree of protein cross-linking, the harder flow of NR molecules. Mooney viscosity of NR has generally depended on molecular weight and molecular weight distribution, molecular structure such as stereochemistry and branching, and nonrubber constituents [30]. It should be noted that Maillard cross-linking of proteins could affect the rheological property of raw NR.

The effect of Maillard cross-linking of proteins in fresh NR latex on the tensile strength of NR is represented in terms of tensile stress at 100%, 300%, and 500% modulus, as shown in Figure 8. Obviously, the tensile strength was found to increase with increasing $C_5H_8O_2$ concentration. It was thought that two or more NR molecules might be connected together through cross-linking of proteins at chain-ends of NR molecules (as evidenced by SDS-PAGE technique), resulting in an increase of molecular weight of NR. It is well-known that tensile strength of NR is mostly dependent on its molecular weight [31–33]. The higher molecular weight values, the more tensile strength of NR. From this finding, it should be reported that the Maillard cross-linking of proteins could provide NR with more resistance to stretching out. Normally, sulfur cross-linking of NR molecules is needed for the NR product manufacturing (such as tires, shoe-soles, and conveyor belts) in order to obtain NR products with high performance. Unfortunately, the chemicals used in the sulfur cross-linking (such as sulfur, accelerators, and activators) are quite toxic to human and environment. Since the Maillard cross-linking of proteins in fresh NR latex could increase the partial strength of solid NR, it might help to reduce the amount of toxic chemicals used for sulfur cross-linking in the product manufacturing.

4. Conclusions

Maillard reaction was successfully done in fresh NR latex by using $C_5H_8O_2$ as Maillard reagent. SDS-PAGE technique and ATR-IR spectroscopy confirmed the protein cross-linking in NR. The disappearance of major protein bands was noticed in treated NR samples, suggesting cross-linking of proteins to products with higher molecular weight. Reduction of the peak area of amide II band (C-N stretching and N-H bending) at wavenumber 1,561 cm^{-1} in ATR-IR spectra of NR films was observed. Physical properties of solid NR were changed after applying the Maillard reaction in fresh NR latex. In contrast to untreated NR, treated NR substances provided more gel content values. Initial Wallace plasticity (P_0) and plasticity retention index (PRI) values of treated NR substances were higher than those of untreated NR. Mooney viscosity of NR increased continuously with increasing $C_5H_8O_2$ concentration. Tensile stress at 100%, 300%, and 500% modulus was obviously found to increase with increasing $C_5H_8O_2$ concentration. From these findings, it could be inferred that Maillard cross-linking of proteins in fresh NR latex had the remarkable effect on bulk NR properties, that is, solvent resistance, hardness, resistance to oxidation, rheological behavior, and resistance to stretching out.

Competing Interests

The authors declare that they have no competing interests.

Acknowledgments

The authors would like to thank Cluster and Programme Management Office (CPMO) for financial support and National Metal and Materials Technology Center (MTEC) for instrumental support.

References

[1] D. C. Blackley, *Polymer Latices: Science and Technology*, Chapman & Hall, London, UK, 2nd edition, 1997.

[2] B. L. Archer and B. C. Sekhar, "The proteins of *Hevea brasiliensis* latex: I. Protein constituents of fresh latex serum," *The Biochemical Journal*, vol. 61, no. 3, pp. 503–508, 1955.

[3] P. W. Allen and G. F. Bloomfield, *Chemistry and Physics of Rubber-Like Substance*, Edited by L. Bateman, Maclaren and Sons, London, UK, 1963.

[4] J. T. Sakdapipanich and P. Rojruthai, "Molecular structure of natural rubber and its characteristics based on recent evidence," in *Biotechnology—Molecular Studies and Novel Applications for Improved Quality of Human Life*, R. Sammour, Ed., InTech, 2012.

[5] J. Tangpakdee and Y. Tanaka, "Characterization of sol and gel in hevea natural rubber," *Rubber Chemistry and Technology*, vol. 70, no. 5, pp. 707–713, 1997.

[6] S. Tuampoemsab, J. Sakdapipanich, and Y. Tanaka, "Influence of some non-rubber components on aging behavior of purified natural rubber," *Rubber Chemistry and Technology*, vol. 80, no. 1, pp. 159–168, 2007.

[7] C. Nakason, A. Thitithammawong, and S. Wisunthorn, "Effect of non-rubber components on properties of sulphur crosslinked natural rubbers," *Advanced Materials Research*, vol. 844, pp. 345–348, 2013.

[8] J. E. Hodge, "Dehydrated foods: chemistry of browning reactions in model systems," *Journal of Agricultural and Food Chemistry*, vol. 1, no. 15, pp. 928–943, 1953.

[9] J. A. Gerrard, P. K. Brown, and S. E. Fayle, "Maillard crosslinking of food proteins I: the reaction of glutaraldehyde, formaldehyde and glyceraldehyde with ribonuclease," *Food Chemistry*, vol. 79, no. 3, pp. 343–349, 2002.

[10] J. A. Gerrard, P. K. Brown, and S. E. Fayle, "Maillard crosslinking of food proteins II: the reactions of glutaraldehyde, formaldehyde and glyceraldehyde with wheat proteins in vitro and in situ," *Food Chemistry*, vol. 80, no. 1, pp. 35–43, 2003.

[11] J. A. Gerrard, P. K. Brown, and S. E. Fayle, "Maillard crosslinking of food proteins III: the effects of glutaraldehyde, formaldehyde and glyceraldehyde upon bread and croissants," *Food Chemistry*, vol. 80, no. 1, pp. 45–50, 2003.

[12] S. Chabba and A. N. Netravali, "'Green' composites part 1: characterization of flax fabric and glutaraldehyde modified soy protein concentrate composites," *Journal of Materials Science*, vol. 40, no. 23, pp. 6263–6273, 2005.

[13] S. Chabba and A. N. Netravali, "'Green' composites part 2: characterization of flax yarn and glutaraldehyde/poly(vinyl alcohol) modified soy protein concentrate composites," *Journal of Materials Science*, vol. 40, no. 23, pp. 6275–6282, 2005.

[14] T. M. Goh, *Film behaviour of natural rubber latex treated with banana skin [Ph.D. thesis]*, Universiti Malaysia Pahang, 2012, http://iportal.ump.edu.my/lib/item?id=chamo:76344&.

[15] A. Nimpaiboon and J. Sakdapipanich, "A model study on effect of glucose on the basic characteristics and physical properties of natural rubber," *Polymer Testing*, vol. 32, no. 8, pp. 1408–1416, 2013.

[16] U. K. Laemmli, "Cleavage of structural proteins during the assembly of the head of bacteriophage T4," *Nature*, vol. 227, no. 5259, pp. 680–685, 1970.

[17] W.-Q. Wang, Y.-H. Bao, and Y. Chen, "Characteristics and antioxidant activity of water-soluble Maillard reaction products from interactions in a whey protein isolate and sugars system," *Food Chemistry*, vol. 139, no. 1–4, pp. 355–361, 2013.

[18] F.-L. Gu, J. M. Kim, S. Abbas, X.-M. Zhang, S.-Q. Xia, and Z.-X. Chen, "Structure and antioxidant activity of high molecular weight Maillard reaction products from casein-glucose," *Food Chemistry*, vol. 120, no. 2, pp. 505–511, 2010.

[19] Q. Liu, B. Kong, J. Han, C. Sun, and P. Li, "Structure and antioxidant activity of whey protein isolate conjugated with glucose via the Maillard reaction under dry-heating conditions," *Food Structure*, vol. 1, no. 2, pp. 145–154, 2014.

[20] J. Kong and S. Yu, "Fourier transform infrared spectroscopic analysis of protein secondary structures," *Acta Biochimica et Biophysica Sinica*, vol. 39, no. 8, pp. 549–559, 2007.

[21] J. Tangpakdee and Y. Tanaka, "Characterization of sol and gel in Hevea natural rubber," *Rubber Chemistry and Technology*, vol. 70, no. 5, pp. 707–713, 1997.

[22] L. Tarachiwin, J. Sakdapipanich, and Y. Tanaka, "Gel formation in natural rubber latex: 1. Effect of $(NH_4)_2HPO_4$ and TMTD/ZnO additives," *Rubber Chemistry and Technology*, vol. 76, no. 5, pp. 1177–1184, 2003.

[23] P. W. Allen and G. M. Bristow, "The gel phase in natural rubber," *Journal of Applied Polymer Science*, vol. 7, no. 2, pp. 603–615, 1963.

[24] B. C. Sekhar, "Abnormal groups in rubber and microgel," in *Proceedings of the 4th Rubber Technology Conference*, pp. 460–469, London, UK, 1962.

[25] I. Migneault, C. Dartiguenave, M. J. Bertrand, and K. C. Waldron, "Glutaraldehyde: behavior in aqueous solution, reaction with proteins, and application to enzyme crosslinking," *BioTechniques*, vol. 37, no. 5, pp. 790–802, 2004.

[26] M. J. Spotti, M. J. Perduca, A. Piagentini, L. G. Santiago, A. C. Rubiolo, and C. R. Carrara, "Gel mechanical properties of milk whey protein-dextran conjugates obtained by Maillard reaction," *Food Hydrocolloids*, vol. 31, no. 1, pp. 26–32, 2013.

[27] H. C. Baker, "Effects of process aids on plasticity," *Journal of Applied Polymer Science*, vol. 13, pp. 1201–1207, 1980.

[28] P. Argos, M. G. Rossmann, U. M. Grau, H. Zuber, G. Frank, and J. D. Tratschin, "Thermal stability and protein structure," *Biochemistry*, vol. 18, no. 25, pp. 5698–5703, 1979.

[29] J. A. Gerrard, "Protein-protein crosslinking in food: methods, consequences, applications," *Trends in Food Science and Technology*, vol. 13, no. 12, pp. 391–399, 2002.

[30] J. Maláč, "Mooney viscosity, mooney elasticity and procesability of raw natural rubber," *Journal of Materials Science and Engineering with Advanced Technology*, vol. 3, no. 1, pp. 67–87, 2011.

[31] F. H. Axtell, P. Phinyocheep, and P. Monthachitra, "The influence of viscosity ratio on the morphology and mechanical properties of polyamide 6/natural rubber blends," *Journal of the Science Society of Thailand*, vol. 18, pp. 195–215, 1992.

[32] J. S. Dick, *Compounding Materials for the Polymer Industries*, Noyes Publications, Upper Saddle River, NJ, USA, 1987.

[33] S. Matsuoka, *Relaxation Phenomena in Polymers*, Hanser, Munich, Germany, 1992.

20

Electrospun Polyurethane/Loess Powder Hybrids and Their Absorption of Volatile Organic Compounds

Jun Cong Ge, Jeong Hyeon Kim, and Nag Jung Choi

Division of Mechanical Design Engineering, Chonbuk National University, 567 Baekje-daero, Jeonju-si, Jeollabuk-do 561-756, Republic of Korea

Correspondence should be addressed to Nag Jung Choi; njchoi@jbnu.ac.kr

Academic Editor: Antonio Riveiro

We investigated the effects of composite electrospun polyurethane (PU)/loess powder (LP) fibers for absorption of volatile organic compounds (VOCs) from air. PU films containing different amounts of LP (0, 10, 30, and 50 wt% LP with respect to PU) nanoparticles (NPs) were analyzed using FE-SEM, FTIR, and XRD experimental analyzers. Electron microscopy and spectroscopy indicated that the proper content of LP NPs is homogeneous dispersion in a polymer matrix. In addition, 50 wt% of LP results in a higher concentration spinning nanosuspension that leads to some agglomeration on the film surface. The chloroform, benzene, and toluene (VOCs) absorption capacity of PU/LP composite films showed a trend of toluene > benzene > chloroform. The highest VOC absorption capacity was found with PU composite mats containing 30 wt% LP NPs. The high VOC absorption capacity of PU/LP composite films can be used to remove VOCs from a new house or car.

1. Introduction

Electrospinning is an important method used to produce nanofibers with polymer solutions [1–4]. In the last 10 years, electrospinning has become one of the most important academic techniques in the fields of materials and technology. Because of its easy manufacturing process, low pollution, and low cost, electrospinning has been widely used in many fields. For example, electrospinning has been used with artificial skin, bandages, bulletproof clothing, battery electrolytes, sensors, and other biomaterials [5–11]. The simple one-step electrospinning process is fast developing in the past two decades. In one way, different types of complicated electrospinning processes include coaxial electrospinning [12], side-by-side electrospinning [13], and triaxial electrospinning [14]. In another way, different kinds of working fluids have been tried on these processes, including polymeric solutions, codissolving solutions of polymers and the functional ingredients, microemulsions, and also nanosuspensions. Electrospinning is a fiber spinning technology that injects charged polymer solution through a syringe pump under a high-voltage electrostatic field to form fibers. Under voltage in the nozzle, macromolecule droplets are affected by droplet surface tension. Surface tension direction is directed opposite to the electric field force, and the electric field induces charge at the macromolecule droplet surface. A Taylor cone is created when voltage slowly increases, causing macromolecule droplets to stretch from spherical to long and tapered. When the electric field strength increases to a critical value, the surface tension of the liquid is overcome by electrostatic force, and the liquid is ejected from the Taylor cone. With a strong electric field, the ejected liquid fluid is elongated, and a large amount of injected solvent evaporates, thereby causing a nanoscale fiber to form on the collector [15–17].

The nanometer fibers and nanometer fiber membranes produced by electrospinning technology have large specific surface areas and porosity, enabling good performance in many fields. These nonwoven fabrics have recently emerged as effective membranes for removing harmful water and airborne contaminants from the environment [18–21]. Filter media can protect people from both natural and man-made contaminants by filtering or absorbing hazardous materials. The main outdoor sources of volatile organic compounds (VOCs) are fuel combustion, transport of industrial waste gas such as car exhaust, and photochemical pollution. Indoors, VOCs mainly come from coal and natural gas combustion products, such as smoking, heating, and cooking, building and decoration materials, furniture, and cleaners [22, 23]. The harm of VOCs is obvious, as they cause headache, nausea, vomiting, and limb weakness. In severe cases when VOC concentration exceeds a certain threshold, convulsions, coma, and memory loss can result. Many researchers are interested in the absorption of VOCs by electrospun nanofibers. Scholten et al. [24] reported that electrospun polyurethane (PU) fiber can be effectively used to absorb VOCs from air. PU fibers demonstrated completely reversible absorption and desorption, with desorption obtained by simple purging with nitrogen at room temperature. Kim et al. [23] showed that PU fibers with 30 wt% fly ash had the highest VOC absorption capacity, which was 2.52–2.79 times greater than that of pristine PU fibers.

Porous clay materials are attractive candidate adsorbents for VOC removal [25-27]. Loess powder (LP) is used to purify water and can be used to adsorb many kinds of heavy metals and organic compounds from water. LP mainly consists of silicate, ferric oxide, and aluminum oxide; is a low cost, nontoxic, environmentally friendly material; and can be easily obtained, making it suitable for use as an adsorbent [28–32]. Some research has shown that clay can be used as an additive to load into polymeric fibers by electrospinning, but information on the absorption of VOCs is limited. In this study, LP is used as an adsorbent loaded into PU fibers by electrospinning to adsorb VOCs. This study also aims to investigate the possibility of LP utilization as an adsorbent for the removal of VOCs from air. This presents a new direction in the fields of environmental science and technology.

2. Experimental Details

2.1. Materials. PU pellets (Estane® Skythane® X595A-11, Lubrizol Advanced Materials, Inc., USA) and loess powder (chemical composition: 58.8% SiO_2, 21.0% Al_2O_3, 8.34% Fe_2O_3, 0.04% CaO, 0.28% MgO, 1.00% TiO_2, 2.35% K_2O, and 0.11% Na_2O; ig. loss: 8.1%) were used to produce composite nanofibers. The loess powder was obtained from HONGIK BIO TECH Co., Ltd. (Damyang-gun, Jeollanam-do, Republic of Korea). N,N-Dimethylformamide (DMF; Showa Chemical Co., Ltd., Japan) and methyl ethyl ketone (2-butanone) (MEK, extra pure; Junsei, Japan) were used as received without further purification. In VOC absorption experiments, three different kinds of gas (chloroform, benzene, and toluene) (purity 99.9%, AR grade) were analyzed.

2.2. Fabrication of Composite Electrospun Mats. PU pellets were dried for at least 3 h in a dry oven at 80°C prior to dissolution in solvents. A 10 wt% PU solution in DMF/MEK (50 : 50 by weight) was prepared using a magnetic stirrer at 300 rpm for 12 h at room temperature, and different amounts of LP (0, 10, 30, and 50 wt% with respect to PU) were added to create a spinning nanosuspension using ultrasonication for 2 h. This was followed by 2 h magnetic stirring at 300 rpm under room temperature. Electrospinning was carried out using the setup shown in Figure 1. A total of 15 kV of high-voltage electricity (New Era Pump Systems, Inc., NY, USA)

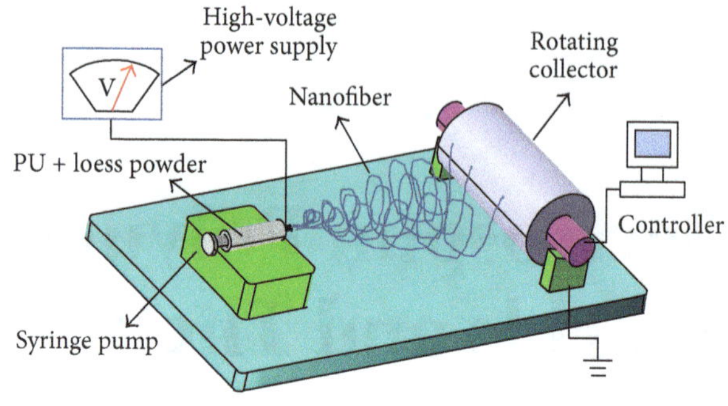

FIGURE 1: Schematic diagram of the electrospinning setup.

was applied to the PU/LP spinning nanosuspension in a syringe for electrospinning. The tip-to-collector distance was 18 cm, and the nanosuspension feed rate was 1 mL/h. Finally, nanofibrous mats were dried under vacuum for 12 h at 60°C after electrospinning.

2.3. Characterization. The morphological structures of nanofibrous mats were characterized using field emission scanning electron microscopy (FE-SEM, JIB-4601F, JEOL, Japan) and biotransmission electron microscopy (Bio-TEM, H-7650, Hitachi, Japan). Fiber diameter was determined on 100 filaments or particles using ImageJ (NIH, USA) software. The Fourier transform infrared spectroscopy (FTIR, Spectrum GX, Perkin Elmer, USA) spectra of samples were obtained using a Paragon 1000 spectrometer. X-ray diffraction (XRD) analysis was carried out using a Multi-Purpose High Performance X-ray Diffractometer (X'pert Powder, PANalytical, Netherlands). Nanosuspension viscosity was measured using an EC meter (Brookfield, USA). An AMETEK tensile test machine (AMETEK, USA) measured the tensile properties of samples.

2.4. VOCs Absorption Experiment. For the VOC absorption experiment, pristine and composite electrospun PU matrices (0 wt% LP, 10 wt% LP, 30 wt% LP, and 50 wt% LP) were composed of 8 cm × 8 cm fibers. The absorption experiment was implemented in a closed container with two valves. For every experiment, a rotary pump was used to remove impurities and other contaminants on the fiber surface and container. First, 100 mg each of chloroform, benzene, and toluene were mixed. Then, methanol was used to dilute the nanosuspension to result in 2 mg of VOCs. A syringe was used to inject 2 mg of VOCs into the container and onto fiber surfaces, which were then allowed to evaporate for 1 h at room temperature. After that, unabsorbed VOCs were placed into a Tenax absorber (Tenax-GR; Japan Analytical Industry, Tokyo, Japan) using a nitrogen stream for 5 minutes. In this study, quantitative and qualitative analyses were performed using GC/MS (gas chromatography-mass spectroscopy, XEVO TQ-S, Waters, USA). A MS/FID in a purge and trap analyzer evaluated the Tenax-GR at 280°C. VOCs were purged and trapped using Automated Purge & Trap analyzer JTD-505II (Japan

TABLE 1: Physical properties of PU/LP spinning solutions.

Spinning solution	Viscosity (cP)	Fiber diameter (nm) Distribution	Mean	BET area (m^2/g)	Tensile strength (MPa)
PU + LP 0 wt%	547	290–987	636	5.9933	7.2 ± 2.31
PU + LP 10 wt%	552	290–993	526	1.3481	1.1 ± 0.05
PU + LP 30 wt%	595	184–618	343	2.6032	0.5 ± 0.03
PU + LP 50 wt%	830	172–711	368	2.2095	—

Analytical Industry, Japan). Gas composition was analyzed by AGC/MS QP 2010 Plus (Shimadzu, Kyoto, Japan). More specific VOC absorption experimental methods are available in our previous article [23].

3. Results and Discussion

3.1. Morphological Characteristics of PU/LP Composite Fibers. In this study, the loess consisted of typical elements, such as 50% Si, 25% Al, 15% Fe, and 5% K, and small amounts of Ti, Mg, and Ba. Figure 2(a) shows FE-SEM images of LP, which was found to have irregular shapes with different sizes and diameters. Table 1 shows the physical properties of PU solution/mats containing different amounts of LP. Table 1 shows that the fibrous membranes had a wide diameter distribution. With increased content of LP, the fiber diameter clearly decreased. The smallest fiber diameter was produced when LP content increased to 30 wt%, and the average diameter was about 343 nm. Similar results have also been observed by other researchers [33–35]. However, when the LP content exceeded 30 wt%, there was an increase in fiber diameter. This might be because the LP highly dissolved into PU solution to reach saturation when LP content increased to 30 wt%.

FE-SEM images of PU/LP nanofibrous membranes with increased LP content are shown in Figure 2. Figures 2(b)–2(e) show PU electrospun fibers containing 0, 10, 30, and 50 wt% LP, respectively. Figure 2(b) is an FE-SEM image of PU solution without LP, showing a smooth and nonporous surface compared with those in Figures 2(c)–2(e). Different LP content dissolved in PU produced smooth or not smooth fiber surfaces. The smooth fiber surfaces produced by LP did not incorporate through electrospinning. However, the not smooth fiber surfaces produced by LP were highly disordered with agglomerated LP. Figure 2 also shows that fiber diameter decreased as LP content increased from 0 wt% to 30 wt%. When LP content exceeded 30 wt%, as shown in Figure 2(e), many LP particles gathered together and connected with PU, disturbing the fiber morphology. The increased viscosity of spinning solutions and gathering of LP particles at the nozzle results in increased fiber diameter and can lead to nozzle blockage during electrospinning. Figures 2(b)–2(e) show that PU solution with LP up to 30 wt% was spinnable, and 30 wt% LP dissolved in PU solution was a suitable amount for effective particle incorporation without disturbing fiber morphology through electrospinning. These figures also clearly show that the smoothness of fibers and homogeneous distribution of LP were excellent when LP content increased to 30 wt%.

TEM images of PU/LP nanofibrous membranes are shown in Figure 3. Figure 3(a) is a pristine PU nanofiber that is beadless, cylindrical, and smooth. When adding the LP NPs into PU solution to produce composite fibers, shown in Figures 3(b)–3(d), LP NPs were either inside PU nanofibers (see Figure 3(c)) or on the surfaces of fibers (see Figures 3(b) and 3(d)). LP NPs appeared inside the fiber because the LP content was less than the 50 wt% of LP in the PU solution, where there was a good degree of mixing, and the viscosity was not high. That is, the nanosuspension did not require significant energy during electrospinning. However, when increasing LP content to 50 wt%, the viscosity of the PU/LP electrospinning nanosuspension became high. Electrospinning requires more energy to overcome surface tension, leading to more agglomeration of LP NPs and bead formation on PU nanofibers [20]. Agglomeration can also reduce the mechanical properties of composite nanofibers.

3.2. Mechanical Properties of PU/LP Composite Fibers. Loess has a good humidity-adjusting function and a far infrared ray radiation function, making it a very beneficial material for the human body. In cosmetics, loess has a whitening effect and strong affinity for the skin. It also can increase skin flexibility and waste discharge. In this study, loess was chosen for absorption of VOCs as a research material by adding it to electrospun fibers. We were interested in whether loess could improve the mechanical properties of PU fibers and act as a commercial air filter textile nanofibrous membrane with high VOC absorption efficiency. Table 1 and Figure 4 show the tensile strength of electrospun nanofibrous mats. The tensile strength of pristine PU nanofibrous mat was 7.2 ± 2.31 MPa, while composite PU electrospun fibers contained 10 and 30 wt% LP with 1.1 ± 0.05 and 0.5 ± 0.03 MPa, respectively. However, there is no measured tensile strength of composite PU electrospun fibers containing 50 wt% LP. This is because its tensile strength was very small, and the fibers broke when performing the tensile strength test. It is clear from these results that tensile strength drastically decreased with increased amounts of LP in fibers. The tensile strength of composite PU electrospun fibers containing 10 and 30 wt% LP decreased by 85% and 93% compared with pristine PU electrospun fibers, respectively. With increased LP content in PU fibers, the mechanical properties of PU/LP composite mats were greatly reduced. Under the existing state without adding any tensile strength additives, PU/LP composite fibers produced by electrospinning would be difficult to use as a commercial air filter.

To analyze the composition of electrospun nanofibers and investigate possible interactions between PU and LP

(a) LP

(b) PU + LP 0 wt%

(c) PU + LP 10 wt%

(d) PU + LP 30 wt%

(e) PU + LP 50 wt%

Figure 2: FE-SEM images of PU/LP electrospun fibers containing 0, 10, 30, and 50 wt% LP.

NPs, the XRD and FTIR spectra of nanofibrous mats were investigated. The XRD patterns of nanofibrous mats are shown in Figure 5, which shows that, with increasing amount of LP in fibers, the relative intensity gradually increased. This finding demonstrates effective loading of LP NPs on the fiber surface. The patterns of all nanofibrous mats had different crystalline characteristic peaks, as shown in Figure 5. The γ-phase crystalline characteristic peaks of all nanofibrous mats are shown at a 2θ of about 21°. The γ-phase crystalline characteristic peaks appeared at almost the same location, indicating very little interaction between PU and LP NPs, similar to findings reported by Pant et al. [20]. The FTIR patterns of nanofibrous mats are shown in Figure 6, which shows many peaks. Bands at 3347, 2973, 2920, 1752, 1588, 1232, and 964 cm^{-1} of neat PU nanofibrous mats were assigned to stretching of the N–H group, CH_2 asymmetric vibration, CH_3 symmetric vibration, stretching of the carbonyl group in hard segments, urethane amide II band, C(O)–O–C stretching of the hard segment, and bending vibration in a benzene ring, respectively [36]. These results are similar to those of our previous research [23]. In addition, the IR characteristic absorption peaks of PU/LP composite nanofibrous mats with different LP contents appeared in almost the same positions as those of neat PU nanofibrous mats. However, with increasing amount of LP in fibers, the bands shifted toward higher values, indicating molecular interactions between PU and LP NPs [34, 37, 38].

3.3. Absorption Characteristics of Different Nanofibrous Mats. This study mainly investigates PU/LP composite nanofibrous mats produced by electrospinning for VOCs adsorption. Loess is an easy to obtain material with low cost that can be utilized for various applications. With excellent VOC absorption capacity, it can be widely used to remove harmful VOCs in new houses and new cars. The VOC absorption characteristics of composite PU mats containing different amounts of LP are shown in Figure 7. With increasing amount of LP, the VOC absorption capacity of PU/LP composite nanofibrous

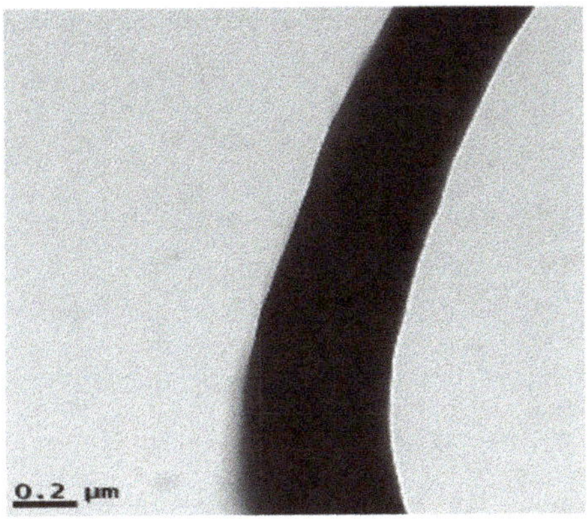

(a) PU + LP 0 wt%

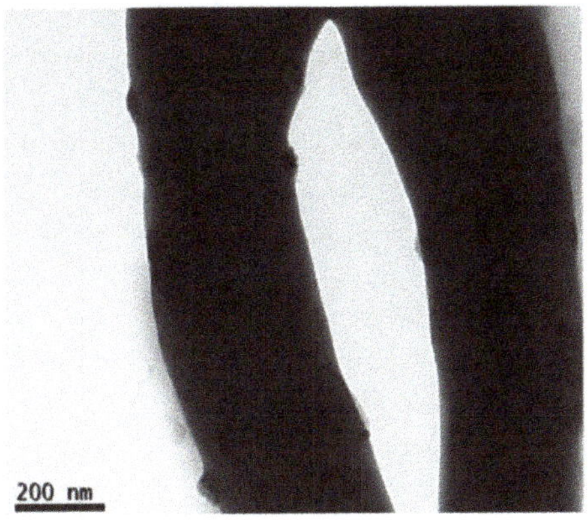

(b) PU + LP 10 wt%

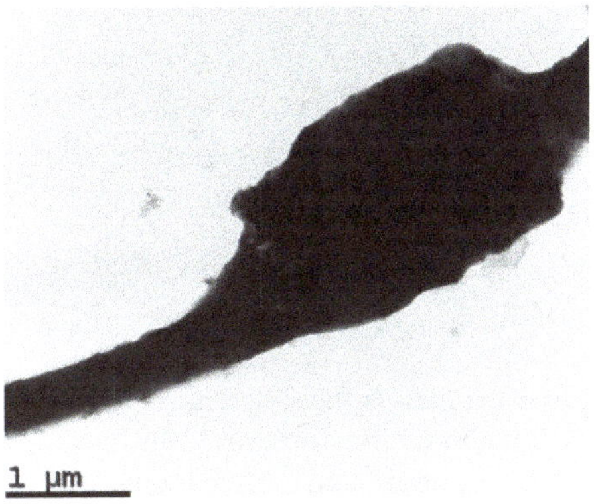

(c) PU + LP 30 wt%

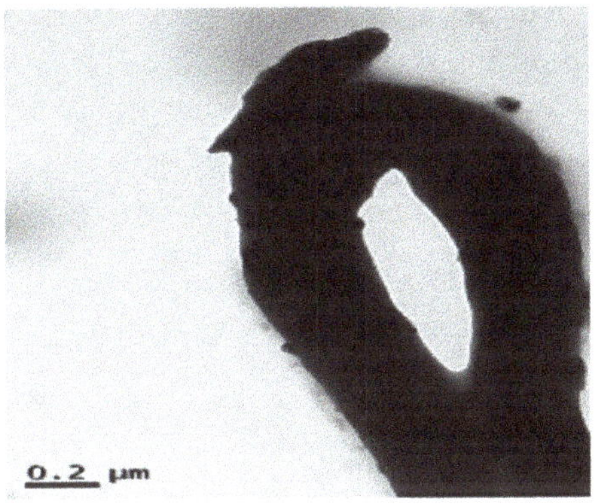

(d) PU + LP 50 wt%

FIGURE 3: TEM images of PU/LP electrospun fibers containing 0, 10, 30, and 50 wt% LP.

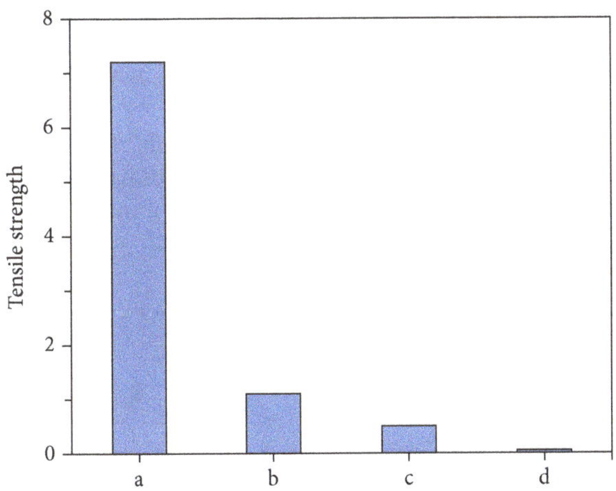

FIGURE 4: Tensile strength of electrospun PU mats with different amounts of LP.

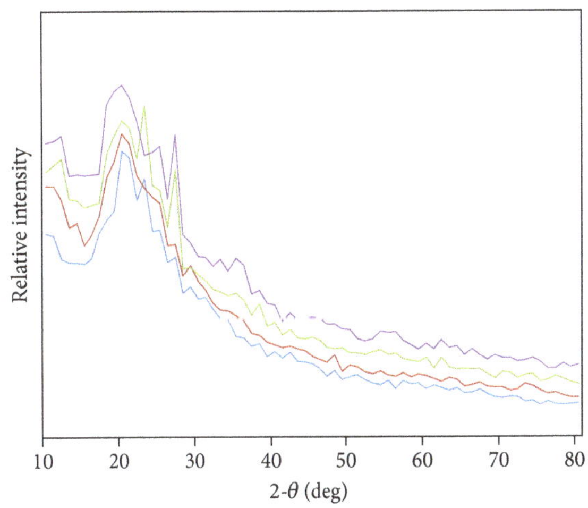

FIGURE 5: XRD patterns of different nanofibrous mats.

mats increased compared to that of neat PU composite nanofibrous mats. These results can be explained by the decreasing fiber diameter of composite fibers with increasing amount of LP, increasing the surface area of composite fibers. The LP NPs present on fiber surfaces increased the absorption capacity. Neat PU composite nanofibrous mats were also able to absorb VOCs [24]. The VOC absorption capacity of PU/LP nanofibers containing 10 wt%, 30 wt%, and 50 wt% LP NPs increased by 55%, 300%, and 192%, respectively. The

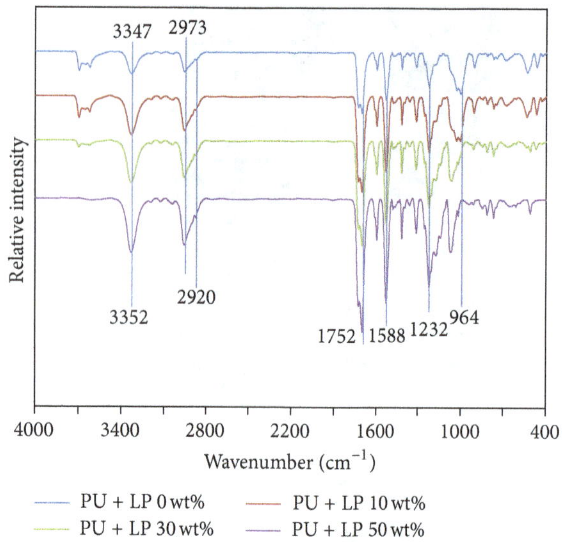

FIGURE 6: FTIR spectra of different nanofibrous mats.

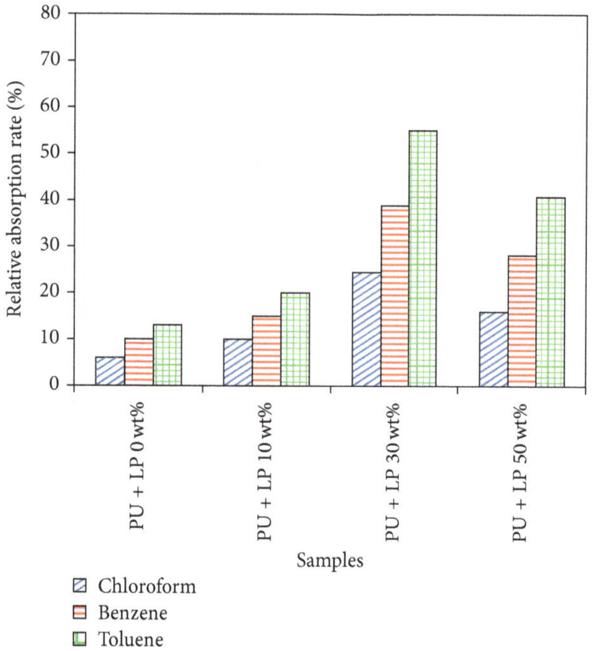

FIGURE 7: VOC absorption capacity of different nanofibrous mats.

VOC absorption capacity of PU mats containing 30 wt% LP NPs was the highest because this concentration produced maximum LP-loaded smooth fibers and high BET surface area [23]. The BET surface area is shown in Table 1. When the concentration of LP in PU spinning nanosuspension was sufficiently large, extensive agglomeration of LP particles occurred, preventing homogeneous LP NP distribution in PU film. Therefore, a high concentration of LP will decrease VOC absorption capacity [22]. The VOC absorption capacity of PU/LP composite nanofibrous films showed a trend of toluene > benzene > chloroform. Kim et al. [23] reported similar results. PU/LP composite nanofibrous films have

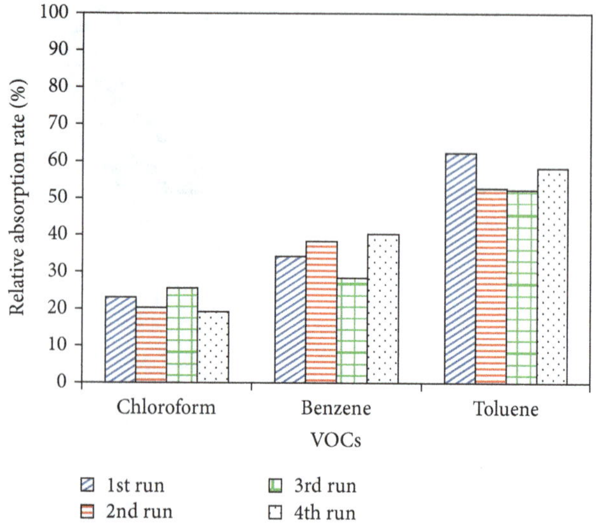

FIGURE 8: VOC absorption capacity of 30 wt% LPs containing PU film in different cycles.

different trends for different VOCs, which is related to the formation of π-complexes [39]. If the π-complex of PU/LP films is larger and more stable, the adsorption capacity of the surface for VOCs increases. Chloroform is not an aromatic compound, and its lack of π-complex formation leads to decreased VOC absorption capacity. Thus, its VOC absorption capacity is the smallest of the tested compounds. In addition, the adsorption capacity of a surface for VOCs is related to the ionization potential of the donor molecule. High ionization potential prevents charge transfer complexes from forming [40, 41]. This property also explains the trend of toluene > benzene > chloroform.

To further demonstrate the reliability of the VOC absorption capacity of PU/LP films, this VOC absorption experiment was repeated four times using PU films containing 30 wt% LP. Figure 8 shows the results, which are nearly the same for all four cycles. This reliability also indicates that the PU/LP composite film had reversible absorption and desorption behavior.

4. Conclusions

In this study, we report that PU/LP composite films can be produced by electrospinning. FE-SEM and TEM images indicated that 10 and 30 wt% of LP NPs homogeneously disperse in a polymer matrix, while 50 wt% of LP forms some agglomeration on the film surface. With increased amount of LP, the viscosity of the spinning nanosuspension increased, also increasing the possibility that the nozzle will be blocked. With increased amount of LP, nanofiber diameter decreased and BET surface area increased. The highest VOC absorption capacity of PU/LP films was that of a film containing 30 wt% LP. Due to its high VOC absorption capacity, this film can be used for VOC absorption in new houses and cars.

Competing Interests

The authors declare that there are no competing interests regarding the publication of this paper.

Acknowledgments

This research was supported by the Basic Science Research Program through the National Research Foundation of Korea (NRF) funded by the Ministry of Education (Project no. 2014059723).

References

[1] Z.-M. Huang, Y.-Z. Zhang, M. Kotaki, and S. Ramakrishna, "A review on polymer nanofibers by electrospinning and their applications in nanocomposites," *Composites Science and Technology*, vol. 63, no. 15, pp. 2223–2253, 2003.

[2] S. Ramakrishna, K. Fujihara, W.-E. Teo, T. Yong, Z. Ma, and R. Ramaseshan, "Electrospun nanofibers: solving global issues," *Materials Today*, vol. 9, no. 3, pp. 40–50, 2006.

[3] D. H. Reneker, A. L. Yarin, H. Fong, and S. Koombhongse, "Bending instability of electrically charged liquid jets of polymer solutions in electrospinning," *Journal of Applied Physics*, vol. 87, no. 9, pp. 4531–4547, 2000.

[4] A. Greiner and J. H. Wendorff, "Electrospinning: a fascinating method for the preparation of ultrathin fibers," *Angewandte Chemie International Edition*, vol. 46, no. 30, pp. 5670–5703, 2007.

[5] N. Bhardwaj and S. C. Kundu, "Electrospinning: a fascinating fiber fabrication technique," *Biotechnology Advances*, vol. 28, no. 3, pp. 325–347, 2010.

[6] H. J. Kim, H. R. Pant, C. H. Park, L. D. Tijing, N. J. Choi, and C. S. Kim, "Hydrothermal growth of mop-brush-shaped ZnO rods on the surface of electrospun nylon-6 nanofibers," *Ceramics International*, vol. 39, no. 3, pp. 3095–3102, 2013.

[7] Y. Dzenis, "Spinning continuous fibers for nanotechnology," *Science*, vol. 304, no. 5679, pp. 1917–1919, 2004.

[8] S. J. Kim, Y. J. Kang, and N.-J. Choi, "Optical-fiber electronic speckle pattern interferometry for quantitative measurement of defects on aluminum liners in composite pressure vessels," *Journal of the Optical Society of Korea*, vol. 17, no. 1, pp. 50–56, 2013.

[9] K. Saeed, S. Haider, T.-J. Oh, and S.-Y. Park, "Preparation of amidoxime-modified polyacrylonitrile (PAN-oxime) nanofibers and their applications to metal ions adsorption," *Journal of Membrane Science*, vol. 322, no. 2, pp. 400–405, 2008.

[10] P. Raghavan, X. H. Zhao, J.-K. Kim et al., "Ionic conductivity and electrochemical properties of nanocomposite polymer electrolytes based on electrospun poly(vinylidene fluoride-co-hexafluoropropylene) with nano-sized ceramic fillers," *Electrochimica Acta*, vol. 54, no. 2, pp. 228–234, 2008.

[11] S. G. Kumbar, S. P. Nukavarapu, R. James, L. S. Nair, and C. T. Laurencin, "Electrospun poly(lactic acid-co-glycolic acid) scaffolds for skin tissue engineering," *Biomaterials*, vol. 29, no. 30, pp. 4100–4107, 2008.

[12] H.-F. Wen, C. Yang, D.-G. Yu, X.-Y. Li, and D.-F. Zhang, "Electrospun zein nanoribbons for treatment of lead-contained wastewater," *Chemical Engineering Journal*, vol. 290, pp. 263–272, 2016.

[13] D. G. Yu, C. Yang, M. Jin et al., "Medicated Janus fibers fabricated using a Teflon-coated side-by-side spinneret," *Colloids and Surfaces B: Biointerfaces*, vol. 138, pp. 110–116, 2016.

[14] D.-G. Yu, X.-Y. Li, X. Wang, J.-H. Yang, S. W. A. Bligh, and G. R. Williams, "Nanofibers fabricated using triaxial electrospinning as zero order drug delivery systems," *ACS Applied Materials and Interfaces*, vol. 7, no. 33, pp. 18891–18897, 2015.

[15] I. G. Kim, J.-H. Lee, A. R. Unnithan, C.-H. Park, and C. S. Kim, "A comprehensive electric field analysis of cylinder-type multi-nozzle electrospinning system for mass production of nanofibers," *Journal of Industrial and Engineering Chemistry*, vol. 31, pp. 251–256, 2015.

[16] C. Burger, B. S. Hsiao, and B. J. Chu, "Nanofibrous materials and their applications," *Annual Review of Materials Research*, vol. 36, pp. 333–368, 2006.

[17] M. M. Demir, I. Yilgor, E. Yilgor, and B. Erman, "Electrospinning of polyurethane fibers," *Polymer*, vol. 43, no. 11, pp. 3303–3309, 2002.

[18] D. Aussawasathien, C. Teerawattananon, and A. Vongachariya, "Separation of micron to sub-micron particles from water: electrospun nylon-6 nanofibrous membranes as pre-filters," *Journal of Membrane Science*, vol. 315, no. 1-2, pp. 11–19, 2008.

[19] H. J. Kim, H. R. Pant, C. H. Park et al., "Electrical properties of ZnO/nylon-6 spider-wave-like nanonets prepared via electrospinning," *Digest Journal of Nanomaterials and Biostructures*, vol. 8, pp. 385–393, 2013.

[20] H. R. Pant, M. P. Bajgai, K. T. Nam et al., "Electrospun nylon-6 spider-net like nanofiber mat containing TiO_2 nanoparticles: a multifunctional nanocomposite textile material," *Journal of Hazardous Materials*, vol. 185, no. 1, pp. 124–130, 2011.

[21] H. R. Pant, D. R. Pandeya, K. T. Nam, W.-I. Baek, S. T. Hong, and H. Y. Kim, "Photocatalytic and antibacterial properties of a TiO_2/nylon-6 electrospun nanocomposite mat containing silver nanoparticles," *Journal of Hazardous Materials*, vol. 189, no. 1-2, pp. 465–471, 2011.

[22] H. J. Kim, H. R. Pant, N. J. Choi, and C. S. Kim, "Fly ash/polyurethane thin film for the adsorption of volatile organic compounds (VOCs) from air," *Fibers and Polymers*, vol. 15, no. 7, pp. 1393–1398, 2014.

[23] H. J. Kim, H. R. Pant, N. J. Choi, and C. S. Kim, "Composite electrospun fly ash/polyurethane fibers for absorption of volatile organic compounds from air," *Chemical Engineering Journal*, vol. 230, pp. 244–250, 2013.

[24] E. Scholten, L. Bromberg, G. C. Rutledge, and T. A. Hatton, "Electrospun polyurethane fibers for absorption of volatile organic compounds from air," *ACS Applied Materials and Interfaces*, vol. 3, no. 10, pp. 3902–3909, 2011.

[25] I. Jarraya, S. Fourmentin, M. Benzina, and S. Bouaziz, "VOC adsorption on raw and modified clay materials," *Chemical Geology*, vol. 275, no. 1-2, pp. 1–8, 2010.

[26] F. Qu, L. Z. Zhu, and K. Yang, "Adsorption behaviors of volatile organic compounds (VOCs) on porous clay heterostructures (PCH)," *Journal of Hazardous Materials*, vol. 170, no. 1, pp. 7–12, 2009.

[27] L. Zhu, S. Tian, and Y. Shi, "Adsorption of volatile organic compounds onto porous clay heterostructures based on spent organobentonites," *Clays and Clay Minerals*, vol. 53, no. 2, pp. 123–136, 2005.

[28] J. H. Park and D. I. Jung, "Removal of total phosphorus (TP) from municipal wastewater using loess," *Desalination*, vol. 269, no. 1-3, pp. 104–110, 2011.

[29] S. S. Tahir and N. Rauf, "Thermodynamic studies of Ni(II) adsorption onto bentonite from aqueous solution," *Journal of Chemical Thermodynamics*, vol. 35, no. 12, pp. 2003–2009, 2003.

[30] B. Koumanova and P. Peeva-Antova, "Adsorption of p-chlorophenol from aqueous solutions on bentonite and perlite," *Journal of Hazardous Materials*, vol. 90, no. 3, pp. 229–234, 2002.

[31] T. An, H. Chen, H. Zhan, K. Zhu, and R. Berndtsson, "Sorption kinetics of naphthalene and phenanthrene in loess soils," *Environmental Geology*, vol. 47, no. 4, pp. 467–474, 2005.

[32] T. B. İyim and G. Güçlü, "Removal of basic dyes from aqueous solutions using natural clay," *Desalination*, vol. 249, no. 3, pp. 1377–1379, 2009.

[33] F. A. Sheikh, N. A. M. Barakat, M. A. Kanjwal et al., "Electrospun antimicrobial polyurethane nanofibers containing silver nanoparticles for biotechnological applications," *Macromolecular Research*, vol. 17, no. 9, pp. 688–696, 2009.

[34] L. D. Tijing, M. T. G. Ruelo, A. Amarjargal et al., "Antibacterial and superhydrophilic electrospun polyurethane nanocomposite fibers containing tourmaline nanoparticles," *Chemical Engineering Journal*, vol. 197, pp. 41–48, 2012.

[35] W. K. Son, J. H. Youk, T. S. Lee, and W. H. Park, "The effects of solution properties and polyelectrolyte on electrospinning of ultrafine poly(ethylene oxide) fibers," *Polymer*, vol. 45, no. 9, pp. 2959–2966, 2004.

[36] A. S. Khan, Z. Ahmed, M. J. Edirisinghe, F. S. L. Wong, and I. U. Rehman, "Preparation and characterization of a novel bioactive restorative composite based on covalently coupled polyurethane-nanohydroxyapatite fibres," *Acta Biomaterialia*, vol. 4, no. 5, pp. 1275–1287, 2008.

[37] C. Zhou, R. Chu, R. Wu, and Q. Wu, "Electrospun polyethylene oxide/cellulose nanocrystal composite nanofibrous mats with homogeneous and heterogeneous microstructures," *Biomacromolecules*, vol. 12, no. 7, pp. 2617–2625, 2011.

[38] G. P. Dong, X. D. Xiao, X. F. Liu et al., "Preparation and characterization of Ag nanoparticle-embedded polymer electrospun nanofibers," *Journal of Nanoparticle Research*, vol. 12, no. 4, pp. 1319–1329, 2010.

[39] J. Völter, "A π-complex mechanism for catalytic hydrogenation of the benzene ring," *Journal of Catalysis*, vol. 3, no. 3, pp. 297–298, 1964.

[40] T. T. Phuong, J. Massardier, and P. Gallezot, "Competitive hydrogenation of benzene and toluene on group VIII metals: correlation with the electronic structure," *Journal of Catalysis*, vol. 102, no. 2, pp. 456–459, 1986.

[41] S. B. Lee, K. W. Chung, I. K. Hong, and H. J. Kim, "Selective catalytic oxidation of volatile organic solvent mixtures," *Journal of Industrial and Engineering Chemistry*, vol. 7, pp. 193–198, 2001.

21

Reduction of Offensive Odor from Natural Rubber Using Zinc-Modified Bentonite

Bryan B. Pajarito, Kayla C. Castañeda, Sofia Denise M. Jeresano, and Dominique Ann N. Repoquit

Polymer Research Laboratory, Department of Chemical Engineering, University of the Philippines, Diliman, Quezon City 1101, Philippines

Correspondence should be addressed to Bryan B. Pajarito; bryan.pajarito@gmail.com

Academic Editor: Peter Majewski

Natural bentonite was modified with zinc ions and was used as a filler to reduce the offensive odor from raw natural rubber (NR). Characterization of filler shows the limited exchange of inherent calcium ions and the preferred adsorption of zinc ions on the surface of natural bentonite during modification. The modification process was also accompanied by bassanite formation due to sulfate ions brought by the zinc salt. Zinc-modified bentonite demonstrates an antimicrobial effect against microorganisms that are known to degrade or produce offensive odor from raw NR. Due to its antimicrobial and adsorption properties, zinc-modified bentonite significantly reduces the offensive odor from raw NR at loadings of 2.5–7.5 phr as shown by the olfactometry test. In terms of vulcanization characteristics, the loading of zinc-modified bentonite at 7.5 phr caused a slight decrease in elastic torque and increase in scorch and curing times of NR. The hardness and tensile properties of vulcanized NR are maintained after compounding with zinc-modified bentonite at 7.5 phr.

1. Introduction

The offensive odor from raw natural rubber (NR) has been a long-standing problem of manufacturers [1]. The obnoxious odor released during processing and mastication of NR is undesirable to workers and a nuisance to the public residing near rubber manufacturing facilities [2, 3]. The volatile compounds produced during microbial degradation of carbohydrates, proteins, and other nonrubber components of NR generate the offensive odor [2–5]. Currently, water scrubbers are used to control the offensive odor from NR [1, 5, 6], but the expense of installation and maintenance makes it unattractive to rubber manufacturers. Several methods based on the application of solid adsorbents and antimicrobial agents to reduce the offensive smell from raw and vulcanized NR were previously studied. Hoven et al. [1] showed the effectiveness of zeolite, chitosan, and carbon black as solid adsorbents in lowering the offensive odor from vulcanized NR. Perng et al. [7] utilized a column of granulated activated carbon to adsorb residual ozone and volatile organic compounds from an ozonation reactor which treated the odorous gas from rubber processing. Nornanysya et al. [2] soaked coagulated NR latex in dilute aqueous solutions of antimicrobial agents (sodium hypochlorite and formaldehyde) and undiluted HISKA (a coagulant and density enhancer for NR latex) to decrease the unpleasant smell. In these methods, adsorption by solid adsorbents was used to capture and limit the emission of odorous compounds, while the antimicrobial characteristics of selected liquid chemicals were employed to inhibit the microbial growth and degradation of NR.

In this study, natural bentonite was utilized as a carrier of zinc ions which are known to possess antimicrobial properties [8–11]. As a clay mineral, bentonite is also commonly used as an adsorbent [12]. Solid adsorbents such as montmorillonite (which is the main component of natural bentonite) and zeolite were previously employed as carriers or supports of heavy metal ions of silver, copper, and zinc

[13–19]. Silver is the preferred antimicrobial agent, but its relatively high cost prompts researchers to explore the oligodynamic effect of other heavy metals such as copper and zinc. Copper seems to be a feasible choice for antimicrobial application in NR, but previous research [20] showed the susceptibility of vulcanized NR to auto-oxidation and degradation when in contact with copper ions. Therefore, this study focused on the modification of natural bentonite with zinc ions and its capability to reduce the offensive odor from raw NR.

Appropriate techniques were used to characterize the phase and elemental composition, chemical structure, and surface morphology of natural and zinc-modified bentonite in order to understand the modification process. Antimicrobial analyses were also performed on selected microorganisms to measure the antimicrobial properties of natural and zinc-modified bentonite. The reduction of offensive odor from NR using zinc-modified bentonite was investigated using an olfactometry method. After determining the optimum loading in NR for odor reduction, the effect of zinc-modified bentonite on curing and mechanical properties of NR composite was also studied and discussed.

2. Materials and Methods

2.1. Materials. Natural bentonite powder (industrial grade, Saile Industries Inc.) was used as received. The cation exchange capacity of natural bentonite is 134.66 meq per 100 g according to the datasheet provided by the supplier. Zinc sulfate heptahydrate ($ZnSO_4{\cdot}7H_2O$; analytical grade, Loba Chemie Pvt. Ltd.) and deionized water were used to prepare the zinc solution. Standard Philippine rubber (SPR 10, from Rhodeco Processing Services Inc. and MJ Rubber Agricultural Products Inc.) was used for the olfactometry test and preparation of vulcanized NR samples. Compounding ingredients added to NR for vulcanization such as stearic acid, zinc oxide, sulfur, and *N*-tert-butyl-2-benzothiazyl sulfonamide (Rhodeco Processing Services Inc.) were also used as received.

The antimicrobial analysis and the required microorganisms/materials such as *M. luteus* (BIOTECH 1793), *X. oryzae* (BIOTECH 1792), sterile saline, agar plates, 10% phenol solution, and distilled water were provided by the National Institute of Molecular Biology and Biotechnology (Laguna, Philippines).

2.2. Zinc Modification of Bentonite. Natural bentonite powder was first sieved (400 mesh, Taylor) before treatment with 0.2 M zinc solution (58 g $ZnSO_4{\cdot}7H_2O$ in 1000 ml deionized water). Constant stirring of 150 g natural bentonite in zinc solution was performed at room temperature for 24 h using a magnetic stirrer (MSH-30A, Wisd Laboratory Instruments) at 1500 rpm. The clay slurry was centrifuged (Rotofix 32A, Hettich), and the collected solid was dried in a laboratory hot air oven (DSO-500D, Digisystem Laboratory Instruments) at 90°C for 100 h. Zinc-modified bentonite was ground with mortar and pestle and then sieved before characterization and further tests.

2.3. Characterization of Natural and Zinc-Modified Bentonite. Phase composition of natural bentonite before and after zinc modification was obtained using a portable X-ray diffraction machine (Terra, Olympus). The instrument used coradiation ($\lambda = 1.79$ Å) while recording intensity data from $2\theta = 5$–$55°$ with the increment of 0.05°. Basal spacing at specific 2θ was computed using Bragg's law. Elemental composition was measured by X-ray fluorescence analysis (XRF; BTX II, Olympus) with a resolution of 200 eV at the 5.9 keV peak of cobalt and an energy range of 3–25 keV.

Chemical structure of samples was studied using the Fourier transform infrared spectrometer (FTIR; Nicolet 6700, Thermo Scientific) coupled with a macrodiamond attenuated total reflectance (ATR) crystal plate accessory. The resolution of the spectrum was 2 cm^{-1}, and the number of coadded scans was 32. Background spectra were obtained without samples prior to actual measurement of samples.

Morphological features of samples were examined using a scanning electron microscope (SEM; S-3400N, Hitachi) operated at 5 kV and 1000x magnification.

2.4. Antimicrobial Analyses. Cylinder-plate assays were performed for antimicrobial analysis of natural and zinc-modified bentonite against selected microorganisms known to degrade or produce offensive odor from raw NR. *Micrococcus luteus* is a Gram-positive microorganism previously classified by Nornanysya et al. [3] to produce foul smell from coagulated NR latex. *Xanthomonas oryzae* is a Gram-negative microorganism known to degrade NR [21]. Natural or zinc-modified bentonite sample was diluted in deionized water to produce 2 ml solution of 0.1 $g{\cdot}ml^{-1}$. The test microorganism (*M. luteus* or *X. oryzae*) from a freshly grown culture was suspended in 3 ml sterile saline. The saline suspension was spread into the surface of an agar plate. The culture was incubated for 24 h at 35°C until growth was apparent. The inoculum was obtained from the incubated culture and added into 50 ml sterile saline. The microorganism was then spread evenly on a new agar plate until a layer of uniform depth was formed.

Three cylinders containing 0.1 ml of the sample solution, 10% phenol (positive control), and distilled water (negative control) were placed on the agar plate. After 18 h, the cylinders were removed from the plate and the diameter of the inhibition zone was measured to the nearest 0.1 mm. Mean diameter of the inhibition zone was reported using triplicate samples.

2.5. Olfactometry Test. The olfactometry method used by Hoven et al. [1] was adapted to assess the effectiveness of zinc-modified bentonite in odor reduction. Although the chromatographic analysis is usually employed to measure the concentration of volatile compounds from NR [1–3, 22], Hoven et al. [1] demonstrated the capability of the technique in describing the odor reduction in NR where human testers judged and rated the odor of rubber samples. Moreover, there is no known relationship between the concentration of volatile compounds and its perceived odor [7]. To obtain statistically reliable data, 50 persons (25 men and 25 women) with ages of 18–25 years participated in the olfactometry test. References 1–5 ranging from low to high intensity of rubber odor were generated by filling 1, 5, 10, 15, and 20 g of raw NR in 250 ml plastic containers, respectively. Rubber samples were prepared by masticating raw NR with natural or zinc-modified bentonite (0–10 phr) in a laboratory mixing mill (BMM-6, Bharaj Machineries Pvt. Ltd.) at room temperature. The odor of 10 g of the rubber sample in a 250 ml container was compared with each reference. Each person specified the reference number that matches the odor intensity of the rubber sample. The results of the olfactometry test were reported as the percentage of people who selected each reference. The mean odor rating (minimum value = 1; maximum value = 5) was also computed for each rubber sample. Every participant had to smell the odor of coffee beans (also in 250 ml container) before each comparison to erase their previous experience from other rubber samples and to minimize odor confusion [1].

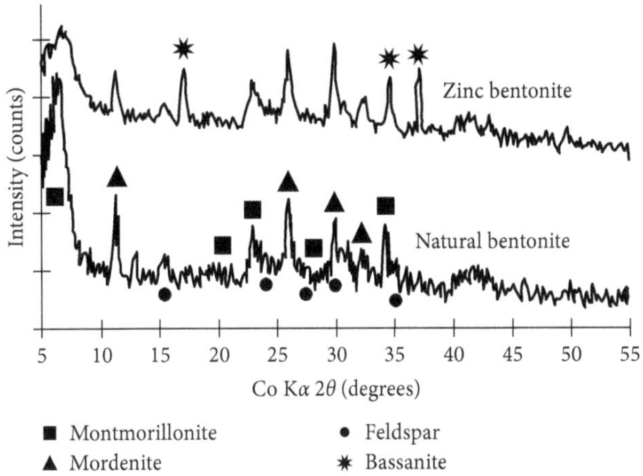

FIGURE 1: X-ray diffractogram of natural bentonite and zinc-modified bentonite.

2.6. Curing and Mechanical Tests. The raw NR sample containing zinc-modified bentonite at proportion found optimum for odor reduction was compounded with 2 phr stearic acid, 5 phr zinc oxide, 2.5 phr sulfur, and 0.6 phr *N*-tert-butyl-2-benzothiazyl sulfonamide using the laboratory mixing mill. After mastication, the vulcanization parameters of the rubber sample were determined at 160°C using a rubber process analyzer (RPA 2000, Alpha Technologies) for 20 min. The equipment was operated at a frequency of 100 cpm and strain of 0.5°. The obtained parameters were minimum and maximum elastic torques (ML and MH), scorch times (time until 1 or 2 dNm rise above ML; ts1 and ts2), and curing times (time required for the sample to reach 50% or 90% of the total state of cure; tc50 and tc90) [23].

The NR sample compounded with curing additives was vulcanized in a compression moulding press (Chemplast, S. V. Industries) at 160°C and duration time indicated by it is tc90. The Shore A hardness of vulcanized sample was measured using a durometer (Kori Seiki) following ASTM D2240. The vulcanized sample was also cut into dumbbell shape (GT-7016-A, Gotech) and subjected to a tensile test using a universal testing machine (AGS-5kNX, Shimadzu) following ASTM D412. The crosshead speed used during the tensile test was 500 mm·min^{-1}. Due to limited tensile stroke of the equipment and failure of vulcanized sample to break during the test, the tensile stress at 100, 200, and 300% elongation was recorded instead of the tensile strength. Tensile set due to permanent deformation was also measured at 140% elongation following ASTM D412. The sample was strained for 10 min and was allowed to rest for another 10 min before measurement of the distance between bench marks. Five replicates were tested to report the mean hardness of the sample, while 3 replicates were used to report the mean values of tensile properties.

The Shore A hardness and tensile properties of the vulcanized sample were also obtained after air oven ageing following ASTM D573. The vulcanized sample was aged in the laboratory hot air oven at 70°C for 168 h. The aged sample was then cooled at room temperature and maintained at 23°C for 13 h before testing for hardness and tensile properties.

The curing characteristics and mechanical properties (before and after air oven ageing) of the vulcanized NR sample without zinc-modified bentonite were also determined and reported for comparison.

3. Results and Discussion

3.1. Characterization of Filler. The natural bentonite used in this study is mainly composed of montmorillonite (smectite) with traces of mordenite and feldspar (Figure 1). The previous study in Philippine natural bentonite [24] detected montmorillonite, mordenite, and hectorite. The measured composition of natural bentonite (Table 1) suggests calcium as the main exchangeable cation. The modification process lowers and broadens the reflection of montmorillonite at $2\theta = 6.35^{\circ}$ (d_{001} basal spacing = 16.15 Å) due to the adsorption of $[Zn(H_2O)_6]^{2+}$ complex on the surface of clay [25–27]. However, the expected shift of d_{001} reflection to lower 2θ due to ion exchange and intercalation of zinc ions in the clay galleries [25–27] is not observed. The lower ionic concentration and larger size of Zn^{2+} than that of Ca^{2+} have limited the intercalation and swelling of bentonite [28].

The diffraction pattern of zinc-modified bentonite also shows the presence of bassanite $2CaSO_4{\cdot}(H_2O)$ with intense peaks at $2\theta = 17.2$, 34.6, and 37.2° [29]. As suggested by Wilson and Bish [30], some of Ca^{2+} reacted with SO_4^{2-} to form bassanite. The presence of calcium (9.89 wt.%) and sulfur (9.19 wt.%) in zinc-modified bentonite (Table 1) further indicates bassanite formation. Moreover, the high concentration of calcium and zinc (32.30 wt.%) in modified bentonite denotes limited ion exchange of Ca^{2+} and preferred adsorption of Zn^{2+} into the clay surface.

Zinc modification also results in dealumination of bentonite [31] (Table 1) due to the weak acidity of zinc sulfate solution [32]. Zinc- and sulfur-free composition of zinc-modified bentonite shows considerable enrichment in silicon due to the dissolution of aluminum. Meanwhile, other elements such as iron and calcium are retained in bentonite after zinc modification.

TABLE 1: Composition of natural bentonite and zinc-modified bentonite by XRF analyses.

Analyte (wt.%)	Natural bentonite	Zinc bentonite	Zinc bentonite (Zn- and S-free)
Si	46.0	31.5	53.8
Fe	25.7	14.5	24.8
Al	8.74	—	—
K	1.32	0.81	1.38
Ti	1.42	0.77	1.32
Ca	15.5	9.89	16.9
Zn	0.05	32.3	—
Cu	0.08	0.07	0.12
Sr	0.24	0.17	0.29
Mn	0.85	0.48	0.82
Zr	0.08	—	—
V	0.06	—	—
S	—	9.19	—
Ba	—	0.36	0.61

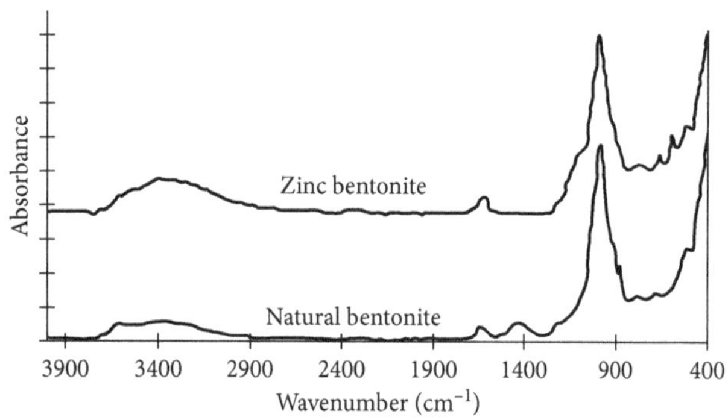

FIGURE 2: ATR-FTIR spectrum of natural bentonite and zinc-modified bentonite.

The infrared bands of as-received natural bentonite (Figure 2) show OH stretching and deformation of water (3390 and 1630 cm^{-1}), Si-O (989, 782, and 686 cm^{-1}), AlAlOH (872 cm^{-1}), and Al-O-Si deformation (518 cm^{-1}) [33]; organic impurity is also detected (1430 cm^{-1}). Zinc modification eliminates the organic impurity and AlAlOH while maintaining the original chemical structure of bentonite. For zinc-modified bentonite, the band broadening from 1070 to 1270 cm^{-1} along with the appearance of new bands at 603 and 669 cm^{-1} suggests the presence of bassanite [34]. Moreover, the slight increase in band intensity of OH stretching indicates hydrated Ca^{2+} and Zn^{2+} ions [27].

SEM micrograph (Figure 3) shows no evident change in surface morphology of bentonite after zinc modification. The particles have an irregular shape with sizes of less than 37 μm due to 400-mesh sieving before and after zinc modification. Sieving keeps the particle or aggregate size of bentonite small before incorporating to NR.

3.2. Antimicrobial Properties. Zinc-modified bentonite demonstrates antimicrobial properties against *M. luteus* and *X. oryzae* (Figure 4). The modified bentonite achieves 81.9 and 52.1% effectivity of 10% phenol against *M. luteus* and *X. oryzae* in terms of the size of the inhibition zone (Table 2). The result shows zinc-modified bentonite to be more effective against *M. luteus* than *X. oryzae* which confirms the susceptibility of Gram-positive bacteria to zinc-induced inhibition than that of Gram-negative bacteria [8, 11]. Zn^{2+} ions liberated by modified bentonite have intrinsic antimicrobial properties attributed to the depletion of antioxidant reserves, protein dysfunction, and enzyme inactivity of microbial cells [9–11].

Natural bentonite without zinc modification also shows microbial inhibition against *X. oryzae* (35.9% effectivity of 10% phenol). Xie and Yang [35] recently reported the antimicrobial activity of Ca^{2+} and Mg^{2+} ions against cardiolipin-rich *Staphylococcus aureus*. Both Ca^{2+} and Mg^{2+} form complexes with cardiolipin which disrupt the cell membrane and kill *S. aureus*. The higher content of cardiolipin in *Xanthomonas* sp. (31% for *X. campestris* [36]) than that in *M. luteus* (5% [37]) results to the selective antimicrobial effect of released Ca^{2+} ions from natural bentonite.

3.3. Reduction of Offensive Odor. Zinc-modified bentonite at 7.5 phr shows the highest effect in reducing the offensive odor of masticated NR (Table 3). The majority (84% of the participants) indicated the odor intensity of raw NR filled with 7.5 phr of zinc-modified bentonite was equal or less than to reference 2 (5 g NR per 250 ml). The remaining participants described the odor intensity to be equal or greater than reference 3 (10 g NR per 250 ml). In terms of average odor rating, zinc-modified bentonite at 7.5 phr improves the odor of unfilled NR by 56.8%. The antimicrobial effect of Zn^{2+} ions released by modified bentonite lowers the microbial growth in raw NR, thereby reducing the microbial degradation of nonrubber components and the emission of offensive odor.

The results of olfactometry show improvement in the odor of NR as the proportion of zinc-modified bentonite is increased from 0 to 7.5 phr. The odor improvement drops at 10 phr which could be due to the poor dispersion and aggregation of modified bentonite in the rubber matrix at high loading. At low filler loading, the homogenous dispersion of modified bentonite in NR allows the creation of microbial inhibition zones distributed within the rubber matrix which maximizes the antimicrobial effect of Zn^{2+}. However, the presence of water and hydrated ions in modified bentonite makes the filler hydrophilic and incompatible with hydrophobic NR during solid mixing [38, 39]. Filler aggregates formed from the poor dispersion limit the distribution and availability of Zn^{2+} within the rubber matrix.

Natural bentonite also contributes to the reduction of offensive odor, as illustrated by the steady decrease in the average odor rating of NR with the increase in the proportion of natural bentonite. The reduction in offensive odor only becomes significant at 10 phr with odor improvement of 37.8% relative to unfilled NR and with 64.2% of participants describing the odor to be equal or less than reference 2. The obnoxious smell of NR is composed of volatile chemicals coming from different reactions: alcohols, aldehydes, carboxylic acids and hydrocarbons (from lipid

(a) (b)

Figure 3: SEM micrograph of natural bentonite (a) and zinc-modified bentonite (b). Scale bar is 50 μm.

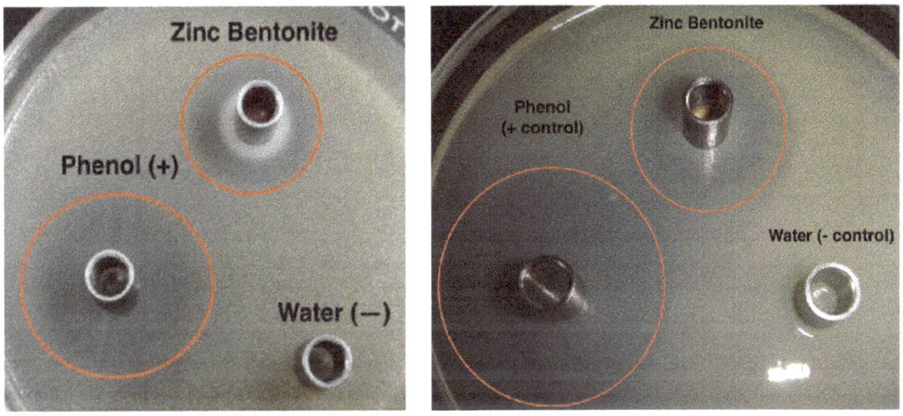

(a) (b)

Figure 4: Antimicrobial activity of zinc-modified bentonite against *M. luteus* (a) and *X. oryzae* (b). Performance is compared with 10% phenol and distilled water as positive and negative control, respectively.

Table 2: Mean diameter of inhibition zone of samples against *M. luteus* and *X. oryzae*.

Sample (mm)	*M. luteus*	*X. oryzae*
Natural bentonite	0.0 ± 0.0	14.0 ± 1.2
Zinc bentonite	21.3 ± 1.2	20.3 ± 1.4
10% phenol (+ control)	26.0 ± 1.0	39.0 ± 1.0
Distilled water (− control)	0.0 ± 0.0	0.0 ± 0.0

oxidation), free fatty acids (from carbohydrate fermentation), esters (from microbial esterification), and sulfur-containing compounds (from amino acid degradation) [22]. Adsorption of these volatile components occurs in the interlayer space of clay and within the secondary pores formed by the clay particles [40]. Adsorption sites become more available as the proportion of natural bentonite increases in NR. However, the saturation of adsorption sites will eventually lead to the release of offensive smell due to the availability of microorganisms and nonrubber components.

Raw NR is more odorous than compounded and vulcanized NR. Compounding additives like zinc oxide and accelerators inhibit the microbial growth in rubber [41]; the high temperature applied to rubber during vulcanization also contributes to this effect. Therefore, much of the offensive odor of raw NR is decreased after vulcanization due to the low microbial population. Reducing the offensive odor via microbial inhibition has more effect on raw NR than vulcanized NR. Hoven et al. [1] showed that an antimicrobial agent-like benzalkonium chloride provides little improvement in the emitted odor of vulcanized NR. On the contrary, Nornanysya et al. [2] demonstrated that the soaking of raw NR in aqueous solutions of the antimicrobial agent eliminates most of the volatile compounds. Reducing the offensive odor using solid adsorbents such as carbon black, zeolite, and chitosan is more effective to vulcanized NR [1]. The treatment of bentonite with Zn^{2+} ions combines antimicrobial inhibition and adsorption in a single filler which can reduce the offensive odor of both raw and vulcanized NR.

3.4. Curing and Mechanical Properties. The incorporation of zinc-modified bentonite at 7.5 phr slightly decreases the elastic torque and prolongs the scorch and cure times of NR (Table 4). Moisture carried by modified bentonite plasticizes and softens the rubber matrix which lowers the elastic torque. Moreover, the adsorption of curing agents into bentonite delays sulfur vulcanization and extends the curing of NR. Moisture-induced plasticization by zinc-modified bentonite also marginally lowers the tensile stress response of vulcanized NR (Tables 5 and 6)

Table 3: Results of the olfactometry test.

Filler	Proportion (phr)	Percentage of olfactometry test					Mean odor rating
		Reference 1	Reference 2	Reference 3	Reference 4	Reference 5	
None	0.0	9.4	17.0	5.7	30.2	37.7	3.7 ± 1.4
Natural bentonite	2.5	3.8	15.1	24.5	39.6	17.0	3.5 ± 1.1
Natural bentonite	5.0	3.8	24.5	34.0	15.1	22.6	3.3 ± 1.2
Natural bentonite	7.5	13.2	30.2	26.4	20.8	9.4	2.8 ± 1.2
Natural bentonite	10.0	34.0	30.2	15.1	13.2	7.5	2.3 ± 1.3
Zinc bentonite	2.5	44.0	32.0	14.0	6.0	4.0	1.9 ± 1.1
Zinc bentonite	5.0	48.0	24.0	12.0	12.0	4.0	2.0 ± 1.2
Zinc bentonite	7.5	62.0	22.0	12.0	2.0	2.0	1.6 ± 0.9
Zinc bentonite	10.0	42.0	26.0	14.0	8.0	10.0	2.2 ± 1.3

Table 4: Vulcanization parameters of unfilled NR and NR composite at 160°C.

Sample	ML (dNm)	MH (dNm)	ts1 (min)	ts2 (min)	tc50 (min)	tc90 (min)
Unfilled NR	0.68	6.82	2.54	3.02	3.49	5.41
NR + 7.5 phr zinc bentonite	0.60	5.45	3.33	4.11	4.47	7.57

Table 5: Mean hardness and tensile properties of unfilled NR and NR composite before air oven ageing.

Sample	Shore A hardness	Tensile stress at specified elongation (MPa)			Tensile set (%)
		100%	200%	300%	
Unfilled NR	30.60 ± 0.89	0.74 ± 0.06	1.13 ± 0.11	1.55 ± 0.17	1.33 ± 1.15
NR + 7.5 phr zinc bentonite	30.40 ± 0.55	0.68 ± 0.01	1.04 ± 0.01	1.43 ± 0.03	0.67 ± 1.15

Table 6: Mean hardness and tensile properties of unfilled NR and NR composite after air oven ageing.

Sample	Shore A hardness	Tensile stress at specified elongation (MPa)			Tensile set (%)
		100%	200%	300%	
Unfilled NR	30.48 ± 0.50	0.79 ± 0.03	1.27 ± 0.04	1.81 ± 0.06	1.67 ± 0.58
NR + 7.5 phr zinc bentonite	31.32 ± 0.84	0.74 ± 0.01	1.23 ± 0.02	1.78 ± 0.04	1.33 ± 1.15

before and after air oven ageing. Meanwhile, the rigidity of bentonite allows sustenance of hardness and resistance to permanent deformation of vulcanized NR. One-way analysis of variance at 95% confidence level shows no statistically significant differences between the mean hardness (before ageing: $p = 0.68$; after ageing: $p = 0.07$), tensile stress (before ageing: $p = 0.13$–0.31; after ageing: $p = 0.06$–0.49), and tensile set (before ageing: $p = 0.52$; after ageing: $p = 0.68$) of unfilled NR and NR composite. Thus, the hardness and tensile properties of vulcanized NR remain uncompromised with the use of zinc-modified bentonite as a filler.

4. Conclusions

Modification of natural bentonite is achieved by the adsorption of Zn^{2+} ions on the clay surface. This modification process is also accompanied by the formation of hydrated calcium sulfate minerals such as bassanite. The controlled release of Zn^{2+} from zinc-modified bentonite has antimicrobial properties against *M. luteus* and *X. oryzae*.

The use of zinc-modified bentonite as a filler reduces the offensive odor from raw NR. The magnitude of odor reduction depends on the loading of zinc-modified bentonite in NR. At the optimum loading, zinc-modified bentonite slightly reduces the elastic torque and extends the vulcanization of NR. The mechanical properties of vulcanized NR composite remain comparable with unfilled NR. These results suggest the suitability of zinc-modified bentonite as a deodorant during mastication of raw NR.

Conflicts of Interest

The authors declare that there are no conflicts of interest regarding the publication of this paper.

Acknowledgments

The authors acknowledge the support provided by the following: Saile Industries Inc. for the natural bentonite; Rhodeco Processing Services Inc. and MJ Rubber

Agricultural Products Inc. for SPR 10 and compounding ingredients; Department of Mining, Metallurgical, and Materials Engineering (University of the Philippines) for the use of laboratory mixing mill and rubber process analyzer; and Cytek Solutions Inc. for X-ray diffraction. This work was supported by the Department of Science and Technology-National Capital Region (DOST-NCR), Philippines.

References

[1] V. P. Hoven, K. Rattanakarun, and Y. Tanaka, "Reduction of offensive odor from natural rubber by odor-reducing substances," *Journal of Applied Polymer Science*, vol. 92, no. 4, pp. 2253–2260, 2004.

[2] F. K. Nornanysya, S. M. Sharifah Aminah, S. Azemi, K. Dzaraini, and W. Y. Wan Mohd Fathil, "Treatment of odor from natural rubber by using selected antimicrobial agent," in *Proceedings of the 2012 IEEE Symposium on Humanities, Science and Engineering Research (SHUSER)*, pp. 559–564, Kuala Lumpur. Malaysia, June 2012.

[3] F. K. Nornanysya, S. M. Sharifah Aminah, and S. Azemi, "Odourless natural rubber (ONR)," *Journal of Rubber Research*, vol. 16, no. 1, pp. 15–35, 2013.

[4] O. H. Nautiyal, "Fragrant rubber sole with new formulation employing EVA crump," *American Journal of Polymer Science*, vol. 2, no. 2, pp. 14–21, 2012.

[5] N. H. Kamarulzaman, N. F. Idris, and Z. Mohd Nor, "Characteristics of odour concentration from raw rubber processing factories via olfactometry technique," *Chemical Engineering Transactions*, vol. 30, pp. 121–126, 2012.

[6] N. F. Idris, N. H. Kamarulzaman, and Z. Mohd Nor, "Determination of volatile fatty acids from raw natural rubber drying activity by thermal desorption-gas chromatography," *Chemical Engineering Transactions*, vol. 30, pp. 175–180, 2012.

[7] C. H. Perng, I. L. Cheng, I. C. Wang, and M. S. Chou, "Ozonation of odorous compounds in gases emitted from rubber processing industries," *Aerosol and Air Quality Research*, vol. 11, no. 1, pp. 51–58, 2011.

[8] S. Atmaca, G. Ü. L. Kadri, and R. Cicek, "The effect of zinc on microbial growth," *Turkish Journal of Medical Sciences*, vol. 28, no. 6, pp. 595–598, 1998.

[9] J. A. Lemire, J. J. Harrison, and R. J. Turner, "Antimicrobial activity of metals: mechanisms, molecular targets and applications," *Nature Reviews Microbiology*, vol. 11, no. 6, pp. 371–384, 2013.

[10] J. Pasquet, Y. Chevalier, J. Pelletier, E. Couval, D. Bouvier, and M. A. Bolzinger, "The contribution of zinc ions to the antimicrobial activity of zinc oxide," *Colloids and Surfaces A: Physicochemical and Engineering Aspects*, vol. 457, pp. 263–274, 2014.

[11] A. Sirelkhatim, S. Mahmud, A. Seeni et al., "Review on zinc oxide nanoparticles: antibacterial activity and toxicity mechanism," *Nano-Micro Letters*, vol. 7, no. 3, pp. 219–242, 2015.

[12] H. H. Murray, "Current industrial applications of clays," *Clay Science*, vol. 12, pp. 106–112, 2006.

[13] M. Rivera-Garza, M. T. Olguin, I. Garcia-Sosa, D. Alcantara, and G. Rodriguez-Fuentes, "Silver supported on natural Mexican zeolite as an antibacterial material," *Microporous and Mesoporous Materials*, vol. 39, no. 3, pp. 431–444, 2000.

[14] A. Top and S. Ulku, "Silver, zinc, and copper exchange in a Na-clinoptilolite and resulting effect on antibacterial activity," *Applied Clay Science*, vol. 27, no. 1-2, pp. 13–19, 2004.

[15] P. Kaali, M. M. Perez-Madrigal, E. Stromberg, R. E. Aune, G. Czel, and S. Karlsson, "The influence of Ag^{+}, Zn^{2+} and Cu^{2+} exchanged zeolite on antimicrobial and long term in vitro stability of medical grade polyether polyurethane," *Express Polymer Letters*, vol. 5, no. 12, pp. 1028–1040, 2011.

[16] K. Malachova, P. Praus, Z. Rybkova, and O. Kozak, "Antibacterial and antifungal activities of silver, copper and zinc montmorillonites," *Applied Clay Science*, vol. 53, no. 4, pp. 642–645, 2011.

[17] J. Hrenovic, J. Milenkovic, T. Ivankovic, and N. Rajic, "Antibacterial activity of heavy metal-loaded natural zeolite," *Journal of Hazardous Materials*, vol. 201-202, pp. 260–264, 2012.

[18] K. K. Krishnani, Y. Zhang, L. Xiong, Y. Yan, R. Boopathy, and A. Mulchandani, "Bactericidal and ammonia removal activity of silver ion-exchanged zeolite," *Bioresource Technology*, vol. 117, pp. 86–91, 2012.

[19] S. Demirci, Z. Ustaoglu, G. A. Yilmazer, F. Sahin, and N. Bac, "Antimicrobial properties of zeolite-X and zeolite-A ion-exchanged with silver, copper, and zinc against a broad range of microorganisms," *Applied Biochemistry and Biotechnology*, vol. 172, no. 3, pp. 1652–1662, 2014.

[20] Y. Ohtake, "Degradation of vulcanised rubber products–problems and solutions (degradation by residual chlorine in tap water, metals and ozone due to deterioration in the global environment)," *Nippon Gomu Kyokaishi*, vol. 79, no. 11, pp. 529–536, 2006.

[21] H. B. Bode, K. Kerkhoff, and D. Jendrossek, "Bacterial degradation of natural and synthetic rubber," *Biomacromolecules*, vol. 2, no. 1, pp. 295–303, 2001.

[22] J. Sakdapipanich and K. Insom, "High-resolution gas chromatography-mass spectrometry," *Kautschuk und Gummi Kunststoffe*, vol. 59, no. 7-8, pp. 382–387, 2006.

[23] J. Dick, *Basic Rubber Testing: Selecting Methods for a Rubber Test Program*, ASTM International, West Conshohocken, PA, USA, 2003.

[24] E. Olegario-Sanchez and J. C. Felizco, "Investigation of the structural properties of amorphous Philippine bentonite clay and its potential use for topical applications," *Key Engineering Materials*, vol. 737, pp. 401–406, 2017.

[25] Q. Shi, S. Tan, Q. Yang, Z. Jiao, Y. Ouyang, and Y. Chen, "Preparation and characterization of antibacterial Zn^{2+}-exchanged montmorillonites," *Journal of Wuhan University of Technology-Materials Science Edition*, vol. 25, no. 5, pp. 725–729, 2010.

[26] O. Kozak, P. Praus, V. Machovic, and Z. Klika, "Adsorption of zinc and copper ions on natural and ethylenediamine modified montmorillonite," *Ceramics–Silikaty*, vol. 54, no. 1, pp. 78–84, 2010.

[27] L. Jiao, F. Lin, S. Cao et al., "Preparation, characterization, antimicrobial and cytotoxicity studies of copper/zinc-loaded montmorillonite," *Journal of Animal Science and Biotechnology*, vol. 8, p. 27, 2017.

[28] R. P. Singh and K. Kumari, "Calcium-zinc exchange equilibria on montmorillonite," *Colloids and Surfaces*, vol. 20, no. 3, pp. 239–246, 1986.

[29] J. W. Anthony, R. A. Bideaux, K. W. Bladh, and M. C. Nichols, *Handbook of Mineralogy*, Mineralogical Society of America, Chantilly, VA, USA, http://www.handbookofmineralogy.org/.

[30] S. A. Wilson and D. L. Bish, "Formation of gypsum and bassanite by cation exchange reactions in the absence of free-liquid H_2O: implications for Mars," *Journal of Geophysical Research*, vol. 116, p. E9, 2011.

[31] I. M. Kolthoff and T. Kameda, "The hydrolysis of zinc sulfate solutions, solubility product of hydrous zinc oxide and the composition of the latter precipitated from zinc sulfate solutions," *Journal of the American Chemical Society*, vol. 53, no. 3, pp. 832–842, 1931.

[32] G. Jozefaciuk and G. Bowanko, "Effect of acid and alkali treatments on surface areas and adsorption energies of selected minerals," *Clays and Clay Minerals*, vol. 50, no. 6, pp. 771–783, 2002.

[33] J. Madejova and P. Komadel, "Baseline studies of the Clay Minerals Society source clays: infrared methods," *Clays and Clay Minerals*, vol. 49, no. 5, pp. 410–432, 2001.

[34] Y. Liu, A. Wang, and J. J. Freemen, "Raman, MIR, and NIR spectroscopic study of calcium sulfates: gypsum, bassanite, and anhydrite," in *Proceedings of the 40th Lunar and Planetary Science Conference*, League City, TX, USA, March 2009.

[35] Y. Xie and L. Yang, "Calcium and magnesium ions are membrane-active against stationary-phase staphylococcus aureus with high specificity," *Scientific Reports*, vol. 6, article 20628, 2016.

[36] R. Moser, M. Aktas, and F. Narberhaus, "Phosphatidylcholine biosynthesis in *Xanthomonas campestris* via a yeast-like acylation pathway," *Molecular Microbiology*, vol. 91, no. 4, pp. 736–750, 2014.

[37] R. M. Epand and R. F. Epand, "Lipid domains in bacterial membranes and the action of antimicrobial agents," *Biochimica et Biophysica Acta*, vol. 1788, no. 1, pp. 289–294, 2009.

[38] B. B. Pajarito, W. F. F. Mangaccat, M. R. M. Tigue, and M. T. Tipton, "Moisture diffusion in natural rubber/bentonite nanocomposites: effect of clay filler treatments," *Defect and Diffusion Forum*, vol. 379, pp. 124–132, 2017.

[39] D. E. O. Santiago, B. B. Pajarito, W. F. F. Mangaccat, M. R. M. Tigue, and M. T. Tipton, "Effect of bentonite modification on hardness and mechanical properties of natural rubber nanocomposites," *AIP Conference Proceedings*, vol. 1736, no. 1, article 020047, 2016.

[40] G. Morozov, V. Breus, S. Nekludov, and I. Breus, "Sorption of volatile organic compounds and their mixtures on montmorillonite at different humidity," *Colloids and Surfaces A: Physicochemical and Engineering Aspects*, vol. 454, pp. 159–171, 2014.

[41] Y. Li, S. Zhao, L. Zhang, Y. Wang, and W. Yu, "The effect of different Fe^{2+} concentrations in culture media on the recycling of ground tyre rubber by *Acidithiobacillus ferrooxidans* YT-1," *Annals of Microbiology*, vol. 63, no. 1, pp. 315–321, 2013.

22

Rice Husk Filled Polymer Composites

Reza Arjmandi,[1] Azman Hassan,[1] Khaliq Majeed,[2] and Zainoha Zakaria[3]

[1] *Department of Polymer Engineering, Faculty of Chemical Engineering, Universiti Teknologi Malaysia (UTM), 81310 Skudai, Johor, Malaysia*
[2] *Department of Chemical Engineering, COMSATS Institute of Information Technology, 54000 Lahore, Pakistan*
[3] *Faculty of Science, Universiti Teknologi Malaysia (UTM), 81310 Skudai, Johor, Malaysia*

Correspondence should be addressed to Azman Hassan; azmanh@cheme.utm.my and Zainoha Zakaria; zainoha@kimia.fs.utm.my

Academic Editor: Mahbub Hasan

Natural fibers from agricultural wastes are finding their importance in the polymer industry due to the many advantages such as their light weight, low cost and being environmentally friendly. Rice husk (RH) is a natural sheath that forms around rice grains during their growth. As a type of natural fiber obtained from agroindustrial waste, RH can be used as filler in composites materials in various polymer matrices. This review paper is aimed at highlighting previous works of RH filled polymer composites to provide information for applications and further research in this area. Based on the information gathered, application of RH filled composites as alternative materials in building and construction is highly plausible with both light weight and low cost being their main driving forces. However, further investigations on physical and chemical treatment to further improve the interfacial adhesion with polymeric matrix are needed as fiber-polymer interaction is crucial in determining the final composite properties. Better understanding on how the used polymer blends as the matrix and secondary fillers may affect the properties would provide interesting areas to be explored.

1. Introduction

The increasing demand for eco-friendly materials, increasing depletion rate, and soaring prices of petroleum based plastics and pressing environmental regulations have all triggered a growing interest towards the field of composites [1, 2]. Composites materials are formed by combining two or more materials to improve properties of their original components. When one or more of the materials used are derived from biological origins, they are then defined as biocomposites (ASTM D7075-04) [3].

In general, polymer composites consist of a polymer resin as the matrix and one or more fillers are added to serve specific objectives or requirements. For example, composites for aerospace and sports applications require high mechanical and thermal properties. Traditionally synthetic fibers such as carbon or glass fibers have been used to reinforce composites and are able to produce such properties. However, with the growing global environmental concerns, their slow biodegradability is a disadvantage. Therefore researchers are finding other viable approaches to enhance or accelerate the biodegradability of polymeric composites. For this reason natural fibers provide good prospective as reinforcements fillers in thermosets, thermoplastics, and elastomers. Some main advantages of using natural fibers in composites are low cost, sustainability, light weight, and being nonabrasive and nonhazardous and more importantly they can accelerate biodegradability of the polymeric composites [4, 5].

Rice (*Oryza sativa* L.) is a primary source of food for billions of people and one of the major crops in the world. It covers around 1% of the earth's surface [6]. Statistics show that, during the period of 2010–2013, the average annual global production of rice was 725 million metric tons with Asia region alone producing over 90% of the total global rice production [7]. Rice husk (RH) is an inexpensive byproduct of rice processing and is separated from rice grain during the rice milling process. It is reported that, for every ton of rice produced, about 0.23 tons of RH is formed [8]. Rice milling is one of the most important industries in countries such as China, India, Indonesia, Malaysia, and Bangladesh [9].

Table 1: (a) Components and (b) physical properties of RH [21].

Property	RH
(a) Components (%)	
Cellulose	25–35
Hemicellulose	18–21
Lignin	26–31
SiO_2 (silica)	15–17
Solubles	2–5
Moisture content	5–10
(b) Physical properties	
Particle size (μm)	26.64
Surface area (m^2/g)	0.92
Density (g/cm^3)	1.00

Table 2: (a) Chemical and (b) physical properties of typical RHA fillers [34].

Property	WRHA	BRHA
(a) Chemical composition (%)		
CaO	0.360	0.120
MgO	0.160	0.078
Fe_2O_3	0.041	0.022
K_2O	0.690	0.950
Na_2O	0.034	0.018
Al_2O_3	0.025	0.023
P_2O_5	0.570	0.270
SiO_2 (silica)	92.200	53.880
Loss of ignition (LOI)	1.620	44.480
(b) Physical properties		
Particle size (μm)	6.60	19.50
Surface area (m^2/g)	1.40	26.80
Density (g/cm^3)	2.20	1.80

The incineration of RH is discouraged, which in turn produces ash, fumes, and toxic gases, leading to serious air pollution. Moreover, silica is intrinsic to RH and has been successfully used to enhance the mechanical properties of composites [10]. In the paddy plants of Malaysia, with a land area of approximately 680,000 hectares, a total of 840,000 tons of RH is produced every year [11, 12].

RH is a cellulose-based fibrous material with a wide range of aspect ratios [13]. The major components and physical properties of RH are tabulated in Table 1 [14–21]. Due to its high availability, low bulk density (90–150 kg m^{-3}), toughness, abrasiveness in nature, resistance to weathering, and unique composition, a variety of applications have been proposed in the literature. RH has the potential to be utilized as an insulating material, in the production of organic chemicals [22], panel boards and activated carbon [23], and supplementary cementing material [9]. Potential of RH as source of power generation and its financial viabilities have also been studied and have produced encouraging results [24, 25]. Even though some of this husk is converted into the above-mentioned end products, however, like many other agricultural byproducts, the industrial applications of this biomass are still limited with little economic value. Therefore, it is very important to find pathways to fully utilize the RH and an intense research scrutiny is currently undertaken worldwide to identify potential applications and to develop economically feasible processes for these applications on a commercial scale.

Various researchers have reported on a variety of applications that involve RH. However, it has also been reported that, in the manufacture of composites panels with the use of RH, there is a poor interaction between the RH and matrix materials, which results in weak particle-matrix adhesion [26]. Although previous studies reported that RH particleboard could be used in the manufacture of furniture and interior fitments, the physical and mechanical properties of the particleboard were lower than those of the particleboards made from wood particles [27–29]. The main reasons for lower physical and mechanical properties of the RH particleboards are low aspect ratio and waxy/silica layer of the RH particles [18]. The incorporation of RH into polymer matrices provides advantageous characteristics, such as biodegradability, light weight, toughness, and resistance to weathering, and also makes the final products more economically competitive [30, 31]. Compared with wood-based composites, the RH filled polymer composites have higher resistance to termite and biological attack and also better dimensional stability upon exposure to moisture. Thus, these composites are increasingly being used in building construction, such as frames for windows and doors, slidings, decks, and interior panels, and in the automotive industry for interior parts like door panels and trims [32].

RH, when burnt in open air outside the rice mill, yields two types of ash that can serve as fillers in plastics materials, namely, white rice husk ash (WRHA) and black rice husk ash (BRHA). The upper layer of the RHA mound is subjected to open burning in air and yields BRHA in the form of a carbonized layer. On the other hand, the inner layer of the mound being subjected to a higher temperature profile results in the oxidation of the carbonized ash to yield WRHA that consists predominantly of silica [33]. Table 2 shows the chemical and physical properties of typical RHA fillers [34]. As can be seen in Table 2, WRHA has been analyzed and found to have approximately 95% silica content, while BRHA has typically only approximately 54% silica content and a substantial carbon content that is about 44%.

Compression molding, injection molding, and extrusion are the most commonly used processing techniques for natural fiber reinforced thermoplastic composites. Enhancement of processability of natural fiber reinforced thermoplastics is an important issue, as the addition of interphase modifiers reduces the flow properties of the composites considerably by increasing the interaction between the fiber and the matrix [19, 20]. On the other hand, incorporation of processing aid improves processability by reducing the viscosity of the molten composites. Processability enhancement, while retaining and/or improving the performance properties, is more economical and can be an added advantage for the natural fiber composites industry [20]. RH has been

(a) (b)

FIGURE 1: SEM micrographs of the fractured surface of composites films with varying MMT content: (a) 0 and (b) 6 wt% of MMT content [42].

incorporated into a wide variety of materials such as low-density polyethylene (LDPE) [18], high-density polyethylene (HDPE) [35], polylactic acid (PLA) [36], polypropylene (PP) [37], and polyvinyl chloride (PVC) [38]. In general, the resulting composites exhibit improved mechanical, thermal, chemical, physical, and other properties with the addition of RH. However, RH was shown to have varying degree of reinforcing effect on the different polymer matrices. In this respect, the aim of the present study is to describe the possibility for utilization of RH as fillers in the different polymer composites.

2. Rice Husk Composites

Many studies on the physical, mechanical, and thermal properties of RH filled polymeric composites and how these incorporation affects various thermosets, thermoplastics, and elastomers have been attempted [18, 35–38]. The composite properties are largely dependent on matrix properties and hence selection of the matrix is based on desired end properties of the composites. The literature review reveals that polyethylene (PE), PP, PVC, and PLA are among popular choices for RH filled polymer composites. The subsequent sections begin with some general characteristics of the polymers along with their particular applications and then report the influence of incorporating RH on the properties. In addition, properties of RH filled polymer composites are also summarized to gain a deep intuitive understanding on the subject of developing multifunctional material.

2.1. Rice Husk/Polyethylene Composites. PE is a synthetic thermoplastic polymer which is made from the monomer ethylene. It is an important family member of the polyolefin resins. It is the most widely used polymer in the world and predominates in plastics owing to a number of reasons such as high elongation at break, good chemical resistance, excellent water barrier and good oxygen barrier properties, light weight, and lower cost. PE has contributed greatly to the development of high performance materials and their production and usage have been enormously increased [39]. PE is being made into products ranging from packaging films to bottles and automobile fuel tanks. PE is classified into different categories based mostly on its density and branching. It is worth mentioning that its mechanical and barrier properties are greatly influenced by the extent and type of branching. The long, chainlike molecules, in which hydrogen atoms are connected to a carbon backbone, can be produced in linear or branched forms. Branched versions are known as low-density polyethylene or linear low-density polyethylene; linear versions are known as high-density polyethylene and ultrahigh-molecular-weight polyethylene. Researchers have extensively studied the mechanical properties of RH filled PE and properties of some RH based composites are given in Table 3.

In our recent study [42], rice husk/maleic anhydride-modified polyethylene/montmorillonite (RH/MAPE/MMT) filled LDPE nanocomposites films (containing 0 to 6 wt% of MMT content) were prepared by extrusion blown film. The films were characterized by morphological, mechanical, oxygen (O_2) barrier, and thermal properties. The delamination of MMT layers evidenced from X-ray diffraction (XRD) results revealed the formation of intercalated nanocomposites, which was confirmed by the increase in the interlayer spacing and substantial reduction in their peak intensity of MMT. Scanning electron microscopy (SEM) confirmed that adding MMT did not alter the adhesion features between RH and LDPE (Figure 1). Furthermore, addition of RH/MAPE/MMT to the LDPE has improved the mechanical and O_2 barrier properties. Tensile and tear properties were improved with the addition of MMT and the maximum improvement of 8% in tensile strength and 5% in tear strength was observed at 3 wt% of MMT content. Concurrently, the modulus continues to increase while elongation at break decreases with increasing the MMT content. However, the most significant finding of this research was the large increase in O_2 barrier, in which the maximum barrier was observed at 4 wt% of MMT content (more than twofold). Additionally, thermogravimetric analysis (TGA) and differential scanning

TABLE 3: Mechanical properties of RH filled PE composites.

Matrix	RH (wt%)	Secondary filler (wt%)	Coupling agent or compatibilizer (wt%)	Tensile strength (MPa) (% incr.)	Tensile modulus (GPa) (% incr.)	Flexural strength (MPa) (% incr.)	Flexural modulus (GPa) (% incr.)	Impact strength (Jm^{-1}) (% incr.) Not.	Impact strength (Jm^{-1}) (% incr.) Unnot.	Reference
HDPE	65	—	—	13.5	2.39	33.5	2.90	26	38	[19]
HDPE	65	—	E-GMA (2.5)	22.5	2.37	49.06	3.32	28	65	[19]
HDPE	65	—	ethylene-acrylic ester-maleic anhydride (2.5)	20.9 (55)	2.42 (1.30)	40.2 (13.5)	2.81 (−3.1)	—	80 (111)	[20]
HDPE	30	—	Ultra-Plast TP10 (1)	—	—	—	—	—	75 (−32)	[40]
HDPE	50	—	MAPE (2)	25.64	2.5	—	—	18.39	—	[41]
HDPE	50	MMT* (2)	MAPE (2)	29.83	3.2	—	—	17.82	—	[41]
rHDPE	45	Sawdust (17.1)	MAPP (7.9)	7.5	0.6	—	—	15.20	—	[35]
rHDPE	44.5	Sawdust (17.1)	MAPP (7.9)	11.6	1.02	—	—	33.20	—	[35]
rHDPE/rPET (75/25)	60	—	MAPE* (3)/E-GMA (5)	23 (15)	1.55 (89)	35 (17)	3.5 (192)	25 (−69)	—	[12]

% incr. = percentage increase; not. = notched; unnot. = unnotched.
*phr.

TABLE 4: DSC results of the nanocomposites film [42].

Samples	T_{onset}	T_{pm}	T_m	T_c	X_c
LDC-0	511	551	107	95	19.1
LDC-2	519	553	107	96	19.5
LDC-3	524	559	108	95	19.8
LDC-4	525	559	108	95	20.7
LDC-5	520	557	107	96	20.3
LDC-6	514	552	108	95	19.6

LDC: LDPE/RH/MAPE (90 : 4 : 6 wt%) composites film reinforced with different MMT content (0–6 wt%); T_{onset}: onset temperature of degradation; T_{pm}: degradation temperature at peak maximum; T_m: melting temperature; T_c: crystalline temperature; and X_c: degree of crystallinity.

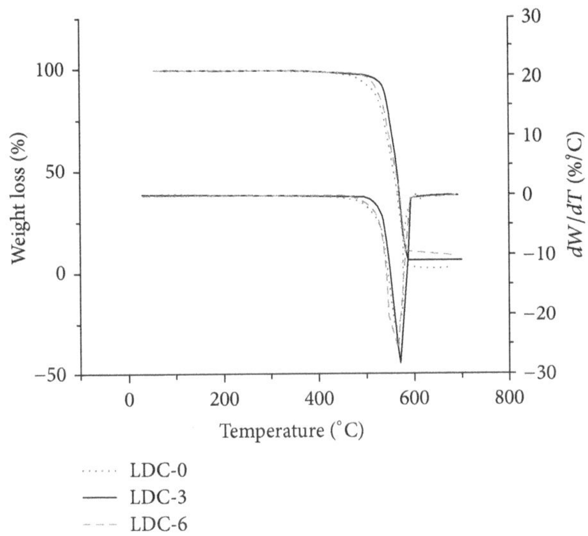

FIGURE 2: TGA and DTG thermograms of the nanocomposites film. LDC: LDPE/RH/MAPE (90 : 4 : 6 wt%) composites film reinforced with different MMT content (0, 3, and 6 wt%) [42].

calorimetry (DSC) results are shown in Figure 2 and Table 4, respectively. As can be seen, the degree of crystallinity (X_c) and thermal stability of LDPE/RH/MAPE/MMT nanocomposites are better than those of LDPE/RH/MAPE composites and the maximum improvement was at 4 wt% of MMT content. The unique combination of O_2 barrier, biodegradation, and mechanical properties for the LDPE/RH/MAPE/MMT nanocomposites films showed that these nanocomposites films provide potential candidates for a variety of food packaging.

The adhesion between the reinforcing fiber and the matrix plays an important role in the final mechanical properties of the materials. In another study by Majeed et al. [43], RH flour (having particle size of less than 75 μm) and nanoclay filled LDPE composites were prepared by extrusion blown film. In this study, MAPE was used as compatibilizer in various concentrations ranging from 0 to 8 parts per hundred (PPH), and the effect of MAPE compatibilizer on the morphological, tensile, and oxygen barrier properties was examined. The XRD results showed the addition of 2, 4, 6, and 8 PPH of MAPE increased the interlayer spacing of MMT by 20, 33, 36, and 38% compared to the uncompatibilized composites, respectively. From morphological observations, it can be concluded that the compatibilizer helps polymer molecules to penetrate and delaminate the clay platelets to realize large filler aspect ratio. In addition, it was found that adding compatibilizer enhances natural fiber dispersion in the continuous LDPE phase, and the RH aggregates started disappearing with increased compatibilizer concentration, which in turn improves homogeneity, as confirmed by SEM (Figure 3). Furthermore, the compatibilizer improved the tensile properties. It was observed that adding only 2 PPH of MAPE to the composites system improved the film tensile strength by approximately 22% compared to uncompatibilized films. The tensile properties of the composites films continued to increase as the MAPE concentration increased. This improvement in tensile properties is due to improved interfacial interactions between filler and matrix, owing to the MAPE. Therefore, composites films with high compatibilizer content showed better tensile strength. Moisture absorption usually deteriorates the mechanical properties of composites due to debonding of reinforcements from the polymer matrices. However, in this particular study, moisture absorption increased the tensile strength of uncompatibilized composites, and the authors attributed this increase in tensile strength to the fiber swelling, which could fill the gaps between fiber and polymer matrix and eventually could lead to an increased strength. It is notable that the increase in tensile strength is not significant for compatibilized composite samples with no significant change observed between the properties of dry and water aged samples. On the other hand, moisture absorption decreased the tensile properties for the compatibilized composites. Indeed, the most significant finding of this research was the large increase in oxygen barrier. Oxygen barrier property of the composite films improved remarkably by the presence of MAPE compatibilizer (more than twofold by 4 PPH MAPE). The decrease in permeability also confirms that the impermeable clay platelets are well dispersed as the extent of dispersion of clay platelets plays a major role in creating a tortuous path for the permeating molecules.

Recently, Kwon et al. [18] reported the effect of adhesive type and content on the dimensional stability and mechanical properties of three-layer particleboards. The composition layers were made by the mixture of wood particles (face layer, 30 wt%) and RH particles (core layer, 70 wt%) under laboratory conditions (Figure 4). Two types of thermosetting adhesives, liquid urea-formaldehyde (UF) and phenol-formaldehyde (PF), and thermoplastic adhesive (LDPE) powder were used as binder in the experiments. They found that the dimensional stability and mechanical properties of particleboards with RH core were significantly improved by the incorporation of LDPE into the core layer. It was studied that at the amounts of UF and PF adhesives decreased from 8 to 4 wt% at the same LDPE content (10 wt%); the thickness swelling (TS) and water absorption (WA) of the particleboards increased but were significantly lower than those of the control specimens. The reason was that a hydrophilic material (RH) was replaced by a hydrophobic material (LDPE). They explained that although the internal bond (IB) strength of particleboards containing 10 wt% LDPE

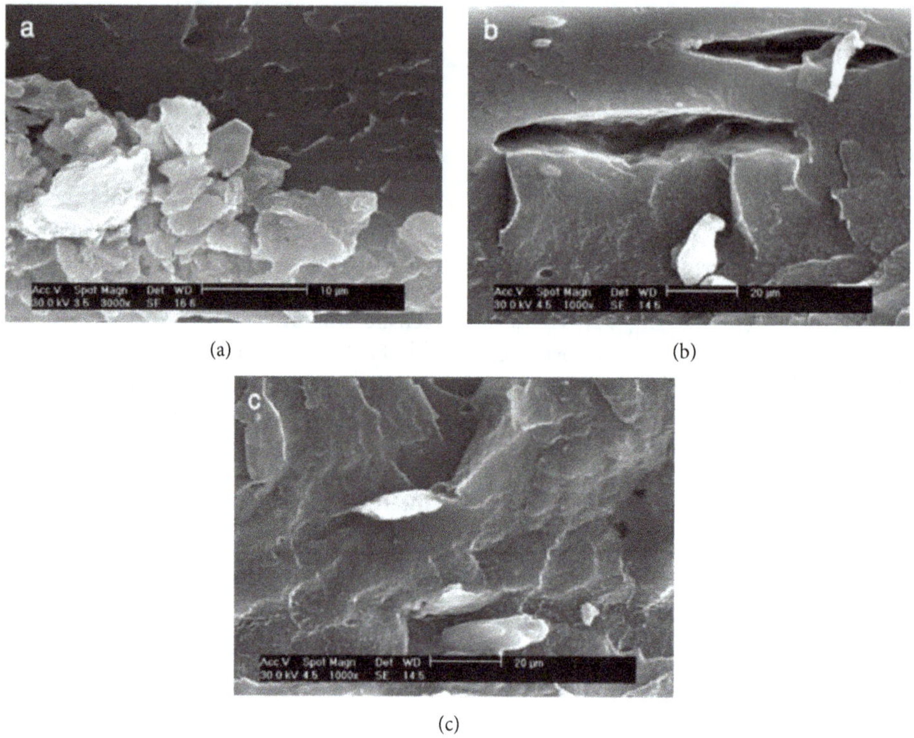

Figure 3: SEM micrographs of the fractured surface of composites films with varying compatibilizer concentrations: (a) 0, (b) 2, and (c) 6 PPH [42].

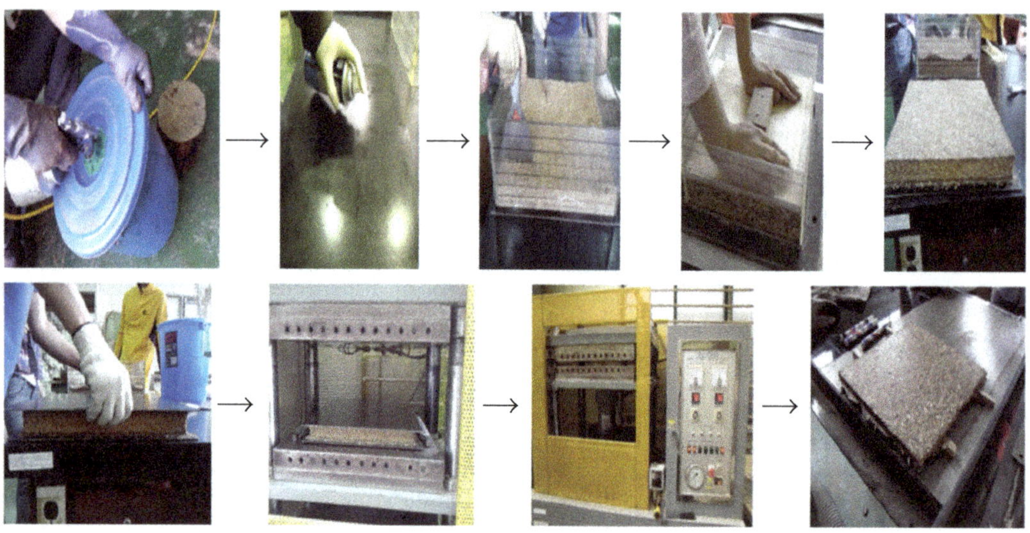

Figure 4: Laboratory-scale production of three-layer particleboards [18].

decreased with decreasing UF adhesive content in the core layer, it was higher than that of the control particleboards. Different fracture modes were observed during the IB tests. The fracture modes of particleboard samples with and without LDPE were presented in Figure 5 [18]. This study also demonstrated that the bending strength (MOR) and modulus of elasticity (MOE) of the particleboards containing 10 wt% LDPE decreased with decreasing the amounts of UF and PF adhesives. However, there were no statistically significant differences in the MOR and MOE values of the particleboard types as the adhesive content decreased from 8 to 4 wt% in the core layer. Based on the findings obtained from Kwon et al.'s [18] study, it can be suggested that the RH particleboards containing LDPE (in particular above 20 wt% of LDPE) are suitable for use in damp places, such as bathrooms, toilets, kitchens, and laundries.

Chuayjuljit et al. [44] successfully obtained silica (SiO_2) from RHA. They explored the possibility of using silica

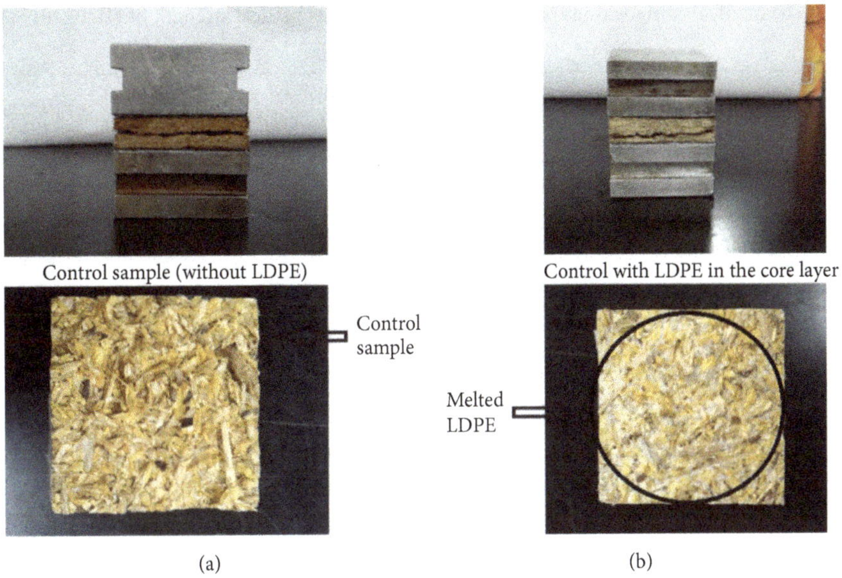

FIGURE 5: Fracture modes and fracture surface of the IB test samples without LDPE (a) and with LDPE (b) [18].

TABLE 5: Properties of RHA silica and commercial silica [44].

Properties	RHA silica	Commercial silica
Average particle size (micron)	4.27	6.61
Specific surface area ($m^2 g^{-1}$)	131.35 ± 2.34	511.97 ± 4.70
Bulk density ($g cm^{-3}$)	0.677	0.319
pH	5.7	6–8
Oil absorption ($g g^{-1}$)	1.92	2
Silica content (% SiO_2)	99.6	99.0

from RHA (RHA silica) as antiblocking agent in LDPE film. Properties of RHA silica were compared with commercial silica, as can be seen in Table 5. In addition, the appropriate amount of silica to be used as an antiblocking agent in LDPE film was also investigated. The results indicated that silica prepared from RHA, although composed of finer particles, showed less porosity than that of commercial silica, resulting in a lower specific surface area. Nevertheless, it is possible to use the obtained RHA silica as an antiblocking agent for LDPE film. They found that the addition of RHA silica to LDPE film modifies the film blocking behavior by inducing film roughness. However, the blocking is reduced not as much as in the case of commercial silica. It is investigated that, to the same antiblocking capability, 2000–3000 ppm of RHA silica is needed as opposed to 500–1500 ppm of commercial silica normally used in the plastic film industry. It was found that, as the amount of silica increased, certain mechanical properties (tensile strength, elongation at break, and tear strength) were decreased markedly. This is probably because silica has a lower bulk density and larger average particle size.

Panthapulakkal et al. [19] explored the potential of extruded RH filled HDPE composites profiles for structural applications. In their study, four different coupling agents based on ethylene-(acrylic ester)-(maleic anhydride) terpolymers and ethylene-(acrylic ester)-(glycidyl methacrylate) terpolymers were used to enhance the performance properties of the composites. The chemical composition and molecular weight of these coupling agents, and also tensile and flexural strength of HDPE/RH (32.5 : 65 wt%) composites are given in Table 6. They studied the effects of various coupling agents (2.5 wt%) on various performance properties of RH filled HDPE composites. The results indicated that these coupling agents significantly enhanced the tensile and flexural strength of the HDPE/RH composites, and the extent of the coupling effect depends on the nature of the interface formed. It is notable that the tensile and flexural strength of HDPE/RH composites without present of any coupling agent are 13.5 and 33.5 MPa, respectively. Their results showed that RH filled HDPE composites gain enough strength and rigidity by incorporating coupling agents and can be used for making structural materials. They found that incorporation of coupling agents enhanced the resistance to thermal deformation and the water absorption properties of the composites, whereas it reduced the extrusion rate significantly. Among the four coupling agents used, the coupling agent with a glycidyl methacrylate functional group and without any methyl acrylate pendant group on the polymer backbone (EGMA1) was found to be the best coupling agent for the RH filled HDPE composites. They concluded that the extrusion rate of these composites is reduced by the incorporation of the coupling agents and has to be enhanced by proper use of process aids.

In another study by Panthapulakkal et al. [20], the effect of coupling agent and processing aid on the performance properties of RH filled HDPE composites was studied. In their study, HDPE composites filled with 65% RH were extruded using a single screw extruder with die dimensions of 3 × 8 mm. They investigated the processability and performance properties of the HDPE composites, which were

TABLE 6: Interrelation between chemical compositions of coupling agents and mechanical strength of HDPE/RH (32.5 : 65 wt%) composites [19].

Coupling agent (2.5 wt%)	Butyl acrylate (wt%)	Maleic anhydrate (wt%)	Methyl acrylate (wt%)	Glycidyl methacrylate (wt%)	Molecular weight ($g\,mol^{-1}$)	Tensile strength MPa	Flexural strength MPa
EBMA1	6	3.1	—	—	120000	20.9	40.2
EBMA2	17.7	3.1	—	—	120000	19.3	42.7
EGMA1	—	—	—	8	100000	22.5	49.6
EGMA2	—	—	15	9	—	15.7	33.8
HDPE/RH*	—	—	—	—	—	13.5	33.5

*Without coupling agent.

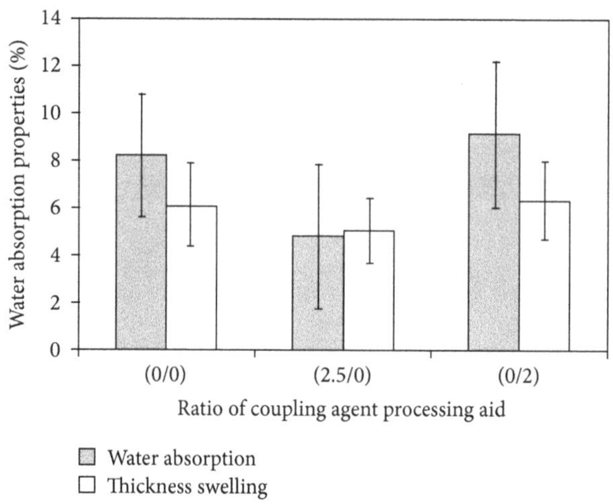

FIGURE 6: Tensile and flexural modulus of the composites with and without process additives [20].

highly dependent on the concentration of the coupling agent and processing aid in the composites formulation. In their study, attempt was made to optimize the composites formulation with respect to both coupling agent and processing aid to achieve optimum mechanical and water absorption properties with an optimum extrusion rate. It was observed that incorporation of a terpolymer (ethylene-acrylic ester-maleic anhydride) based coupling agent enhanced the tensile strength, flexural strength (Figure 6), and resistance to water absorption properties of the composites with a significant reduction in the processability. However, this study revealed that the addition of processing aid reduced the strength and stiffness of the composites and the effect was at a higher content of processing aid in the formulation. They found that the extrusion rate of the composites improved by the incorporation of processing aid and processing aid adversely affected water absorption resistance of the composites. They observed that the composites with a coupling agent to processing aid ratio of 0.73 : 0.59 by weight showed an optimum combination of performance properties and extrusion rate.

In an interesting study, the compounding of RH and HDPE was investigated using corotating twin-screw extruder [40]. Four sizes of RH were studied at various compositions. The size ranged from 500 μm and below, while the content of RH in the composites varied from 30 to 50 wt%. Also, a fixed amount of Ultra-Plast TP10 as a compatibilizer and Ultra-Plast TP 01 as lubricant was added to the composites compound. It was noticed that among the four RH sizes (A, 250–500; B, 125–250; C, 63–125, and D, <63 μm), RH with the size of 250–500 μm which has the largest particle was considered to be the most appropriate size as the composites filler based on thermal stability test. The melt flow rate of RH/HDPE decreased with the increasing in RH filler and apparent viscosity also increased with composition for all fillers sizes. They found that, when RH was incorporated into the HDPE matrix, the melt flow rate of RH size A (250–500 μm) composites performed well above the control limit set. Although composites with RH size C and D had better melt flow at 30 wt%, the fillers tended to degrade at high processing temperature, such as 190°C or when longer residence time was used. Melt flow rate above 4 g 10 min^{-1} was found to be the lower limit for injection molding process. It is observed that RH size A has the highest impact resistance among the other sizes. The large particles of RH have low agglomeration and dispersed better in the composites matrix. Filler incorporated at 30% of RH size A was found to be suitable for processing and to have optimum rheological properties with respect to impact strength compared to 40 and 50 wt% of RH.

Kord (2011) fabricated nanocomposites based on HDPE and rice husk flour (RHF) blend (ratio of 50 : 50 wt%) with different content of nanoclay using an internal mixer [41]. In his study, the influence of nanofiller at four levels, 0, 2, 4, and 6 parts per hundred compounds (phc), on the thermal and rheological behavior was studied. The morphology of nanoclay was determined by XRD, and the effect of morphology on the thermal and dynamic mechanical properties was also considered. It was indicated that the crystallization temperature, crystallization enthalpy, and crystallinity level increased with increasing of nanoclay up to 2 phc and then decreased. Additionally, the dynamic mechanical behavior of nanocomposites was improved by the addition of nanofiller and XRD patterns revealed that the nanocomposites formed were intercalated. It was also reported that the sample containing 2 phc of nanoclay had higher order of intercalation and better dispersion, as confirmed by XRD. It was found that with increasing of coupling agent (MAPE) a fully exfoliation morphology could be obtained. Therefore, he concluded

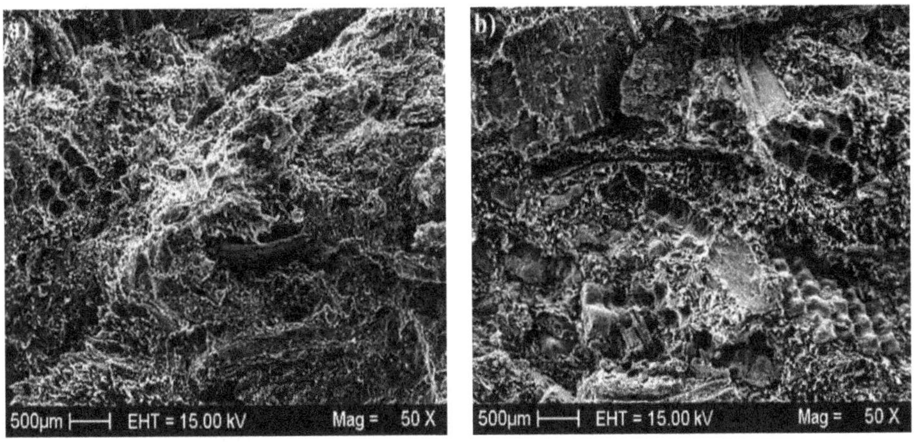

(a) (b)

FIGURE 7: SEM micrographs of fracture surfaces of sample from composites at 50x: (a) 0.5% AOs and (b) 20% FRs [35].

that thermal and rheological properties of the HDPE/RHF nanocomposites were improved by increasing of MAPE content.

Hamid and coworkers [35] studied the physical and mechanical and thermal properties of recycled high-density polyethylene (rHDPE) reinforced with a high content of hybrid RH and sawdust. They also studied the effect of antioxidants (AOs) and fire retardants (FRs) as additives. The rHDPE and natural fibers were compounded into pellets by a compounder; then the pellets were extruded using a corotating twin-screw extruder and the test specimens were prepared using hot and cold press processes. Also, maleic anhydride-grafted polypropylene (MAPP) was used as a coupling agent to increase the adhesion of the fiber-matrix interface. Results showed that the samples with 0.5 wt% of AOs and 20 wt% of FRs produce the most reasonable strength and elasticity of composites. Furthermore, the effect of the presence of AOs and FRs on water uptake was minimal and microstructure analysis of the fractured surfaces of composites confirmed improved interfacial bonding. As can be seen in Figure 7, composites with (a) 0.5% AOs and (b) 20% FRs content show the good dispersion of hybrid fibers in the matrix. Therefore, good dispersion of fibers in composites leads to high performance on physical and mechanical properties. There is no clear gap between hybrid fibers and matrix (Figure 7), indicating the good interface bonding. The fibers were perfectly attached to the matrix and it is also evident that the fibers were strongly imbedded in the matrix, indicating the efficiency of the mixing, which was attributed to a good interfacial interaction between hybrid fibers and matrix. It was concluded from the results that the addition of AOs and FRs as mineral fillers improves not only the aging and fire resistance aspects but also the durability performance of rHDPE reinforced with a hybrid of RH and sawdust. As a result, various parameters influence the mechanical properties of fiber-reinforced composites: fiber aspect ratio, fiber-matrix adhesion, stress transfer at the interface, mixing temperatures, and the addition of a coupling agent.

More recently, Chen et al. [12] fabricated RHF composites based on uncompatibilized (UPB) and compatibilized (CPB) recycled high-density polyethylene/recycled polyethylene terephthalate (rHDPE/rPET) with ethylene-glycidyl methacrylate (E-GMA) copolymer by a two-step extrusion and hot pressing with fiber content of 40, 60, and 80 wt%. They noted that the tensile and flexural properties of the composites were significantly increased with increasing RHF content, especially the tensile and flexural modulus. In contrast, as expected, they observed that the addition of RHF to the polymer blends (PB) matrices decreased the elongation to break and impact strength of the composites. However, the results indicated that the compatibilization of PB matrices by E-GMA compatibilizer could further increase the overall mechanical properties of CPB-based composites. In their research, WA test was examined in distilled and seawater. In long-term water immersion test, Chen et al. [12] found that the WA and TS increase with RHF content and immersion time for all types of composites irrespective of the types of polymer matrix (UPB and CPB) and immersion conditions (distilled and seawater). They also observed that the diffusion coefficient, thermodynamic solubility, and permeability obtained from the Fickian model and the swelling rate parameter obtained from swelling model increased with the RHF content. However, all these parameters decreased with the compatibilization of rHDPE/rPET by E-GMA, especially at lower content matrix. SEM analysis of the compatibilized matrix composites confirmed the improved interfacial bonding of matrix-matrix and filler-matrix phases (Figure 8). As can be seen in Figures 8(a) and 8(a′), the UPB displayed an incompatibility morphology of an obvious phase segregation structure between rHDPE and rPET, whereas the CPB exhibited a finer dispersion of rPET component inside the HDPE matrix. Figures 8(b) and 8(c) show a coarse morphology in the polymer matrices. From Figure 8(b), it can be clearly observed that the larger particle size shows no evidence of interfacial interaction and adhesion within the matrix phase, which confirmed the incompatibility of the two matrix individual components. The black arrows in the

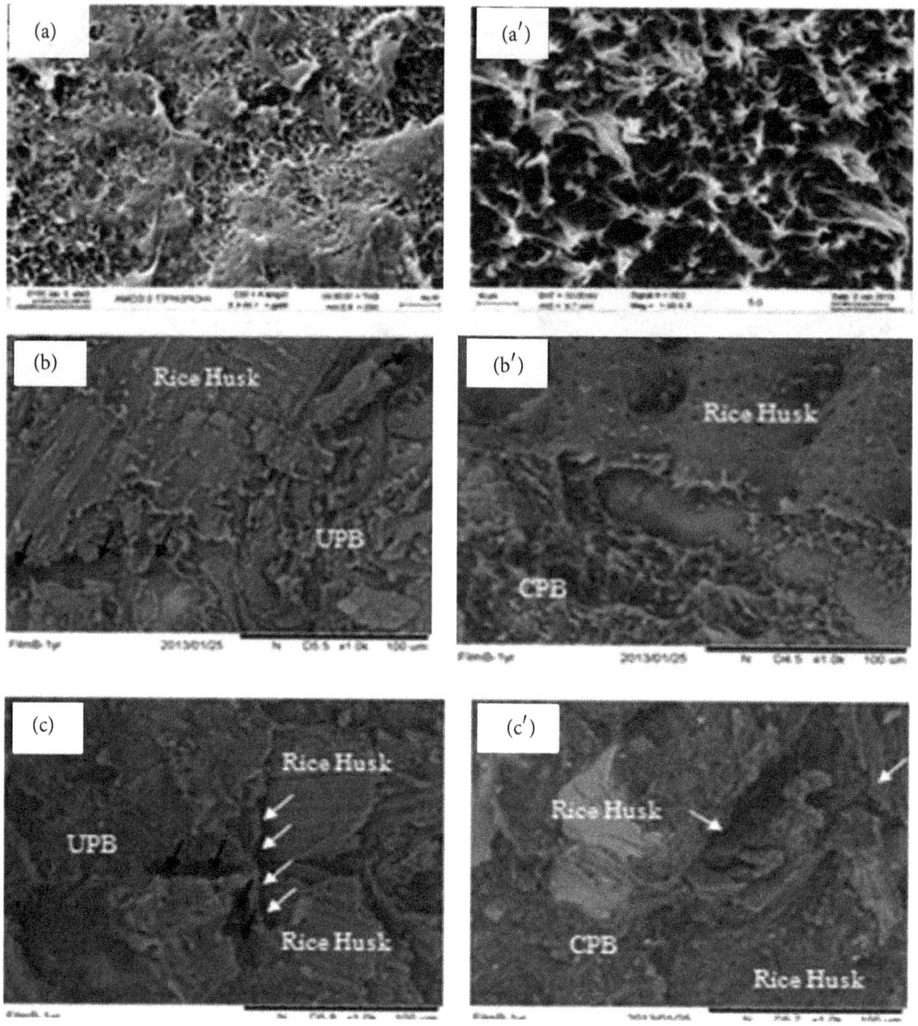

FIGURE 8: SEM micrographs of (a) 0 wt% UPB, (a′) 0 wt% CPB, (b) 40 wt% RHF-UPB composite, (b′) 40 wt% RHF-CPB composite, (c) 80 wt% RHF-UPB composite, and (c′) 80 wt% RHF-CPB composite (magnification, 1000x) [12].

Figures 8(b) and 8(c) show that the available spaces or sites for the water molecules diffuse and form hydrogen bonds with the polar rPET by breaking the existing bonds between the hydroxyl groups of the rPET chain. However, RHF-reinforced composites with CPB matrices exhibited a finer surface morphology (Figures 8(b′) and 8(c′)) with smaller size matrix domains, which indicates the good dispersion or rPET particles inside the rHDPE matrix and the better interfacial adhesion within the matrix. By comparing Figures 8(b/b′) and (c/c′), it can be observed that the RHF fillers at the concentration of 40 wt% were well embedded in the PB matrix, which indicates the relatively good interface adhesion between the fibers and matrix. Meanwhile, the white arrows in the latter figure show a clear gap between the polymer matrix and the RHF fillers, which enhances the water diffusion. This indicated the poor filler-matrix interfacial bonding, as a result of insufficient amount of coupling agent at high RHF content (80 wt%). Based on the results and findings in this work [12], they proposed that the agrofiller/recycled plastics composites are suitable for outdoor applications in terms of low WA and TS.

2.2. Rice Husk/Polylactic Acid Composites. PLA is a synthetic thermoplastic polymer and one of the most promising candidates for biodegradable polymers with excellent mechanical properties [45]. PLA is produced on a large scale from fermentation of cornstarch to lactic acid and subsequent chemical polymerization. As PLA is obtained from renewable resources, therefore, it represents a good candidate to produce disposable packaging in addition to its application in the textile and automotive industries and clinical uses [46]. Pure PLA can degrade to carbon dioxide, water, and methane in the environment over a period of several months to two years, compared to other petroleum plastics needing a very long period. Although PLA is an eco-friendly plastic with good biocompatibility, poor hardness, slow degradation, hydrophobicity, and lack of reactive side-chain groups limit

TABLE 7: Mechanical properties of RH filled PLA composites.

Matrix	RH (wt%)	Coupling agent or compatibilizer (wt%)	Tensile strength (MPa) (% incr.)	Tensile modulus (GPa) (% incr.)	Flexural strength (MPa) (% incr.)	Impact strength (Jm^{-1}) (% incr.) Unnot.	Reference
PLA	AT RHF (25)	—	7.35 (−82)	—	13.9 (−84)	—	[48]
PLA	WT RHF (25)	—	39.47 (−21)	—	52.4 (−39)	—	[48]
PLLA/starch (42/42)	5	TC/MDI (10/1)	8.58	0.062	—	35.23	[49]
PLLA/starch (37/37)	15	TC/MDI (10/1)	14.58	0.225	—	22.31	[49]

AT RHF = alkaline-treated rice husk fiber; WT RHF = water-treated rice husk fiber; TC = triethyl citrate; MDI = diphenyl-methane-4,4′-diisocyanate.

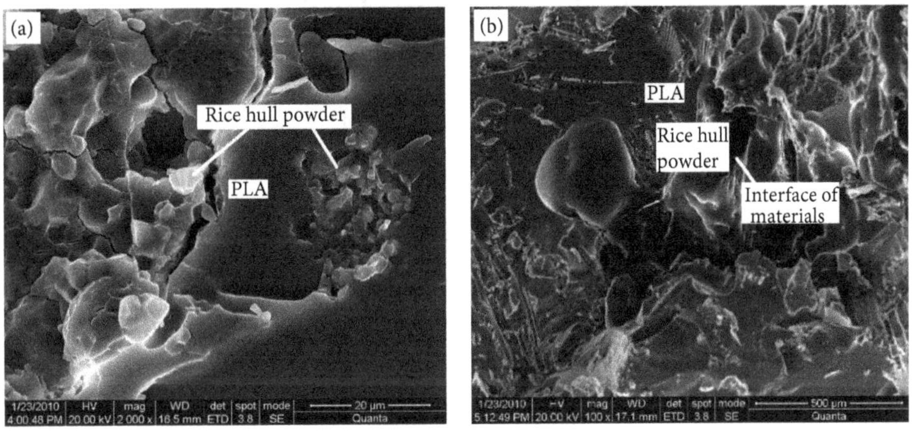

FIGURE 9: SEM micrographs of composites: (a) 20 g of RHP and (b) 20 g of RHP + 2 g of MAPP [50].

its application [47]. The mechanical properties of RH filled PLA composites are illustrated in Table 7.

In research by Hua et al. [50], the rice husk powder (RHP) was modified by PLA using melt-blending technique, and MAPP was used as coupling agent. The mechanical properties, water absorbability of RHP/PLA composites, and the relations between these properties and microscopic characteristics were investigated. Their result showed that, with the increasing content of RHP, the maximum tolerable tension and tensile strength increased. The water absorptivity of materials increased with the increasing of RHP content (after 24 hours). They also found that coupling agent could significantly reduce the water absorbability of composite materials. They observed that the hygroscopicity was slow in the beginning and it became strong after a certain period of time because the characteristics will change remarkably after the RHP soaking in the water for a long time. It is also reported that, with the increasing of RHP, the hardness of composite materials increased the tensile strength enhancement, and the maximum bearable tension increased too. As shown in Figure 9(b), there is no obvious change in the content density and the porosity of the RHP in the composites with MAPP compared to composites without MAPP (Figure 9(a)). But it shows that the pore becomes obviously small and smoother. In addition, when the adhesion between PLA and RHP became closer, the agglutination performance of materials enhanced, the composite materials had lower water absorptivity, and the tensile strength and the most high tension all obtained enhancement obviously when the MAPP was added to RH composites. Therefore, the usage of coupling agent is one of the effective ways to strengthen the performance of composite materials.

Zawawi et al. [51] did a comparison study on microwave properties of natural fiber reinforced PLA composites in order to recognize their potential as alternatives to common printed circuit board (PCB) for electronic communication industries. Thus, they reported on measured dielectric properties for two composites (kenaf/PLA and RH/PLA) and their results were compared. The samples were prepared in equal wt% (50 : 50 wt%), and the complex dielectric permittivity ($\varepsilon = \varepsilon' - j\varepsilon''$) and loss tangent ($\tan\delta$) of both composites have been studied in the frequency range of 500 MHz to 10 GHz. It was observed that, from 500 MHz to 3.32 GHz, real permittivity (ε') values are consistent throughout the wide frequency range, at approximately 3.3. However, the permittivity seems to decrease at higher frequencies starting from 3.35 GHz for both composites, down to 2.5 at 10 GHz. From their observation, real permittivity (ε') parameter has shown significant variation with frequency range. It is also observed that kenaf/PLA composites with the same weight content have higher permittivity (ε') value compared to RH/PLA composites. Meanwhile, they noticed that loss tangent ($\tan\delta$) is low and remains similar for both composites. The results showed that both kenaf/PLA and RH/PLA composites can

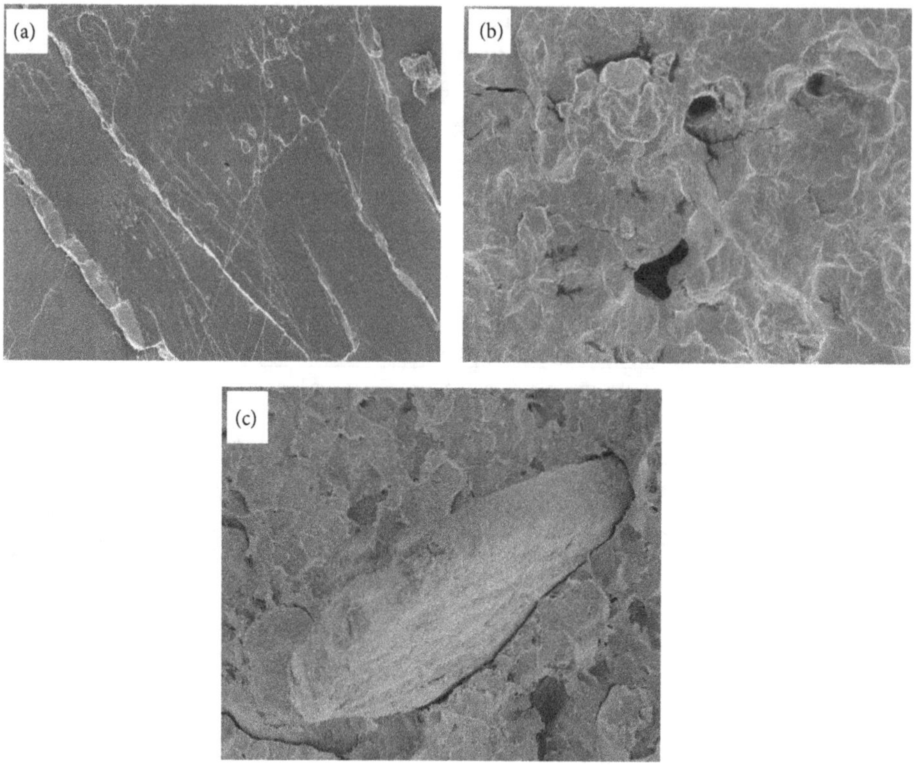

FIGURE 10: FESEM micrographs of flexural fractured surfaces of (a) pure PLA, (b) PLA/AT RHF, and (c) PLA/WT RHF composites [48].

be used for microwave applications due to their permittivity range and low loss tangent comparable to existing commercial substrates, such as epoxy-based dielectrics.

Most recently, a research group in (2014) focuses on the influence of surface treatment and fiber sizes on mechanical behavior, physical properties, and morphology of RH fiber reinforced PLA [48]. Modified RH fiber was prepared by using 6 wt% sodium hydroxide (NaOH) and distilled water. PLA composites reinforced with modified RH fiber (25 wt%) was mixed using the internal mixer and fabricated by the mini injection molding. Farah Dina et al. (2014) found that the tensile and flexural strength of PLA composites with 100, 200, and 500 μm particles sizes of water treated fiber are much higher than those of alkaline treated ones. DSC and TGA analyses were performed and indicated that the water and alkaline treatments reduced the thermal stability of the PLA composites (Table 8). The authors claimed that surface treatment by NaOH on RH fiber could not successfully improve the interfacial adhesion between fibers and PLA matrix. Field emission scanning electron microscopy (FESEM) micrographs of flexural fractured surfaces of composites are shown in Figure 10. Microcrack and pores are indicated by circle and arrow, respectively (Figure 10(b)). This behavior is due to the brittle fracture of PLA matrix adjacent to the alkaline treated RH fiber (AT RHF). The interface between the alkaline treated RH fiber and PLA is loose. It may be due to the loss of fiber-matrix interfacial adhesion and the failure of the fibers in the load distribution which are the evidence of poor fiber-matrix adhesion. From Figure 10(c), it is clear that the PLA/water treated RH fiber (WT RHF)

TABLE 8: DSC and TGA results of WT RHF and AT RHF filled PLA composites.

Samples	T_g (°C)	T_m (°C)	ΔH_m (J/g)	Decomposition temp. of the different weight loss (wt%)		
				5	50	75
PLA	64	171	99	364	395	407
PLA/WT RHF	66	169	77	347	382	400
PLA/AT RHF	57	160	52	296	339	—

fracture surface is rougher than the PLA/alkaline treated RH fiber (Figure 10(b)), which facilitates the mechanical interlocking. Additionally, two phases can be clearly seen and RH fibers were pulled out from the matrix in the fracture process, with large voids thereby being created, as indicated by arrow (Figure 8(c)).

Working in another direction, Battegazzore et al. [52] extracted silica powder from RH by a simple extraction procedure (Figure 11) for preparing PLA composites with differing filler contents (5, 10, 20, and 30 wt%) using melt compounding. The thermal, mechanical, and morphological properties have been investigated. Indeed, the presence of extracted silica has induced significant improvements of Young's modulus as well as a slight reduction of the oxygen permeability of PLA. The collected values at 10 and 30 wt% have been compared with prepared samples using a commercial silica and they found that the achieved mechanical properties turned out to be higher than those exhibited by the homologous samples prepared by using a commercial

(c)

FIGURE 11: Typical images of (a) RH, (b) RHA, and (c) extracted silica [52].

silica, as can be seen in Figure 12. The economic analysis of the materials and whole process (silica extraction and composites preparation) highlighted that the production of a formulation containing 20 wt% silica can be considered economically sustainable, when the energy recovered from the RH burning is reused.

A group in France [53] investigated specific surface treatments, which aimed to improve the adhesion between PLA, RH, and Einkorn wheat husk. In the first part of their study, husks (RH and Einkorn wheat husk) were treated by alkaline solutions at three concentrations (2, 5, and 10%) at room temperature for different times between 6 and 48 hours. Results revealed that the alkaline treatment dissolved a fraction of waxes, lignin, and hemicellulose, so that the polarity of their surface was increased and therefore alkaline treated husks were more sensitive to moisture and had lower surface energies. Moreover, the alkaline treatments had more influence on wheat than on RH, as confirmed by chemical mechanism determination (Figure 13), FTIR (Figure 14), and SEM observations (Figure 15). In the second part of the Tran et al. (2014) study, different surface treatments including alkaline treatments with NaOH solutions at different concentrations and silane agents (γ-aminopropyltriethoxysilane (APS) and γ-glycidoxypropyltrimethoxysilane (GPS)) on the characteristics of RH and wheat husk were investigated. These silane treatments were applied either alone or on previously alkaline treated husks (NaOH 5% for 24 h). The results showed that silane treatments reduced the moisture sensitivity and raise the energy surface of the husks. In the last part of their study, treated husks were incorporated into PLA to produce composites. Nevertheless, alkaline treatments did not improve the mechanical properties of composites. However, an increase in the mechanical properties and the thermal stability of composites was obtained. Better results were obtained with APS agent compared to GPS agent because of a higher reactivity; the group amine of APS showed more effect on PLA than group epoxy of GPS. When silane treatments were combined with an alkaline pretreatment, composites showed higher mechanical properties and thermal stability than composites with only silane treatment. Alkali treatment cleaned the surface of husk and activated the hydroxyl groups. They found that these active hydroxyl groups reacted with the silanes more easily than the hydroxyl group on the surface of untreated husks. It was also given in evidence that wheat husks (with higher cellulose content) lead to better performances when they are incorporated within PLA compared to RH. They proposed future works on the influence of the treatment surface on the aged resistance of composites.

In another work, Sareeladdanon and Potiyaraj [54] prepared polymer composites of PLA and linear low-density polyethylene (LLDPE) with the addition of different amount of rice husk silica (RHS), 0.5 and 1 wt%, using melt-mixing in a twin-screw extruder and processed into film by a chill roll cast extruder. The mechanical properties of PLA/RHS and LLDPE/RHS composites were investigated through tensile testing and tear testing. The results showed that the tensile modulus and tensile strength of PLA increased with the addition of 0.5 wt% RHS and dropped with further addition of RHS (1 wt%); this may arise from the agglomerates of silica particles in the PLA matrix. However, the elongation at break of PLA increased with the addition of silica at the higher content. On the other hand, the tensile modulus of LLDPE decreased with the addition of silica because of nonuniform dispersion of silica particles.

In a recent publication, Mohammadi-Rovshandeh et al. [49] investigated the use of RH and bleached rice husk (BRH) in the form of powder and the effect of lignin removal on mechanical and thermal properties of prepared composites with PLA and starch. They confirmed that the lignin has been removed from RH using FTIR technique, as indicated by disappearing of lignin aromatic C=O peak (1648 cm^{-1}) in the FTIR spectra of BRH (Figure 16). The tensile, Charpy impact, and hardness tests were implemented to study the

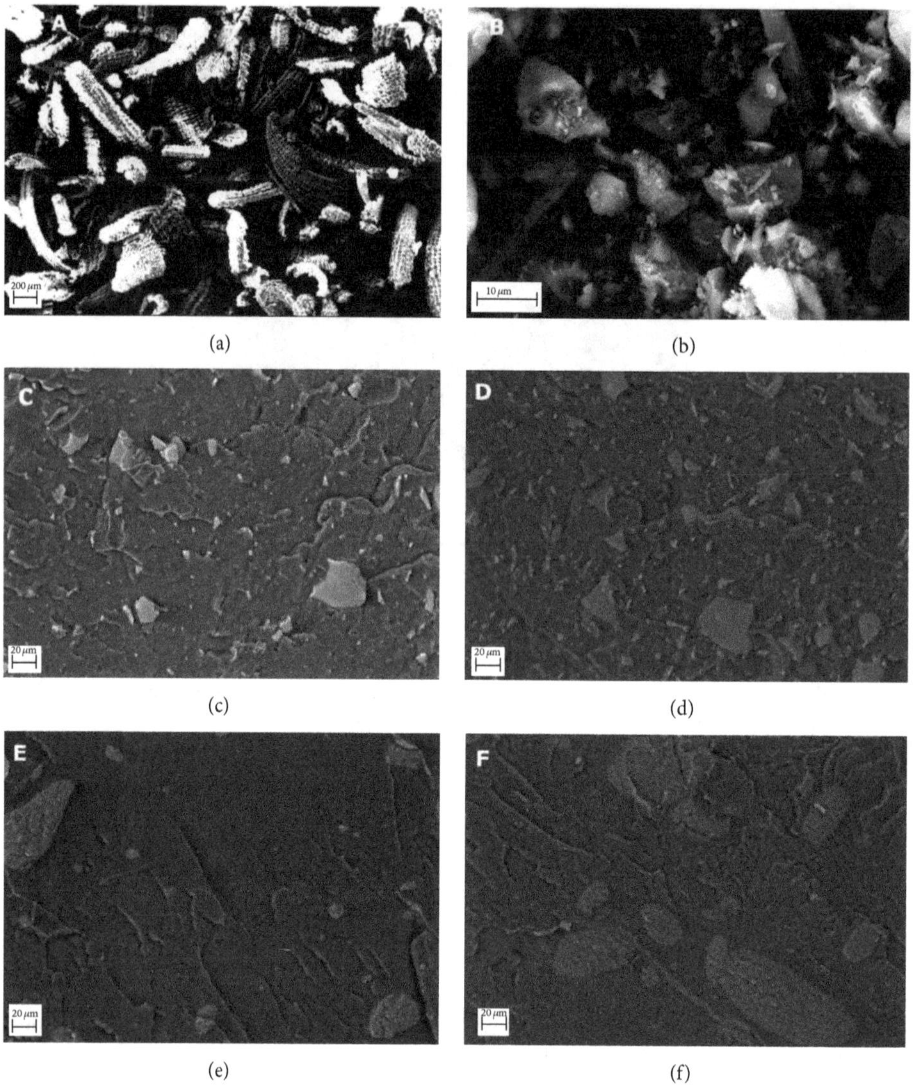

FIGURE 12: SEM micrographs of silica powder (a and b) and composites containing 10 and 30 wt% of extracted (c and d) and of commercial silica (e and f) [52].

mechanical properties. It was shown that the filler content in PLA/starch blends could improve the mechanical properties of composites. It was concluded that filler caused an increase in the toughness and Young's modulus of composites. It was also found that filler improved the tensile strength and hardness of PLA/starch blend. The mechanical properties revealed that 5–10% of BRH or RH may be used in composites which result in improvement of the tensile properties as well as hardness. It was clearly observed that the BRH samples show larger modulus, Charpy impact strength, and hardness compared to those composites filled with RH. Despite larger values for the elongations at break of RH composites, the mechanical properties showed a significant improvement (25–50%) on lignin removal. In addition, the differential scanning calorimetry thermograms revealed an improvement in crystallization behavior and reduction of T_g from 47 to 31°C in the case of BRH filled polymer composites. Finally, they reported that, due to the removal of lignin in the samples and modification of mechanical properties, the BRH could have the ability of wide use in the bioplastics and food packaging industries.

In an interesting research in Italy, Battegazzore et al. [36] focused on the extraction of cellulose from RH through a multistep process (Figure 17) and used different extracted cellulose contents (5, 10, 20, and 30 wt%) to fill PLA composites by melt-blending. The crystalline structure and morphology of the extracted cellulose have been investigated by using XRD and SEM, as can be seen in Figure 18. Four main peaks have been found at 16°, 17°, 22°, and 34° (Figure 18(a)). Indeed, the extracted cellulose has a crystalline structure with respect to hemicellulose and lignin, which are amorphous in nature. More specifically, cellulose has a crystalline structure due to the hydrogen bond interactions and Van der Waals forces between adjacent macromolecules. The thermal, mechanical, and oxygen barrier properties of the resulted PLA composites have been investigated and correlated with the observed

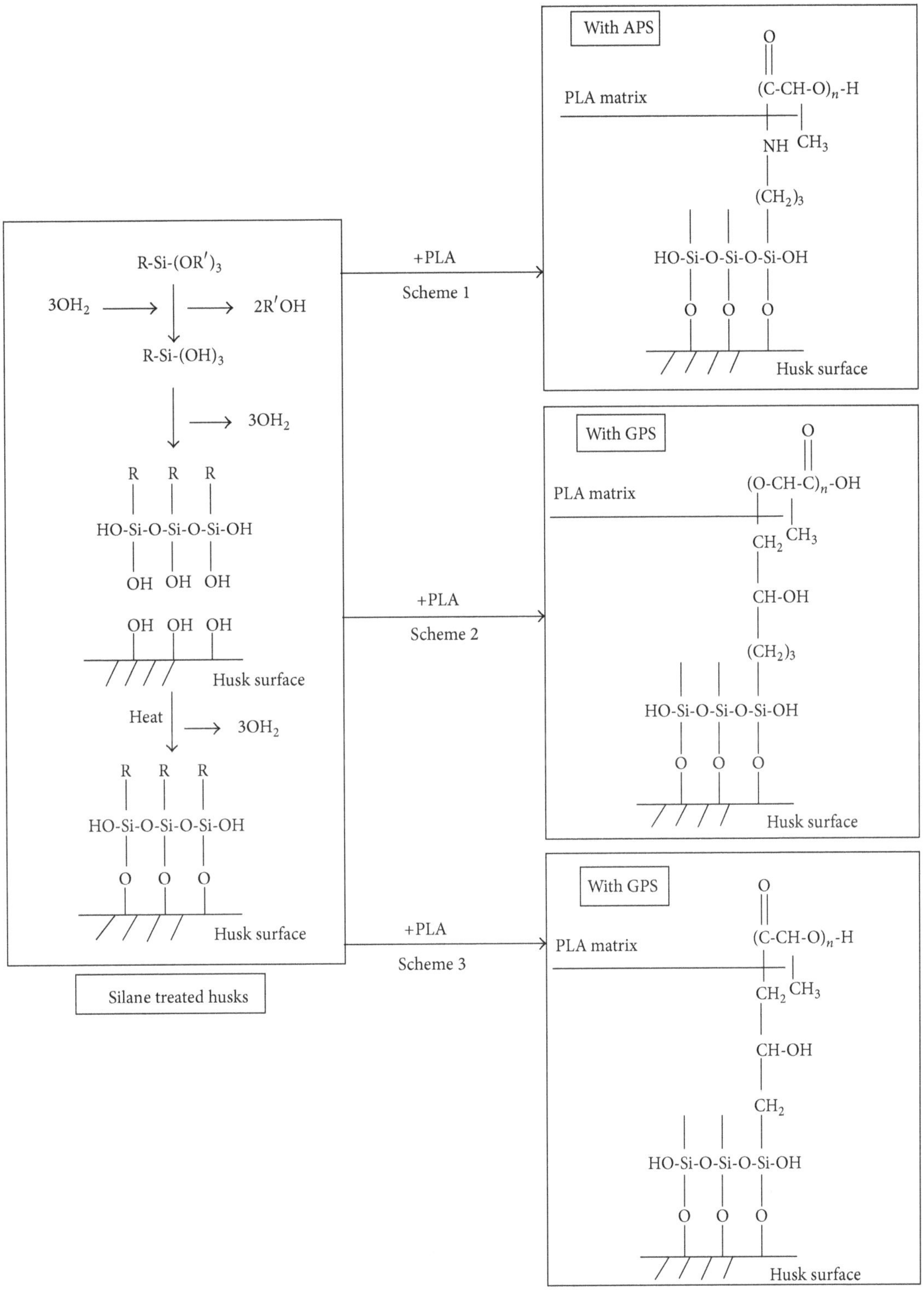

FIGURE 13: Chemical mechanisms at the interfacial area between silane (Scheme 1: APS and Schemes 2 and 3: GPS) treated natural fibers and PLA matrix [53].

morphologies (Figure 19). They reported that, in spite of a broad micrometric distribution of the cellulose particle size (as indicated by arrow in Figures 19(b) and 19(d)), the mechanical properties turned out to be strongly improved and the oxygen permeability properties have proven to be slightly reduced as well. The most interesting aspect of this work was that similar mechanical and oxygen permeability properties have been achieved for both commercial cellulose

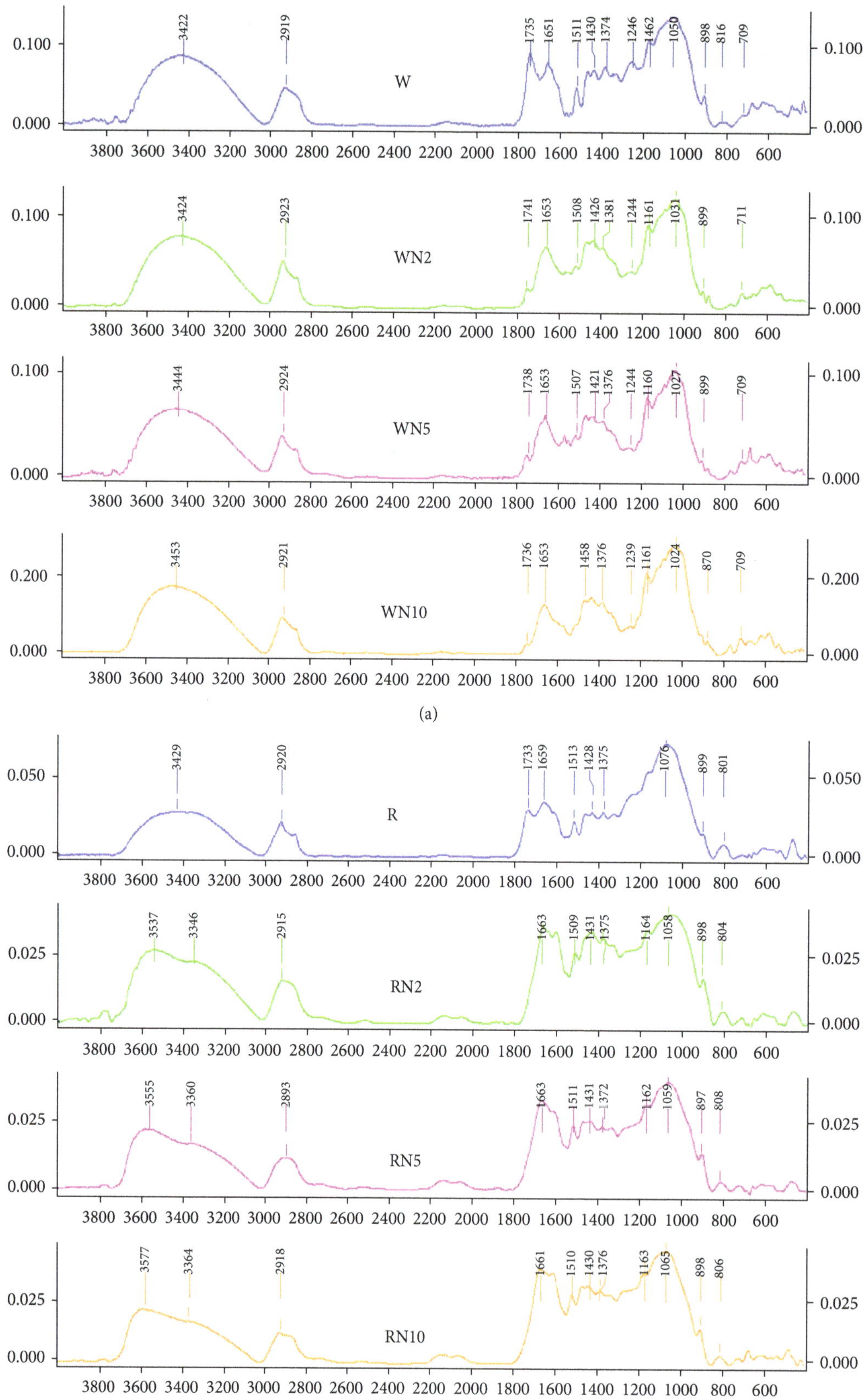

FIGURE 14: FTIR spectra of untreated and alkaline treated (a) wheat husk and (b) RH at different NaOH concentrations (2, 5, and 10%). Einkorn wheat husk (W), einkorn wheat husk + NaOH (WN), long-grain RH (R), and long-grain RH + NaOH (RN) [53].

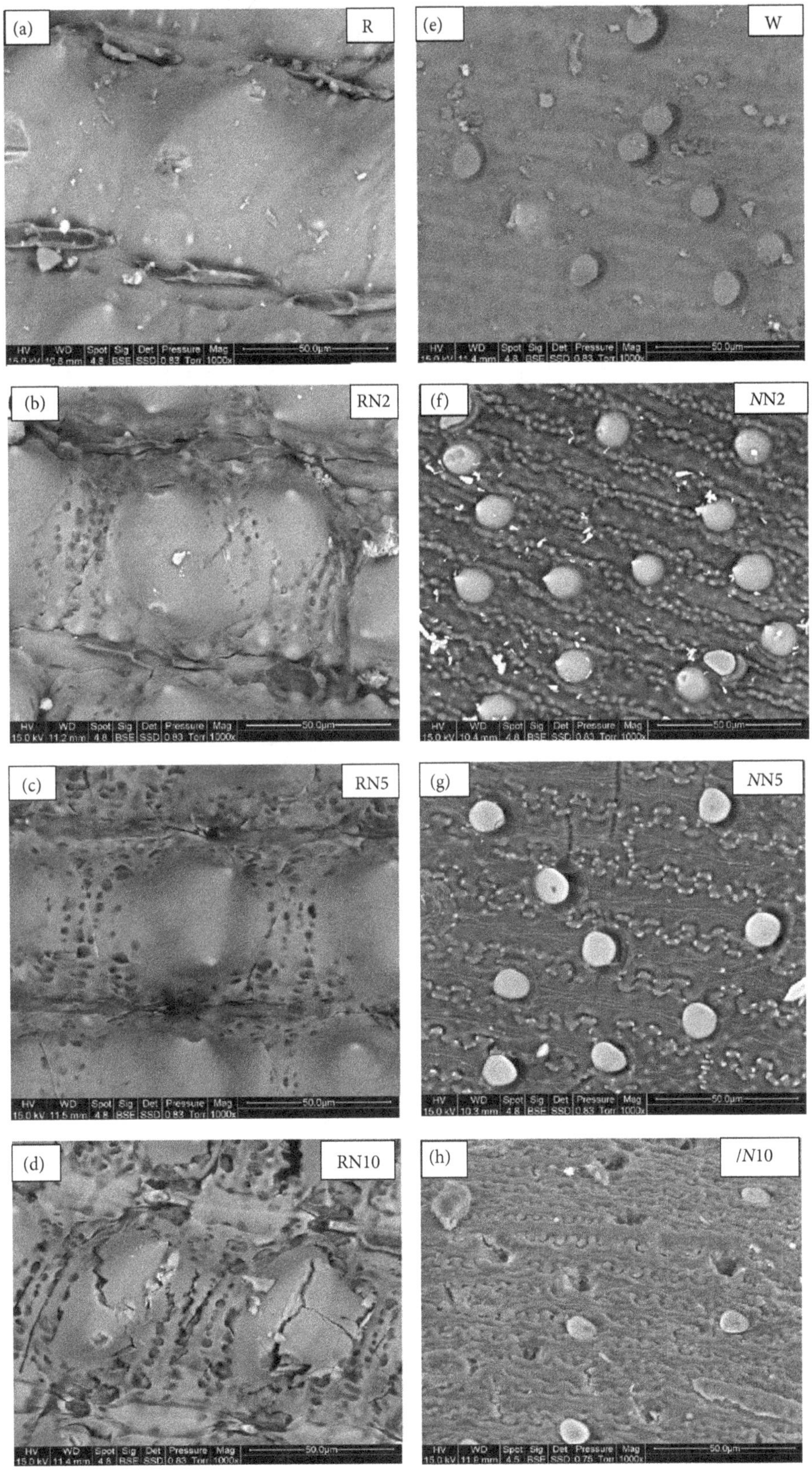

FIGURE 15: SEM micrographs of husks before and after treatment by NaOH: (a) untreated RH; (b–d) treated RH with NaOH concentration of 2%, 5%, and 10%, respectively; (e) untreated wheat husk; and (f–h) treated wheat husk with NaOH concentration of 2%, 5%, and 10%, respectively [53].

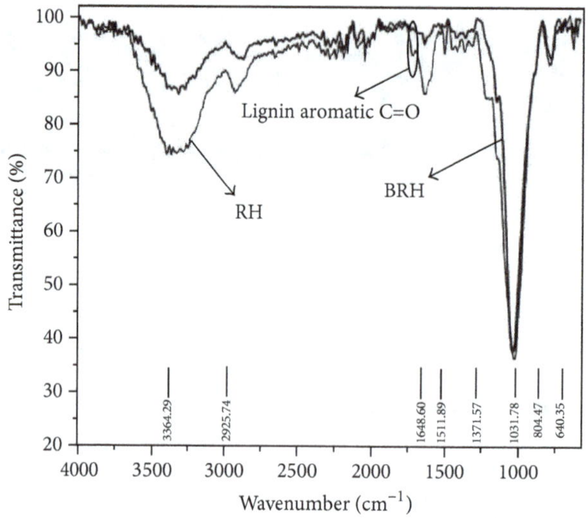

FIGURE 16: FTIR spectra of RH and BRH [49].

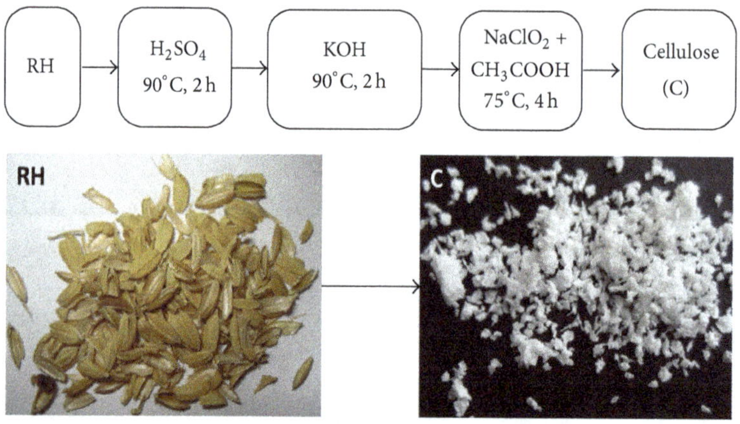

FIGURE 17: Schematic representation of the cellulose extraction from RH [36].

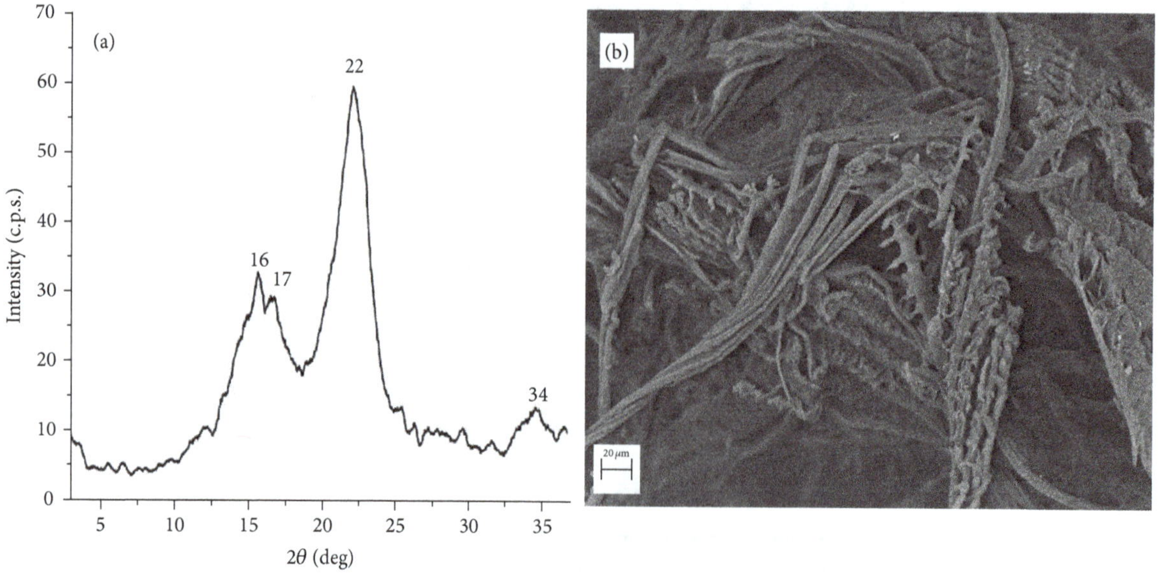

FIGURE 18: (a) XRD patterns and (b) SEM micrograph of the extracted cellulose [36].

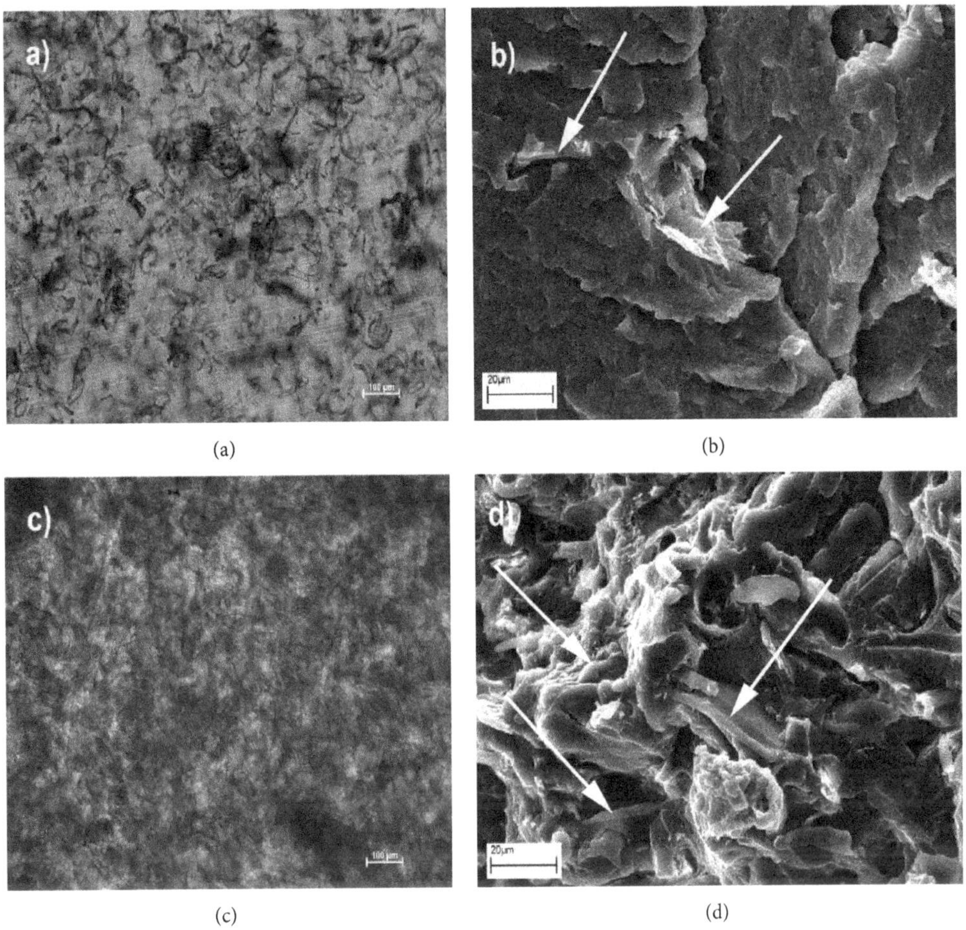

FIGURE 19: Optical and SEM macrographs of PLA filled with 5 wt% cellulose (a and b) and PLA filled with 30 wt% cellulose (c and d), respectively [36].

and cellulose derived from RH. Above all, Battegazzore et al. [36] demonstrated that an industrial waste as RH can be recycled for conferring enhanced final properties to PLA.

2.3. Rice Husk/Polypropylene Composites. PP is an addition polymer chemically synthesized from the monomer propylene. It is a thermoplastic polymer mainly used in packaging, laboratory equipment, automotive components, yogurt containers, and margarine tubs. Its transparency and resistance to heat and chemicals nature make it suitable for a variety of applications. Melt processing of PP can be achieved through extrusion and molding. However, the most common shaping technique used is injection molding. Table 9 shows experimental results from different researches in terms of mechanical properties of RH composites by employing PP as matrix material.

Many researchers have reported comparative studies on different fibers based composites. Premalal et al. [21] compared two types of fillers (RHP and talc) to prepare PP composites using a Brabender Plasti-Corder internal mixer at 180°C, which contained 0, 15, 30, 45, and 60 parts per hundred parts of polymer (php) of fillers. The processability and mechanical properties of the composites with respect to the filler type and filler content were investigated. It was found that talc composites are easier to process than RHP filled polymer composites. They reported that, in terms of mechanical properties, Young's modulus and flexural modulus increased, whereas yield strength and elongation at break decreased with the increasing in filler contents for both types of composites. They observed that the PP/RHP composites have lower yield strength, Young's modulus, flexural modulus, and higher elongation at break than PP/talc composites. SEM studies of tensile fracture surfaces of the composites are shown in Figure 20. The voids which can be seen in Figures 20(a) and 20(b) clearly indicated the poor interaction between RHP and matrix. As reported by Premalal et al. (2002), the polar nature of RHP and nonpolar nature of the matrix can be cited as one of the possible causes for this observation. By increasing the RHP content (Figure 20(b)), significant filler aggregation has also appeared, which contributed to the poor stress transfer from matrix to filler resulting in poor properties. On the other hand, the talc filled polymer composites (Figures 20(c) and 20(d)) show brittle-type failure. They do not exhibit a tendency towards formation of large agglomerates. Compared to the RHP composites, a stronger talc-matrix interface is

TABLE 9: Mechanical properties of RH filled PP composites.

Matrix	RH (wt%)	Coupling agent or compatibilizer (wt%)	Tensile strength (MPa) (% incr.)	Tensile modulus (GPa) (% incr.)	Flexural strength (MPa) (% incr.)	Flexural modulus (GPa) (% incr.)	Impact strength (Jm^{-1}) (% incr.)		Reference
							Not.	Unnot.	
PP	30*	—	24 (−21)	2.1 (80)	—	1.9 (27)	—	13 (−64)	[21]
PP	WRHA* 40	—	23	1.18	—	2.48	—	—	[33]
PP	10	—	29 (−15)	—	—	—	15 (−21)	23 (−45)	[55]
PP	30	—	—	—	26	1.6	40	—	[56]
PP/NBRr (70/30)	10*	—	13 (−33)	1.1 (2)	—	—	—	—	[57]
PP	50	MAPP (4)	37.5 (5)	0.8 (78)	34 (143)	2.28 (61)	—	—	[37]

*phr.

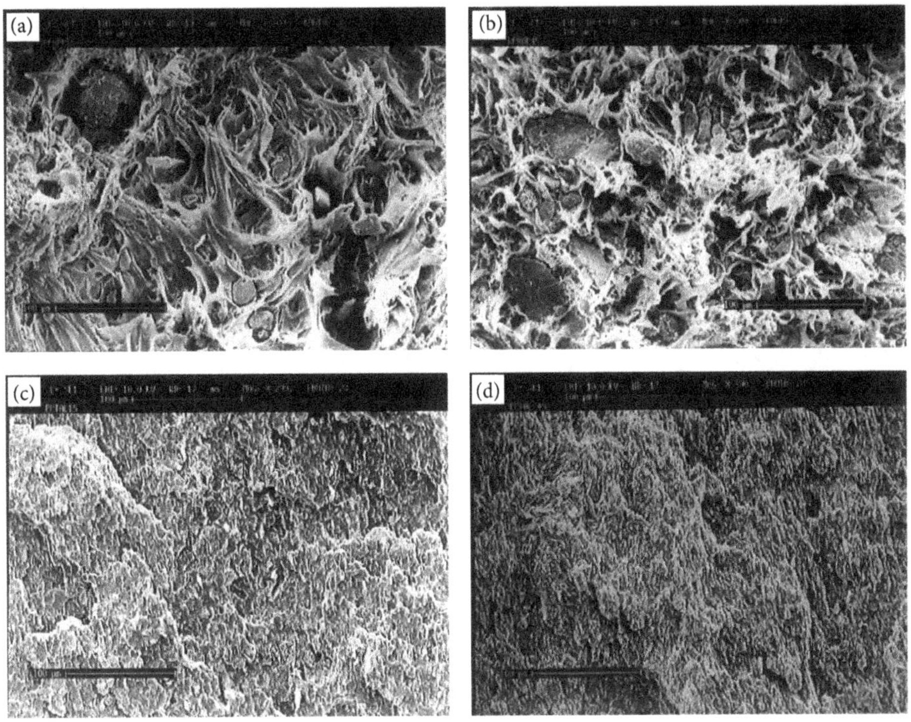

FIGURE 20: SEM micrographs of the tensile fracture surface of (a) and (b) PP/RHP composites and (b) and (c) PP/talc composite at 15 and 45 php of fillers content, respectively (mag. ×300) [21].

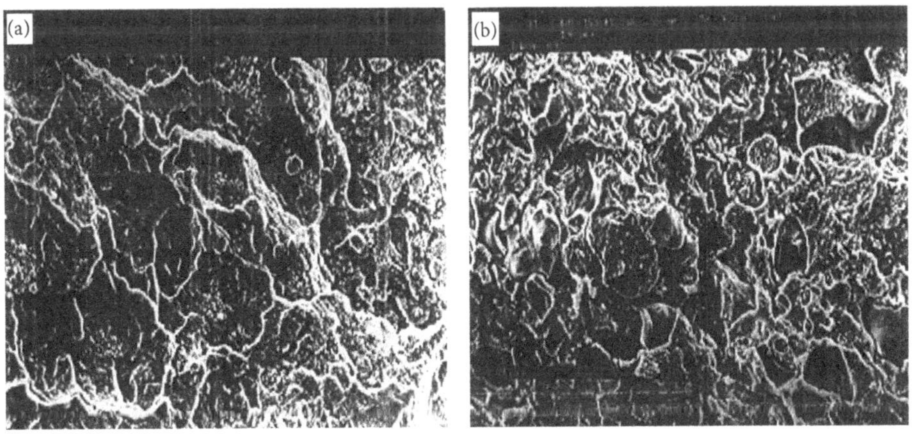

FIGURE 21: SEM micrographs of tensile fracture surface of (a) 40/0 and (b) 20/20 silica/WRHA (wt/wt) of PP/silica/WRHA hybrid composites, respectively (mag. ×300) [33].

evident from the SEM micrographs. Furthermore, it seems that better filler distribution is achieved in talc composites than RHP composites.

As reported by Ismail et al. [33], PP hybrid composites were prepared using silica and WRHA fillers as reinforcing agents. In their study, both fillers were used at the similar total filler content (40 wt%). Ismail et al. (2003) found that the flexural modulus and tensile modulus of the composites increased with increasing of WRHA content in silica/WRHA weight ratio of PP/silica/WRHA hybrid composites, which was attributed to the higher aspect ratio of WRHA compared to silica. However, they noted that the tensile strength, elongation at break, stress at yield, and water absorption of the composites increased with increasing of silica content in silica/WRHA weight ratio of PP/silica/WRHA hybrid composites. It was reported that the smaller particle size of silica could affect on the interfacial adhesion between silica and PP matrix. Degradation study showed that percentage loss of above properties increases with increasing of WRHA content in silica/WRHA weight ratio of composites. SEM micrographs of tensile fracture surfaces of composites are shown in Figure 21. It can be observed from Figure 21(a) that the fracture surface is rough with many tear lines. The rougher surface structure indicates that failure occurred with higher tensile strength. Silica fillers were well embedded in PP matrix and good adhesion occurred between silica and PP matrix. In contrast, Figure 21(b) shows the filler agglomerates pull-out, which is responsible for the poor tensile strength and poor dispersion of WRHA in the PP matrix. It can be concluded that the silica has better adhesion with PP matrix compared to WRHA.

Another study presented by Yang et al. [55] showed that RHF-reinforced PP composites were prepared using twin-screw extruder and physical, mechanical, and morphological properties of PP/RHF were investigated. In their study, four levels of filler contents (10, 20, 30, and 40 wt%), six levels of test temperature (−30, 0, 20, 50, 80, and 110°C), and five levels of crosshead speed (2, 10, 100, 500, and 1500 mm min^{-1}) were designed. By increasing the RHF loading and crosshead speed of test samples, the composite became more brittle. The authors attributed this increase in brittleness with increasing filler loading to the decrease in fraction of thermoplastic polymer and increased interfacial area. Concerning increase in brittleness with increasing crosshead speed, the frictional resistance between filler and matrix polymer decreased at higher crosshead speeds, and this increased the brittleness. The tensile strengths according to the filler loadings presented the same tendency at each test temperature. At the lower test temperatures (−30 and 0°C), the composites exhibited strong and brittle properties like glass, but the tensile strength drastically decreased as the test temperature was increased from (0 to 20°C) due to the glass transition of the polymer matrix. The matrix polymer became more ductile and softened as the test temperature increased. Therefore, they suggested that this problem could be reduced by using a compatibilizing or coupling agent. In this experiment, Yang et al. described that by increasing of test temperature the thermoplastic polymer was softened and the composites showed plastic matrix deformation, which decreased the tensile strength and modulus. They reported that, by increasing of filler contents, morphological study revealed more filler particles and increased the numbers of holes where filler particles have pulled out traces (Figure 22).

In the past ten years (2005), our research group [56] reported on the effects of coupling agent and impact modifier on the mechanical properties of PP/RH composites. In our study, as suggested by Yang et al. [55], ethylene 1-octene copolymers (EOC) and MAPP were used as the impact modifier and coupling agent, respectively. These compounds were melt blended using twin-screw extruder and then injection-molded into standard test samples. An improvement of 35% in flexural strength with the addition of 4 wt% of MAPP into the composites with 30 wt% of RH was observed. However, the flexural modulus and impact strength were decreased. Significant enhancement was also found in the impact properties by the incorporation of 20 wt% of EOC into the composites, but the flexural modulus reduced.

In a research by Santiagoo et al. [57], the properties of RHP filled PP/recycled acrylonitrile butadiene rubber (NBRr) composites with and without coupling agent (APS) were investigated. The composites with different RHP filler content (0 to 30 phr) were prepared using a Haake internal mixer. They studied the mechanical properties, swelling behavior, and water absorption of PP/NBRr/RHP composites. They demonstrated that the addition of RHP to PP/NBRr/RHP composites increased processing torque, tensile modulus, water absorption, and swelling in oil but decreased the tensile strength and elongation at break of the composites. It was also found that, due to the strong bonding between APS treated RHP filler and PP/NBRr matrices, the APS treated RHP composites have higher processing torque, tensile strength, tensile modulus, and lower elongation at break when compared to untreated RHP composites. Furthermore, the APS treatments on RHP improved the adhesion between hydrophilic RHP fiber and the hydrophobic PP/NBRr polymer matrices, which led to reduction of water and oil absorption into PP/NBRr/RHP composites.

Recently, Duy Tran et al. [37] focused on preparation of composites based on PP and Vietnamese RHF (size distribution, 100–125 μm) with the presence of MAPP as a compatibilizer. The researchers have investigated the effect of RHF and MAPP content on properties of PP composites. The results showed that the addition of MAPP to PLA/RHF composites not only did improve tensile and flexural strength but also improved toughness of composites, as confirmed by SEM micrographs (Figure 23). In the absence of MAPP, the composites displayed a rough morphology with appearance of the phase separation between the RHF particles and PP matrix (Figure 23(a)), which clearly showed the poor interaction between filler and matrix. However, in the presence of 4 wt% MAPP (Figure 23(b)), the morphological observation demonstrates that the compatibilizer reduces the separating positions and covered the free spaces between filler and the matrix, which indicated improve adhesion at the interfacial region between filler and matrix. Besides, they found that the thermal stability of composites was also enhanced in the presence of compatibilizer. Their results showed that MAPP acted as the effective factor to improve interfacial adhesion between RHF filler and PP matrix. Duy Tran et al. [37] claimed that Vietnamese RHF is a promising renewable resource to be applied for polymer composites.

2.4. Rice Husk/Polyvinyl Chloride Composites. PVC is a thermoplastic polymer. It is a vinyl polymer constructed of repeating vinyl groups having one of their hydrogens replaced with a chloride group. It is the third most widely produced plastic after polyolefin and is medium strong and transparent material, resistant to chemicals, grease, and oil. It is used for sewage pipe lines and other pipe line applications owing to its biologically and chemically resistant characteristics. It is also used for windows, door frames, and such other building materials by adding impact modifiers and stabilizers. It becomes flexible on addition of plasticizers and can be used in cabling applications as wire insulator. Mechanical properties of RH/PVC composites have been studied by various researchers to understand the effect of RH content on strength of the resulting composites and their ability to resist deformation and are shown in Table 10.

In one set of our experiments [62], RHA fillers were added to acrylonitrile butadiene styrene (ABS) impact modified unplasticised PVC (PVC-U) in order to reduce the cost and

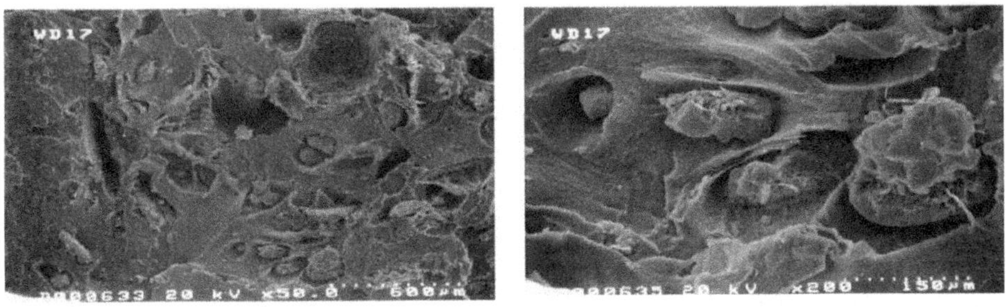

(a)

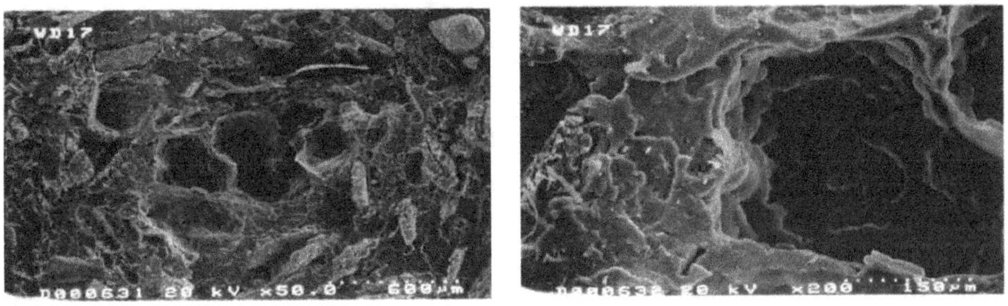

(b)

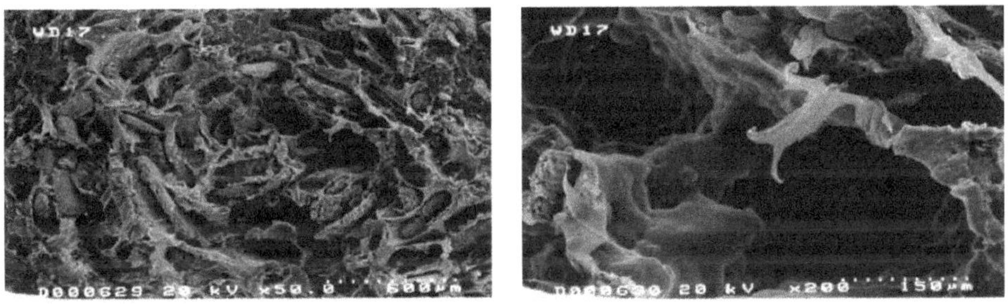

(c)

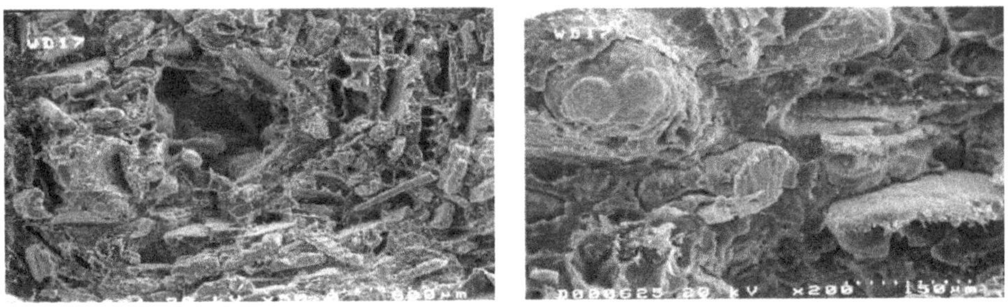

(d)

FIGURE 22: SEM micrographs of the tensile fracture surfaces of the PP/RHF composites with different filler contents: (a) 10, (b) 20, (c) 30, and (d) 40 wt% of RHF contents [55].

maintain or improve properties. A processability study was conducted to investigate the fusion behavior of filled impact modified PVC-U using Brabender Plasti-Corder with fitted mixing head. The processability study showed that as the RHA content increased the fusion time increases. In such cases, attempts to analyze and compare the effect of RHA content and various types of coupling agents on the fusion behavior of impact modified PVC-U were carried out. Our work was carried out on the preprepared RHA, with different content levels. It was found that 10 phr RHA filled impact modified PVC-U samples has shorter fusion time and higher fusion torque compared to other levels of filler contents. Upon the addition of 20 phr RHA, the end torque value decreased compared to unfilled sample. It was also found that the zirconate-treated samples have longer fusion time and lower processing torque compared to other types of coupling agents. Our experimental results also showed that the low contents of the filler have the faster fusion during processing.

Unmodified and modified RHP filled PVC composites with different amounts of RHP content (0, 10, 20, 30,

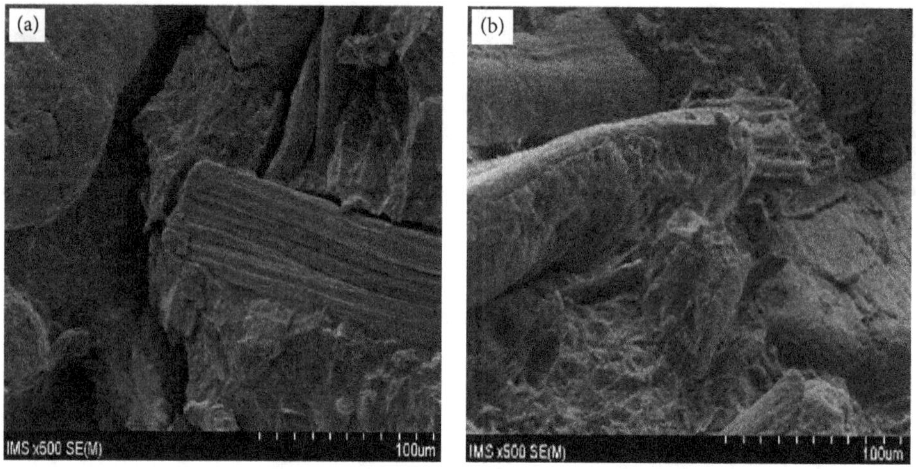

Figure 23: SEM micrographs of (a) PP/RHF and (b) PP/RHF with 4 wt% of MAPP [37].

and 40 wt%) were developed by Chand and Jhod [58]. Mechanical, thermal, and electrical properties of these composites were determined. Density of PVC/RHP composites decreased with increasing RHP content in the PVC composites. The tensile strength of PVC/RHP composites was found to decrease with increasing of RHP in the composites (from 33.9 to 19.5 MPa). However, adding of MAPP modified RHP improved the tensile strength of PVC/RHP composites. In addition, the researchers found that the flexural strength and flexural modulus of composites also increased on treatment of RHP due to the improved bonding between RHP and PVC matrix. They reported that Arc-resistance of PVC/RHP composites decreased with increasing of RHP content in the composites. Furthermore, they found that, due to the presence of impurities and water molecules, volume resistivity and surface resistivity decreased with increasing of RHP content in the composites. They confirmed that Vicat softening temperature of PVC/RHP composites increased in both treated and untreated RHP filled PVC composites.

In an intensive study, Xu et al. [59] fabricated PVC and natural fiber composites by melt compounding and compression molding. The influence of fiber type (i.e., bagasse, rice straw, RH, and pine fiber) and content level of styrene-ethylene-butylene-styrene (SEBS) block copolymer on composites properties were investigated. Mechanical analysis showed that storage modulus and tensile strength increased with fiber contents at the 30 wt% level for all composites. SEBS showed moderate effect on tensile strength of PVC/natural fiber composites, but no obvious contributions to impact strength of the composites were observed. However, among the four natural fibers used, rice straw offered better mechanical properties than bagasse, RH, and pine fiber. They reported that the addition of natural fibers decreases the thermal stability of neat PVC but also weakened the self-accelerating effect during degradation of PVC. The researchers also found that the SEBS had little effect on the thermal stability of the composites. After being immersed in water for four weeks, PVC/RH composites showed lower water absorption rate and better dimensional stability in water compared to the other three types of composites. The results showed that the addition of SEBS increased the water absorption rate of PVC/natural fiber composites. They concluded that PVC composites filled with agricultural fibers had properties comparable with those of PVC/wood composites. They proposed future work with effective coupling agents for PVC (virgin and recycled) and natural fibers.

In the study by Crespo et al. at Polytechnic University of Valencia [60], new composites materials were fabricated, based on lignocellulosic fillers filled thermoplastic composites. In their study, carboxylate plasticizer-di-isononyl-cyclohexane-1,2-dicarboxylate were used as a plasticizer. Filler amount influenced the PVC/RH composites regardless of particle size (150, 500, or 1000 μm). Mechanical properties and SEM analysis of composites were investigated. They found that although the mechanical properties decreased, the resulted composites became more rigid with increasing filler content in all of the particle formulations. They reported that the behavior of the composites at different plasticizer concentrations was very similar to that of plasticized PVC without filler and also resistance values decreased as plasticizer concentration increased regardless of particle size or filler amount. They concluded that the particle distribution had a considerable influence on the properties of the PVC/RH composites, and it was seen that smaller particles equipped the material with improved mechanical properties due to the better dispersion and alignment of the particles within the PVC polymer matrix, as indicated by SEM micrographs (Figure 24).

In 2010, our research group [38] investigated the effects of RH and acrylic impact modifiers on the mechanical properties of PVC-U composites. The composites were prepared using a two-roll mill at 165°C before being hot-pressed at 185°C. In our study, it was found that the incorporation of RH fillers from 10 to 40 per hundred resins (phr) has increased the flexural and tensile modulus of the unmodified and modified (8 phr impact modifier) PVC-U composites. The flexural strength for both unmodified and modified PVC-U composite was observed to increase until 20 phr of RH content. However, the tensile and impact strength of PVC-U composite decreased with increasing RH content. The SEM

TABLE 10: Mechanical properties of RH filled PVC composites.

Matrix	RH (wt%)	Coupling agent or compatibilizer (wt%)	Tensile strength (MPa) (% incr.)	Tensile modulus (GPa) (% incr.)	Flexural strength (MPa) (% incr.)	Flexural modulus (GPa) (% incr.)	Impact strength (Jm^{-1}) (% incr.) Not.	Unnot.	Reference
PVC	10	—	19.5 (−43)	—	95 (−12)	3.1 (−16)	—	—	[58]
PVC	30	SEBS (2.5)	33 (61)	—	—	—	15.5 (−38)	—	[59]
PVC	20	Carboxylate plasticizer* (40)	6.2 (−63)	0.035 (335)	—	—	—	—	[60]
PVC	20	Acrylic impact modifier* (4)	58 (−2)	1.8 (20)	83 (−8)	4.1 (2.5)			[38]
PVC	10	—	19.4 (−43)	—	92.4 (−13)	3.07 (−16)	—	116 (−30)	[13]
PVC	TRH (10)	—	27.6 (−19)	—	102.2 (−4)	3.52 (−3.5)	—	140 (−15)	[13]
rPVC	35	—	12.84 (−15)	1.55 (25)	—	—	—	—	[61]

THR = treated RH using maleic anhydride and benzylperoxide.
*phr.

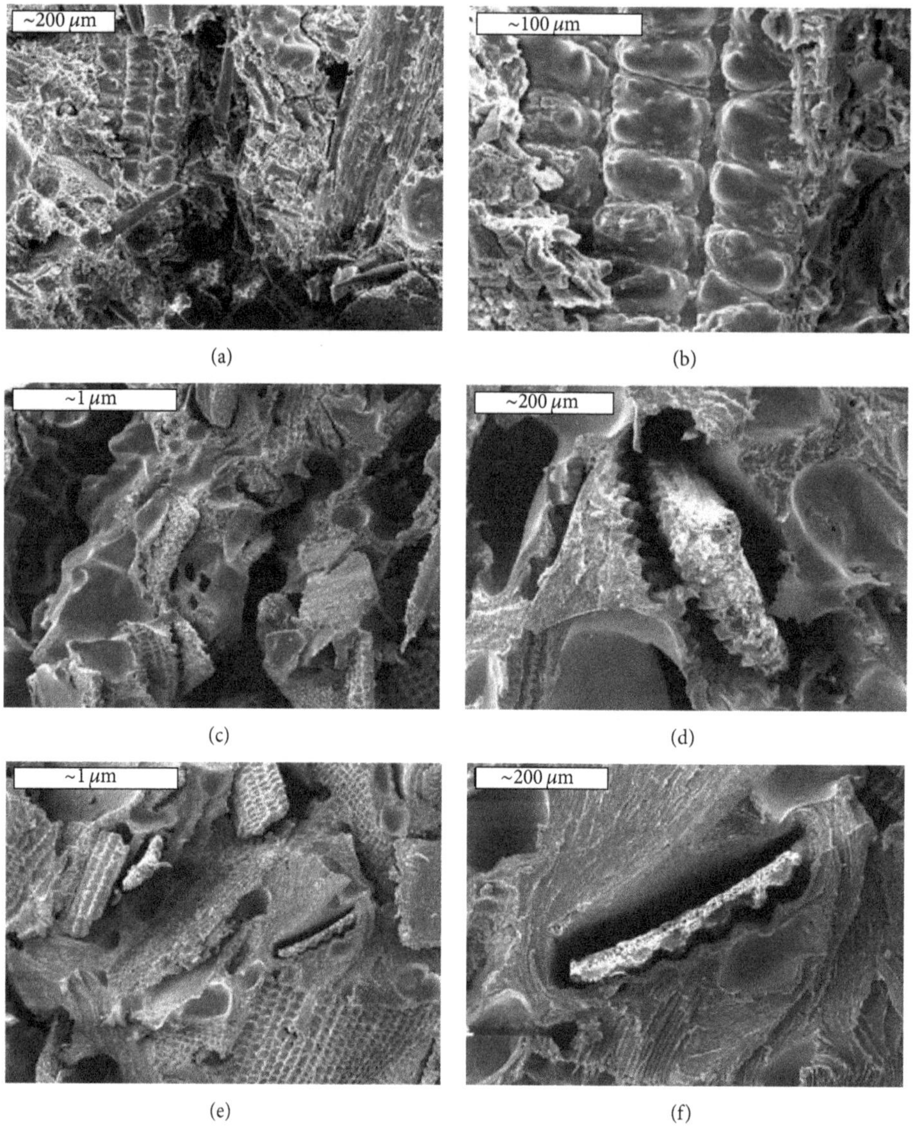

Figure 24: SEM micrographs of plasticized PVC containing 20 wt% RH filler with different plasticizer and particle size: (a) 40 phr, 150 μm, ×150; (b) 40 phr, 150 μm, ×500; (c) 60 phr, 500 μm, ×50; (d) 60 phr, 500 μm, ×150; (e) 60 phr, 1000 μm, ×50; and (f) 60 phr, 1000 μm, ×150 [60].

micrographs showed that the RH fillers agglomerated and unevenly distributed throughout the matrix, as indicated by circles (Figure 25). The result showed that the impact strength of the filled PVC-U composites (20 phr filler) increased, but the tensile and flexural properties decreased with increasing impact modifier content from 0 to 8 phr. It was also found that the formulation containing 8 phr of acrylic impact modifier and 20 phr of RH content has the best balance of stiffness and toughness properties.

In the study by Chand et al. [13], a good interfacial interaction between RH and PVC has been achieved by modifying the surface of the former using MAPP as a compatibilizing agent. In their study, the positive effect of RH reinforcement as well as surface treatment has been reflected in the improved tribological and mechanical properties of the resulted PVC composites. According to Chand et al. [13], the filler concentration should be kept to a minimum if optimum improvement is desired. Therefore, low filler concentration of RH (10 wt%) was proved to be the optimum concentration to achieve maximum strength as well as tribological properties. However, they reported that the larger filler concentration results in larger filler-filler interaction, inhomogeneous dispersion, and inadequate wetting of filler by polymer resin. They noted that the wear resistance of RH filled PVC composites can be increased if the surface of RH is treated suitably with a coupling agent. They observed that, by increasing abrasive grit size, the wear rate increased. SEM micrographs were used to understand the wear resistance of MAPP modified RH filled PVC composites (Figure 26). It is also reported that, with increasing sliding distance, wear rate gradually decreased and attains an almost steady state in multipass condition. Size effect was also observed

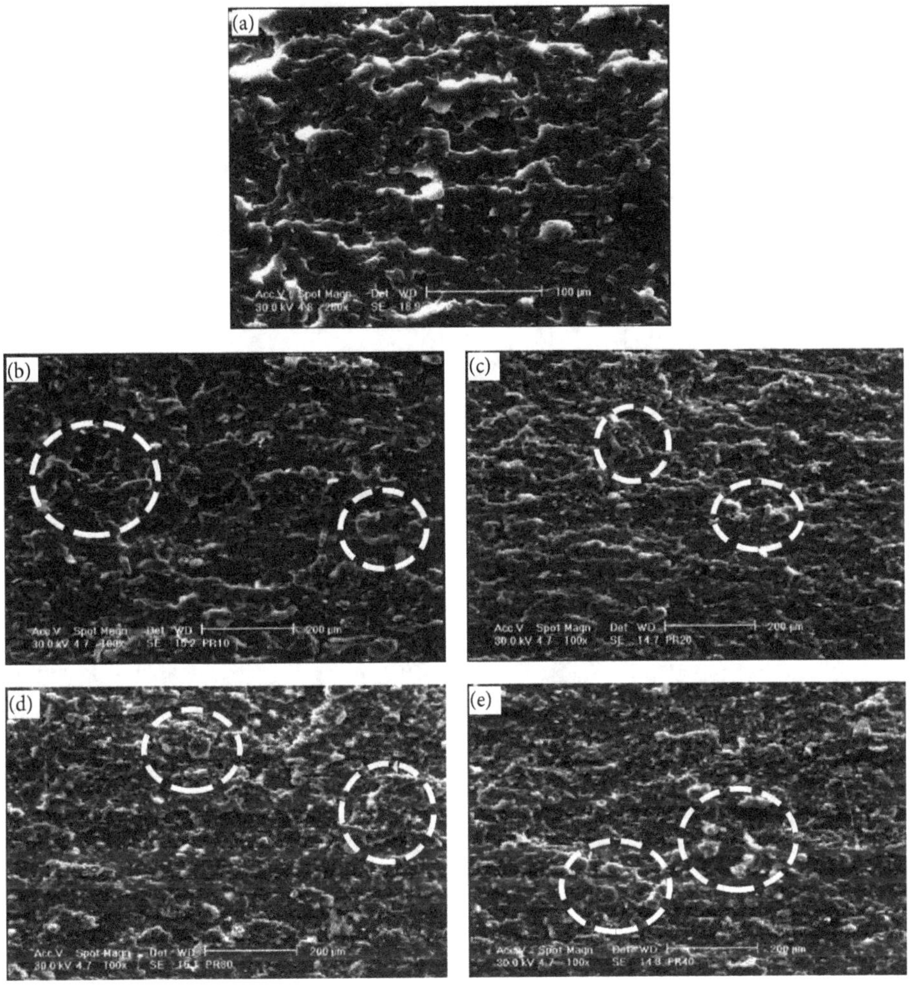

FIGURE 25: SEM micrographs of impact fracture surfaces of (a) unfilled unmodified PVC-U; (b) RH filled impact modified PVC-U composites at 10, (c) 20, (d) 30, and (e) 40 phr of RH content [38].

for composites having 40 wt% of modified filler. Lancaster-Ratner correlation was found to be more linear for MAPP-treated RH filled PVC composites compared to untreated RH/PVC composites. They concluded that probably this is due to increased filler-matrix adhesion upon treatment in the case of the former.

In another study by Chand et al. [63], the influence of a foaming agent on the mechanical and wear behavior of surface modified RH filled PVC composites has been studied. In their study, the surface of RH was modified using MAPP and benzylperoxide for better coupling with the PVC polymer. In Chand et al.'s [63] research, six composites having 10, 20, and 30 wt% of unmodified and surface modified RH having a small percent of a foaming agent have been investigated and compared with foamed PVC. It was observed that with an increase in RH content wear resistance as well as compressive strength of composites increased. They found that wear resistance of PVC foamed composites further improved with surface treatment of RH by MAPP, due to improvement in compatibility between RH and PVC compound, as confirmed by SEM analysis (Figure 27). They proposed a correlation between compressive strength and wear resistance and also they reported that foaming agent reduces the density of composites.

In another work by Chand and coworkers [64], the effect of addition of RH in PVC composites on the dielectric properties at different frequency and temperature has been investigated. Measurements have been performed in the frequency range of 1–10 kHz and temperature range of 32–80°C. Chand et al. [64] proposed a new reasonable model based on existing equations which was experimentally verified. In their investigation, frequency and temperature dependent parameters were experimentally obtained and taken as contribution to the proposed model. The dielectric constant for composites with various fibers to matrix ratio was determined with this model. They reported that dielectric constant of the composite was highly influenced by the fiber composition. The experimental results showed that dielectric constant (ε') increased with the addition of RH in PVC and decreased with increasing frequency, which indicated that the major contribution to the polarization came from orientation polarization. Additionally, they demonstrated that the

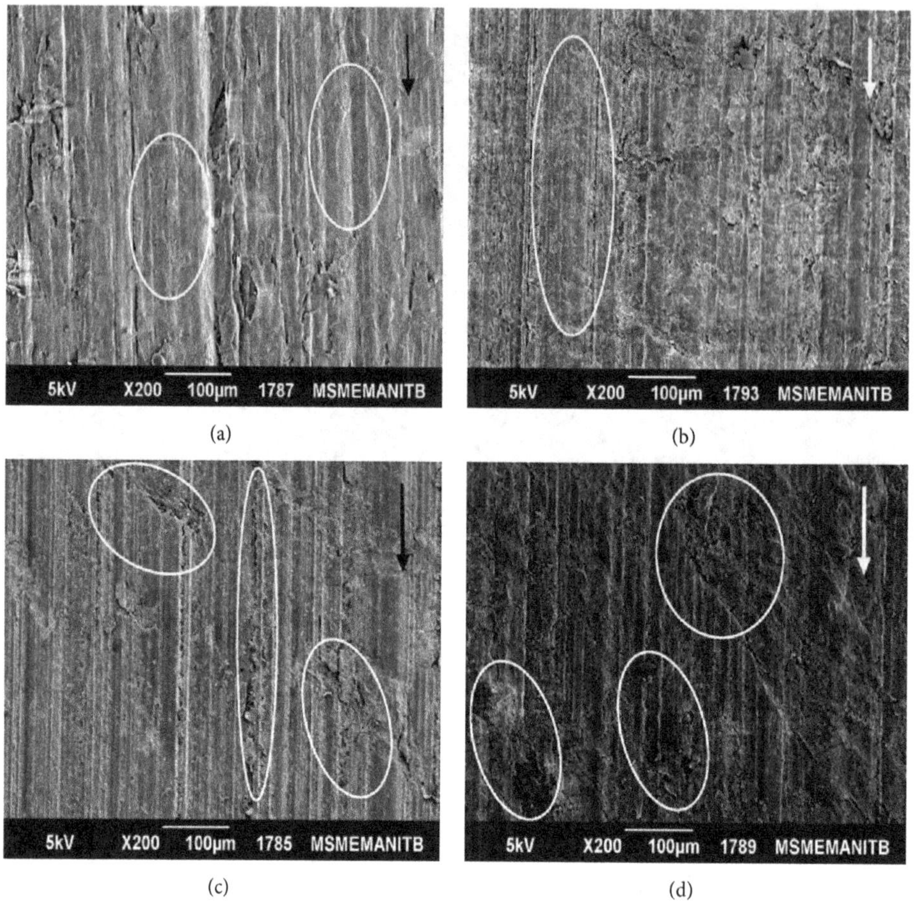

FIGURE 26: SEM micrographs of maleic anhydride modified RH filled PVC composites: (a) PTRH10 and (b) PTRH30; unmodified RH filled PVC composites: (c) PRH20 and (d) PRH40; with the sliding speed of 10 s, sliding distance of 6.28, and applied load of 10 N. (P = PVC; T = treated (modified) and downward arrow shows sliding direction) [13].

dielectric constant (ε') increases with increasing RH content temperature due to greater freedom of movement of dipoles within PVC at higher temperatures. Researchers reported that the proposed frequency and temperature dependent model to calculate the dielectric constant (ε') of the RH filled PVC composites matches the experimental results. They also found that Tan δ peaks at different frequencies are dominant in case of 30 and 40 wt. % of RH composite.

A recent study by Ramle et al. [61] demonstrated good interfacial interactions between RH and recycled PVC (rPVC) by incorporation of aminosilane as coupling agent. The aminosilane-treated rPVC was mixed with RH and then fabricated using hot compression molding. Three different ratios of RH : rPVC (65 : 35, 55 : 45, and 45 : 55) were used to prepare RH/rPVC composites. The positive effect of RH reinforcement as well as surface treatment has been reflected in the improved tensile properties of the resulted rPVC composite. However, the tensile properties of RH/rPVC composites decreased with the increasing RH content. They claimed that this is due to the poor adhesion between RH and rPVC matrix. Ramle et al. (2013) reported that the flexural stress of the treated composite decreased with increasing RH content. However, the incorporation of aminosilane as coupling agent caused to significantly increase the tensile stress of RH/rPVC composites with 45 wt% of RH content. SEM micrographs of RH/rPVC composites are shown in Figure 28. The incompatibility between RH and PVC can be clearly observed, as indicated by the circle (Figure 28(a)), resulting in low tensile strength. However, better adhesion and bonding between the RH and PVC were achieved after the addition of aminosilane, as indicated by the circle (Figure 28(b)). Based on these findings, it is beneficial to carry out further study in the development of RH filled rPVC waste composites.

3. Conclusion and Future Prospects

This review aims to highlight previous works on RH filled polymer composites and provides relevant data for several applications as well as further research areas. RH has been used in combination with various polymeric matrices including PE, PP, PVC, and PLA to form polymer composites. RH filled polymer composites were prepared by different

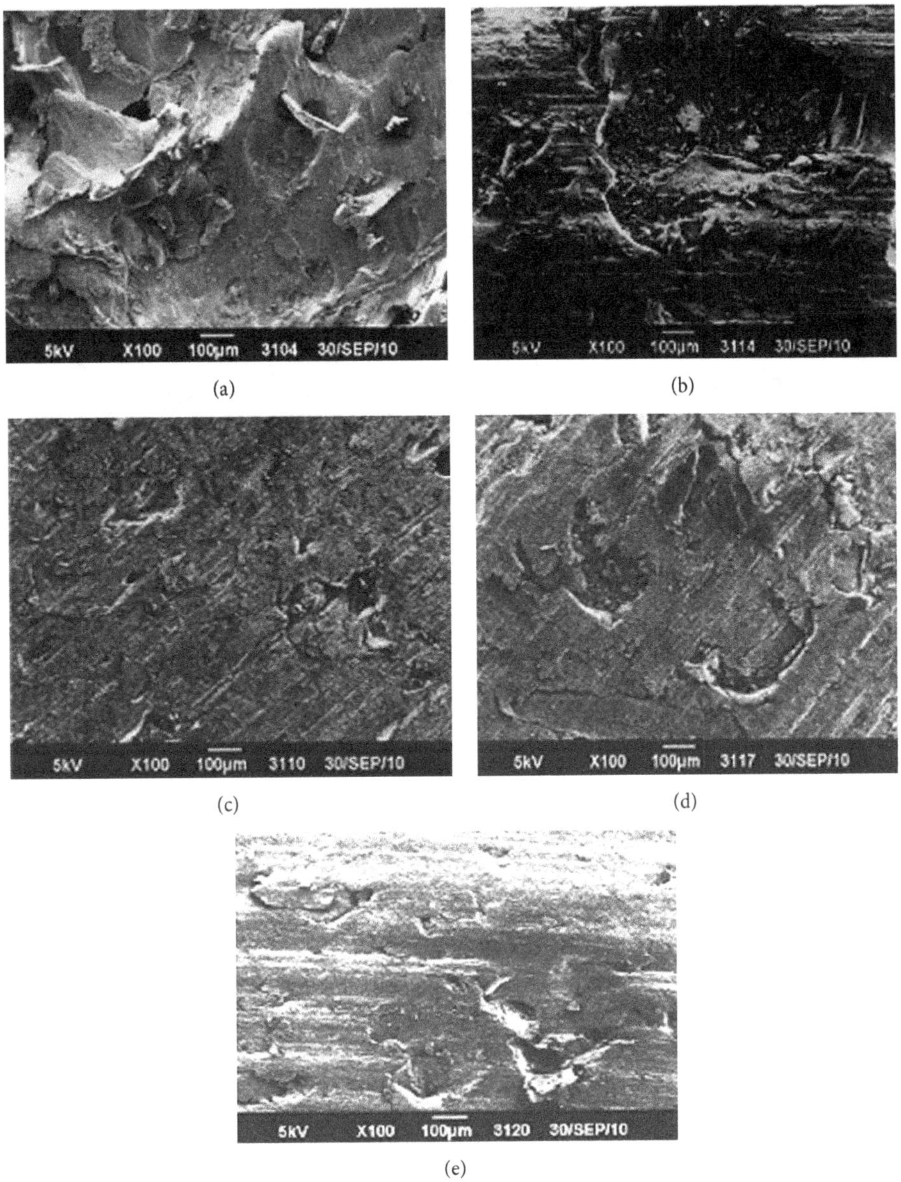

FIGURE 27: SEM micrographs of worn surfaces of foamed PVC without RH: (a) FPRH0; RH filled foamed PVC: (b) FPRH2 and (c) FPRH3; MAPP modified RH filled PVC composites: (d) FPTRH2 and (e) FPTRH3. (P = PVC; T = treated; F = foaming agent) [63].

methods, for example, single screw extruder, corotating twin-screw extruder, internal mixer, and compression molding. The different preparation methods have resulted in composites with different mechanical, physical, thermal, and morphological properties. The addition of RH resulted in reduced tensile strength of composites. Therefore, the use of various types and amount of coupling agent/compatibilizer (approximately $\leq$ 5 wt%) have been reported in order to improve the mechanical properties of composites. Furthermore secondary fillers such as MMT along with compatibilizers have been used to improve mechanical properties of composites. Interestingly, RH filled PE composites have been found to possess better overall mechanical properties compared to other composites such as PLA, PP, and PVC.

Based on this review, the application of RH filled polymer composites as alternative composites materials in building and construction is highly plausible with both lightweight and low cost as its main driving forces. For such applications, biodegradability and weathering properties of the composites need to be evaluated. Although many studies have reported on the use of coupling agent and compatibilizers, further investigations on physical and chemical treatment to further improve the interfacial adhesion with polymeric matrix are needed. The reason is that fiber-polymer interaction is important in determining the final composite properties. So far only single polymer has been used as the matrix in RH filled composites. Limited work has been reported on the use of secondary fillers. Therefore, better understanding of the

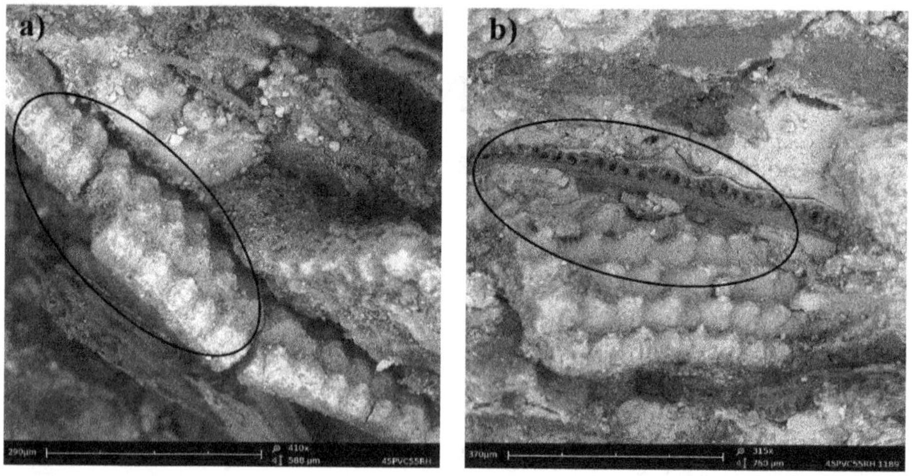

(a) (b)

FIGURE 28: SEM micrographs of RH/rPVC composites (55% RH + 45% rPVC) (a) without the addition of aminosilane and (b) with addition of aminosilane [61].

use of polymer blends as the matrix and secondary fillers in controlling the properties of polymer composites would provide interesting areas to be investigated.

Conflict of Interests

The authors declare that there is no conflict of interests regarding the publication of this paper.

Acknowledgments

The authors wish to acknowledge the Universiti Teknologi Malaysia (UTM) and Research University Grant 05H22, subcode: Q.J130000.2509.05H22, for financial support.

References

[1] A. N. Netravali and S. Chabba, "Composites get greener," *Materials Today*, vol. 6, no. 4, pp. 22–29, 2003.

[2] H. P. S. Abdul Khalil, I. U. H. Bhat, M. Jawaid, A. Zaidon, D. Hermawan, and Y. S. Hadi, "Bamboo fibre reinforced biocomposites: a review," *Materials & Design*, vol. 42, pp. 353–368, 2012.

[3] P. A. Fowler, J. M. Hughes, and R. M. Elias, "Biocomposites: technology, environmental credentials and market forces," *Journal of the Science of Food and Agriculture*, vol. 86, no. 12, pp. 1781–1789, 2006.

[4] K. P. Kumar and A. S. J. Sekaran, "Some natural fibers used in polymer composites and their extraction processes: a review," *Journal of Reinforced Plastics and Composites*, vol. 33, no. 20, pp. 1879–1892, 2014.

[5] N. S. Egute, P. L. Forster, D. F. Parra, D. M. Fermino, S. Santana, and A. B. Lugao, "Mechanical and thermal properties of polypropylene composites with curaua fibre irradiated with gamma radiation," in *Proceedings of the International Nuclear Atlantic Conference*, 2009.

[6] F. Adam, J. N. Appaturi, and A. Iqbal, "The utilization of rice husk silica as a catalyst: review and recent progress," *Catalysis Today*, vol. 190, no. 1, pp. 2–14, 2012.

[7] J. FAO and M. H. I. FOODS, Food and agriculture organization of the United Nations, Rome, Italy, 2004, http://faostat.fao.org.

[8] S. L. Lim, T. Y. Wu, E. Y. S. Sim, P. N. Lim, and C. Clarke, "Biotransformation of rice husk into organic fertilizer through vermicomposting," *Ecological Engineering*, vol. 41, pp. 60–64, 2012.

[9] E. Aprianti, P. Shafigh, S. Bahri, and J. N. Farahani, "Supplementary cementitious materials origin from agricultural wastes—a review," *Construction and Building Materials*, vol. 74, pp. 176–187, 2015.

[10] S. Kumagai and Y. Matsuo, "Composite produced from rice husk and chopped carbon fiber without using any binders," *Industrial Crops and Products*, vol. 43, no. 1, pp. 640–647, 2013.

[11] K. M. Isa, S. Daud, N. Hamidin, K. Ismail, S. A. Saad, and F. H. Kasim, "Thermogravimetric analysis and the optimisation of bio-oil yield from fixed-bed pyrolysis of rice husk using response surface methodology (RSM)," *Industrial Crops and Products*, vol. 33, no. 2, pp. 481–487, 2011.

[12] R. S. Chen, M. H. Ab Ghani, M. N. Salleh, S. Ahmad, and M. A. Tarawneh, "Mechanical, water absorption, and morphology of recycled polymer blend rice husk flour biocomposites," *Journal of Applied Polymer Science*, vol. 132, no. 8, pp. 1–12, 2015.

[13] N. Chand, P. Sharma, and M. Fahim, "Tribology of maleic anhydride modified rice-husk filled polyvinylchloride," *Wear*, vol. 269, no. 11-12, pp. 847–853, 2010.

[14] A. K. Rout and A. Satapathy, "Development and characterization of rice husk-filled glass fiber-reinforced epoxy biocomposites," *Composites: Mechanics, Computations, Applications*, vol. 3, no. 2, pp. 95–106, 2012.

[15] L. Ludueña, D. Fasce, V. A. Alvarez, and P. M. Stefani, "Nanocellulose from rice husk following alkaline treatment to remove silica," *BioResources*, vol. 6, no. 2, pp. 1440–1453, 2011.

[16] I. A. Rahman, J. Ismail, and H. Osman, "Effect of nitric acid digestion on organic materials and silica in rice husk," *Journal of Materials Chemistry*, vol. 7, no. 8, pp. 1505–1509, 1997.

[17] M. M. Hessien, M. M. Rashad, R. R. Zaky, E. A. Abdel-Aal, and K. A. El-Barawy, "Controlling the synthesis conditions for silica nanosphere from semi-burned rice straw," *Materials Science and Engineering B: Solid-State Materials for Advanced Technology*, vol. 162, no. 1, pp. 14–21, 2009.

[18] J. H. Kwon, N. Ayrilmis, and T. H. Han, "Combined effect of thermoplastic and thermosetting adhesives on properties of particleboard with rice husk core," *Materials Research*, vol. 17, no. 5, pp. 1309–1315, 2014.

[19] S. Panthapulakkal, M. Sain, and S. Law, "Effect of coupling agents on rice-husk-filled HDPE extruded profiles," *Polymer International*, vol. 54, no. 1, pp. 137–142, 2005.

[20] S. Panthapulakkal, S. Law, and M. Sain, "Enhancement of processability of rice husk filled high-density polyethylene composite profiles," *Journal of Thermoplastic Composite Materials*, vol. 18, no. 5, pp. 445–458, 2005.

[21] H. G. B. Premalal, H. Ismail, and A. Baharin, "Comparison of the mechanical properties of rice husk powder filled polypropylene composites with talc filled polypropylene composites," *Polymer Testing*, vol. 21, no. 7, pp. 833–839, 2002.

[22] A. M. Das, A. A. Ali, and M. P. Hazarika, "Synthesis and characterization of cellulose acetate from rice husk: eco-friendly condition," *Carbohydrate Polymers*, vol. 112, pp. 342–349, 2014.

[23] P. Mishra, A. Chakraverty, and H. D. Banerjee, "Studies on physical and thermal properties of rice husk related to its industrial application," *Journal of Materials Science*, vol. 21, no. 6, pp. 2129–2132, 1986.

[24] T. Kapur, T. C. Kandpal, and H. P. Garg, "Electricity generation from rice husk in Indian rice mills: potential and financial viability," *Biomass and Bioenergy*, vol. 10, no. 5-6, pp. 393–403, 1996.

[25] C. Sookkumnerd, N. Ito, and K. Kito, "Financial viabilities of husk-fueled steam engines as an energy-saving technology in Thai rice mills," *Applied Energy*, vol. 82, no. 1, pp. 64–80, 2005.

[26] B. S. Ndazi, *The performance of hydrolyzed tannin-CNSL resin in rice husk particleboards [M.S. thesis]*, University of Dar es Salaam, 2001.

[27] Y. K. Lee, S. M. Kim, H. H. Yang, and H. J. Kim, "Mechanical properties of rice husk flour-wood particleboard by urea-formaldehyde resin," *Journal of the Korean Wood Science and Technology*, vol. 31, no. 3, pp. 42–49, 2003.

[28] J. O. Osarenmwinda and J. C. Nwachukwu, "Effect of particle size on some properties of Rice Husk Particleboard," *Advanced Materials Research*, vol. 18-19, pp. 43–48, 2007.

[29] D. García, J. López, R. Balart, R. A. Ruseckaite, and P. M. Stefani, "Composites based on sintering rice husk-waste tire rubber mixtures," *Materials & Design*, vol. 28, no. 7, pp. 2234–2238, 2007.

[30] Q. Zhao, B. Zhang, H. Quan, R. C. M. Yam, R. K. K. Yuen, and R. K. Y. Li, "Flame retardancy of rice husk-filled high-density polyethylene ecocomposites," *Composites Science and Technology*, vol. 69, no. 15-16, pp. 2675–2681, 2009.

[31] J. H. Kwon, N. Ayrilmis, and T. H. Han, "Enhancement of flexural properties and dimensional stability of rice husk particleboard using wood strands in face layers," *Composites Part B: Engineering*, vol. 44, no. 1, pp. 728–732, 2013.

[32] H.-S. Kim, B.-H. Lee, S.-W. Choi, S. Kim, and H.-J. Kim, "The effect of types of maleic anhydride-grafted polypropylene (MAPP) on the interfacial adhesion properties of bio-flour-filled polypropylene composites," *Composites Part A: Applied Science and Manufacturing*, vol. 38, no. 6, pp. 1473–1482, 2007.

[33] H. Ismail, H. B. Hong, C. Y. Ping, and H. P. S. Abdul Khalil, "Polypropylene/silica/rice husk ash hybrid composites: a study on the mechanical, water absorption and morphological properties," *Journal of Thermoplastic Composite Materials*, vol. 16, no. 2, pp. 121–137, 2003.

[34] G. Pritchard, *Plastics Additives: An AZ Reference*, vol. 1, Chapman & Hall, London, UK, Springer, 1998.

[35] M. R. Y. Hamid, M. H. Ab Ghani, and S. Ahmad, "Effect of antioxidants and fire retardants as mineral fillers on the physical and mechanical properties of high loading hybrid biocomposites reinforced with rice husks and sawdust," *Industrial Crops and Products*, vol. 40, no. 1, pp. 96–102, 2012.

[36] D. Battegazzore, S. Bocchini, J. Alongi, A. Frache, and F. Marino, "Cellulose extracted from rice husk as filler for poly(lactic acid): preparation and characterization," *Cellulose*, vol. 21, no. 3, pp. 1813–1821, 2014.

[37] T. Duy Tran, M. Dang Nguyen, C. N. H. Thuc, H. H. Thuc, and T. Dang Tan, "Study of mechanical properties of composite material based on polypropylene and Vietnamese rice husk filler," *Journal of Chemistry*, vol. 2013, Article ID 752924, 6 pages, 2013.

[38] M. Ahmad, A. R. Rahmat, and A. Hassan, "Mechanical properties of unplasticised PVC (PVC-U) containing rice husk and an impact modifier," *Polymers & Polymer Composites*, vol. 18, no. 9, pp. 527–536, 2010.

[39] S. Kayaisang, S. Saikrasun, and T. Amornsakchai, "Potential use of recycled PET in comparison with liquid crystalline polyester as a dual functional additive for enhancing heat stability and reinforcement for high density polyethylene composite fibers," *Journal of Polymers and the Environment*, vol. 21, no. 1, pp. 191–206, 2013.

[40] W. A. W. A. Rahman, N. M. Isa, A. R. Rahmat, N. Adenan, and R. R. Ali, "Rice husk/high density polyethylene bio-composite: effect of rice husk filler size and composition on injection molding processability with respect to impact property," *Advanced Materials Research*, vol. 83-86, pp. 367–374, 2010.

[41] B. Kord, "Nanofiller reinforcement effects on the thermal, dynamic mechanical, and morphological behavior of HDPE/rice husk flour composites," *BioResources*, vol. 6, no. 2, pp. 1351–1358, 2011.

[42] K. Majeed, A. Hassan, A. A. Bakar, and M. Jawaid, "Effect of montmorillonite (MMT) content on the mechanical, oxygen barrier, and thermal properties of rice husk/MMT hybrid filler-filled low-density polyethylene nanocomposite blown films," *Journal of Thermoplastic Composite Materials*, 2014.

[43] K. Majeed, A. Hassan, and A. A. Bakar, "Influence of maleic anhydride-grafted polyethylene compatibiliser on the tensile, oxygen barrier and thermal properties of rice husk and nanoclay-filled low-density polyethylene composite films," *Journal of Plastic Film and Sheeting*, vol. 30, no. 2, pp. 120–140, 2014.

[44] S. Chuayjuljit, C. Kunsawat, and P. Potiyaraj, "Use of silica from rice husk ash as an antiblocking agent in low-density polyethylene film," *Journal of Applied Polymer Science*, vol. 88, no. 3, pp. 848–852, 2003.

[45] Y. Lemmouchi, M. Murariu, A. M. D. Santos, A. J. Amass, E. Schacht, and P. Dubois, "Plasticization of poly(lactide) with blends of tributyl citrate and low molecular weight poly(d,l-lactide)-b-poly(ethylene glycol) copolymers," *European Polymer Journal*, vol. 45, no. 10, pp. 2839–2848, 2009.

[46] E. A. Jaffar Al-Mulla, "Preparation of new polymer nanocomposites based on poly(lactic acid)/fatty nitrogen compounds

modified clay by a solution casting process," *Fibers and Polymers*, vol. 12, no. 4, pp. 444–450, 2011.

[47] E. A. Al-Mulla, A. H. Suhail, and S. A. Aowda, "New biopolymer nanocomposites based on epoxidized soybean oil plasticized poly(lactic acid)/fatty nitrogen compounds modified clay: preparation and characterization," *Industrial Crops and Products*, vol. 33, no. 1, pp. 23–29, 2011.

[48] A. Farah Dina, S. Siti Zaleha, B. Noor Najmi, and I. Nor Azowa, "The influence of alkaline treatment on mechanical properties and morphology of rice husk fibre reinforced polylactic acid," *Advanced Materials Research*, vol. 911, pp. 13–17, 2014.

[49] J. Mohammadi-Rovshandeh, P. Pouresmaeel-Selakjani, S. M. Davachi, B. Kaffashi, A. Hassani, and A. Bahmeyi, "Effect of lignin removal on mechanical, thermal, and morphological properties of polylactide/starch/rice husk blend used in food packaging," *Journal of Applied Polymer Science*, vol. 131, no. 22, pp. 1–8, 2014.

[50] J. Hua, Z. Zhao, W. Yu, and B. Wei, "Hydroscopic and mechanical properties performance analysis of rice husk powder/PLA composites," *Advanced Materials Research*, vol. 230-232, pp. 1231–1235, 2011.

[51] N. A. Zawawi, A. Ismail, K. Abdan, and M. A. Mahdi, "Investigation on the microwave properties of kenaf and rice-husk fiber reinforced pla composite utilizing one-port coaxial transmission line reflection method," *Key Engineering Materials*, vol. 471-472, pp. 868–873, 2011.

[52] D. Battegazzore, S. Bocchini, J. Alongi, and A. Frache, "Rice husk as bio-source of silica: preparation and characterization of PLA–silica bio-composites," *RSC Advances*, vol. 4, no. 97, pp. 54703–54712, 2014.

[53] T. P. T. Tran, J.-C. Bénézet, and A. Bergeret, "Rice and Einkorn wheat husks reinforced poly(lactic acid) (PLA) biocomposites: effects of alkaline and silane surface treatments of husks," *Industrial Crops and Products*, vol. 58, pp. 111–124, 2014.

[54] S. Sareeladdanon and P. Potiyaraj, "Mechanical properties of PLA/LLDPE films reinforced with silica from rice husk," *Advanced Materials Research*, vol. 1025-1026, pp. 221–226, 2014.

[55] H.-S. Yang, H.-J. Kim, J. Son, H.-J. Park, B.-J. Lee, and T.-S. Hwang, "Rice-husk flour filled polypropylene composites; mechanical and morphological study," *Composite Structures*, vol. 63, no. 3-4, pp. 305–312, 2004.

[56] M. S. E. Mustapa, A. Hassan, and A. R. Rahmat, "Preliminary study on the mechanical properties of polypropylene rice husk composites," in *Proceedings of the Simposium on Polimer Kebangsaan (Ke '05)*, pp. 185–191, 2005.

[57] R. Santiagoo, H. Ismail, and K. Hussin, "Mechanical properties, water absorption, and swelling behaviour of rice husk powder filled polypropylene/recycled acrylonitrile butadiene rubber (pp/nbrr/rhp) biocomposites using silane as a coupling agent," *BioResources*, vol. 6, no. 4, pp. 3714–3726, 2011.

[58] N. Chand and B. D. Jhod, "Mechanical, electrical, and thermal properties of maleic anhydride modified rice husk filled PVC composites," *BioResources*, vol. 3, no. 4, pp. 1228–1243, 2008.

[59] Y. Xu, Q. Wu, Y. Lei, F. Yao, and Q. Zhang, "Natural fiber reinforced poly(vinyl chloride) composites: effect of fiber type and impact modifier," *Journal of Polymers and the Environment*, vol. 16, no. 4, pp. 250–257, 2008.

[60] J. E. Crespo, L. Sánchez, D. García, and J. López, "Study of the mechanical and morphological properties of plasticized PVC composites containing rice husk fillers," *Journal of Reinforced Plastics and Composites*, vol. 27, no. 3, pp. 229–243, 2008.

[61] M. S. Ramle, A. Z. Romli, and M. H. Abidin, "Tensile properties of aminosilane treated rice husk/ recycled PVC composite," *Advanced Materials Research*, vol. 812, pp. 151–156, 2013.

[62] A. Hassan and K. Sivaneswaran, "The effect of rice husk ash fillers on processability of ABS impact modified PVC-U," in *Proceedings of the 3rd National Symposium on Polymeric Materials*, pp. 174–179, 2002.

[63] N. Chand, M. Fahim, P. Sharma, and M. N. Bapat, "Influence of foaming agent on wear and mechanical properties of surface modified rice husk filled polyvinylchloride," *Wear*, vol. 278-279, pp. 83–86, 2012.

[64] N. Chand, J. Sharma, and M. N. Bapat, "A new proposed model for dielectric behavior of PVC/rice husk composites," *Journal of Applied Polymer Science*, vol. 126, no. 3, pp. 1105–1111, 2012.

Permissions

All chapters in this book were first published in AMSE, by Hindawi Publishing Corporation; hereby published with permission under the Creative Commons Attribution License or equivalent. Every chapter published in this book has been scrutinized by our experts. Their significance has been extensively debated. The topics covered herein carry significant findings which will fuel the growth of the discipline. They may even be implemented as practical applications or may be referred to as a beginning point for another development.

The contributors of this book come from diverse backgrounds, making this book a truly international effort. This book will bring forth new frontiers with its revolutionizing research information and detailed analysis of the nascent developments around the world.

We would like to thank all the contributing authors for lending their expertise to make the book truly unique. They have played a crucial role in the development of this book. Without their invaluable contributions this book wouldn't have been possible. They have made vital efforts to compile up to date information on the varied aspects of this subject to make this book a valuable addition to the collection of many professionals and students.

This book was conceptualized with the vision of imparting up-to-date information and advanced data in this field. To ensure the same, a matchless editorial board was set up. Every individual on the board went through rigorous rounds of assessment to prove their worth. After which they invested a large part of their time researching and compiling the most relevant data for our readers.

The editorial board has been involved in producing this book since its inception. They have spent rigorous hours researching and exploring the diverse topics which have resulted in the successful publishing of this book. They have passed on their knowledge of decades through this book. To expedite this challenging task, the publisher supported the team at every step. A small team of assistant editors was also appointed to further simplify the editing procedure and attain best results for the readers.

Apart from the editorial board, the designing team has also invested a significant amount of their time in understanding the subject and creating the most relevant covers. They scrutinized every image to scout for the most suitable representation of the subject and create an appropriate cover for the book.

The publishing team has been an ardent support to the editorial, designing and production team. Their endless efforts to recruit the best for this project, has resulted in the accomplishment of this book. They are a veteran in the field of academics and their pool of knowledge is as vast as their experience in printing. Their expertise and guidance has proved useful at every step. Their uncompromising quality standards have made this book an exceptional effort. Their encouragement from time to time has been an inspiration for everyone.

The publisher and the editorial board hope that this book will prove to be a valuable piece of knowledge for researchers, students, practitioners and scholars across the globe.

List of Contributors

Xiaoqing Xu, Bohan Liu and Yibing Li
State Key Laboratory of Automotive Safety & Energy, Department of Automotive Engineering, Tsinghua University, Beijing 100084, China

Takashi Sasaki and Masaaki Nakagiri
Department of Materials Science and Engineering, University of Fukui, 3-9-1 Bunkyo, Fukui 910 8507, Japan

Satoshi Irie
Department of Frontier Fiber Technology and Science, University of Fukui, 3-9-1 Bunkyo, Fukui 910 8507, Japan

Yanqing Wang and Beicheng Wu
School of Materials Science & Engineering, China University of Mining & Technology, Xuzhou 221116, China

Yu Guo
School of Mines, China University of Mining & Technology, Xuzhou 221116, China

Haifeng Yang
School of Mechatronic Engineering, China University of Mining & Technology, Xuzhou 221116, China

Zhi-peng Zhong
Department of Civil Engineering, Shanxi University, Taiyuan, Shanxi 030013, China
Key Laboratory of C&PC Structures of the Ministry of Education, Southeast University, Nanjing 210096, China

Hong Liu
Department of Civil Engineering, Shanxi University, Taiyuan, Shanxi 030013, China

Han Ki Lee and Dae Sik Kim
New Technology Development Team, Research & Development Division, Hyundai-ngv, Seoul 151-742, Republic of Korea

Jong Sung Won, Da Young Jin, Hyun Jae Lee and Seung Goo Lee
Department of Organic Materials Engineering, Chungnam National University, Daejeon 34134, Republic of Korea

P. Xiao, J. J. Zhang, Z. Q. Wang and H. Lin
School of Textile and Material Engineering, Dalian Polytechnic University, Dalian 116034, China

Dandan Yin and Dongfeng Zhao
School of Petrochemical and Energy Engineering, Zhejiang Ocean University, Zhoushan 316022, China
United National-Local Engineering Laboratory of Harbor Oil & Gas Storage and Transportation Technology, Zhejiang Ocean University, Zhoushan 316022, China

Fadila Benlizidia Lalam and Faouzi Djemmal
Quantum Chemistry and Physics Laboratory, Mouloud Mammeri University, Tizi-Ouzou, Algeria

Akram R. Jabur, Laith K. Abbas and Saja A. Moosa
Department of Materials Engineering, University of Technology, Baghdad, Iraq

Il-Ju Ko, Hyun-Chang Oh, Yong-Jun Cha and Chae Hyeok Han
Department of Chemistry and Biology, Korea Science Academy of KAIST, Busan, Republic of Korea
Department of Chemistry, KAIST, Daejeon, Republic of Korea

Eun-Young Choi
Department of Chemistry and Biology, Korea Science Academy of KAIST, Busan, Republic of Korea

Alfredo Alán Rey Calderón and Alberto Díaz Díaz
Centro de Investigación en Materiales Avanzados, S. C. Miguel de Cervantes 120, Complejo Industrial Chihuahua, 31136 Chihuahua, Chihuahua, Mexico

Anca Aldea
Department of General Chemistry, University Politehnica Bucharest, 1-7 Gheorghe Polizu Sector 1, 011061 Bucharest, Romania

Ana-Maria Albu
Department of Bioresources and Polymer Science, University Politehnica Bucharest, 1-7 Gheorghe Polizu Sector 1, 011061 Bucharest, Romania

Alina Nicolescu
Centre for Organic Chemistry Costin D. Nenitescu, Romanian Academy, 202B Splaiul Independentei, CP 35-108, 060023 Bucharest, Romania
Institute of Macromolecular Chemistry "Petru Poni" Iasi, 41A Aleea Grigore Ghica Voda, 700487 Iasi, Romania

Victorita Tecuceanu
Centre for Organic Chemistry Costin D. Nenitescu, Romanian Academy, 202B Splaiul Independentei, CP 35-108, 060023 Bucharest, Romania

Zhao Yang and Kun Wu
Institute of Urban Construction, Wuhan University of Science and Technology, Wuhan 430065, China

Youmin Wang, Zhichao Yan and Xuejun Shan
School of Mechanical and Automotive Engineering, Anhui Polytechnic University, Wuhu, China

Rubén Sierra-Ávila, Janett Valdez-Garza, Carlos Alberto Avila-Orta, Enrique Javier Jiménez-Regalado and Gregorio Cadenas-Pliego
Centro de Investigación en Química Aplicada, Blvd. Enrique Reyna 140, 25294 Saltillo, Coahuila, Mexico

Marissa Pérez-Alvarez and Enrique Soto-Castruita
CONACyT Research Fellow, Instituto Mexicano del Petróleo, Eje Central Lázaro Cárdenas Norte 152, 07730 Ciudad de México, Mexico

José M. Mata-Padilla
CONACyT Research Fellow, Centro de Investigación en Química Aplicada, 25294 Saltillo, Mexico

Sivaldo Leite Correia and Denilso Palaoro
State University of Santa Catarina (UDESC), Center of Technology Sciences (CCT), 89223-100 Joinville, SC, Brazil

Ana Maria Segadães
Department of Materials and Ceramics Engineering (CICECO), University of Aveiro, 3810-193 Aveiro, Portugal

Shiyong Jiang, Weilai Yao, Jin Chen and Tao Cai
Department of Military Infrastructure Engineering, Army Logistics University of PLA, Chongqing, China

José Ricardo Carneiro and Maria de Lurdes Lopes
Construct-Geo, Faculty of Engineering, University of Porto, Rua Dr. Roberto Frias, 4200-465 Porto, Portugal

Paulo Joaquim Almeida
Requimte, Faculty of Sciences, University of Porto, Rua do Campo Alegre, 687, 4169-007 Porto, Portugal

S. Montha, P. Suwandittakul and C. Kongkaew
MTEC, National Science and Technology Development Agency, 114 Thailand Science Park, Phahonyothin Road, Khlong Nueng, Khlong Luang, PathumThani 12120, Thailand

A. Poonsrisawat
BIOTEC, National Science and Technology Development Agency, 113 Thailand Science Park, Phahonyothin Road, Khlong Nueng, Khlong Luang, PathumThani 12120, Thailand

P. Oungeun
Department of Chemistry, Faculty of Science, Naresuan University, Phitsanulok 65000, Thailand

Jun Cong Ge, Jeong Hyeon Kim and Nag Jung Choi
Division of Mechanical Design Engineering, Chonbuk National University, 567 Baekje-daero, Jeonju-si, Jeollabuk-do 561-756, Republic of Korea

Bryan B. Pajarito, Kayla C. Castañeda, Sofia Denise M. Jeresano and Dominique Ann N. Repoquit
Polymer Research Laboratory, Department of Chemical Engineering, University of the Philippines, Diliman, Quezon City 1101, Philippines

Reza Arjmandi and Azman Hassan
Department of Polymer Engineering, Faculty of Chemical Engineering, Universiti Teknologi Malaysia (UTM), 81310 Skudai, Johor, Malaysia

Khaliq Majeed
Department of Chemical Engineering, COMSATS Institute of Information Technology, 54000 Lahore, Pakistan

Zainoha Zakaria
Faculty of Science, Universiti Teknologi Malaysia (UTM), 81310 Skudai, Johor, Malaysia

Index

www.ingramcontent.com/pod-product-compliance
Lightning Source LLC
LaVergne TN
LVHW060535210926
848485LV00005B/1204